Historical Essays on Meteorology
1919–1995

The Diamond Anniversary History Volume
of the American Meteorological Society

edited by

James Rodger Fleming

Boston
American Meteorological Society
1996

The American Meteorological Society
45 Beacon St.
Boston, MA 02108

Library of Congress Cataloging-in-Publication Data

Historical Essays on Meteorology, 1919–1995
Edited by James Rodger Fleming
Foreword by Warren M. Washington
(The diamond anniversary history volume of the American
Meteorological Society)
p. 618 cm.
ISBN 1-878220-17-9 (alk. Paper)
1. Meteorology—History
2. Atmospheric science—History

CIP 96-83535

Printed in the United States of America
by Braun-Brumfield, Inc., A Sheridan Group Company

The paper used in this book meets the minimum requirements of
American National Standard for Information Sciences—Permanence
of Paper for Printed Library Materials, ANSI Z39.48-1984.

Contents

Foreword

As part of its 75th Anniversary, the American Meteorological Society (AMS) initiated a history book. This volume of essays chronicles achievements in the field of meteorology in many specialized areas, including basic and applied research, the private sector, and education. From the beginning, the society appealed to a wide spectrum of interests. In addition to meteorologists (both professional and amateur), the membership included people involved in the practical applications of weather and climate information to business, commerce, and agriculture. That tradition remains today. The first issue of the *Bulletin of the American Meteorological Society (BAMS)* captured this sentiment when it noted that the birth of the AMS, "marked the beginning of a movement not only to push forward investigations of weather processes and climate conditions, but also to widen the valuable applications of the knowledge already at hand . . . into almost every line of human endeavor." As this history book well demonstrates, the founders were futuristic. Meteorology has strongly progressed, both as a basic and applied science and as a major influence in the development of electronic computing, remote sensing, and aviation technologies. Information about weather and climate has also found its way into every sector of the economy. "Surely, the applications of meteorology are manifold" (*BAMS* 1, 1920, 1).

The AMS has grown from about 600 members at its beginning to approximately 11 000 at the present. The society, through its extensive committee structure, meetings and publications programs, and helpful headquarters staff, has also grown in terms of its service to the members. There are many significant milestones that could be mentioned; among the most significant are the following.

- 1919–1920: The AMS was organized on 29 December 1919 at a meeting of the American Association for the Advancement of Science in St. Louis, Missouri. It was incorporated three weeks later, on 21 January 1920, in Washington, D.C. The original constitution and by-laws are quite similar to today's, providing for different grades of membership, the election of officers, the governing Council, the meeting structure, the publications, and the various committees. The first volume of *BAMS* was published in 1920 to keep the society informed of new developments. (See *BAMS* 26, 1945, 163–170.)

- 1938: The awards program began with the Clarence Leroy Meisinger Award.
- 1944: The society was reorganized to give increased emphasis to being a scientific, professional organization. This was undertaken by presidents C.-G. Rossby and H.G. Houghton. As part of the process of professionalization the *Journal of Meteorology* was initiated.
- 1946: Because of the rapid growth of the field during and after World War II, the society required a formal directorate to coordinate its many activities. This was established in Boston with Ken Spengler as the first executive secretary.
- 1950–1951: *Meteorological Abstracts and Bibliography* was started under the editorship of Malcom Rigby with the purpose of making available citations to all the literature in meteorology and related fields. In 1951 the *Compendium of Meteorology* was published by a committee chaired by H.G. Houghton. The editor was Tom Malone, who stated that the purpose was to "take stock of the present position of meteorology . . . and to indicate the avenues of further study and research which needed to be explored in order to extend the frontiers of our knowledge."
- 1957: The Certified Consulting Meteorologist program was established as was the AMS Seal of Approval for broadcast meteorologists.
- 1959: The *Glossary of Meteorology* was published to standardize the definitions of terms used in the field. A new version of the Glossary is currently being prepared by the society.
- 1960: The AMS moved into their historic home at 45 Beacon Street, overlooking Boston Common. The house was designed by Charles Bulfinch, architect of the Massachusetts State House and the Capitol Building in Washington, D.C. It was originally the residence of Harrison Gray Otis, who served as mayor of Boston and then as a U.S. senator. The society raised funds from its members to help renovate the building and restore it to its

former elegance. (See *BAMS* 41, 1960, 507–517 and *BAMS* 62, 1981, 493–497). The society was further reorganized under the leadership of president T. Malone with the initial draft prepared by P.D. McTaggart-Cowan. Committee and commission structures were revamped to keep the society in step with the times. By 1968, Council membership was expanded to allow for representation from hydrology, aeronomy, and oceanography.

- 1962: The *Journal of the Atmospheric Sciences* and the *Journal of Applied Meteorology* were formed from the *Journal of Meteorology*.

- 1964–1965: The Council developed a series of extensive education programs including films, popular books, a visiting foreign scientist program, career guidance materials, support for science fairs, and a special program to introduce high school students to the field. A year later a curriculum guide for students in the atmospheric sciences was developed. This guide is still being published biannually (now in conjunction with the University Corporation for Atmospheric Research), and educational outreach programs are still a prominent part of AMS.

- 1971–1988: The *Journal of Physical Oceanography* was established in 1971, and the hundred-year-old *Monthly Weather Review* was converted from a government publication to an AMS journal in 1974. The growth of AMS journals continued in the 1980s with the addition of the *Journal of Atmospheric and Oceanic Technology* (1984), *Weather and Forecasting* (1986), and the *Journal of Climate* (1988). In 1988 Richard Hallgren became the society's second executive director.

The essays in this 75th anniversary history book do not focus on the history of the AMS. Rather, they concern the overall intellectual and institutional development of meteorology in the twentieth century. They constitute a rich sampling of what has been learned, where we stand, and where we might be going—in research, in education, and in the private sector. Each was written by a recognized authority in the field. The AMS was fortunate to obtain the editorship of James R. Fleming for the 75th Anniversary history volume. His professional background in the history of science and technology contributed both rigor and clarity to the historical essays. He has worked closely with the contributors to help tie the chapters together and make the various historical aspects more accurate. As you read this book, keep in mind that this is a book of meteorological discovery and innovation. Important factors for continued discovery and innovation are to be found in understanding the history of our field. *Historical Essays on Meteorology, 1919–1995,* The Diamond Anniversary Volume of the American Meteorological Society, is a welcome addition to that history, and the AMS is proud to sponsor it.

Warren M. Washington
President of the American Meteorological Society, 1994

ACKNOWLEDGMENTS

Many people were involved in producing a book such as this. Special thanks are due Richard Hallgren, executive director of AMS; Warren Washington, president of AMS in 1994; and the members of the Council who supported this project from beginning to end. I also thank the AMS Committee on the History of the Atmospheric Sciences, chaired by Professor Robert Fleagle, and the program chairs of the individual conferences for inviting speakers to the AMS diamond anniversary meeting and for generating an enormous amount of goodwill and enthusiasm for the book project. Appreciation is due each of the individual authors and their support networks for their perseverance and goodwill through the editorial process.

Friends at AMS—Ken Spengler, Evelyn Mazur, and Barry Mohan—provided valuable support and encouragement for this project, as did William Kellogg, the new chair of the Committee on the History of the Atmospheric Sciences.

Preparation of the manuscript was facilitated by Alice Ridky, Beverly Eaton, and my research students Simone Kaplan and Amy Lyons at Colby College. The volume's final form is a result of my hard-working and most professional colleagues in the publications department at AMS—Keith Seitter, Melissa Weston, Marianne Simas-Valcich, Patricia English, Irwin Abrams, Gary Gorski, and Julie Burba.

I would be remiss if I failed to acknowledge the sacrifices made by my wife Miyoko and my sons Jamitto and Jason while I was preoccupied with manuscript and pencil.

Introduction

It has been my privilege, honor,—and challenge!—to serve the American Meteorological Society (AMS) as editor of its 75th anniversary history volume. The Society was fortunate to engage a number of eminent pioneers and leading practitioners to write about the fields they helped develop. They were joined by several professional historians of science and technology. The resulting book, *Historical Essays on Meteorology, 1919–1995*, constitutes a substantial sampling of what has been learned since 1919 in the atmospheric sciences and services.[1]

The volume is not complete in its coverage—no such collection is. No book in history is ever granted the status of being considered the last word on a subject; rather, it is sufficient if some consider it the *latest* word. I sincerely believe that these 20 essays accomplish this goal and constitute important benchmark contributions to the history of the atmospheric sciences. The historical foreword by Warren Washington and the photographic essay compiled by Julie Burba at AMS Headquarters also serve to enhance the book's value. This volume should be of interest to weather professionals and enthusiasts, to historians of science, and to students of science and history. It will help us calibrate where we are, where we have been, and where we might be going as a discipline. Hopefully it will inspire others to value the past and to dig into it more deeply.

These historical essays are meant to celebrate—without being overly celebratory—a period of disciplinary formation and remarkable growth in the field of meteorology; an era of expanding theoretical, observational, and institutional horizons; an era in large part shaped and documented by the organizational activities of the American Meteorological Society.

---

[1]Many of the essays in this book were originally presented at the AMS diamond anniversary meeting in Dallas in January 1995. Tapes of many of the speeches are available through AMS.

Dynamic meteorology and numerical weather prediction

The book begins with a lucid account by Ed Lorenz of "The Evolution of Dynamic Meteorology." Lorenz discusses his own first exposure to the subject in 1942 from his teacher Bernhard Haurwitz. He traces the evolution of dynamic meteorology as a "new species of investigation" that developed in response to a changing "scientific environment," namely, the advent of computing, while an older "species," classical dynamics, continued to thrive. Three concepts new to the twentieth century—potential vorticity, baroclinic instability, and chaos—receive extended treatment. Concerning chaos, Lorenz modestly remarked, "I have personally devoted much attention" to this concept.

George Cressman, former head of the Joint Numerical Weather Prediction Unit and former director of the National Weather Service, describes "The Origin and Rise of Numerical Weather Prediction." His survey ranges from early efforts to compute atmospheric changes numerically by Felix Exner, Vilhelm Berknes, and L.F. Richardson in the first decades of this century, through the development of electronic computers in the 1950s, to the implementation, circa 1965, of daily forecasting with primitive equation models.

Observational tools and computational devices

Radar, satellites, and computers are without question the tools of choice for modern meteorologists. Bob Serafin, director of the National Center for Atmospheric Research, provides a broad perspective, both topically and chronologically in his chapter, "The Evolution of Atmospheric Measurement Systems." He documents the ever-expanding and ever more sophisticated set of tools for measuring the traditional meteorological variables—wind, pressure, moisture, and temperature, at the surface and aloft—and the new systems that measure trace chemical species and characteristics of the deep oceans. His focus is on operational observing systems: from the earliest instruments and networks to the current "modernization" revolution taking place in the forecasting system used by the National Weather Service.

Roddy Rogers and Paul Smith, both established radar experts and students of the pioneers of the field, have written "A Short History of Radar Meteorology." Viewing radar as fundamental to the development of both mesoscale meteorology and modern atmospheric remote sensing, they document its utility—in forecasting, aviation, storm research, and precipitation physics, and the interdisciplinary breadth of its appeal—to physicists, electrical engineers, applied mathematicians, and others in both the military and civilian sectors. Radar meteorology, although relatively young as a scientific field, has generated a rich literature and holds the record for the longest series of conferences sponsored by AMS. This review documents the emergence of the field, from the first revolutionary observations of rain storms using military radars to today's wind profilers, Doppler radars, and efforts to integrate radar observations and numerical models.

The "Evolution of Satellite Observations in the United States and Their Use in Meteorology" is examined by satellite specialists James Purdom and Paul Menzel. Atmospheric scientists have always desired to have an observational platform aloft. Earlier efforts included manned balloon flights, kite measurements of the upper atmosphere, and rocket probes into the stratosphere. Since 1960 meteorological satellites have provided a "weather eye in the sky" from both polar and geostationary orbits. The authors document the interplay between the hardware developed by the space program and the operational and research needs of the meteorological community. Their comprehensive review of satellite capabilities extends over 35 years, from *TIROS-1* to *GOES-8,* and points out new challenges and opportunities just over the horizon.

Historian of science and technology Frederik Nebeker has written "A History of Calculating Aids in Meteorology" that surveys the wide variety of calculating aids used in meteorology from the earliest times. Such devices, include numerical tables, graphical techniques, computing forms, special-purpose slide rules, punched-card and adding machines, and special-purpose analog computers. He describes their various applications and their replacement by the electronic digital computer—a general-purpose computational device that could produce tables and charts and, significantly, was able to generate useful numerical weather forecasts.

Cloud microphysics and dynamics

Roscoe Braham, a pioneer in the field of cloud physics, has prepared an authoritative chapter, "Formation of Rain: A Historical Perspective," documenting the modern development and multidisciplinary nature of the field. He discusses the various theories of rain formation condensation, collision, coalescence, and ice crystal processes that have been developed in recent decades. His view of the future of cloud physics is a bright one, illuminated by new technologies and interdisciplinary opportunities.

The influence of cloud microphysics on the dynamics of clouds is a central theme in Harold Orville's chapter, "A History of Research in Cloud Dynamics and Microphysics." Orville employs the third equation of motion, describing the vertical component of airflow, as a framework for discussing the interactions of dynamics and microphysics in clouds. Observational studies of clouds and thunderstorms and theories of plumes and thermals highlight the work done before 1960. More recently, although field studies continue, the focus has been on the development of numerical models of clouds and severe storms.

Hurricanes, convective storms, and lightning

"A History of Hurricane Forecasting for the Atlantic Basin, 1920–1995," is the subject of the chapter by Mark DeMaria. Using case studies of specific storms as a way of illustrating the problems involved in forecasting hurricane tracks and intensities, he reviews technological advances that influenced the field, including the implementation of the upper-air network in the late 1930s, the establishment of routine aircraft reconnaissance in the 1940s, the advent of numerical weather prediction in the 1950s, and the availability of satellite observations since the 1960s. He also describes organizational changes in the hurricane forecast offices and the interaction between the research and forecast communities.

Over the years tornado outbreaks have, with distressing regularity, caused numerous deaths, untold suffering, and massive amounts of property damage. Meteorologists have, with varying degrees of success, tried to understand and predict these outbreaks and provide timely warnings to the public. Ken Crawford and Ed Kessler, both from the University of Oklahoma, have combined efforts to document the history of these efforts in their chapter, "Severe Convective Storms: A Brief History of Science and Practice." Their account begins in the 1880s with the tornado studies of Signal Sergeant John P. Finley, surveys over a century of research in severe storms and the growing capacity of the National Weather Service to issue timely warnings, and concludes with a review of the technological foundation upon which more skillful forecast and warning decisions are being made in the 1990s.

Phil Krider, an international authority on atmospheric electricity, traces the development of research on the physical properties of lightning from 1919 to the present. His thorough review, "75 Years of Research on the Physics of a Lightning Discharge," discusses the luminous, electrical, and magnetic aspects of natural and triggered flashes between cloud and ground. He also reviews techniques used to detect and locate lightning discharges—a field in which he was a pioneer.

Climatology and hydrology

"Steps in the Evolution of Climatology: From Descriptive to Analytic," by John Kutzbach, traces two paths in our understanding of the climate. The first outlines ideas about the earth's energy budget from the time of Halley (1693) to the present; the second sketches concepts of the general circulation and hydrologic cycle. These paths have led to today's energy budget and dynamic climate models that are being used to reconstruct past climates, understand ocean–biosphere–atmosphere feedbacks, and evaluate the effects of human intervention in the climate system.

In his chapter, "Applied Climatology: A Glorious Past, An Uncertain Future," Stanley Changnon assesses the accomplishments and potential of this ever-changing interdisciplinary realm. The author defines applied climatology broadly as "any useful endeavor" that "describes, defines, interprets, and/or explains" the relationships between climate conditions and thousands of weather-sensitive activities; he locates it within a complex network bounded by fundamental climate research, climate data, and the users of climate information. After reviewing the field's distinguished past, Changnon concludes that major challenges remain, especially in applying climate information to key economic sectors such as agriculture, engineering, and energy production; in maintaining the integrity of the historic climate database; and in educating a new and more sophisticated generation of applied climate scientists.

Edwin Engman's chapter, "Hydrology in the Twentieth Century," traces the history of hydrology from its foundations as an engineering discipline for solving water resource problems through its development as a distinct and interdisciplinary geoscience. Changing technology, especially in computing and remote sensing, and the needs of other disciplines, including meteorology, have shaped the development of the field. The author provides a periodization of important ideas and individuals and discusses the challenges and opportunities currently facing the hydrologic sciences.

The private sector

In "A History of Private Sector Meteorology," David Spiegler traces the rise of the field since 1945, when some of the currently existing private weather forecasting companies began operations. He identifies 10 segments of the private sector—weather forecasting services, instrumentation companies, data and graphics providers, staff meteorologists in industrial companies, general meteorological consulting services, forensic meteorologists, specialty companies, environmental consultants, weather system developers, and media meteorology—that today generate a billion dollars worth of business annually. Using surveys, interviews, and his own extensive experience in the field, Spiegler's account includes memorable individuals, important companies, and significant events and factors in the development of the private sector.

Gordon Cartwright and Charlie Sprinkle have combined efforts to prepare a dynamic chapter, "A History of Aeronautical Meteorology: Personal Perspectives, 1903–1995," that clearly demonstrates the symbiosis between weather and flying. They divide their hybrid history of meteorology and aviation into three eras: before 1939, World War II and the emergence of global aviation, and the contemporary era. The authors' account is based on their extensive personal experiences in aviation meteorology, notably their lifelong involvement in the development of international institutions and accords. They conclude that the services generated in support of the aviation industry, both military and commercial, have had a profound impact on the whole field of meteorology and the development of weather services in all nations.

Roy Leep's career spans 40 years as a professional weathercaster. His chapter, "The American Meteorological Society and the Development of Broadcast Meteorology," spans over 70 years of history, from the radio broadcasts of E.B. Rideout in the 1920s on WEEI Boston, through the latest developments in TV weathercasting technology as exemplified by his own station, WTVT Tampa. Leep highlights the role of the AMS in setting professional standards through its Seal of Approval program and documents both the notable achievements and the current challenges facing the broadcast industry.

Education

"From Geo- to Physical Science: Meteorology and the American University, 1919–1945," by historian and geographer William Koelsch, examines alternative approaches to the institutionalization of advanced meteorological education. In the late nineteenth century, meteorology was not recognized as a distinct discipline. Introductory courses on weather and climate, if they were offered at all, were largely descriptive and were incorporated into university departments of geology, geography, physics, or even agriculture. In the 1920s and 1930s increasing interest in aeronautics helped meteorology find a tentative home in engineering departments. By the early 1940s the introduction of Norwegian methods of analysis, the war-induced influx of new students, and the rise of graduate programs at five major universities resulted in meteorology being recognized as a university discipline focused on the fundamental principles underlying the behavior of the atmosphere.

David Houghton, the current president of the American Meteorological Society, has prepared an account of "Meteorology Education in the United States after 1945." This chapter documents the support provided by AMS, the National Academy of Sciences, and a number of federal agencies for meteorological education from K–12 through the Ph.D. level. Starting with a handful of graduate programs just after World War II, the number of meteorology programs at universities now has reached 109, including 77 offering the Ph.D. degree. Membership in the University Corporation for Atmospheric Research, originally 14 in 1960, now stands at 61. Today's efforts are increasingly interdisciplinary, involving education in meteorological applications, oceanography, earth system science, and global environmental issues.

Special essays

The volume concludes with "Historical Writing on Meteorology: An Annotated Bibliography," a discussion of over 70 books and dissertations on the history of meteorology by myself and Colby College student Simone Kaplan. The references and annotations are meant to serve as a working "historian's bookshelf" and guide to further reading in the field.

This is followed by Julie Burba's "Historical Photo Gallery," which depicts meteorologists and meteorological equipment of bygone eras.

The history of meteorology is growing in popularity. In the last academic year alone (1994–1995), sessions on the history of meteorology were held at meetings of the American Meteorological Society, the American Geophysical Union, the History of Science Society, the Royal Meteorological Society, and the Society for the History of Technology. This newfound popularity may be due in part to increased media coverage of severe weather events, in part to growing awareness of the scientific and human dimensions of global change, in part to the retirement of a large number of meteorologists trained during and just after World War II, and in part to greater diversification among professional historians of science and technology. Whatever might be the proximate sources of this interest, we need and deserve to know more about the rich heritage of meteorology. There is much known and much yet to be learned. Indeed there are great opportunities here, especially for collaborations between enterprising scientist–historians and historians of science. It is a necessary step in the maturation of a scientific discipline.

James Rodger Fleming
Waterville, Maine
12 October 1995

Dynamic Meteorology and Numerical Weather Prediction

CHAPTER 1

The Evolution of Dynamic Meteorology

EDWARD N. LORENZ

Introduction

Philosophers have pondered the vicissitudes of the weather for millenia, but it is only during the twentieth century and part of the nineteenth that our understanding of the physical laws that govern the atmosphere has been thorough enough to enable us to account for what we observe. As meteorologists well know, our eventual grasp of the laws merely made it *possible* to account for things; the explanations themselves were not immediately forthcoming, and some of them elude us even today. In this exposition I shall be examining how the search for these explanations—the work of the dynamic meteorologist—has advanced during the lifetime of the American Meteorological Society (AMS).

What constitutes the *state* of an evolving scientific discipline—dynamic meteorology or something else—at a particular moment in history? Does it encompass the ideas taking shape in the minds of the most forward-looking scientists? Does it include only those ideas that have found their way into the refereed literature? Is it the knowledge that is regularly imparted in the classroom in institutions of higher learning, available to all who have the opportunity to enroll? Is it limited to the material appearing in textbooks, available to a still greater audience?

More generally, does the state depend upon the activities of a handful of leaders or a multitude of followers? Is it also partly determined by rather different factors, such as the existence and availability of intricate and often expensive equipment—giant telescopes for astronomers, instrumented seagoing vessels for oceanographers, and supercomputers for almost everybody?

I would maintain that in addition to describing the tools of the trade and noting such details as the physical locations where the relevant activities take place, a comprehensive account of a scientific discipline ought to cover the specific activities of the innovators and the typical activities of the ordinary practitioners. Some qualifying comments are in order. Even recently published scientific textbooks are often sadly out of date, if their authors are not currently active in the special fields about which they write, and errors sometimes seem to propagate without viscous dissipation from one book to the next. Nevertheless, textbooks, good or mediocre, and also programs of study, exert a distinct influence on what is taking place within a discipline; surely they are part of its state. At the other extreme, a leader's advanced idea that has yet to reach a journal may not influence anyone at all, unless it has been presented in a lecture or offered to a protégé as a possible topic for a dissertation. Even published works can lie dormant, but does this mean that they are not part of the state? Perhaps a satisfactory assessment of the state of a discipline needs to include some combination of ideas, publications, classes, and texts.

Next, what constitutes dynamic meteorology? Clearly the term alludes to methods of attack rather than specific atmospheric phenomena. For example, one can learn much about extratropical cyclones through careful examination of the extensive observational data that have been accumulating for many years. The task of the dynamic meteorologist, however, is to account for the observed characteristics of these cyclones by means of the physical laws that govern their behavior. More generally, the laws of hydrodynamics and thermodynamics constitute the foundation of dynamic meteorology.

There is no rule that says mathematical equations must be used in a dynamical study; witness, for example, Victor Starr's lucid equationless discussion of nearly 50 years ago on the controlling influence of angular-momentum transport on the general circulation (Starr 1948). Nevertheless, the more involved problems are often rather unapproachable if properly formulated equations are not put to quantitative use; some problems, in fact, have proven intractable even with equations.

The intricate laws of absorption and emission of radiation are often considered to be in the realm of physical rather than dynamic meteorology. Indeed, there is no universal agreement as to where dynamic meteorology stops and another branch takes over. In introducing a recent volume of the well-established journal *Tellus,* the newly installed editor has written (Sundqvist 1992), "the theme of *Tellus A* will continue to be *Dynamic meteorology and oceanography.* The series will encompass [all of] dynamic meteorology and oceanography, including numerical modelling, synoptic meteorology and weather forecasting, and dynamic climatology." As a dynamicist, I naturally think that synoptic meteorologists should be proud to be included among us, but I suspect that many synopticians do not share my feelings.

Finally, no account of dynamic meteorology can be complete without acknowledging that some of those who tackle and solve its problems do not even consider themselves to be meteorologists; often they are fluid dynamicists or more general applied mathematicians. Likewise, a complete account would have to recognize the field of dynamic oceanography, both because the ocean and the atmosphere have rather similar dynamics, so that results established for one system often apply to the other, and because the two systems are so strongly coupled physically.

As for the remaining word in my title, "evolution" is indeed an old term with a fairly general meaning, but it signifies, to many people, the specific process through which new living species come into being. In the case of our own species, early primates did not change continually until they all became human beings; there are still plenty of apes and monkeys around. More generally, older species sometimes survive even as new ones evolve from them. Sometimes the appearance of new environmental conditions will favor a mutated form of an established species, while, if the original conditions are still to be found at nearby geographical locations, the earlier form can continue to flourish.

This is what appears to have happened in dynamic meteorology. A new species of investigation has developed in response to the change in the scientific environment brought about by the advent of the computer, while those dynamicists who have preferred to make minimal use of computers have had no difficulty in keeping the older species alive. I shall presently consider this matter in greater detail.

The early years

For an overview of the state of dynamic meteorology in the early years of the AMS, let me turn to my own first experiences with the subject. The time was 1942, a few months after the United States had entered World War II, and I suddenly found myself transplanted from the Mathematics Department at Harvard University, a school located about two miles up the Charles River from the Massachusetts Institute of Technology (MIT), to MIT itself. The program at the MIT was the regular graduate course in meteorology, but, to accommodate the armed services in their effort to acquire a large number of weather forecasters within a short time, two ordinary academic years had been crowded into a fraction of one. Lectures filled the mornings, with map analysis and forecasting in the afternoons.

One subject that I found particularly inviting, with my mathematical background, was called "Dynamic Meteorology." Our teacher was the late Bernhard Haurwitz, already one of the universally acknowledged experts in the field. The textbook that he chose was also called *Dynamic Meteorology,* and, in fact, he was the author. Certainly he was in a position to write an up-to-date book, and he evidently took great pains to do so. Published in 1941, the book even contains a derivation of Rossby's celebrated formula for the speed of large-scale waves superposed on a westerly wind current, announced only two years previously (Rossby et al. 1939). It also mentions the even more recent suggestion of Starr and Neiburger (1940) that potential vorticity might be used as a tracer of atmospheric motions.

Much of the more recent material of the book was, however, offered only in more advanced courses, and most of the dynamics that we learned could have been taught when the AMS first came into being. We learned the equations of motion in their Eulerian form in a coordinate system rotating with the earth—the form still favored in the bulk of today's studies. We learned how to apply the equations to account for specific features—for example, the manner in which the wind typically varies with elevation through the lowest kilometer of the atmosphere in extratropical latitudes. This is of course the familiar Ekman spiral, a structure first deduced in an oceanic context (Ekman 1902). I had been looking forward to the moment when we would learn how to use the equations for forecasting—the task to which our accelerated program supposedly was primarily dedicated. That moment never arrived. Had I looked ahead in the book more carefully, I would have guessed that this would be so, for, at the opening of a chapter devoted to a rather different topic, Haurwitz (1941, 180) states, almost as an aside, "an attempt to compute the future weather by direct application of the equations of thermodynamics and dynamics seems at present not promising." After mentioning the well-known attempt by Richardson (1922), Haurwitz concludes the paragraph by saying that "a computation of the future weather by dynamical methods will be possible only when it is known more definitely which factors have to be taken into account under given conditions and which may be neglected." Here Haurwitz anticipates what would be realized a decade later, when the first moderately successful numerical forecasts would be produced by neglecting most of the conceivably important factors (see Charney et al. 1950).

Extratropical cyclones are arguably the most prominent features of sea level weather maps, and explaining their existence constitutes as fundamental a problem in dynamic meteorology as forecasting their behavior. The most commonly accepted hypothesis before World War II, championed by Vilhelm Bjerknes and his collaborators, was that they originated as small growing disturbances on the polar-frontal surface (see Bjerknes et al. 1933). Acceptance was often tempered by some reservations, since some of the observations suggested that a frontal surface and an accompanying cyclone would appear simultaneously rather than the former preceding the latter.

Haurwitz emphasized to us that the equations of motion were nonlinear, with the nonlinearity, in the absence of external heating and internal dissipation, arising from advective processes, which show up as quadratic terms in the equations. He devotes a chapter to the perturbation method, where linear equations are derived from the nonlinear ones, and he applies the method to such problems as the instability of a horizontal surface of discontinuity, but subsequently, in discussing the wave theory of cyclones, involving the postulated instability of a sloping surface of discontinuity, he states (p. 307): "the mathematical problems arising out of a study of these instability conditions are very complicated, and it cannot be claimed that a complete solution has been reached."

Certain nonlinear equations are readily solved analytically, but the equations of motion are not among them. Indeed, no workable method for solving the equations in their full nonlinear form was currently in use.

The great mutation

Long before World War II came to an end, meteorologists became aware that automatic electronic computers were in the process of development and that there would undoubtedly be attempts to use them to forecast the weather by numerically determining particular solutions of the dynamic equations. As to whether such attempts could ever succeed, there was no unanimity of opinion.

It would be easy to conclude that once that war had ended, meteorologists finally had time to become seriously involved with the problem, so that, soon afterward, numerical weather forecasting became a fact, but this does not seem to be the way that things generally happen. If during wartime a new development is perceived as being important to the war effort, it will probably progress faster rather than slower than it would in peacetime; witness the sudden implementation of radar. Numerical weather forecasting, whatever its importance, had to await the digital computer, which, even with such influential champions as John von Neumann, probably could not have been developed much sooner. Most likely it was coincidence that the first numerical integration of the barotropic vorticity equation (Charney et al. 1950) followed the war by only a few years.

The story of this integration, and of the continual advances in numerical forecasting that have filled the 45 years since that time, will be told elsewhere in this volume. What most of us may not have anticipated in the early days of computers, when they were a luxury to which few meteorologists had access, was the role that they would play in ordinary research. This possibility was forcefully revealed when Norman Phillips (1956) performed his first numerical experiment, where his numerical integration, extending over a simulated month, produced a general circulation, from which he could compile climatological statistics. In the years that followed, theoretical studies built around numerical solutions of the dynamic equations became more and more abundant as computers became available to more and more potential users. Gradually a new standard format for a dynamical investigation took shape.

First, in such a study, one formulates a specific hypothesis. Next, one constructs a model; this is typically a considerably simplified form of the dynamic equations, arranged for numerical solution. The model must not be so highly simplified that it cannot possibly represent the anticipated outcome, and in fact, it should be general enough to be able to represent other possible outcomes as well. Next, one obtains one or several numerical solutions of the model equations; often these are time-dependent solutions originating from selected initial states. Finally, one interprets the results and observes whether the hypothesis is supported.

Of course the format has a number of variants. In particular, despite what is sometimes maintained, a working hypothesis is not essential; one may simply ask a question. It seems perfectly legitimate, for example, to seek the stability of a specific flow pattern without first hypothesizing that it is stable or unstable.

Questions may be of various types. One that refers to the real atmosphere presumably cannot be answered unequivocally by a study that uses a model. However, a question may refer to what is in fact a model; for example, it might ask about an adiabatic frictionless atmosphere, or flow on a beta plane; it might also ask about a model that has been so simplified that analytic solutions are easily found.

Sometimes the computer is simply a convenience (or an inconvenience) rather than a necessity. If a model is moderately simple and only a steady-state solution is sought, the needed computations might require a few milliseconds on today's computers, but they might also be performed by hand in a week or so. Since several weeks would presumably be needed to write up the results afterward, the absence of a computer would not greatly increase the total effort. If instead the model is rather large, and if a number of time-dependent solutions are needed, what the computer could do in a few hours might take a lifetime to do by hand. Effectively the study would be impossible without the computer. Even if one were willing to devote a lifetime to a computation, the relevance of many questions does not endure for a lifetime.

The computer has thus given rise to a new type of investigation seldom encountered in yesterday's dynamic meteorology. In effect, the change in the scientific environment produced by the computer has enabled a mutation—the substitution of numerical for analytical procedures—to produce a flourishing species.

Meanwhile, the original species still thrives. Particularly when computers were rather new, there was a widespread feeling among mathematicians—von Neumann was the most notable exception—that numerical computation was not a legitimate method of solving problems. This attitude extended to a number of mathematically minded meteorologists and other fluid dynamicists.

We have seen that the perturbation method—linearizing the dynamic equations, or some simplification of them, about a known steady solution—will often produce a system of equations that may be solved analytically. Ordinarily it will offer a sound means for testing a flow pattern for stability. As time progressed, however, it became widespread practice to treat the solutions of perturbation equations as approximate solutions of the nonlinear equations themselves, regardless of whether their departures from any known steady solutions were in some sense small. Indeed, for many dynamicists who found solving the equations the most satisfying way of putting them to use, linearization became the method of choice.

The method acquired a number of refinements. One can formally express each dependent variable of a system of equations as a power series in some quantity, say ϵ; this may be any scalar that is representative of the amplitude of the perturbations. These series may be substituted into the equations, whereupon the coefficients of separate powers of ϵ form a sequence of systems of linear equations. The known steady solution satisfies the zero-order system, and with this solution, the first-order system becomes the usual set of perturbation equations. Having solved these, we can in principle solve in turn the second-order system, the third-order system, etc., thereby obtaining corrections to the perturbation solution. It appears that in many cases the series will converge for small values of ϵ. The possibility that they may not converge for the values of principal physical interest has not appeared to disqualify the studies for publication.

In a variant of this procedure, no basic flow is assumed, and the quantity ϵ is simply some quantity that may in some sense be considered small. A favorite choice is the Rossby number. Here the first-order system need not be linear, and instead of solving it one often chooses it as a new model of the original system (which may itself have been a model). Various quasigeostrophic systems, regularly used in the early days of numerical forecasting, and still extensively used in research, may be derived in this manner.

Numerous more elaborate techniques for handling the dynamic equations have been introduced, and I shall mention just one, which typifies the ingenuity that dynamicists often exhibit in their attempts to pursue an analytical approach. It is used when the solutions are expected to contain oscillations with contrasting timescales. It is known as two-timing, and it consists of replacing the independent variable, time, by two independent variables: fast time and slow time. Some dependent variables, perhaps divergence, are treated as functions of fast and slow time, while others, such as vorticity, may be treated as functions of slow time only. The procedure leads to logical complications, which appear, however, to have been reasonably well resolved.

The two species of investigation that I have described, along with their subspecies, are distinguished from one another by their methodology rather than by the particular meteorological problems they aim to solve. The upshot of the many developments is that, for many problems, linear and nonlinear systems now stand a more or less equal chance of yielding to solution by one method or another. It might be anticipated that linear theorists and nonlinear theorists would occupy opposing camps and denounce each other at scientific meetings, but this is by no means the rule. There have been numerous papers, often by single authors, that develop, in separate sections, the linear theory and the nonlinear theory of one and the same problem.

One may legitimately ask what is to be gained by working out the linear theory of a problem if the supposedly more accurate nonlinear theory is to be worked out in any case. There are actually a number of potential benefits. Linear and nonlinear methods agree fairly well when applied to certain problems and rather poorly when applied to others. If, by applying both types of method when this is feasible, one can get some idea of the sort of problem where linear methods do work well, one will have some idea of how much confidence to place in them when the nonlinear approach is not convenient. Perhaps more importantly, inspection of a linear solution, obtained by analytical procedures, often leads to some understanding of what is taking place to produce a particular effect, whereas observing a sequence of numerical steps may yield little insight. On the practical side, any particular numerical solution applies to only one set of values of the parameters of a system, while a single analytic solution can cover a wide range of values. I suspect, however, that many dynamicists continue to pursue analytical methods, which are likely to demand a linear approach, because they hold these methods in higher esteem than numerical ones.

New concepts

Wholly irrespective of the influence of the computer, which has brought about such remarkable changes in methodology, dynamic meteorology has, both before and after the advent of the computer, witnessed the appearance of a continual stream of new concepts. Some of these have been properties or processes that were waiting to be discovered. Some have been the dynamicists' own creations.

Let me offer, in approximate chronological order of their appearance, a rather incomplete list of terms denoting concepts that were unknown in the dynamic meteorology of 1920 but are much in evidence today. Some are now so familiar that it is hard not to think of them as always having belonged to the meteorological language. Time will tell whether some of the newer terms will acquire the same status. The list follows:

- isentropic analysis
- Rossby waves
- the beta plane
- potential vorticity
- baroclinic instability
- vacillation
- the balance equation
- available potential energy
- quasi-biennial oscillation
- low-order models
- CISK (conditional instability of the second kind)
- chaos
- global circulation models
- semigeostrophy
- enstrophy
- waveguides
- wave overreflection
- Eliassen–Palm flux
- gravity wave drag
- the surf zone.

I have purposely limited the list to 20 entries; it could easily have been made twice as long. Any discussion that I could present here, covering even 20 items, would have to do injustice to all or most of them. I have therefore chosen a sample of three for special consideration.

High on anybody's list of concepts that have altered the course of dynamic meteorology must be baroclinic instability—the instability of a flow possessing a continuous poleward decrease of temperature, and an accompanying continuous upward increase of westerly wind speed, as opposed to a flow with a sloping discontinuity in temperature and wind. It is hard to imagine any phenomenon that, following its discovery, has formed the subject of more papers and dissertations. The original works of Charney (1947) and Eady (1949), which first identified the phenomenon, used simple models and chose basic-flow patterns whose decreases in temperature occupied the entire width of the region involved, while the increases in wind occupied the entire depth. Subsequent studies have applied numerous models of varying physical complexity to numerous temperature and wind profiles. Other studies have attempted to assess the importance of baroclinic instability for the workings of the real atmosphere.

Dynamically, polar-frontal instability and baroclinic instability are perhaps equivalent phenomena, the former being a limiting form of the latter as the zone of transition shrinks to a single surface. It is noteworthy that in separate laboratory experiments, involving flow in rotating containers (see Fultz et al. 1959), the breakdown of a surface of discontinuity separating two homogeneous fluids of different densities, rotating at different rates, closely resembles the breakdown of a baroclinic zone. The mathematical details, however, are much more simply handled in the case of baroclinic instability—a situation that has undoubtedly favored the proliferation of studies of that phenomenon. Meteorologically, the phenomena are different in that polar-frontal instability, if it is present at all, is most pronounced in the lower troposphere and produces cyclones of limited horizontal extent, while baroclinic instability is more prevalent in the upper troposphere and gives rise to longer waves. A set of cyclones and a set of waves, if the latter is present above the former, are necessarily coupled in agreement with the hydrostatic relationship, but individual waves and their associated cyclones may move at different speeds and may even become decoupled as the intervening temperature field undergoes continual deformation. It was undoubtedly the meteorological rather than the mathematical distinction that Charney had in mind when, in recalling his early work, he emphatically maintained that the two phenomena were not the same (see Platzman 1990).

For idealized, zonally symmetric flow patterns, baroclinic instability is virtually an established fact, but it is likely not the answer to the origin of all large-scale tropospheric disturbances, even in temperate and polar latitudes. One seldom sees, even temporarily, essentially symmetric flow on which waves can proceed to grow—at any chosen time the waves are already present. Do new waves form because of the instability of flow patterns already endowed with waves? Do they instead form because the natural distortion of existing patterns produces details that are wavelike in structure? Are these two alternatives really the same thing looked at from different perspectives?

Contrasting with baroclinic instability in its nature, but rivaling it in its influence on dynamical thinking, is the concept of potential vorticity. Indeed, a recent paper (Holopainen and Kaurola 1991) even opens with the sentence, "potential vorticity is the backbone of dynamic meteorology." Whereas baroclinic instability is a *quality*, which the whole atmosphere, or a substantial section of it, may or may not possess, potential vorticity is a *quantity,* whose value, like temperature, varies from point to point throughout the atmosphere. It was identified by Rossby (1940) and in a more general form by Ertel (1942) as a quantity whose value at individual parcels of air does not vary under adiabatic flow. The proposal of Starr and Neiburger (1940) that it might prove useful as a tracer of atmospheric motions actually bore fruit more than a decade later when Reed and Sanders (1953) investigated an upper-level frontal surface and found that within the narrow transition zone the potential vorticity had high values typical of the stratosphere, while, on either side, the values were lower and typical of the troposphere. They came to the rather striking conclusion that the two boundaries of the transition zone were actually a section of the tropopause sharply folded over and enclosing air of strato-spheric origin.

It might appear that an outpouring of papers centered on potential vorticity, like the outpouring of papers on baroclinic instability that later followed Charney's and Eady's discoveries, should have quickly followed Rossby's and Ertel's findings, but evidently this was not the case. Whereas it was easy to test additional and presumably more relevant basic-flow patterns for instability, one had to think of specific problems where potential vorticity could be put to use. Not the least of the obstacles was the difficulty of determining actual detailed distri-butions of potential vorticity on any regular basis with the data then available, especially in the stratosphere where the flow was most likely to be nearly adiabatic.

Although potential vorticity is a fairly complicated function of the observable variables—wind, pressure, and temperature—it assumes a simpler form in an isentropic coordinate system. An important finding has been that a field of potential vorticity can be inverted; that is, if one knows the distribution of potential vorticity on each isentropic surface, together with the total atmospheric mass above each isentropic sur-face, and the temperature distribution at the base of the atmosphere, one can, under the assumption of hydrostatic and quasigeostrophic balance, determine the complete fields of wind, pressure, and temper-ature. The actual inversion process may be awkward to implement but it becomes fairly simple in a quasigeostrophic model where the potential vorticity assumes a rather simple form and can even serve as the basic prognostic variable.

Once the routine use of large global circulation models, extending to considerable heights, had become standard practice in operational weather forecasting, it became possible to evaluate the potential vorticity on a point-by-point and day-by-day basis from the analyses constructed by the models for use as initial states. This in turn made it possible to construct sequences of synoptic charts showing the distribution of potential vorticity on various isentropic surfaces.

The individual studies that have benefited from the availability of these charts are too numerous to list, and the reader is referred to the comprehensive discussion by Hoskins et al. (1985), with its 176-entry bibliography. Here I shall mention only one of the more spectacular applications, that of McIntyre and Palmer (1983). By examining a succession of charts, they have detected the breaking of Rossby waves in the stratosphere. That is, the waves become irreversibly deformed, like ocean waves as they approach the shore, and ultimately bring about an injection of air from lower latitudes into the polar vortex; in doing so they dissipate a substantial amount of kinetic energy.

A final concept to which I have personally devoted much attention is what is popularly called chaos. A chaotic system is one that exhibits sensitive dependence on initial conditions; that is, a small alteration of the state at one time will lead eventually to a state differing considerably from the state that would have occurred if the alteration had not been made. The atmosphere cannot be examined directly for chaos, but numerical studies with models with various degrees of complexity leave little doubt that the atmosphere is chaotic. Indeed, the nonlinearity introduced by the presence of advective processes—the only nonlinearity in some of the simpler models—is by itself quite sufficient to produce chaos.

The most obvious and most familiar consequence of atmospheric chaos is the limitation that it places on the possibility of forecasting most aspects of the weather pattern at long range, say two weeks or more in advance, in view of the impossibility of starting from a perfect analysis or using a perfect extrapolation procedure. However, there are more fundamental changes in dynamical thinking that the recognition of chaos has brought about.

For example, we now realize that many of the equations that we would like to solve—even some rather simples ones—do not possess general solutions that can be expressed in analytic form. If we have succeeded in analytically determining particular solutions, presumably steady or periodic ones, and if our equations are realistic enough to have captured the atmosphere's chaotic nature in their general solutions, our solutions will be highly specialized ones, and their relevance to the real atmosphere will at least be suspect.

Should we conclude, for example, at least in the context of a model that we are using, that the long-term average transport of angular momentum across middle latitudes is poleward, if a particular solution that we have found analytically should show a poleward transport? The real atmosphere does not exhibit a continual poleward transport but instead possesses periods of poleward and also of equatorward transport, with the former dominating. An extended solution of any realistic model should be expected to behave likewise. One property of chaotic solutions is that if one can identify a segment, and there should be many such segments, where the initial and final states are very much alike, a slight alteration of the initial state will produce a segment where they are exactly alike, that is, a periodic solution, albeit an unstable one. Hence, we might have happened upon an analytic solution in which equatorward transport prevailed, rather than the solution that we did find. This being the case, can we show that the solutions with poleward transport are in some sense more representative? Are they, although unstable, perhaps less unstable? Here are plenty of problems left for the dynamicist to think about.

It also appears that, more generally, we must meticulously avoid obtaining several numerical solutions for any problem and then concluding without further inquiry that the ones that support a previously conceived hypothesis are the more representative ones. Certainly we must avoid acknowledging only these solutions in our write-up, even though we could do so without falsifying any results. Similarly, if we have been forced to base our conclusions on a single chaotic solution, we should be aware that another chaotic solution of the same equations might have led us to conclude something else.

So far I have been stressing the possibility that a system once thought to be behaving regularly may be found to be chaotic. We should not ignore the possibility that some systems once thought to be *random* may be found to be chaotic, in which case we may be able to discover their underlying dynamics. Chaos indeed has its positive side as well as its negative one.

Concluding remarks

In summary, the aim of dynamic meteorology—the determination and explanation of the observed or observable properties of the atmosphere through application of the physical laws—has not changed during the past three-quarters of a century. The total effort has undergone some shift from explanation to mere determination, if we regard forecasting as the determination of the course of individual states of the atmosphere. The forecasting problem has introduced subsidiary dynamical problems, such as parameterization and initialization.

The most evident change has been in methodology. The construction of models and the subsequent determination of time-dependent numerical solutions, using the computer, has not simply dominated the forecasting process; it has become a favorite research method. Meanwhile, new analytical techniques, as well as new numerical ones, are continually being devised.

Where will dynamic meteorology go in the years to come? I cannot visualize any single new event that will so profoundly affect dynamic meteorology as the development of the computer, although, had I been a meteorologist rather than a small boy 75 years ago, I presumably would not have visualized the coming of the computer. Perhaps the best that I can do is to extrapolate current trends.

What dynamic meteorology seems likely to do is to shift toward regions and aspects of the atmosphere that have been somewhat neglected in the past. Already more attention is being paid to the stratosphere and the tropical troposphere, where typical processes seem to be less nonlinear, and to mesoscale and smaller-scale systems, where they may be even more nonlinear. As for longer-period behavior, we do not yet know to what extent the progressive changes of climate, other than those associated with changing external conditions, are determined by the climate itself and to what extent they are mere statistical residuals of shorter-period fluctuations. I feel confident that dynamic meteorology will provide the answer in the coming years.

REFERENCES

Bjerknes, V., J. Bjerknes, H. Solberg, and T. Bergeron, 1933: *Physikalische Hydrodynamik.* Springer, 797 pp.

Charney, J.G., 1947: The dynamics of long waves in a baroclinic westerly current. *J. Meteor.*, **4**, 135–162.

——, R. Fjörtoft, and J. von Neumann, 1950: Numerical integration of the barotropic vorticity equation. *Tellus*, **2**, 237–254.

Eady, E.T., 1949: Long waves and cyclone waves. *Tellus*, **1**, 33–52.

Ekman, V.W., 1902: Om jordrotationens inverkan paa vindströmmar i hafnet. *Nyt. Mag. Naturv.*, **40**, 1–18.

Ertel, H., 1942: Ein neuer hydrodynamischer Wirbelsatz. *Meteor. Z.*, **59**, 271–281.

Fultz, D., R.R. Long, G.V. Owens, W. Bohan, R. Kaylor, and J. Weil, 1959: *Studies of Thermal Convection in a Rotating Cylinder with Some Implications for Large-Scale Atmospheric Motions. Meteor. Monogr.*, No. 4, Amer. Meteor. Soc., 104 pp.

Haurwitz, B., 1941: *Dynamic Meteorology.* McGraw-Hill, 365 pp.

Holopainen, E., and J. Kaurola, 1991: Decomposing the atmospheric flow using potential vorticity framework. *J. Atmos. Sci.*, **48**, 2614–2625.

Hoskins, B.J., M.E. McIntyre, and A.W. Robertson, 1985: On the use and significance of isentropic potential vorticity maps. *Quart. J. Roy. Meteor. Soc.*, **111**, 877–946.

McIntyre, M.E., and T.N. Palmer, 1983: Breaking planetary waves in the stratosphere. *Nature*, **305**, 593–600.

Phillips, N.A., 1956: The general circulation of the atmosphere: A numerical experiment. *Quart. J. Roy. Meteor. Soc.*, **82**, 123–164, 535–539.

Platzman, G.W., 1990: Charney's recollections. *The Atmosphere—A Challenge*, R.S. Lindzen, E.N. Lorenz, and G.W. Platzman, Eds., Amer. Meteor. Soc., 11–85.

Reed, R.J., and F. Sanders, 1953: An investigation of the development of a midtropospheric frontal zone and its associated vorticity field. *J. Meteor.*, **10**, 338–349.

Richardson, L.F., 1922: *Weather Prediction by Numerical Process.* Cambridge University Press, 236 pp.

Rossby, C.-G., 1940: Planetary flow patterns in the atmosphere. *Quart. J. Roy. Meteor. Soc.*, **66** (Suppl.), 68–87.

——, and Collaborators, 1939: Relation between variations in the intensity of the zonal circulation of the atmosphere and the displacements of the semipermanent centers of action. *J. Mar. Res.*, **2**, 38–55.

Starr, V.P., 1948: An essay on the general circulation of the earth's atmosphere. *J. Meteor.,* **5,** 39–43.

———— , and M. Neiburger, 1940: Potential vorticity as a conservative property. *J. Mar. Res.,* **3,** 202–210.

Sundqvist, H., 1992: Editorial. *Tellus,* **44A,** 1.

The Origin and Rise of Numerical Weather Prediction

GEORGE P. CRESSMAN

Introduction

The first known effort to compute atmospheric changes by physical methods was recorded by Felix M. Exner (1908). Exner was a professor of geophysics at the University of Vienna and the author of a widely circulated book on dynamic meteorology. His first calculations of this type were made at a time when very little was known about atmospheric circulations above the ground and they were restricted to low levels and a small area. Exner's computations, previously almost completely unknown, are still being investigated in Germany and, it is hoped, may be reviewed and published in detail again within the next few years.

Modern scientific weather prediction, called "numerical weather prediction" followed earlier work by Vilhelm Bjerknes in Norway and experimental calculations by Lewis F. Richardson in France and in England. Whether or not Bjerknes or Richardson knew of Exner's calculations is not known and probably not important, since Bjerknes's and Richardson's research diverged from Exner's work.

In this century, a historic divide in meteorology was crossed with the development of the electronic computer. By this time a sufficient basis in theory was available and could be put to use following the successful development of the electronic computer, first at the Aberdeen Proving Grounds in Maryland and then at the Institute for Advanced Study in Princeton, New Jersey. There John von Neumann and Jule Charney led the work to combine modern meteorological theory with the means to compute a forecast. The successes at the Institute for Advanced Study quickly led to the establishment of operational numerical prediction, where theory and facts were sometimes in agreement and at other times in disagreement, with ensuing reconciliation and benefits to weather forecasting. Numerical weather prediction became the accepted approach to the weather forecast problem and is still yielding improvements and many benefits to those who must enjoy or endure the weather.

Earliest efforts—Felix M. Exner, Vilhelm Bjerknes, and Lewis F. Richardson

Exner's calculations, long forgotten and now rediscovered, were appropriate to the available data and to the then-current knowledge of atmospheric motions as a function of elevation above the earth's surface. Exner assumed that the nonperiodic air pressure variations extend only as high as 5-km elevation and that the motion of the air at and above 5 km is parallel to the motion of the air at the earth's surface. He then computed the lower-level temperature changes due to horizontal advection, as modified by diabatic and adiabatic effects, yielding a map of forecast pressure changes at the surface of the ground. Some of his results were fairly realistic. Together with Alfred Defant, Exner calculated forecast charts daily over a period of several weeks. His method of calculation and one of his experiments were published in *Meteorologische Zeitschrift* in 1908 and most likely came to the attention of Bjerknes and, later, Richardson. Exner was well aware of problems arising from his upper-boundary condition. He wrote "as our method cannot take into account special events in the upper air layers, errors cannot be avoided in such circumstances."

The origin of the principles used in modern numerical weather prediction can be traced back to 1898 when, according to Friedman (1989), Bjerknes presented a lecture on the circulation theorem, suggesting its use to clarify the nature of atmospheric motions. In 1901 Bjerknes expanded the theorem to its now-classic formulation. Wanting to incorporate physics into weather forecasting, Bjerknes envisioned starting with a complete set of initial conditions of the three-dimensional atmosphere and then using graphical methods for solution of the equations and determination of the rates of changes in the atmosphere. While sufficient initial data were not available and a clearly laid-out procedure eluded him, his vision was far ahead of his time. Beginning on page 36 of his book *Appropriating the Weather, Vilhelm Bjerknes and the Construction of a Modern Meteorology,* Friedman mentions Bjerknes's interest in a three-dimensional analysis of the atmosphere for forecast purposes and he emphasizes there and elsewhere Bjerknes's interest in considering atmospheric motions through the circulation theorem.

Carl-Gustav Rossby was a student of Bjerknes and spent several of his younger years under Bjerknes's influence. Given Bjerknes's interest in and use of the circulation theorem, it should not seem surprising that the young Rossby followed his interest in this direction. Rossby subsequently produced his work on long waves in the westerlies by looking at large-scale atmospheric circulations through a vorticity theorem. As is well known, this approach eventually led to the barotropic forecast, to the use of equations from which gravity waves were filtered, and to an early success in numerical weather prediction. In chapter 5 of his book (1989), Friedman describes the establishment, beginning in 1917, of a program of research and weather prediction in Bergen, in which Bjerknes and his assistants used hydrodynamic equations in an effort to calculate a 3-h forecast of the wind field. Its failure, according to Bjerknes, was because their wind data were unrepresentative of true atmospheric conditions. Eliassen (1956), reviewing this matter much later, discovered that Bjerknes did not use the continuity equation to obtain the initial vertical component of the wind, for which there was no way of direct measurement. The result of the ensuing inconsistent initial wind field was the "unrepresentative" data that spoiled his calculations.

Richardson, making a similar effort at about the same time, used the continuity equation to obtain initial vertical velocities but failed for other reasons. Richardson's strategy was to use the undifferentiated equations of state and motion (the primitive equations) in a direct assault on the forecast problem. His principal occupation at the time was as an ambulance driver in the British Army during World War I, a serious preoccupation that certainly interfered with his scientific interests. Richardson's failure to achieve credible results has often been attributed to a lack of sufficient and accurate initial data. Perhaps so, but the task he undertook was much too large for one person doing arithmetic by hand, which completely excluded the possibility of experimentation with numerical methods because of the enormous amount of time that it would require. Richardson lacked a method of initialization of the wind and mass fields and was planning to use large time steps, on the order of many hours, for continuing the integration, as described in his book *Weather Prediction by Numerical Process*

(1922).[1] If his initial data did not destroy the calculation at the beginning, computational instability would have done so later. [The computational stability criterion of Courant et al. (1928), which would have required much shorter time steps to advance the calculation, had not yet been published.] According to Brunt (1939), Richardson concluded in 1926 that "barotropy is not even an approximation to the truth in the atmosphere." The conclusion was wrong but consistent with Richardson's thorough and comprehensive approach to an attempted forecast calculation. Thus, Richardson could find no way to simplify the calculation to a level where doing arithmetic by hand might have yielded some kind of answers.

Oliver Ashford (1985), a personal friend of the Richardson family, has written a biography of Richardson and his many interests. According to Ashford, upon the first publication of Richardson's book in 1922, Exner wrote to Richardson in a quite critical tone, advising Richardson to publish a work on theoretical meteorology instead of concerning himself with an immediate application to prediction (Ashford 1985, 9).

Carl-Gustaf Rossby

Here our attention must shift to Carl-Gustaf Rossby. In 1939, at a joint meeting of the Royal and American Meteorological Societies in Toronto, Rossby presented a paper of fundamental importance, "Planetary Flow Patterns in the Atmosphere" (Rossby 1940). Rossby's conclusion was that, *the factors determining the stationary or progressive character of the motion are to be found in the vorticity distribution and that the displacement of the pressure field is a secondary effect."* Taking absolute vorticity as a conserved quantity, Rossby derived an equation for the movement of the deep atmospheric waves aloft as a function of wavelength and the speed of the large-scale zonal flow. If he was correct, it would follow that many of the large-scale (or planetary) waves could be described as essentially barotropic in nature. This was a rather incredible notion to meteorologists whose education emphasized fronts, cyclones, and a description of conversions from potential to kinetic energy. In 1939, even fronts were a radical notion to many. It

---

[1]Richardson's book gives full details of his forecast calculation.

seems quite likely that Rossby either was not understood or not believed. The minutes of the meeting, published in the same issue of the *Quarterly Journal of the Royal Meteorological Society* as Rossby's paper, contain no mention of any discussion following his presentation except that "Prof. C.-G. Rossby presented the results of his own studies and of some investigations by Mr. R.A. Allen, whose paper also appears in this issue." Throughout the 1940s, considerable forecast skill was found in this relationship, giving credence to Rossby's insight (Namias and Clapp 1944; Cressman 1948).

In his 1939 paper, without giving it a name, Rossby introduced the concept of "potential vorticity" [p. 71, Eq. (10)]. Potential vorticity was formally introduced in a paper by Ertel and Rossby (1949) during or following a visit Rossby made to Ertel in Berlin. The mathematical elegance of this paper suggests a dominance by Ertel in writing it. The origin of the concept of potential vorticity is a matter that seems to be a delicate one and has aroused some passion in private correspondences.

The first barotropic forecasts

In 1949 Jule Charney published "On a Physical Basis for Numerical Prediction of Large Scale Motions in the Atmosphere," a paper of fundamental importance. In the paper Charney showed that since the large-scale motions of the atmosphere are approximately geostrophic and hydrostatic, they can be described in terms of a geostrophic potential vorticity. A time integration of the prognostic equations can thus be accomplished using the proper combinations of geopotential. This brought numerical prediction within the range of possibility as soon as a suitably powerful computer became available. In a 1950 letter to George Platzman, Rossby wrote, "I am completely convinced that the pilot project discussed last summer and carried further by Charney in a recent memo to von Neumann, namely the computation of the state of motion in an 'equivalent' barotropic atmosphere (with enough convergence and divergence to keep the signal velocity finite) is not only psychologically important to the Princeton group but could also lead to highly significant practical results" (Platzman 1979).

A feasibility test was successfully accomplished with a barotropic model in 1950 by Charney et al. (1950) on the ENIAC (Electronic Numerical Integrator and Calculator) computer at the Aberdeen Proving Grounds. ENIAC was sponsored by U.S. Army Ordinance and was hundreds of times faster than the fastest electromechanical computers. The computer program was hardwired. ENIAC could multiply 10 two-digit numbers in 1/300 s (Aspray 1990). The forecast was not a good one but it looked meteorological. The important thing was that the computation proceeded as planned, stayed within scale, and produced a plausible-looking 500-mb chart for a forecast. The accounts by Platzman (1979) and Aspray (1990) taken together give a very detailed account of the ENIAC and the 1950 computation, which should be of interest to those interested in both computer and meteorological history.

The earliest barotropic forecasts (and baroclinic forecasts as well) employed the quasigeostrophic approximation. In this, the mass field carried the history of the forecast, and a geostrophic wind field was used in place of the actual winds. (A comprehensive analysis and history of the quasigeostrophic approximation was produced by N.A. Phillips in 1989). From 1958 to 1965, an essentially nongeostrophic barotropic forecast encompassing practically the whole Northern Hemisphere, but containing only enough divergence to allow the mutual adjustment of wind and pressure fields, carried out to two or three days in advance was found to be the best midtropospheric forecast tool available. Part of the reason for the longevity of barotropic forecasting was that the construction of truly better baroclinic models of the atmosphere was a very complex matter requiring rigorous physical and mathematical consistency. In addition, baroclinic forecasting had to be paced by the availability of larger and faster computers than were available for some years. The lesson learned was that *barotropic processes are mostly dominant in large-scale atmospheric motions for a timescale of a few days.*

Ragnar Fjörtoft (1952) scored a significant coup with the publication of a method for one- or two-level forecasts based on the conservation of absolute vorticity and potential temperature. The forecasts could be accomplished in real time by graphical integration methods, thereby bringing the early proposal by Bjerknes to fruition. A two-level forecast for one-third of a hemisphere could be produced with Fjörtoft's method by two or three people in an hour. This method was introduced and used for a while in 1952 and 1953 in several U.S. Air Force forecast centers by K.R. Johannessen and me. It was also taught in several U.S. universities. Sverre Petterssen and Frederick Sanders are said to have used this method in about 1954. In the air force effort, the graphical arithmetic was accomplished with the use of overlaid acetates illuminated from below on a light table. The lines were drawn in various colors of grease pencil. The graphical arithmetic was accomplished by drawing a new chart with lines through intersections of previous lines in one direction or the other. The grease pencil lines tended to rub off on the clothing of the forecasters. Understandably, very few forecasters had both the energy and the interest to do this on a routine basis, and it was never widely used. For those who tried it, it had the effect of removing some of the mystery from numerical forecasting, which could then be anticipated as something the average meteorologist could hope to understand. The use of this method in the United States, and probably abroad as well (at least in Norway), can be counted as the first known operational use of numerical weather prediction.

The beginning of baroclinic forecasts

During the two years after 1950 a larger and faster electronic computer designed by John von Neumann was under construction at the Institute for Advanced Study at Princeton. In 1952, a meteorology group was established at the institute with the objective of applying dynamical laws to the solution of the forecast problem. Charney was the leader of the group. On 10 June 1952, the institute announced the successful development of the new electronic computer, sometimes referred to as the "Johniac," after von Neumann. Its key features were the following: (a) its program and the initial data were stored in an internal electrostatic memory of 1024 40-bit words ("bits" equal binary digits); (b) the program could modify itself as the calculation progressed; and (c) its operation was parallel, that is, on all bits in a word at once. The electrostatic memory of the computer was somewhat volatile. The program had to be checked a number of times, and many parts of the calculation had to be done twice and in different ways where possible. Periodic storage of intermediate results on the computer's magnetic drum was needed to enable recovery in case of a subsequent computer error or "crash" (a catastrophic error).

While I was temporarily assigned to the Institute for Advanced Study at Princeton in 1953, Dr. Bruce Gilchrist taught me how to program and operate the Johniac. The computer had a characteristic called "read-around," in which high-frequency reference to memory at a given address (spot) on a memory tube could contaminate the memory at adjacent spots. I had an unforgettable experience while running a program on the computer. The address of a particular memory location in an instruction erroneously altered itself by an address change of two, which turned that part of the program into a two-word loop at the highest possible frequency. The event happened at night in a dark room. The sudden appearance of a very bright spot on the face of the memory tube could be seen across the small room, a frightening sight that suggested possible imminent damage to the tube. An emergency shutdown of the computer prevented damage.[2]

For a series of tests, Charney chose a previously unpredicted case of sudden and strong cyclogenesis that occurred from 24 to 25 November 1950. The storm, completely unforecast by the traditional methods of the time, had caused major disruptions of all kinds in the mid-Atlantic States, as it delivered torrential rain and strong winds and then flash floods and heavy snow. The storm appeared to be a good test case for Charney's new quasigeostrophic forecast models. Charney's test calculations failed to predict the storm (post hoc, of course) with a one- and then a two-level quasigeostrophic model. Both models were based on the conservation of entropy and of geostrophic potential vorticity. Charney tried another forecast, with the vertical resolution increased to three levels. It produced cyclogenesis in the northeast states, centered about 240 miles north of the correct position. The forecast was better than the position would indicate, because it did develop a very intense storm near the mid-Atlantic coastline but not quite as intense as that that occurred in nature. It was a stunning success by the standards of the day (Charney 1954).

In late 1953 Charney used the same model to compute a successful test forecast for another previously unforecast storm, of 6 November 1952, that had dropped six inches of snow on unsuspecting Washington, D.C. When the test forecast was finished, very late at night, Charney telephoned Harry Wexler, director of research of the U.S.Weather Bureau in Washington, woke him from a sound sleep, and informed him, "Harry, it's snowing like hell in Washington on November 6, 1952." (I know because I was in Charney's office at Princeton at the time.)

Following these and other successes, the heads of the Weather Bureau and the military weather services of the U.S. Air Force and the

---

[2]A detailed history of the development of the computing facility at Princeton is available from Aspray (1990).

U.S. Navy established the Joint Numerical Weather Prediction (JNWP) Unit with the task of producing regular numerical weather forecasts on a daily basis. The heads of meteorological services were F.W. Reichelderfer, Weather Bureau; Thomas S. Moorman, U.S. Air Force Weather Service; and Robert O. Minter, U.S. Naval Weather Service. The three services shared the funding and personnel needed for the JNWP Unit. This could hardly have come at a worse time for the Weather Bureau. Reichelderfer had to close 20 Weather Bureau field stations to find the funds to make up for a retrenchment in the Department of Commerce as well as to fund the Weather Bureau's share for the new numerical prediction activity. Let Reichelderfer speak for himself:

> No one can say at present what the future applications of numerical techniques of weather prediction will turn out to be. That they might turn out to be revolutionary in synoptic practice is well known to those close to the investigations. We desire to take the fullest practical part in contributing to the developments with the modest facilities and staff we are now able to devote to the subject. The paucity of our staff and facilities for participation is no measure of our interest in the development. (In a letter dated 18 November 1952 to F.G. Shuman, assigning him to the new Joint Numerical Weather Prediction Unit.)

Reichelderfer had a reputation in some circles as being too conservative and parsimonious. I knew him well to the end of his life and can say with certainty that this was an undeserved calumny.

The first numerical predictions prepared on an electronic computer and issued in real time (before the weather changes had taken place) were prepared three times a week in Stockholm on the Swedish "BESK" computer by an international group (sometimes known as Rossby's group, since Rossby had returned to his native country, Sweden). Beginning in December 1954, the predictions were prepared with the use of a barotropic model. The forecast domain was limited to the North Atlantic area, giving stabilization of the very long atmospheric waves by the fixed boundaries of the relatively limited forecast domain. This activity was concluded after a successful trial period and resumed again at a later date (Bergthorsson et al. 1955).

The beginning of permanent operational numerical prediction

In the United States, an IBM 701 computer was procured for the initial daily operation. On 6 May 1955, the first of regular and continuing daily numerical weather forecasts was issued. The initial decision was made with substantial expectation of achieving more accurate daily forecasts, but it certainly was not made immediately. A development group was included in the organization of the JNWP Unit for the purpose of increasing the scope and accuracy of the future forecast products. This group was led initially by Philip D. Thompson.

The quasigeostrophic approximation, which brought us to this point, proved to have a serious and unexpected fault. This approximation substantially overestimated the extremes of cyclonic vorticity and was responsible for, among other things, producing overestimation of deepening in cases of cyclonic development. This bias was countered by another bias toward insufficient cyclonic development of storms because of low vertical resolution of the models that could be handled by computers of the time and by not including water vapor and the release of latent heat in the models. These and other complications, not fully appreciated at the time, had to be dealt with later, one at a time.

In fact, it is fair to say that when the first regular use of numerical weather prediction began, the results were not nearly as good as the earlier tests had indicated. It had not occurred to us that we could have success with very difficult baroclinic events and then have severe problems with many routine, day-by-day weather situations. In response, our efforts—and those of Rossby's group, as well—had to shift to attacking the forecast problem with the use of the simpler barotropic model. Mastering that, we felt regular positive benefits of numerical weather prediction, while at the same time we could make more deliberate progress toward realizing the benefits for which we had hoped. In retrospect, the decision to start a permanent numerical forecasting activity must be seen as a wise one, even though insufficient testing of the early Princeton model had been done. Without the daily confrontation of theory with fact and the constant intense efforts to improve the forecasts, several more years would have passed before reliably useful numerical predictions would have been available. Thus, the early decision to begin daily operational numerical forecasts might be seen as either a wise or a lucky decision. More details of this difficult period are described by Shuman (1989).

Another seldom-discussed matter was that weather forecasts in the mid-twentieth century were really not very good. Simple extrapolation and experience of the forecasters were useful for the first 24 hours of a forecast, with some notable exceptions, but the forecasts still left very much room for improvement. Unforecast major storms would seriously inconvenience the public and often led to a cascade of public and congressional criticism. Phillips (1989) quotes R.C. Sutcliffe, a highly respected British meteorologist and forecaster, as writing "I was closely in touch with both the British and American weather services in the late phases of the war and can say with some assurance that the techniques of the day were entirely innocent of dynamic theory." A better way to make forecasts was urgently needed.

Dissenting opinions

In 1956, Norbert Wiener of the Massachussetts Institute of Technology (MIT), published his conclusion that dynamic prediction of atmospheric developments would not work and was a mistake (Weiner 1956). He stated that,

> It is bad technique to apply the sort of scientific method which belongs to the precise, smooth operations of astronomy or ballistics to a science in which the statistics of errors is wide and the precision of observations is small. In the semi-exact sciences, in which observations have this character, the technique must be more explicitly statistical and less dynamic than in astronomy.

Wiener's views had no discernible effect on the continuing rapid development of numerical weather prediction but may have been influential in the subsequent establishment of serious work on the development of statistical forecast methods. According to Aspray (1990), in 1945 Wiener had attempted unsuccessfully to lure von Neumann to chair the Department of Mathematics at MIT. Not all the skepticism about numerical prediction came from academia. Petterssen once remarked to me that it was not realistic to base a forecasting method on the correctness of the calculations of the second horizontal derivative of geopotential on a pressure surface. However, this may have been a sample of the subtle humor he often enjoyed, for Petterssen also supported the initial efforts to develop numerical prediction and then to make it operational.

With the kinds of conflicting advice available from prominent and respected scientists, the heads of the three U.S. weather services, made a correct and forward-looking decision to proceed with the beginning of an operational numerical weather prediction activity. That they did so showed they had confidence in themselves and in their own advisors. From today's viewpoint their decision might be counted as more luck than foresight. But then, didn't Napoleon prefer lucky generals?

Early operational forecasting

The first operational numerical forecasts in the United States began with the once-daily issuance of geostrophic barotropic forecasts out to 72 hours and with the additional preparation and issuance of forecasts made with a three-level quasigeostrophic model. This was essentially the model that Charney used for the test in Princeton. At first the initial data were analyzed by traditional methods and read off at grid points for card punching and entry into the computer. After the first six months of operation, the data were prepared by an objective analysis program beginning in April 1956. This suite of products was essentially a transplant of the developments from the Institute of Advanced Study at Princeton. Regular exposure to these products on a day-by-day basis brought out all sorts of problems, which could have been fatal to the nascent activity except for the low quality of the products of the traditional methods of forecasting. In fact, administrators and forecasters both showed a tolerance toward the initial problems, with the faith that they could be solved. The administrators and forecasters did not expect instant success but they wanted continued progress.

The poor quality of the observational data received over the international communication networks presented an unexpected problem. Many ambiguities and errors in encoding observations and in data identification were a serious handicap to the development of automatic data processing and analysis. Within several years, these were gradually reduced or eliminated through a special effort by the World Meteorological Organization (WMO). The fact that sloppy international data encoding and collection had existed for many years may tell us something about the former use of such worldwide data, namely, that there were many omissions and mistakes in analyses because of misplaced, miscoded, or garbled data (Bedient and Cressman 1957).

The early barotropic forecasts in Sweden and in the United States developed disagreeable symptoms when the area of integration was expanded to a large area, for example, one-half of a hemisphere. The first was a distortion of the large-scale flow beginning in low latitudes.

This was partly a consequence of the divergence of the geostrophic wind (Shuman 1957; Cressman and Hubert 1957). The problem was not noticed in earlier forecasts, because the fixed boundary conditions of a small area severely restricted large-scale changes. This was an interesting situation in that the geostrophic assumption, which had made possible the start of numerical prediction, quickly became a barrier to further progress. The problem was partly resolved by replacing the geopotential in the model by a nondivergent streamfunction. This remedy in turn was subsequently complicated by another error in formulating the finite-difference analog of the balance equation, which had to be found and corrected (see Shuman 1989, 289). Both errors resulted in similar symptoms.

Another problem appeared as a retrogression of the waves of planetary scale (which have the slowest motions of all in the spectrum of observed atmospheric waves). This was diagnosed as a spurious retrogression of the atmospheric waves of wavenumber approximately 1–4. The problem was traced to a difficulty in the mutual adjustment of wind and pressure in the barotropic formulation then being used. This matter had been discussed earlier by Rossby but was not so easily identified as such at the beginning of the numerical forecasts. In some cases these effects were catastrophic, leading to the most urgent efforts to find a remedy. Early remedial efforts included a bizarre method of periodically extracting the initial very long waves and enforcing stationarity before putting them back into the forecast (Wolff 1958). After a rereading of Rossby's earlier papers (Rossby 1940, 1945) and the thesis of Yeh (1949), a much simpler remedy was effected, namely, the introduction of an implied tropopause and stratosphere into the barotropic models (Bolin 1956; Cressman 1958). This allowed for the mutual adjustment of mass and motion fields within the forecast. The change actually sped the forecast since relaxing the resulting Helmholz equation was much faster than relaxing the previously used Poisson equation.

If all this seems rather odd to the reader, we need to remember that a realistic-looking barotropic forecast was usually, in those days, the best midatmospheric forecast available. Some good forecasters used to say that if we could give them a good 500-mb forecast, that was all they needed to produce the forecast at other levels—an exaggeration but it contained much truth. In his 1985 WMO interview, K. H. Hinkelmann made the following comment:

> I had very valuable discussions with Arnt Eliassen and I read papers by Charney and by Eady on how to operate a baroclinic model. But at the 500 hPa level the barotropic model produced pretty good results, and it was going to be a very long time before we developed a baroclinic model which could do as well, let alone better (Taba 1988).

The somewhat unexpected success of the barotropic forecast model with a balanced wind approximation led us to a view of the atmosphere in which *the large-scale daily motions appear to be mostly equivalent barotropic in nature, with occasional bursts of baroclinic activity*. In a 1987 interview Fjortoft, commenting on the success of the barotropic model in forecasting, remarked that,

> ... there is seemingly still a scientific challenge: to find the degree to which barotropic and baroclinic effects can be considered as operating separately and to what extent they must be regarded as coupled phenomena (Taba 1988).

Baroclinic forecasts and the primitive equations

While numerical weather prediction started as an attempt to solve the problem of forecasting baroclinic events, its first few years were occupied with many other kinds of problems, such as very large scale atmospheric divergence; advancing from quasigeostrophic forecasting; expanding the domain of the forecast to near-hemispheric, automatic processing and analysis of data; and reduction or elimination of computational problems. Yet NWP had been launched on a wave of enthusiasm about being able to forecast those relatively rare sudden energy conversions that could make or break the reputation of the profession of meteorology. The three-level baroclinic model of Charney, which so successfully forecast the November 1950 storm, proved unsuitable for day-by-day baroclinic forecasting, leaving us with a void in this important effort. Several two-level models were proposed and tried. Thompson, in an unpublished manuscript (copy available from me), wrote, "By 1953, no less than five major research groups, at the Air Force Cambridge Research Center, the British Meteorological Office, the International Institute of Meteorology at the University of Stockholm, the Deutscher Wetterdienst, and the Japanese Meteorological Agency, had developed and partially tested two-level models based on Charney's quasi-geostrophic equations." At the JNWP Unit considerable use was made of a similar model developed by Thompson and Gates

(1956). These two-level models had much in common with the earlier models by Fjortoft and by Eliassen (1956). Initial data were used from the surface and from the midtroposphere. Energy and vorticity equations were used to compute time derivatives. In general, these two-level models were not capable of forecasting baroclinic development. The reason was that in the type of baroclinic development associated with storm development there is a significant generation of midtropospheric absolute vorticity by horizontal convergence. This is, of course, not possible to model directly when no data levels are used above the midtroposphere. Therefore, an absolute minimum of three data levels had to be carried in a successful baroclinic model (Cressman 1961).

Other factors that could not be ignored were the divergent component of the wind in all advections and latent heat of condensation. Inclusion of these proved to be a severe strain on the computer resources available in the early 1960s if the numerical forecast was to be delivered in time for use. Baroclinic models based on the vorticity equation and using an initially nondivergent wind field obtained from the balance equation proved capable of forecasting areas of storm development but with insufficient strength of the storm. Use of the primitive equations, as attempted by Richardson, was not practical before the 1960s. The short forward time step required in this approach placed such demands on a computer that the forecast could hardly advance faster than the weather itself. Early pioneering efforts at construction of primitive equation models were carried out by Phillips (1959), Hinkelmann (1959), Richtmyer (1963), Shuman (1964), and Robert et al. (1970). In the same WMO interview mentioned earlier, Hinkelmann was quoted by Taba as saying,

> . . . the primitive equations promised to simulate atmospheric dynamics and energetics more realistically, and moreover, they should work in the tropics where the geostrophic approximation is invalid. Finally, the extra computer time required because of time steps in minutes rather than hours would, to some extent, be offset by much simpler solution operations and avoiding time-consuming relaxations and iterations. On my first attempt, using idealized initial data, I got a most encouraging result which reproduced new developments, occlusions and even the kinks of isobars along a front.

In the United States, we proceeded with the development of baroclinic models along two lines. One was to compute the forecast with a full balance between wind and geopotential fields, and the other was to include the latent heat of condensation. A minimum of four data levels in the vertical were included. This project was dropped before operational implementation when Shuman and Hovermale (1968) completed and tested a baroclinic primitive equations model suitable for daily forecasting. This was an entry into a new age of numerical forecasting and of forecasting in general. We should also note that a balanced model of this complexity no longer had a significant operational speed advantage over a primitive equation model of the same resolution. The outcome of the competition between filtered and primitive equation models was settled in favor of the latter by the nearly continuously advancing speed of commercially sold computers.

Shuman (1978) presented a record of a skill score of the 36-hour 500-mb prognoses of the circulation over North America, as issued by the U.S. National Meteorological Center. The *increase* from prenumerical forecast days in 1955 to 1965, which can be designated as the end of the early period of operational forecasts, was 46% of the total skill score of the subjective prognoses at the beginning of numerical prediction. Additional and substantial increases have continued, but that matter exceeds the scope of this paper.

Thus, in 1965 the early days of numerical weather prediction could be considered at a close, with the implementation of daily forecasting with primitive equation models. The modeler's attention had to turn from large-scale dynamics to the physics and continuing numerical problems of modeling. One era came to an end and another was launched. Continuing advances in the inclusion of physics in models were enabled by the flexibility of the models and by the continuing advances in computer development. It is a good thing we did not have supercomputers at the beginning. There was too much to learn before we could have been ready for them.

Postscript

The reader of this account may have noticed the informal awareness and communication of ideas across international boundaries of the western nations. It was an informal club of good friends and colleagues. The leading workers in numerical weather prediction were in frequent communication through letters, telephone calls, and visits, both short and extended. Nothing was secret, and there were no monetary charges to each other for data, computer time, methods, or results. What benefitted one benefitted all. No significant advance was private for more than a very short time. It was an ideal milieu for scientific progress—a Camelot that deserves to be remembered.

Acknowledgments. I gratefully acknowledge the receipt of very helpful reprints from F.G. Shuman, including a copy of the paper by Ertel and Rossby, which I thought I would never find. Anders Persson, Norman Phillips, and George Platzman were very helpful in locating important references and reprints. The work of F. Exner was kindly brought to my attention by Hans Volkert of the Instut für Physik der Atmosphäre, Wessling, Germany. Thanks are due to Mrs. Lisa Shields of the Irish Meteorological Service, who provided an excellent and elegant translation of Exner's *Meteorologische Zeitschrift* paper.

REFERENCES

Ashford, O.M., 1985: *Prophet or Professor? The Life and Work of Lewis Fry Richardson.* Adam Hilger Ltd., 304 pp.

Aspray, W., 1990: *John von Neumann and The Origins of Modern Computing.* The MIT Press, 376 pp.

Bedient, H.A., and G.P. Cressman, 1957: An experiment in automatic data processing. *Mon. Wea. Rev.,* **85,** 333–340.

Bergthorsson, P., B.R. Doos, S. Frylkund, O. Haug, and R. Lindquist, 1955: Routine forecasting with the barotropic model. *Tellus,* **7,** 272–274.

Bolin, B., 1956: An improved barotropic model and some aspects of using the balance equation for three-dimensional flow. *Tellus,* **8,** 61–75.

Brunt, D., 1939: *Physical and Dynamical Meteorology.* Cambridge University Press, 428 pp.

Charney, J.G., 1949: On a physical basis for numerical prediction of the large scale motions in the atmosphere. *J. Meteor.,* **6,** 371–385.

———, 1954: Numerical prediction of cyclogenesis, 1950. *Proc. Natl. Acad. Sci. USA,* **40,** 99–110.

———, R. Fjörtoft, and J. von Neumann, 1950: Numerical integration of the barotropic vorticity equation. *Tellus,* **6,** 309–318.

Courant, R., K. Friedrichs, and H. Lewy, 1928: Uber die partiellen differenzen Gleichungen der mathematischen Physik. *Math Ann.,* **100,** 32–74.

Cressman, G.P., 1948: On the forecasting of the long waves in the westerlies. *J. Meteor.*, **5**, 1273.

______ , 1958: Barotropic divergence and very long atmospheric waves. *Mon. Wea. Rev.*, **86**, 367–374.

______ , 1959: An operational objective analysis system. *Mon. Wea. Rev.*, **87**, 367–374.

______ , 1961: A diagnostic study of midtropospheric development. *Mon. Wea. Rev.*, **89**, 74–82.

______ , and W.E. Hubert, 1957: A study of numerical forecasting errors. *Mon. Wea. Rev.*, **85**, 235.

Eliassen, A., 1956: A procedure for numerical integration of the primitive equations of the two-parameter model of the atmosphere. Sci. Rep. 4, Meteorology Department, University of California, Los Angeles.

Ertel, H., and C.-G. Rossby, 1949: A new conservation—Theorem of hydrodynamics. *Geofis. Pura Appl.*, **XIV**, 3–4.

Exner, F.M., 1908: Uber eine erste Annaherung zur Vorausberechnung synoptischer Wetterkarten. *Meteor. Z.*, **25**, 57–67.

Fjörtoft, R., 1952: On a numerical method of integrating the barotropic vorticity equation. *Tellus*, **4**, 179–194.

Friedman, R.M., 1989: *Appropriating the Weather, Vilhelm Bjerknes and the Construction of a Modern Meteorology.* Cornell University Press, 251 pp.

Hinkelmann, K., 1959: Ein numerisches experiment mit den primitiven Gleichungen. *The Atmosphere and Sea in Motion,* B. Bolin, Ed., Rockefeller Institute Press, 486–500.

Namias, J., and P.F. Clapp, 1944: Studies of the motion and development of long waves in the westerlies. *J. Meteor.*, **1**, 57–77.

Phillips, N.A., 1989: *The Emergence of Quasi-Geostrophic Theory.* Historical Monograph Series, Amer. Meteor. Soc., 321 pp.

Platzman, G.W., 1979: The ENIAC computations of 1950—Gateway to numerical weather prediction. *Bull. Amer. Meteor. Soc.*, **60**, 302–312.

Richardson, L.F., 1922 (1965): *Weather Prediction by Numerical Process.* Dover Publications, 236 pp.

Robert, A.J., F.G. Shuman, and J.P. Gerrity Jr., 1970: On partial difference equations in mathematical physics. *Mon. Wea. Rev.*, **98**, 1–6.

Rossby, C.-G., 1940: Planetary flow patterns in the atmosphere. *Quart. J. Roy. Meteor. Soc.,* **66,** 68–77.

______ , 1945: On the propagation of frequencies and energy in certain types of oceanic and atmospheric waves. *J. Mar. Res.,* **2,** 38–55.

Shuman, F.G., 1957: Predictive consequences of certain physical inconsistencies in the geostrophic barotropic model. *Mon. Wea. Rev.,* **85,** 229–234.

______ , 1978: Numerical weather prediction. *Bull. Amer. Meteor. Soc.,* **59,** 5–17.

______ , 1989: History of numerical weather prediction at the National Meteorological Center. *Wea. Forecasting,* **4,** 289–296.

______ , and J.B. Hovermale, 1968: An operational six-layer primitive equation model. *J. Appl. Meteor.,* **7,** 525–557.

Taba, H., 1988: *The Bulletin Interviews.* World Meteorological Organization, 405 pp.

Thompson, P.D., and W.L. Gates, 1956: A test of numerical prediction methods based on the barotropic and two-parameter baroclinic models. *J. Meteor.,* **13,** 127–141.

Wiener, N., 1956: Nonlinear prediction and dynamics. *Procedings of the Third Berkley Symposium,* J. Neyman, Ed., University of California Press, 247–252.

Wolff, P.M., 1958: The error in numerical forecasts due to retrogression of ultra-long waves. *Mon. Wea. Rev.,* **86,** 1–16.

Yeh, T.-C., 1949: On energy dispersion in the atmosphere. *J. Meteor.,* **6,** 1–16.

Observational Tools and Computational Devices

CHAPTER 3

The Evolution of Atmospheric Measurement Systems

ROBERT J. SERAFIN

Introduction

A wide range of topics can be considered in looking at the history of systems for observing the atmosphere. There is an ever-expanding set of tools for measuring wind, pressure, moisture, and temperature at the surface and aloft. There are sophisticated tools for measuring trace chemical species in the atmosphere and the characteristics of the deep oceans. There are research measurement systems and operational systems, and a revolution is taking place today in the operational forecasting systems used by the National Weather Service (NWS). In keeping with the theme of the AMS 75th anniversary meeting, this chapter focuses on operational observing systems and their relationships to weather services.

Meteorological observations have formed the basis of the science of meteorology for millennia (Frisinger 1977). To measure winds, the ancient Egyptians and Chinese used weather vanes, and the Greeks examined the deformation of trees and the waves in fields of grass. Aristotle wrote about the relationships between cloud formations and winds. Measurements of temperature, humidity, and pressure were advanced by scientific giants, including Galileo, Torricelli, and Hooke.

Weather observational networks in the western world began in Italy in about 1650. Weather service networks in the United States began in the 1800s. The telegraph influenced Joseph Henry, secretary of the Smithsonian Institution, to organize a regular weather observing network in 1849 (Fleming 1990), and the U.S. Weather Service was established in 1870 under President Ulysses S. Grant, as part of the Department of War (Whitnah 1961; Grice 1985).

A principal foundation of modern meteorology was established when Vilhelm Bjerknes (1904) stated that "determination of the atmosphere at the initial time is the task of observational meteorology" and "sufficiently accurate knowledge of the laws according to which one state of the atmosphere develops from another" should be the bases for weather forecasting. Three years earlier, Cleveland Abbe had proposed that weather forecasting should become a mathematical science (Gedzelman 1994). While Bjerknes's hypothesis was simply posed, establishing the initial state of the atmosphere has been no simple task. Early attempts at three-dimensional measurements used human-piloted balloons, kites, and aircraft (Haig and Lally 1958). It was the radiosonde, however, that revolutionized three-dimensional observational measurements of the atmosphere. By the 1940s the radiosonde had become the mainstay of the weather forecasting community.

The postwar years

Radar was invented before World War II. [See chapter 4 by Rogers and Smith, this volume.] Soon after its introduction as an aircraft detection and tracking device, radar operators learned that weather signals could also be detected and, indeed, interfere with the primary mission of detection and warning of aircraft. In the 1940s young meteorologist cadets, such as Dave Atlas and Lou Batten, were being taught the principles of radar so they would be able to exploit the radar systems as a method for detecting and tracking precipitation. In the book *Radar Days,* written by E. G. Bowen (1987), I found some very interesting comments. Bowen refers to the British researcher R. A. Watson Watt's suggestion in 1935 that radiowaves reflected from aircraft could be used to detect the aircraft position. Bowen noted that the energy would be less than a million million millionth part of that sent out; the efficiency would be 10^{-19}. There were many people who thought this could not be done and it would be foolish even to try.

Well, so much for conventional wisdom. We all know what radar can do today. A modern meteorological radar such as a NEXRAD (Next Generation Weather Radar) can detect small concentrations of insects at ranges of 60 miles and greater. Wind-profiling radars receive backscatter from the turbulent atmosphere in order to measure winds through the troposphere in optically clear air.

The first Weather Radar Conference was held at the Massachusetts Institute of Technology in 1947. At that meeting there was a great focus on the quantitative measurement of rainfall with radar. The weather radar conferences of the AMS represent the longest standing continuous series in the Society.

In the late 1950s, the U.S. Weather Bureau installed its network of WSR-57 radars in the United States. These noncoherent radars, or non-Doppler radars (augmented by C-band and S-band local warning systems) have remained for more than 30 years the mainstay of the severe weather warning system in this country.

However, in the 1960s while we were busy operating a noncoherent radar network for operational purposes, there was much work under way in the field of Doppler techniques. People prominent in the field included Roger Lhermitte, Herb Groginski, Dave Atlas, Roddy Rogers, Ralph Donaldson, Dick Doviak, Richard Strauch, and many others, including me. This was an exciting time in radar meteorology since there were many technical questions that had not been answered. These questions related to signal processing theory and the accuracy with which Doppler velocities could be measured. It was in the late 1960s when some of the signal processing work of Groginski (1972), Rummler (1968), Lhermitte (1972), people at the Chicago School (Denenberg et al. 1972), and others finally could quantify the accuracy of the measurements.

At this time, however, in addition to the theoretical questions, there was a second major impediment to developing practical operational or research Doppler radars; it was the inability to process, rapidly enough, the complex radar signals in real time. While the algorithms were well understood, the technology of signal processing hardware had not yet been fully developed. This second constraint was removed when the availability of low-cost, integrated circuits set the stage for a decade of exciting development in the Doppler radar field.

By the late 1970s the joint Doppler studies in Oklahoma (JDOP), involving the National Severe Storms Laboratory and the Air Force Geophysical Laboratory, had definitively established what many of us had believed for many years: Doppler radars through their observations of circulations in severe storms could improve short-term forecasts of tornadoes and other hazardous phenomena (Burgess et al. 1979). In 1978, microbursts were detected for the first time by Ted Fujita (1979), using the National Center for Atmospheric Research (NCAR) Doppler radars. By 1994, Doppler radars were being installed in major airports around the country, and the NEXRAD network was well on its way to completion.

An important adjunct to the Doppler radar research with ground-based systems was the development of the tail Doppler radar systems in the mid-1970s for the National Oceanic and Atmospheric Administration (NOAA) P-3 research aircraft. The work by Hildebrand and Mueller (1985) and Mueller and Hildebrand (1985) clearly established the value of these systems for mapping three-dimensional winds from the moving platform. Today these systems are invaluable tools for monitoring and studying tropical cyclones. NCAR's ELDORA (Hildebrand and Moore 1990) system takes these capabilities one step further by creating a dual-Doppler radar system while flying a straight-line track out of harm's way. The P-3 aircraft now have been modified with this capability as well.

An excellent reference for the history of radar meteorology is found in the book *Radar in Meteorology*, edited by David Atlas (1990) and published by the AMS. This book is a collection of the contributed papers from the Battan Memorial Conference held in honor of Lou Battan, who had recently died, and his family.

The 1960s must also be considered the decade of satellite meteorology. I am sure none of us can forget the day when we learned that the Russians had successfully launched their *Sputnik*. On this date, 4 October 1957, I was an undergraduate electrical engineering student at Notre Dame University. Several of my friends and I were sitting together, listening to the radio, when the incredible news release was broadcast. We wondered how the United States could and would respond. By 1961, in a famous speech before Congress, President Kennedy called upon the United States to place a man on the moon by the end of the decade (Johnson 1994). He also asked Congress for $53 million to fund the earliest possible satellite system for worldwide weather observation. [See chapter 5 by Purdom and Menzel, this volume.]

In 1954 Harry Wexler of the Weather Bureau publically proposed the use of satellites in meteorology. The TIROS (Television and Infrared Observation Satellite) project began in about 1958 and soon found itself under the direction of the newly formed National Aeronautics and Space Administration. The first meteorological satellite, *TIROS-1*, was launched in April 1960. By 1965, *TIROS-9* had been launched and within 20 days of its orbit yielded the first mosaic picture of the entire earth. We were finally able to view our entire planet as a whole. Even today this capability absolutely amazes me. Anytime of the day or night we can tune to the Weather Channel or the local news and have a look at global and national weather systems and their evolution.

In the 1960s the descendants of TIROS, the Environmental Survey

Satellites, were launched. These satellites represented the first joint effort in the field of satellite meteorology between the United States and the Soviet Union. Even during the Cold War, meteorologists on both sides of the Iron Curtain worked together, perhaps primarily because of the global nature of meteorology.

Geostationary satellites were first proposed in 1959, and the first successful launch of the Advanced Technology Satellite was in 1966. The success of the spin scan cloud camera, invented by Vern Suomi, etched into history Vern's enormous contributions to the field. By 1978 we had three GOES (Geostationary Operational Environmental Satellite) satellites in orbit that formed the basis of the Global Atmospheric Research Programme's Global Weather Experiment. After much delay and concern, we have, as of April 1994, successfully launched *GOES-8,* a three-axis stabilized design; *GOES-9* will also soon be operational. [*GOES-9* is now operational.]

Modern weather observations—The future

The very strong technological and observational base established in the postwar years set the stage for modernized weather services. The restructured and modernized U.S. Weather Service will focus upon four major components: 1) the NEXRAD radar systems, approximately 160 of which will be installed throughout the contiguous United States and Alaska and Hawaii; 2) the Automated Surface Observing System, which will provide automated observations of temperature, humidity, winds, pressure, precipitation, ceiling, and visibility (these measurements, made at 1-min intervals, will be available through digital communications channels for use by forecasters throughout the country); 3) the next generation of earth-orbiting satellites and the new GOES that, because of its tri-axis stabilized design, provides for simultaneous imaging and vertical profiling; 4) the Advanced Weather Interactive Processing System (AWIPS) that brings together numerical data and observational data to forecasters at a modern workstation for interactive manipulation and analysis as they perform their daily forecasting tasks.

While these are the four core elements of the modernized National Weather Service, the observations will be augmented by a host of other observational systems. These include the terminal Doppler weather radar, which will be operated by the Federal Aviation Administration

(FAA) to protect about 40 airports; Doppler wind profilers and possibly thermodynamic profilers (which are now still operated in an experimental demonstration mode only); a national lightning network; and aircraft in situ observations made by commercial aircraft.

As part of its restructuring, the National Weather Service is examining the configuration of the sounding system of the future. The principal system upon which weather services have relied for so many years has been the national and global radiosonde measurements. Despite the elegance and enormous value of the radiosonde, its deficiencies are also well recognized. In particular, observational data over the oceans are insufficient in their temporal and spatial resolution. And, twice-daily observations are insufficient for mesoscale forecasting on the regional and local scales. Despite these deficiencies, we have not yet found a satisfactory replacement.

There is now strong consensus that the sounding system of the future will rely upon many technologies. To be sure, there will be a next-generation radiosonde that is likely to use Global Positioning System (GPS) technology for wind finding. This rawinsonde will also be equipped with lightweight next-generation sensors and will transmit all of its data digitally. Such systems are currently being evaluated by the National Weather Service, by several manufacturers, and by NCAR. However, improved spatial and temporal resolution and continuity will depend upon several other technologies. Commercial aircraft in the United States today are producing thousands of observations of winds and temperatures daily (Benjamin et al. 1991), and an operational humidity sensor is currently being tested and evaluated by the Commercial Aircraft Sensing of Humidity program, sponsored by NOAA, the FAA, and the National Science Foundation (NSF)/NCAR (Fleming and Hills 1993). Aircraft observations are already being used to improve numerical weather products currently being produced by the National Centers for Environment Prediction (formerly the National Meteorological Center) in 3-h Rapid Update Cycle.

Yet, while commercial aircraft are likely to play a significant role, their observations will be obtained primarily at flight altitudes, although soundings will also be available in the vicinity of principal airports during approach and departure. Over the oceans, data will still be lacking. It is here that greater reliance must be placed upon observations from both geostationary and orbiting satellites. Indeed, the satellite-based GPS provides a very interesting and new alternative for global sensing of humidity and temperature. It is my impression that the GPS may become one of the more important inventions that will profoundly influence society in the next 20 years. It is likely to revolutionize air traffic control systems and navigation of all kinds in the air, at sea, and on land. It will also be enormously beneficial in search and rescue missions.

The potential of GPS for meteorological applications depends upon the very precise and accurate phase of the GPS signals. The differential phase delay, as GPS signals propagate through the atmosphere, can be measured. This phase delay is related directly to the index of refraction of the atmosphere from which water vapor and/or temperature can be inferred. For ground-based GPS receivers, this phase delay is interpreted as vertically integrated water vapor. Research under way at the University Corporation for Atmospheric Research and NCAR, by Steve Businger at the University of Hawaii, by scientists in Russia, by scientists at the Jet Propulsion Laboratory, and in many other places, is leading to practical applications of this technology.

In 1995 a low earth orbiting satellite with a GPS receiver onboard was launched (Ware et al. 1996). This launch and test are supported by the private sector, NOAA, the FAA, and the NSF. The project, termed GPS/MET, is aimed at establishing the feasibility of measuring profiles of the index of refraction and assimilating these profiles into numerical weather prediction models.

Another interesting sounding concept is the use of adaptive networks. It is now known through adjoint modeling techniques that numerical weather forecasts can be improved if higher-resolution observations can be made in specific regions, since the models are most sensitive to errors in these regions. However, these regions—often associated with strong gradients—are not fixed geographically; they move with the weather. It is therefore being proposed that low-cost pilotless vehicles be used to probe the atmosphere adaptively for improved initialization of numerical models (Langford and Emanuel 1993). Specific examples might include the northeast Pacific Ocean, for improved forecasting of extratropical cyclones, and regions of the Tropics in the vicinity of potentially hazardous tropical cyclones. A region of the globe that is of great interest for the European continent is the North Atlantic Ocean. There are at least two versions of small pilotless vehicles that might qualify as adaptive sensors. One version is that being produced by Aurora Flight Systems, Manassas, Virginia, which would have the capability of carrying several dropwindsondes for many days over remote regions and deploying them upon command. A second interesting technology is the very small aircraft produced by the INSITU Group, Underwood, Washington. The INSITU aircraft itself serves as the probe for measurements of winds, temperature, and humidity and communicates these data back through a satellite to a central point, to be assimilated into numerical models. Commercial aircraft may serve as a third source to an adaptive network, through flight-level measurements of winds, temperature, and humidity and by adding a dropsonde capability to provide profiles.

Wind profiling from the surface is another promising technology. The NEXRAD system itself produces profiles of the horizontal wind several times per hour in its normal scanning modes. These profiles often extend to as high as 10 km and routinely are available within the planetary boundary layer. The longer wavelength wind profilers, however, provide much more reliable and regular observations of the winds. The hybrid sounding system of the future is virtually certain to include some wind profilers distributed throughout the country for the continuous measurements of horizontal winds. Measurements of three-dimensional wind fields over the oceans and remote regions of the globe are possible, using space-based lidar techniques. Baker et al. (1995) suggest that recent advancements in 2-μm lasers make the technology competitive for small satellite missions. Abreu et al. (1992) describe an alternate approach, using shorter-wavelength lidar.

I think you will agree with me that we are on the threshold of a new era. The number of observations available to weather forecasters in time and in space will increase by a factor approaching 100, and these will be available in real time. Perhaps equally important, these observations will be in digital form and of research quality. I envision a time when local radar observations at a weather forecasting office, combined with soundings from an approaching aircraft, will allow for the initialization of a finescale cloud model embedded within a regional-scale mesoscale model, embedded within a global numerical weather prediction model. The future of meteorology should include a four-dimensional weather database that is neither model nor observation but a combination of the two and, moreover, is neither model oriented nor observational system oriented but rather consists of information on the phenomenology of the atmosphere. This four-dimensional database, which Sandy McDonald and the Forecast Systems

Laboratory refer to as the Aviation Gridded-Forecast System for aviation purposes, can and should serve a much broader audience. Its data should be accessible to forecasters, air traffic controllers, pilots, airline dispatch centers, highway management authorities, the agricultural community, the media (including cable television), and the general public.

Concluding remarks

It is worthwhile to consider some of the nontraditional applications of improved weather observations. In the context of weather services, we frequently point to the great damage caused by weather phenomena. Flash floods, lightning, tornadoes, and damaging winds account for the losses of many lives per year, and weather continues to be a factor plaguing aviation. To be sure, a better weather observational and forecasting system will, with appropriate adaptation strategies, reduce losses associated with all weather hazards. However, there are also applications of another nature that are very exciting and that we stumble upon almost by accident. Let me give you a few examples of these.

Recently I visited Texas A&M University and there learned about an interesting application of NEXRAD technology. Mike Biggerstaff, on the faculty in the Department of Meteorology, has been interested for some time in better quantitative precipitation measurements with NEXRAD but had been unable to obtain support for his work. Somewhat accidentally, he became acquainted with people in the Texas agricultural community who needed better precipitation information. Their interest relates to the application of insecticides on crops. Apparently there is a form of weevil that resides in the soil and emerges when soil moisture reduces the soil friction sufficiently for the weevil to squirm out. Because precipitation measurements and soil models are currently not adequate for good forecasts, farmers apply insecticides on a regular and periodic basis. The agricultural community is optimistic about obtaining broad-area, quantitative precipitation measurements from NEXRAD and introducing this into its soil moisture and hydrological models. The potential benefits are substantial. Farmers save money as a consequence of fewer applications. Meteorologists, through the research, obtain better precipitation measurements that translate into improved hydrologic forecasts. The general public benefits through higher quality food at lower cost and a cleaner environment.

Everyone wins in a situation like this. We note that the problem is interdisciplinary in nature, involving specialists in entomology, agriculture, meteorology, land surface processes, and hydrology. Partnerships such as these will lead, I am sure, to many exciting applications of the new technology that will benefit society.

A second interesting application has to do with the electric power industry. On a recent visit to the National Weather Service Forecasting Office in Phoenix, Arizona, I spoke with meteorologists about their experiences with the NEXRAD radars. These experiences have generally been very positive. I also visited with staff of the Salt River Project, which is a water management and electric power generation and distribution organization headquartered in Phoenix and collocated with the local office of the National Weather Service. Staff at the Salt River Project described their interest in thunderstorm forecasting. Of course, they were very interested in quantitative precipitation measurements over their catchment basins that extend over large areas of Arizona and that drain into several reservoirs. However, I learned that they are also interested in the cooling that takes place as a consequence of outflows from convective storms in the summertime. This cooling effect can last for several hours and is used tactically in real time to predict power demand and therefore to decide whether to buy power from or sell power to other regions of the country. Here is another example of how two communities, public and private, can work together to create previously unforseen opportunities to their mutual benefit.

There is another perspective that I would like to share. Improved weather observational systems will produce for us a mesoscale, climatological database of unprecedented detail and quality. This database will be important for many segments of our society but also for those working in the field of climate and climate change. We will have, for the first time, a permanent and quantitative record of the frequency of severe weather, cloudiness, surface variables, upper-level winds, temperatures, and moisture. This record can be used for validation of regional climate simulations and for testing climate change hypotheses. The database will also be useful for the detection of any regional climate changes that might be occurring.

The future of meteorological measurements is likely to be influenced positively by a change in technological emphasis from military applications to public and private sector applications. Satellite technologies that might have been used for military reconnaissance will be used instead to observe the earth's system and its changes. Radar technologies, once designed for tracking aircraft, are now more likely to be used for observations of the atmosphere and the earth's surface while new technologies, such as GPS, will assume a greater role in controlling air traffic. Very sophisticated military systems, previously of a classified nature, are likely to be made available for probing the atmosphere.

None of the technological future would be possible without the rapid advance of computational capability. In each two- to three-year period, we are seeing a doubling, tripling, or quadrupling of the power of computers. We already have workstations available to us that are capable of storing enormous quantities of information and running complex numerical simulations, previously the domain only of the most powerful supercomputers.

Yet, despite all of this technological development, there are still some fundamental sensing problems facing us. Measuring humidity remains a substantial challenge in the often hostile environment of clouds and the upper atmosphere. Single-parameter measurements with radar can take us only so far; we must develop practical multi-parameter systems for improved precipitation measurements for meteorology, hydrology, agriculture, and many other applications. We must place greater emphasis on the integration of spaceborne and earth-based observations for more accurate interpretation of the physical and dynamical aspects of the atmosphere.

From antiquity until now, measurements have been at the core of the science of meteorology. Observations of the atmosphere raised the questions that stimulated the creative minds of meteorologists for millennia. Until the twentieth century, models of how the atmosphere works had been largely conceptual and qualitative. Now, we are experiencing an explosion in the field, with modern measurements and modern computers that have produced quantitative numerical simulations. All of this has occurred during the 75 years of the American Meteorological Society's existence.

This framework is now providing unprecedented opportunities for further advancement. Still, at the core of meteorology will be the much improved observations. Increasingly, however, more creative and accurate interpretative methods will be devised. Satellite remote sensors do not measure temperature or moisture directly but rather measure very accurately the radiance (the electromagnetic radiation) emanating from the atmosphere. Radars measure the reflectivity factor, not the precipitation rate. And, sensors such as GPS receivers measure the index of refraction, while others may measure electromagnetic wave absorption. Researchers have dealt with these issues in the past by attempting to translate nontraditional variables into what meteorologists have traditionally used—namely, temperature, moisture, humidity, winds, pressure, and precipitation. And much remains to be done in this arena. Increasingly, however, we must view the measurement of atmospheric quantities as a multivariate estimation process through which multiple data sources are combined to produce the best estimate possible of the state of the atmosphere.

The merger of numerical modeling with these diverse data sources may lead to the greatest advances. Data assimilation is a rapidly evolving field of research whereby modelers are experimenting with new methods for assimilating the diverse data now available to them into their modeling schemes. Rather than transforming all observations into atmospheric variables, many modeling centers are instead translating the model data into the observed variables (e.g., model temperature profiles are translated into radiance or index of refraction). These model-created variables are then compared to the observations of radiance or index of refraction, and the models are adjusted accordingly.

We are now on the threshold of developing a meteorological database that will be source independent. This database will speak to users in language that they understand through complex language translation methods. The community of users will become much less concerned with radar data, satellite data, surface data, or model data but will work with the products of the merger of these data through assimilation techniques. The result will be more accurate nowcasts and forecasts for all users, be they professional meteorologists, other professionals who need meteorological data, or the general public. This will represent a new quantum state for the field of meteorology. I believe that much of this will occur within the next decade.

Acknowledgments. The author would like to thank Matthew Riel, an undergraduate engineering student at the University of Colorado, for his background literature search. Thanks are also due to Dale Kellogg, Robert Hinson, Judy Brown, and Kevin Geiger for their editorial assistance and typing of the manuscript.

REFERENCES

Abreu, V.J., J.E. Barnes, and P.B. Hays, 1992: Observations of winds with an incoherent lidar detector. *Appl. Opt.,* **31** 4509–4514.

Atlas, D., Ed., 1990: *Radar in Meteorology.* Amer. Meteor. Soc., 806 pp.

Baker, W.E., G.D. Emmitt, F. Robertson, R.M. Atlas, J.E. Molinari, D.A. Bowdle, J. Paegle, R.M. Hardesty, R.T. Menzies, T.N. Krishnamurti, R.A. Brown, M.J. Post, J.R. Anderson, A.C. Lorenc, and J. McElroy, 1995: Lidar-measured winds from space: A key component for weather and climate prediction. *Bull. Amer. Meteor. Soc.,* **76,** 869–888.

Benjamin, S.G., K.A. Brewster, R.L. Brummer, B.F. Jewett, T.W. Schlatter, T.L. Smith, and P.A. Stamus, 1991: An isentropic three-hourly data assimilation system using ACARS aircraft observations. *Mon. Wea. Rev.,* **119,** 888–906.

Bjerknes, V., 1904: Das problem der wettervorhersage, betrachtet vom standpunkte der mechanik und der physik. *Meteor. Z.,* **21,** 1–7.

Bowen, E.G., 1987: *Radar Days.* Adam Hilger-IOP Publishing Ltd., 231 pp.

Burgess, D.W., K.E. Wilk, J.D. Bonewitz, K.M. Glover, D.W. Holmes, J. Hinkleman, D. Sirmans, K. Shreeve, and I. Goldman, 1979: Final report on the Joint Doppler Operational Project (JDOP) 1976–1978: Part I—Meteorological applications; Part II—Doppler radar engineering. NOAA Tech. Memo. ERL/NSSL 86, 84 pp. [Available from the National Technical Information Service, Springfield, VA 22151; Order No. PB80-107188/AS.]

Denenberg, J.N., R.J. Serafin, and L.C. Peach, 1972: Uncertainties in coherent measurement of the mean frequency and variance of the Doppler spectrum from meteorological echoes. Preprints, *15th Radar Meteorology Conf.,* Champaign, IL, Amer. Meteor. Soc., 216–221.

Fleming, J.R., 1990: *Meteorology in America, 1800–1870.* The Johns Hopkins University Press, 264 pp.

Fleming, R.J., and A.J. Hills, 1993: Humidity profiles via commercial aircraft. Preprints, *Eighth Symposium on Meteorology, Observations and Measurements,* Anaheim, CA, Amer. Meteor. Soc., J125–J129.

Frisinger, H.H., 1977: *The History of Meteorology to 1800.* Science History Publications, 148 pp.

Fujita, T.T., 1979: Objectives, operations, and results of Project

NIMROD. Preprints, *11th Conf. on Severe Local Storms,* Kansas City, MO, Amer. Meteor. Soc., 259–266.

Gedzelman, S.D., 1994: Chaos rules. *Weatherwise,* **47,** 21–26.

Grice, G.K., Ed., 1985: *The Beginning of the National Weather Service: The Signal Service Years.* National Weather Service, U.S. Dept. of Commerce, 52 pp.

Groginsky, H.L., 1972: Pulse pair estimation of Doppler spectrum parameters. Preprints, *15th Radar Meteorology Conf.,* Champaign, IL, Amer. Meteor. Soc., 233–236.

Haig, T.O., and V.E. Lally, 1958: Meteorological sounding systems. *Bull. Amer. Meteor. Soc.,* **39,** 401–409.

Hildebrand, P.H., and R.K. Moore, 1990: Meteorological radar observations from mobile platforms. *Radar in Meteorology,* D. Atlas, Ed., Amer. Meteor. Soc., 287–314.

———— , and C.K. Mueller, 1985: Evaluation of meteorological airborne radar. Part I: Dual-Doppler analysis of air motions. *J. Atmos. Oceanic Technol.,* **2,** 362–380.

Johnson, D., 1994: Evolution of the U.S. meteorological satellite program: some reminiscences. *Bull. Amer. Meteor. Soc.,* **75,** 1705–1708.

Langford, J.S., and K. Emanuel, 1993: An unmanned aircraft for dropwindsonde deployment and hurricane reconnaissance. *Bull. Amer. Meteor. Soc.,* **74,** 367–375.

Lherimite, R., 1972: Real-time processing of meteorological Doppler radar signals. Preprints, *15th Radar Meteor. Conf.,* Champaign, IL, Amer. Meteor. Soc., 364–367.

Mueller, C.K., and P.H. Hildebrand, 1985: Evaluation of meteorological airborne radar. Part II: Triple-Doppler analyses of air motions. *J. Atmos. Oceanic Technol.,* **2,** 381–392.

Rummler, W.D., 1968: Introduction of a new estimator for velocity spectral parameters. Tech. Memo. MM-68-4121-5, Bell Telephone Labs, Whippany, N.J.

Ware, R., M. Exner, D. Feng, M. Gorbunov, K. Hardy, B. Hermon, Y. Kuo, T. Meehan, W. Melbourne, C. Rocken, W. Schreiner, S. Sakolovsy, F. Solheim, X. Zou, R. Anthes, S. Businger, and K. Trenberth, 1996: GPS sounding of the atmosphere from low earth orbit: Preliminary results. *Bull. Amer. Meteor. Soc.,* **77,** 19–40.

Whitnah, D.R., 1961: *A History of the United States Weather Bureau.* University of Illinois Press, 267 pp.

CHAPTER 4

A Short History of Radar Meteorology

R.R. ROGERS AND P.L. SMITH

Introduction

For what may seem to be a narrow specialization of atmospheric science, radar meteorology has a surprisingly rich literature. Papers including radar observations have appeared regularly in journals of the American Meteorological Society (AMS) for the past 50 years. Radar meteorology conferences are the oldest series of technical conferences of the AMS and have the longest series of conference proceedings. Critical reviews in the formative years and later (Ligda 1951; Wexler 1951; Marshall et al. 1955; Marshall and Gordon 1957; Atlas 1964) served periodically to bring the subject up to date. The widely used textbooks of Louis J. "Lou" Battan (1959, 1973) did much to delineate the field and to establish radar meteorology as a valid discipline. The text of Doviak and Zrnić, now in its second edition (1993), provides a modern synthesis. Other books on the subject are those of Gossard and Strauch (1983), with its focus on clear-air echoes, and of Rinehart (1991), intended for undergraduate students or operational meteorologists. But it was the publication in 1990 of the monumental *Radar in Meteorology,* conceived as a tribute to Lou Battan by David Atlas and edited by him, that makes it easy to argue that no part of atmospheric science has been better served. And even the history of the subject is available for the reading—the sparkling account of Walter Hitschfeld (1986) and the first 18 chapters (150 pages) of *Radar in Meteorology.* What accounts for the extraordinary attention paid to radar in meteorology? And, perhaps more to the point, what can we hope to say in this short review that has not already been said?

The prominence of radar is explained by its sudden and rather late emergence as part of meteorology and by the breadth of radar applications. The first radar observations of rain storms, early in World War II, made it clear that here was a revolutionary way of observing the atmosphere. Immediate applications included identification of rain areas and short-term prediction of the motion of these areas, with telling consequences for military operations. The association of radar rain patterns with fronts, squall lines, tropical cyclones, and other features not well resolved on weather maps was a valuable aid in synoptic analysis. The speed and direction of weather echo motion gave information on winds aloft that was difficult to obtain otherwise. Before the end of the war, theories were in place to explain the origin of radar echoes from precipitation, the reasons for their fluctuating character, and some of their polarization properties. This fundamental understanding of the observations made it possible to deduce certain information about the raindrops and snowflakes from the echoes they produce, giving new life to the study of cloud physics. Radar was thus instrumental in establishing both mesoscale meteorology and modern atmospheric remote sensing. Because of the usefulness of the observations in forecasting, aviation, storm research, and precipitation physics, radar had captured the imagination of many of the military officers and civilian scientists who contributed to its wartime development. Physicists, electrical engineers, and applied mathematicians, some with brief training as weather officers, were the pioneers. Their publications in the postwar years conveyed the excitement of the observations and gained radar its recognition as a meteorological tool. Radar meteorology took on a life of its own once the range of applications became known and the research challenges were recognized.

The authors of this review are members of the first generation of radar meteorologists who were not initiated into the craft during military service or as part of war-related work. Our perspective is not that of the pioneers but of students of the pioneers. We cannot improve on the historical articles already cited, but because of our own experience and viewpoint we may uncover some details overlooked before. Whatever, we hope the review will awaken enough interest in nonspecialists that some will want to read the more detailed articles in Atlas's *Radar in Meteorology* and the other sources. And we hope those readers already familiar with the background of radar meteorology may draw fresh inspiration from recalling the accomplishments of our predecessors.

The next section of the review outlines the key historic events leading to the emergence of radar meteorology. Later sections focus on contributions of radar to certain areas of research and operational meteorology. We have naturally emphasized the areas of application with which we are most familiar, though these may not be the ones that everyone would agree are most important.

The origins of weather radar

In August 1940 a six-member scientific mission sailed from Liverpool, England, to Halifax, Nova Scotia, Canada, carrying an assortment of new military equipment. The leader of the team, Sir Henry Tizard, had flown ahead to Ottawa, Ontario, Canada, and Washington, D.C., to prepare the way. The purpose of the mission was to share with the Americans and Canadians some of the recent wartime developments in England. In the team's baggage was one of the first production models of the resonant magnetron, invented six months earlier by John Randall and Henry Boot at the University of Birmingham. The crucial importance of the magnetron for the outcome of World War II is well known. It also figures in the history of radar meteorology. Experiments with ground-based radar sets operating at wavelengths of 50 m, and later 26 m, had been under way on the channel coast of England since 1935. Because of the long wavelengths, these radars transmitted very broad beams and had poor directional sensitivity. The trend in radar development was steadily toward shorter wavelengths. By 1937, E.G. "Taffy" Bowen had managed to fit radars with a wavelength of 1.5 m to airplanes, but these were still cumbersome and only barely satisfactory. The magnetron made it possible to transmit high power at wavelengths of a few centimeters. This meant that radars could be made much smaller than before and that radar beams could be narrower. Some progress on radar had already been made in America, independently of the work in England, and also in Germany, but a stumbling block had been the inability to produce useful power at short wavelengths. The magnetron was therefore decisive in the course of radar development. Radar sets employing magnetrons were already being built in England by the time of the Tizard mission. By November 1940 the Radiation Laboratory at the Massachusetts Institute of Technology (MIT) was established as a national facility to develop radar in the United States, and within two months the first American microwave radar was built and operating. There have been many accounts of the early history of radar. Of special interest to meteorologists may be that of Bowen (1987), a member of the Tizard team, who went on to contribute to the postwar development of cloud physics.

A consequence of the move to shorter wavelengths was that rain and snow, previously invisible to radar, now became detectable and interfered with the military job of observing ships and airplanes. It is accepted that on 20 February 1941 a shower was tracked by a 10-cm radar to a range of 7 km off the English coast and that the first precipitation echoes were probably detected at the MIT Radiation Laboratory at about the same time (Atlas and Ulbrich 1990). Meteorologists quickly recognized the differences in radar appearance of showers and widespread rain, the association of precipitation echoes with fronts, and the characteristic echoes of tropical storms. Maynard (1945) gives examples of scope photographs obtained at Lakehurst, New Jersey, and in the Philippine Sea, explaining that "radar offers to the pilot and forecaster an actual 'picture' of storms," which can be analyzed to determine the "length, depth, height, intensity, speed and direction of movement of any storm in the area covered." Another key early paper (Bent 1946) includes a fine example of an airmass thunderstorm from MIT Radiation Laboratory observations of July 1942. Because of wartime secrecy, reports such as these were not published until 1945 and later, but from 1941 onward rapid progress was being made on both data interpretation and the theoretical basis of radar detection of weather echoes.

Paramount for theory was the work of J.W. Ryde of the General Electric Company in Wembley, England, where the first magnetrons were manufactured. Before the invention of radar, he had analyzed the diffusion of light by incandescent lamps by applying the classic (1871) scattering theory of Lord Rayleigh. Ryde now recognized that the same theory could be used to estimate the strength of scattering and attenuation of radar waves by clouds, rain, and other particulate matter in the air. Some weeks before the first actual observations, Ryde concluded from calculations that the energy reflected from the water drops in moderate rain should be strong enough to give a detectable radar echo (Ligda 1951). In his comprehensive analysis, finally published in the open literature in 1946, Ryde included a study of the attenuation of radar waves by atmospheric gases as well as by clouds and precipitation. Equally important, he employed the more general theory of Gustav Mie, dating from 1908, to analyze the scattering and attenuation by precipitation of wavelengths too short for the Rayleigh approximation. Ryde paid more attention to extinction than to backscattering because the main wartime concern was to penetrate the weather echoes and reach the targets of interest. From the start, questions about atmospheric effects on radio propagation have prompted advances in radio methods of observing the atmosphere.

The MIT Radiation Laboratory quickly became a center for research in both theory and experimental techniques. Donald Kerr, Herbert Goldstein, Isadore Katz, and Arnold Siegert, among others, helped to build the foundations of quantitative radar meteorology through a stream of contributions that were later reported in the MIT Radiation Laboratory Series of books, published in 28 volumes by McGraw-Hill in the late 1940s.

Wartime weather radar applications were mainly in support of aviation. Some of the earliest developments were in Panama. Because heavy convective showers were a major flying hazard, the Air Weather Service (AWS) of the U.S. Army Air Corps set up a network of radars for storm detection in the Canal Zone in 1943. Pilots flying in the area were continually informed of the locations of storms and could plan their missions making use of this information. The need for special training to interpret radar data became evident and, on the initiative of Captain Joe Fletcher, led to the sponsorship by the AWS of an intensive seven-month radar course for new weather officers at Harvard and MIT. A group of 10 new students started each month; the first group graduated in October 1944. The importance of this training program for the development of radar meteorology can hardly be overestimated. Some of the roughly 100 weather officers who completed the course continued in meteorology when the war was over and are among the pioneers (e.g., Atlas, Battan, Donaldson, Ligda, Swingle).

Used for weather detection at this time were certain of the existing military radars, either gun-laying radars or modified bombsight radars. Raymond Wexler and others at the Army Signal Corps Laboratories in New Jersey evaluated the different available radars for their suitability and in so doing helped to define the characteristics appropriate for a weather radar set. The first radar designed specifically for meteorological applications was the CPS-9, planned for use by the AWS and produced starting in 1949 by the Raytheon Company. For weather surveillance, the Weather Bureau initially adapted different military radars (denoted WSR-1, -2, -3, etc.) in the postwar years. Only much later, in 1957, was the design established for the radar to be used in the Weather Bureau's operational network. This was the WSR-57, again a product of Raytheon.

Metcalf and Glover (1990) note that, by the end of the war, research in the United States related to weather radar was being conducted at the Signal Corps in Belmar, New Jersey; the Air Force All-Weather Flying Division at Wright Field, Ohio; the University of Texas; the University of Florida; Harvard University; and the Naval Research Laboratory. To this list we can add the MIT Radiation Laboratory and Bell Telephone Laboratories.

Elsewhere, weather radar research had begun in Canada in 1944 in Project Stormy Weather of the Canadian Army Operational Research Group in Ottawa under the direction of Stewart Marshall. Early activities included time-lapse photography of the radar scope so that successive positions of weather echoes could be viewed as movies. The association of the echoes with precipitation was confirmed by phone calls around the Ottawa Valley. In 1945 Marshall joined the Physics Department at McGill University, bringing some of the Ottawa team with him and continuing Stormy Weather Group research at McGill. In England, storm detection was of central interest, but much work was done at the Telecommunications Research Establishment in Malvern on the problem of anomalous propagation. Before the end of the war the U.K. Meteorological Office had established a radar research station, equipped with a 10-cm radar, at a site 30 miles north of London (Probert-Jones 1990).

Important events in the few years following the war helped to secure radar its place in meteorology. In 1946 the MIT Radiation Laboratory was closed, but the weather-related work was transferred to the MIT Meteorology Department as the Weather Radar Project, sponsored by the Signal Corps. Austin and Geotis (1990) explain that in the summer of that year the project consisted of a 16-member research and support staff, a 3-cm TPS-10 radar, a 10-cm SCR-615-B, and an instrumented B-17 airplane with crew, stationed at the Bedford airport. The first report by the project director, Alan Bemis, dated December 1946, provides a useful summary of the status of weather radar. The report also reviews the theory of precipitation development, including the uncertainties about the relative importance of coalescence and the ice crystal process in accounting for the formation of precipitation. Describing the research planned for the project, Bemis wrote: "A better understanding of the propagation of microwaves in the atmosphere is only one of our objectives. We hope through the use of radar to learn more about those meteorological processes associated with precipitation and the internal structure of storms." Thus, from the start of the Weather Radar Project precipitation physics was a major area of research.

Another key event was the Thunderstorm Project, directed by Horace Byers of the University of Chicago. Although much progress had been made in flight safety as a result of wartime needs, thunderstorms remained a serious aviation hazard. On an initiative from the airlines and the Weather Bureau, funding was approved before the end of the war for a major project to study thunderstorms. The timing was lucky, for just when the project was scheduled to begin, the war ended and military equipment and personnel became available to support the project. Four government agencies contributed: the air force, the navy, the Weather Bureau, and the National Advisory Committee for Aeronautics. In what was the largest-scale meteorological investigation up to the time, the project had two observing phases. Phase I was in Orlando, Florida, in the summer of 1946, and Phase 2 was in Wilmington, Ohio, the following summer. Ground-based and airborne radars were an important part of the project, along with balloon and airplane measurements and networks of surface instruments. Byers and Braham (1949) later explained that radar observations were needed to integrate all the other data and obtain a unified picture of the storms. Radar observations were also used on their own to study sizes, shapes, growth rates, and patterns of organization of thunderstorm echoes. [See chapter 7 by Braham, this volume.]

In 1948 a new weather radar research program was established by the Air Materiel Command at its Watson Laboratories in Red Bank, New Jersey, under the direction of David Atlas. In October of that year the laboratories, renamed Geophysics Research Directorate (GRD), moved to the Boston area. Wilbur Paulsen and Vernon Plank were early members of the GRD weather radar group. Major effort at first was on propagation and scattering. Collaboration with the Stormy Weather Group began early, and by 1950 much of the work at McGill was actually supported under contract from GRD.

In summary, by 1950 radar meteorology had outgrown its military roots. Although much remained to be learned about atmospheric effects on radio propagation, the contributions radar could make to precipitation physics and convective storm research had been convincingly demonstrated. At least two universities (MIT and McGill) operated weather surveillance radars as part of their teaching and research programs, helping to legitimize the pursuit. GRD had a vigorous research program, and several universities were active in radio propagation research or fundamental studies of atmospheric scattering. Offices of the army, navy, and air force were engaged in research and operational applications. In a census of radar weather projects, Bemis (1951) lists a total of 32 laboratories in the United States, Canada, England, and Australia where relevant work was being done. Thus, the way was prepared for the achievements described in the next sections.

Storm studies

Storm avoidance and storm prediction are the most important applications of weather radar. Thunderstorms and hurricanes were recognized from the start by their characteristic shapes on the radar screen. Work has continued since then on sharpening the ways of identifying hazardous weather elements such as tornadoes, hail, high winds, and heavy rain. Radar has also been at the center of storm research, from the Thunderstorm Project of the 1940s through the large field programs of the 1960s and 1970s (e.g., BOMEX, NHRE, GATE, SESAME); the more recent projects (DUNDEE, ERICA, STORM-FEST, TOGA COARE . . . , a panorama of acronyms); and finally the current research based on the use of Next Generation Weather Radar (NEXRAD) radars. Radar facilities are a fundamental part of several National Oceanic and Atmospheric Administration (NOAA) laboratories (National Severe Storms Laboratory, Environmental Technology Laboratory, Hurricane Research Division) and the Atmospheric Technology Division of the National Center for Atmospheric Research (NCAR), to cite only examples in the United States.

Storm identification

Radar observations from the Thunderstorm Project led to the concept of the thunderstorm "cell" as the basic structural element of convective storms and to the understanding of the life cycle of the cell as comprising the cumulus, mature, and dissipating stages. Later research showed that many thunderstorms are multicellular, composed of several individual cells in different stages of development and thus larger and longer lasting than a single cell. Early recognized as indexes of storm severity were the reflectivity of echoes and their vertical extent. Donaldson (1965), in a review of radar methods of identifying severe storms, summarizes the work up to the time just before Doppler radar observations were becoming available. His list of useful indicators includes 1) the arrangement of echoes in lines; 2) scalloped echo edges, including "hooks"; 3) echo tops that extend above the tropopause; and 4) reflectivities exceeding 55 dBZ. These criteria are still used, but along with other indicators based on Doppler information. Tornadoes, for example, are indicated by a mesovortex signature—a closely spaced pair of radial velocity peaks having opposite signs—the modern-day equivalent of the squiggles that Armstrong and Donaldson (1969) pointed out on their Plan Shear Indicator in the early Doppler days. The tornado is only one of many features potentially identifiable by a modern Doppler radar of the NEXRAD type. [The WSR-88D (Weather Surveillance Radar–1988 Doppler), the current operational radar of the National Weather Service (NWS), is still referred to by its planning-stage acronym.] Much current research is devoted to the development of automatic methods (computer algorithms) for identifying features of interest. It is also recognized that observers must be trained to interpret the new data, and this is one of the needs that prompted the establishment of the COMET program (Cooperative Program for Operational Meteorology, Education, and Training) of the University Corporation for Atmospheric Research (UCAR) and NOAA.

Prelude to NEXRAD

Establishment of the NEXRAD network has entailed a massive transfer of technology from the research laboratory to operational weather offices. That the transfer was slow in coming should not be surprising in view of the scale of the undertaking and its cost. Certain crucial events and the steady efforts of a few individuals made it possible.

By the early 1970s, the potential of Doppler radar for severe storm identification was widely recognized. Ted Fujita and others had shown that tornadoes were associated with circulations called mesocyclones, intermediate in scale between tornadoes and the synoptic scale. Ralph Donaldson (1970) analyzed the ability of Doppler radar to resolve vortices of different sizes and found that the mesocyclone could often be identified at ranges where the actual tornado vortex would not be resolved. Mike Kraus of the Air Force Cambridge Research Laboratories (AFCRL) identified the Brookline tornado of 1972 on the basis of its vortex signature. Soon after that, Don Burgess and others were frequently observing such signatures with the new Doppler facilities of the National Severe Storms Laboratory (NSSL). But there was still no consensus that techniques developed in the environment of the research laboratory could readily be used operationally. Meanwhile, the FPS-77 radars of the Air Weather Service were wearing out and in need of replacement. Captain Roger Whiton was convinced of the potential operational usefulness of Doppler radar and was a strong advocate for replacing the FPS-77 radars with Doppler radars. David Atlas had been promoting the operational use of Doppler radar for severe storm detection for more than a decade. He used the occasion of the 1975 Severe Storms Conference in Norman, Oklahoma, when he was president of the AMS, to organize a special review session on severe storms before an audience of the U.S. House of Representatives Subcommittee on Environment and the Atmosphere. The first recommendation of the report on this session (Atlas et al. 1976) was for accelerated research and early deployment of Doppler radars for tornado detection. This initiative, plus the continued urging from people like Ed Kessler at NSSL, Roger Whiton at AWS, and Ken Glover at AFCRL, helped to rally support for an experiment to evaluate the operational utility of Doppler radar, the JDOP (Joint Doppler Observational Project) program. Based in Norman, this project operated in the spring storm seasons of 1977 and 1978 under sponsorship of the Air Weather Service, the Federal Aviation Administration, and the National Weather Service. The project succeeded in demonstrating the ability of Doppler radar in an operational environment to improve the detection of severe storms. Identifying the mesoscale vortex of tornadoes extended the warning time from what had been about 2 min, based on earlier kinds of radar data, to more than 20 min. During the second year, a tornado obliterated the town of Piedmont, Oklahoma, killing many horses and cattle. However, the warning time had been sufficient so that all people of the town survived. By 1979, no further convincing was necessary, and the sponsoring agencies approved the development of NEXRAD, with Doppler capability, as the replacement radar for the FPS-77s of the Air Weather Service and the WSR-57s of the National Weather Service. [See chapter 10 by Crawford and Kessler, this volume.]

Storm classification

What Peter Ray (1990) and others have called the taxonomy of convective storms began when it was recognized that many thunderstorms, especially the severe ones, were different from those studied on the Thunderstorm Project. In his often-cited 1964 paper, Keith Browning described the airflow and precipitation trajectories inferred from observations of a long-lasting Oklahoma thunderstorm. The shear of the ambient wind displaced the precipitation from the updraft, creating a circulation that was not suppressed by the precipitation but rather was supported by it, allowing the storm to continue in a quasi-steady state. This analysis explained the hook echo and the tendency of large storms of this kind to move to the right of the ambient wind. Another kind of severe storm, distinct from Browning's "supercell," was the multicell storm, described in 1972 by Alex Chisholm and Jim Renick on the basis of observations in Alberta. A multicell storm is made up of as many as four or five convective elements, each undergoing a life cycle like that of the thunderstorm cell. As one dissipates, another forms on the side of the low-level inflow to the storm, so the storm system is large and long lasting and moves relative to the ambient wind (usually to the right). These models are very helpful in making sense of what can appear to be complex and confused patterns. There are plenty of examples of classic supercell and multicell storms. Yet many storms are in between: long lasting, quasi-steady, but with recognizable evolving elements, as explained by Burgess and Lemon (1990). Consequently, supercell and multicell storms are now regarded as limiting idealizations of a continuous spectrum of storm types.

Storm research

Using Doppler radars, it became possible to observe directly the interactions between precipitation and air motions in convective storms. In one of the earliest examples of Doppler radar research, Lou Battan (1964) estimated vertical air velocities and precipitation sizes in a thunderstorm. Most of the early Doppler applications were with a vertically pointing beam to enable estimates of either vertical air velocities or drop sizes. Partly because of the ambiguity of such observations, but perhaps more because of advances in signal processing and data storage, Doppler research in the 1970s shifted from vertical measurements to near horizontal to enable the study of horizontal air motions. It was only with the recent introduction of radar wind profilers, discussed in the section on wind profiling and clear-air echoes, that attention returned to vertical observations.

Doppler radar is limited to measuring the radial velocity component. It takes two radars some distance apart to measure the horizontal wind vector at a point, and three to measure the total wind vector, without employing extra assumptions. Roland Pilié et al. (1963) first explained the principle of dual-Doppler measurements of the wind, but it was left for Roger Lhermitte (1970) to demonstrate the use of the technique for mapping the wind field in convective storms and for Jay Miller (1975) to make it convincing. Dual-Doppler analysis became standard for field projects on convective storms of the 1970s and 1980s. The power of the technique for revealing previously unmeasurable details of the wind field led Tzvi Gal-Chen, Bob Kropfli, and others to the idea of deriving the thermodynamic field from the wind field, which in turn led to the subject of data assimilation in numerical models, as discussed in the section on retrievals and data assimilation. Thus, the ability to measure the airflow in convective storms accurately and with high spatial resolution has prompted the development of better numerical storm models.

Airborne Doppler radar

Although some of the first microwave radars to be built were installed in airplanes, most of the principles and practices in radar meteorology originated from ground-based equipment. An exception is in hurricane research. To meet the need of observing these storms far from land, unique, tail-mounted, scanning Doppler radars are employed on the NOAA P-3 "hurricane hunter" airplanes, which enable measurement of the wind and reflectivity fields (Jorgensen and Marks 1984). To observe more than one wind component requires flying a pattern around the storm to get more than one view of the same region. During the time for this maneuver, changes can occur in the storm, introducing uncertainties in the wind analysis. To get around this problem, radars have been designed that employ two scanning beams, one pointed slightly forward, the other aft, so that as the airplane flies past a storm it measures in close succession two wind components from every region scanned (Jorgensen and DuGranrut 1991; Dou et al. 1991). The most advanced version of this concept is the ELDORA radar (Electra Doppler Radar) on the NCAR Electra airplane, a joint French–American development. In principle, ELDORA will provide from a portable platform the kind of information on storms that previously required a multiple-Doppler surface network.

Precipitation physics

From the beginning, precipitation physics and radar meteorology have been intimately connected. The proper understanding of weather echoes required some knowledge of precipitation; the echoes in turn provided new information about the precipitation. Fundamental for the reflectivity of rain, and hence for its detectablity, are the numbers and sizes of drops in a given volume of space. In his 1946 paper, Ryde used data that had just been published on raindrop sizes for various rain rates by Otis Laws and Donald Parsons of the Soil Conservation Service in Auburn, Alabama. Their 1943 paper explains that the first systematic measurements of raindrop-size distributions were in Germany at the turn of the century and that the only American investigator to their time had been Wilson Bentley, the remarkable Vermont farmer, whose dough-ball method of drop sizing was adopted for their work. The Laws and Parsons data, incidentally, which figure so prominently in the early quantitative interpretation of radar measurements, were obtained not with radar in mind but to understand soil erosion by rain. The drop-size distribution provides a good example of the close connection between radar meteorology and precipitation physics. Initially, information about drop sizes was essential for understanding the origin of radar weather echoes. Later, it was found that Doppler radar can sometimes be used to measure the drop-size distribution with useful accuracy.

Radar, in brief, has made it possible to observe the initiation, growth, and phase transitions of precipitation and to estimate the sizes and concentrations of the precipitation particles. Some of the consequences for precipitation physics will now be described.

Initial echoes and rain formation

Because the scattering cross section of a drop is proportional to the sixth power of its diameter, radar is much more sensitive to rain than to nonprecipitating clouds. In fact, for most weather radars operating in the wavelength range from 3 to 10 cm, light rain and snow are detectable and clouds are invisible. An important early contribution to precipitation physics was the investigation of the altitude (and hence the temperature) at which echoes from convective clouds first appeared. There was at the time much discussion about the relative importance of two precipitation mechanisms: the Bergeron–Findeisen process, requiring the coexistence of ice crystals and supercooled water, and the coalescence process among water droplets, not requiring the ice phase. Some cloud physicists doubted that significant rain could be produced without the presence of ice; others thought the ice phase unnecessary. Ian Browne may have been the first to suggest that radar could resolve the question (see discussion in Jones 1950). If echoes in developing clouds first appeared at altitudes below and warmer than the freezing level, this would be strong evidence for the coalescence process. Independently, Taffy Bowen in Australia was carrying out coordinated airplane and radar measurements in different kinds of precipitation and reported in 1951 the frequent occurrence of echo tops in growing clouds at temperatures warmer than freezing. Lou Battan and Roscoe Braham meanwhile analyzed radar data from several different geographical locations and found that many initial echoes were at warm temperatures. As summarized in Battan's textbook, these studies showed that rain can be initiated by coalescence, even in clouds that grow to heights above the freezing level. The conclusion is that precipitation may be initiated by either of the two processes and that in many clouds with cold tops both may be acting simultaneously. [See chapter 7 by Braham, this volume.]

Growth rates of precipitation

Battan (1953) analyzed data from the Thunderstorm Project to study not only the locations of initial radar echoes of convective clouds but also their subsequent development. Typically the echoes when first observed have a vertical extent of at least several hundred meters. They then grow rapidly, the tops rising, the bases descending, the reflectivity increasing rapidly at all altitudes. Precipitation in showers thus forms nearly simultaneously over a considerable altitude range and grows at a rapid rate. In widespread precipitation, on the other hand, the radar reflectivity is fairly steady with time at all levels but increases with descending altitude because the drops are growing as they fall. These observations and their differences led Caroline Rigby to the "shower" and "continuous rain" approximations for precipitation development, which provided a way to compute the changes in reflectivity associated with the growth of raindrops as they sweep out cloud droplets. Equations describing the growth of a single raindrop had already been developed. What was new in Rigby et al. (1954) was an extension of the equations to the entire drop-size distribution, which is essential for calculating the changes in reflectivity. Ken Gunn's appendix to the paper showed how the same approach could be used to calculate the depletion of cloud water by growing precipitation. This paper formed a needed bridge between cloud microphysics and radar reflectivity measurements. Observed changes in reflectivity were found to be approximately consistent with drop growth predicted by gravitational sweepout, without invoking turbulence or electrical effects to speed up the growth, a finding that has held up well over the years.

Determination of the thermodynamic phase of precipitation

In the early 1950s work was proceeding in England, Australia, and America on the depolarization of radar waves by precipitation. One of the first collaborations between the Air Force Geophysics Research Directorate and the McGill Stormy Weather Group was a theoretical study of scattering, attenuation, and depolarization by nonspherical ice and water particles. Milton Kerker of Clarkson College was a key collaborator in this research. Results showed that a radar capable of transmitting radiation with different polarization states (e.g., vertical and horizontal linear polarization) could provide information on the shape and orientation of ellipsoidal particles (Atlas et al. 1953). The work was motivated partly by the idea that polarimetric measurements might give a better understanding of the microphysical processes in the radar bright band, the transition layer where falling snow melts to become rain. Similar calculations and findings were reported by N.R. Labrum of the Commonwealth Scientific and Industrial Research Organisation, Sydney, at about the same time. And in England at the Cavendish Laboratories and Malvern, measurements of the cross-polarized component in different kinds of precipitation were confirming that the melting layer was strongly depolarizing (Browne and Robinson 1952; Hunter 1954). These reports led the way for theoretical and experimental investigations that have continued intermittently to the present time and show more promise now than ever before for providing a method based on radar measurements to discriminate between the ice and water phases. Some of the milestones are the experiments by Reginald Newell and colleagues at MIT (Newell et al. 1955); the development by Glen McCormick and Archie Hendry at the Canadian National Research Council of an S-band (10 cm) radar for Alberta and a K-band (1.8 cm) radar for Ottawa, both designed for accurate polarization measurements, exposition of the theory for interpreting the observations (McCormick and Hendry 1975); a focus on one of the measurable quantities; the differential reflectivity for linear polarization (Seliga and Bringi 1976); and successful employment of the differential reflectivity by the group at Rutherford Appleton Laboratory (Hall et al. 1980). The new insight provided by polarimetry was by then proved and prompted the development of more radars of advanced design—for example the DFVLR radar in Germany, the CP-2 at NCAR, and the CHILL radar in Illinois.

V.N. Bringi, in collaboration with John Tuttle and others (Tuttle et al. 1989), and more recently using the CHILL radar newly located at Colorado State University (Bringi et al. 1995), has shown that polarimetry can identify supercooled raindrops swept above the freezing level in large convective clouds and can recognize the onset of freezing. The ability to observe these microphysical details holds great potential for contributing to the understanding of the formation and growth of precipitation in clouds of different kinds.

The melting layer

One of the earliest and surely the most distinctive contribution of radar to precipitation physics was the uncovering of the many processes occurring in the radar bright band. By the late 1940s, although there were still a few dissenters, consensus opinion held that the bright band was explained by the melting layer of snow. Ryde (1946) had correctly identified the two dominant causes of the bright band as the change in refractive index and the increase in fall speed during the melting process. These effects, however, could only explain a reflectivity increase of about 7 dB at the brightband maximum, while observations often indicated an increase of 10 dB or more. Austin and Bemis (1950) and Labrum (1952) suggested that the coalescence of particles while melting could account for part of the additional increase in reflectivity. Atlas et al. (1953) showed that deviations from spherical shape could contribute an additional few decibels. Wexler (1955) summarized these effects and added another one-half decibel of his own to the brightband maximum by considering the growth of the melting snow by condensation. Lhermitte and Atlas (1963) showed how Doppler radar measurements of particle fall speed could give additional insight on the processes in the melting layer, but the continuing attention paid to the bright band since then proves that an unequivocal explanation of all the observed characteristics has not yet been found.

Before a bright band was ever observed, Walter Findeisen (1940) explained that in continuous rain the melting of snow cools the air and can create an isothermal layer at 0°C, below which the air may be made unstable by the cooling from above. From time to time since then, others have given evidence of dynamic effects of the melting layer. For example, Atlas et al. (1969) described mesoscale and smaller-scale wind oscillations near the melting layer observed with Doppler radar, and Willis and Heymsfield (1989) reported that the melting layer serves to decouple the air motions above from those below.

It seems that with every new observing system and every new observer fresh details of the bright band are uncovered. Yet fundamental mysteries remain. Is aggregation of wet snowflakes in the bright band necessary to explain the reflectivity maximum? Is the multiplication of ice crystals in any way essential? Is breakup of the melting particles needed to explain the decrease of reflectivity below the bright band? Are the scattering properties of a soggy mix of air, ice, and water truly understood? How important are the alleged dynamic effects, and under what circumstances? All indications are that the bright band, the quintessential contribution of radar to precipitation physics, will remain a subject of research and controversy for some time to come.

Precipitation measurement

Soon after it was recognized that wartime radars were seeing echoes from precipitation and the basic theory explaining raindrop scattering became available, workers in several locations began to develop quantitative relationships between the strength of the radar echo and the intensity of the rainfall. Wexler and Swingle (1947) were the first to present what would today be recognized as a weather radar equation, developed from basic principles and showing how the parameters of the radar system affect the echo intensity. They also connected the important radar meteorological quantity ND^6, the product of drop concentration times the sixth power of the diameter, with available drop size data. In the same year Wexler (1947) published the first "Z–R" relationship linking reflectivity factor with rain rate—not expressed with these symbols, though remarkably similar to the relationships used today for showery rain. He used his relationship to infer rainfall rate from the echoes at minimum detectable threshold and thus anticipated the "stepped gain" approach of Langille and Gunn (1948) used for many years to develop quantitative estimates of echo intensity and rainfall rate.

It was left for Marshall et al. (1947) of the McGill Stormy Weather Group to develop a weather radar equation on the basis of scattering theory and introduce the symbol Z for ND^6 to radar meteorology (a quantity still pronounced "zed" in deference to its Canadian roots). They also conducted experiments confirming that echo power was indeed related to Z, and made the Z–R relationship explicit. The more recent logarithmic, or dBZ, notation is also a product of the Stormy Weather Group, having come into being during the course of a "Project M" seminar at McGill conducted by one of this chapter's authors (PLS) on 4 November 1969.

Radar–rain gauge comparisons

These pioneering studies, which are reviewed by Ligda (1951) and Atlas and Ulbrich (1990), were soon followed by many others, from which several things became apparent. One was that good quantitative calibration of the radar is essential for achieving accurate rain-rate measurements—a point emphasized most strongly by Austin and Geotis (1960). Another was that even with accurate calibrations there seemed to be a "missing 7 dB" or so (Marshall et al. 1955). This may have been due in part to the use of peak detecting, as opposed to true averaging, for determining the echo intensity (Smith 1966), but Probert–Jones (1962) pointed out that most of the discrepancy was caused by an oversimplified conceptual model of the antenna beam pattern. That, plus a similar factor related to the effect of the receiver response in the along-range direction first noted by Nathanson and Smith (1972), and proper accounting for the clear-atmospheric attenuation along the beam path, removed any remaining systematic discrepancies between the radar and the raindrop measurements of reflectivity factor.

It nevertheless appeared that there was a great deal of variation in these Z–R relationships, as pointed out by Twomey (1953) and as summarized in Battan's textbook. Sampling variability may account for an undetermined, though perhaps large, part of the variation (Smith et al. 1993), but indications of geographic and other systematic variations exist. Austin (1987) discusses the various kinds of uncertainties in comparing radar and rain gauge measurements. Factors affecting the variations, at least from the standpoint of measuring rainfall at the surface, include such things as rain type, up- and downdrafts, wind drift and size sorting–merging, and subcloud evaporation. Various attempts have been made to deal with most of these factors, usually individually, but there are many such factors, each of which contributes roughly equally to the overall variance. Therefore, compensating for any one factor is rather ineffective in reducing the total variance.

In view of these difficulties, the concept gained credence of making adjustments to the radar rainfall estimates, which can cover broad areas, on the basis of coordinated rain gauge observations at a small number of points under the radar umbrella. As early as 1954, Hitschfeld and Bordan had written, "a radar used for rain measurement should be calibrated against a rain gauge, rather than by any other means." Work of this kind went on in many places, good examples being Wilson and Brandes (1979) and Barnston (1991). Experience shows (not surprisingly) that when the radar calibration was not very good, such adjustments could produce marked improvement in the radar rainfall estimates. Since the scatter in the gauge–radar comparisons seems to be random in character, however, it becomes more difficult to achieve significant improvement when the basic agreement is good (on the average) in the first place. Situations, usually convective, even arise in which the adjustments actually degrade the agreement (Barnston 1991). Notable exceptions occur where special factors such as orographic effects on rainfall or the confounding influence of the bright band on reflectivity measurements enter the picture (e.g., Harrold et al. 1974).

In the light of this variability, it was also recognized that better agreement could be obtained by integrating over space or time to smooth out some of the random variations. Stout and Neill (1953) were apparently the first to attempt rainfall measurement by radar over a small area and compare the results with measurements from a dense network of rain gauges. This was soon followed by similar work in Japan and many other places and continues to be a theme of precipitation-measurement research to this day (e.g., Joss and Lee 1995). The study of operational applications of these techniques began at NSSL in the mid-1960s (McCallister et al. 1966). A remarkable characteristic of recent work is that hydrologists, who must employ the data in the context of many uncertainties about such things as infiltration versus surface runoff, seem much less concerned about uncertainties in radar estimates of precipitation than are radar meteorologists, who may be too much aware of all the difficulties these measurements involve. [See chapter 14 by Engman, this volume.]

More recently, Doneaud et al. (1984) expanded the findings of Byers (1948) on the relationship between rainfall production and storm dimensions to introduce the concept of the area–time integral. This quantity is strongly correlated with rainfall amount (for convective storms, at least) and has helped to justify the trading of temporal for spatial sampling in obtaining climatological estimates of rainfall from satellite-borne weather radar for tropical regions poorly covered by ground-based radars (Theon and Fugono 1988).

Drop-size distribution

The Laws and Parsons data mentioned earlier satisfied the initial need for information about the drop-size distribution in rain. Stewart Marshall and his first Ph.D. student, Walter McKinnon Palmer, measured drop-size distributions using the filter-paper method and were able to accurately approximate their observations and the Laws and Parsons data with exponential functions of the form $N_0 \exp(-\Lambda D)$. Subsequent observations by many researchers have supported the findings reported in their short paper (Marshall and Palmer 1948), which must be the most-cited reference in the history of radar meteorology. There may well be local variations from the strict exponential form (e.g., Joss and Gori 1978), and the constancy of the intercept parameter N_0 reported by Marshall and Palmer has not always been substantiated (e.g., Waldvogel 1974). Hence, the Marshall–Palmer drop-size distribution should be regarded more generally as a two-parameter function, one parameter (Λ) being related to some measure of the mean drop size and the other (N_0) to a measure of the total drop concentration.

Use of polarimetry

Because the Marshall–Palmer distribution has two varying parameters, reflectivity observations alone cannot distinguish between variations in one or the other. In their research mentioned earlier, Seliga and Bringi (1976) suggested that the slope parameter Λ might be determined by taking note of the distortion in raindrop shapes with increasing size and measuring the difference in reflectivity between horizontally and vertically polarized echoes. This approach has an advantage over other polarization-based methods, most of which require measuring a very weak cross-polarized component, because the signals on both echo channels (horizontal and vertical polarization) are of nearly equal strength.

The differential reflectivity concept rejuvenated polarization research in radar meteorology, leading to efforts to measure the complete polarimetric scattering matrix and interpret it in terms of hydrometeor types. The basic hope of the initial efforts—improved rainfall measurement—has not yet been realized with polarimetric methods, probably because variations in the $Z\text{–}R$ relationship are only one of the factors causing the variation in radar–rain gauge comparisons. This was pointed out by Zawadzki (1984) and reemphasized by Joss and Waldvogel (1990), who contend that variations in the vertical profile of the reflectivity, particularly at altitudes below those the radar can observe, may be the greatest obstacle to quantitative measurement of rainfall by radar in operational environments. When the radar-observed volume is quite close to the ground where gauges are located, excellent agreement can be readily achieved (e.g., Joss 1968). At close ranges, such things as differential reflectivity measurements may be of some assistance in maintaining fairly good agreement. But at the ranges required for operational measurements (typically 100 km or more), vertical variations tend to become a dominant factor. That, along with efforts to refine the use of polarimetric observations and to devise and implement on-line algorithms for computing rainfall accumulations, promises to be the dominant theme of precipitation measurement research in the immediate future.

Use of attenuation

Among the complicating effects that must be dealt with in quantitative precipitation measurement is attenuation, which can be serious at any wavelength shorter than 10 cm. As early as 1947, Austin suggested that the attenuation might be turned to advantage and used to improve the precipitation measurements. This idea has been advocated from time to time (e.g., Atlas and Ulbrich 1977) but has not really been found practical. There are probably two reasons for this: 1) if the rain is light, the attenuation is weak and difficult to measure accurately; 2) if the rain is heavy, the attenuation is larger but then the signal disappears after traveling some distance through the rain. Moreover, experience has always shown more value in measuring intrinsic scattering effects, such as reflectivity, than cumulative propagation effects like attenuation (Rogers 1984). In spite of these difficulties, attenuation-based algorithms play a prominent part in the forthcoming Tropical Rainfall Measuring Mission satellite program of the National Aeronautics and Space Administration.

Effects of hail

Another complicating factor in rainfall measurement is the presence of hail in many convective storms: the high reflectivity of the hailstones can give erroneously high indications of precipitation rates if the reflectivity measurements are used in typical Z–R relationships. Atlas and Ulbrich (1990) note that results applicable to hail were already contained in the original Ryde calculations. The hail-scattering problem was tackled experimentally by Atlas et al. (1960) and computationally by Herman and Battan (1961). The large radar cross sections of hailstones were determined, but it was not until the work of Probert-Jones (1984) that a physical explanation for the process leading to these large cross sections was obtained. The question of discerning the presence of hail in mixed-phase precipitation has remained as one of the driving factors behind the continuing research on polarimetric techniques.

Wind profiling and clear-air echoes

The echoes observed by microwave radars that arise from scattering by raindrops and snowflakes are the raw material for much of radar meteorology. Under some conditions echoes from the clear air are also detectable. Though sometimes confused with reflections from birds or clouds of insects, there are echoes from the truly clear air, caused by strong gradients or spatial irregularities in the radio refractive index of the air. In the late 1960s clear-air reflections were studied using special microwave radars with high transmitted power and high sensitivity. The observations, summarized by Hardy (1973), revealed several different echo structures. Some echoes were in the form of stratified layers; others were associated with convection in the boundary layer. Separately from these studies, Woodman and Guillen (1974) reported observations of reflections from the stratosphere using a radar of an entirely different kind—a VHF radar designed for ionospheric research, operating at a wavelength of 6 m rather than a few centimeters. Also a Doppler radar, it was able not only to detect what were evidently regions of strong spatial variability in the refractive index but also to measure the drift velocity of the reflective elements from which the wind speed and direction could be inferred. In 1979 Warner Ecklund, David Carter, and Ben Balsley of the NOAA Aeronomy Laboratory showed that VHF Doppler radars could measure winds still lower in the troposphere. Work soon began on constructing radars similar in many ways to the ionospheric radars but designed for measuring wind in the troposphere and lower stratosphere. The success of this effort is evident from the proliferation of radar wind profilers, which are now widely used in operational meteorology and research. This section describes some of the history of wind profilers and gives examples of how they have been used.

Relation to ionospheric studies

Wind profilers are descended partly from ionospheric radars whose roots go back to 1924, the year Breit and Tuve, in America, and Appleton and Barnett, in England, observed reflections of pulsed radio waves from the ionosphere. Villard (1976) notes that there may actually have been rough measurements of the height of the reflecting layer as early as 1912, though based not on pulsed transmission but on the interference patterns along continuous-wave transmission paths. Because of the importance of the ionosphere for long-distance radio communication, the National Bureau of Standards began a program of pulse soundings in 1929 to monitor the height and strength of the so-called E layer (Evans 1969). The principle of these observations is that reflection occurs at the altitude where the plasma frequency of the free electrons equals the transmitted frequency of the radar. Because the plasma frequency is proportional to the square root of the electron density, the vertical profile of electron density may be obtained by sweeping the radar frequency. Required are frequencies in the high-frequency (HF) range of the spectrum, corresponding to wavelengths of hundreds of meters.

It was discovered in World War II that it was possible to detect signals scattered from the ionosphere at higher frequencies than the plasma frequency. This scattering was explained in 1958 by W.E. Gordon of Cornell University as caused by spatial variations in the index of refraction arising from fluctuations in the electron density. Gordon proposed that a special radar be built for ionospheric research based on this scattering mechanism (now called Thomson scattering, though Gordon scattering would be as apt). Work began the following year on the Arecibo Ionospheric Observatory in Puerto Rico. The wavelengths decided on were 7.5 and 0.7 m. Other radars designed for Thomson scattering observations in the ionosphere were later built in France, England, Peru, Canada, and the United States (Evans 1969). Following the 1974 report of echoes from the stratosphere by Woodman and Guillen at the Jicamarca radar in Peru, some of these radars were used increasingly for measurements at altitudes below the ionosphere. The scattering at these levels is not from free electrons but from refractive index variations in the electrically neutral atmosphere caused by local gradients of temperature and humidity. The echoes from such inhomogeneities are weak and are made easier to detect by special signal processing, called coherent integration. This procedure had been developed for ionospheric radars because of the weakness of Thomson scattering. Such processing was not ordinarily employed in microwave weather radars, partly because the reflections from rain and snow at the short wavelengths are usually strong enough to be detectable without special provision and partly because coherent integration imposes tighter limits on the compromise between unambiguous range and unambiguous velocity measurable by a Doppler radar. Given the Doppler spectrum of the scattering inhomogeneities, the wind speed and direction can be estimated by assuming that the inhomogeneities move with the wind. This was done initially with the ionospheric radars by employing a single vertically pointing antenna for transmission and several receiving antennas located in different directions from the transmitter to measure different components of the wind in a common scattering volume. Only later was the practice of beam swinging employed, adopted from the "VAD" (Velocity Azimuth Display) used for wind measurement by microwave Doppler radars. In this approach a single transmitter and receiver are collocated, and the beam is pointed successively in several azimuth directions at a fixed elevation angle. The wind vector is determined by assuming horizontal uniformity of the wind over the area sampled so that each measurement gives information on a different component of the same wind vector.

Early measurements in the "C layer"

Curiously, before the war, before the theory of Thomson scattering, before coherent Doppler processing, there were reports of VHF echoes from the stratosphere and troposphere. Albert Friend and Robert Colwell of West Virginia University, in the course of their ionospheric studies, made such observations as early as 1935. So ubiquitous were the echoes from altitudes below 50 km, well below the ionospheric E layer and D layer, that Friend and Colwell (1937) named the low altitudes the C layer. Over the next few years they developed special equipment to give better resolution and accuracy for the measurements in the C layer. Although radar as we know it had not been invented, their equipment amounted to a bistatic radar with a baseline of several kilometers. A problem was to distinguish real echoes from the "ground wave," the line-of-sight signal from the transmitter to the receiver. To accomplish this, they designed short-pulse transmitters and receivers (4 μs) and a directional loop antenna for reception. By 1939, Friend was able to report that measurements with an instrumented airplane showed the echoes to be at altitudes where there were temperature or humidity discontinuities, associated with inversion layers or frontal transition zones. He proposed that the measurements could be used to monitor the changes in height of such boundaries between the times of regular radiosonde measurements. After the war he contributed to the newly emerging theory of radar scattering by atmospheric layers.

Independently of the work of Friend and Colwell, Watson Watt et al. (1937) also investigated echoes observed at VHF below 50 km—the region they suggested calling the Z layer. (Interestingly, neither name stuck.) Watson Watt and colleagues attributed the echoes to ionized layers, an interpretation strongly challenged by Gish and Booker (1939). With Bill Gordon at Cornell, Henry Booker was later to become a leader in research on scattering by refractive irregularities in the clear air.

Enter the wind profiler

By 1980, methods for radar wind measurement had been perfected by two almost entirely separate communities. Radar meteorologists, using microwave radars scanning at low-elevation angles, had developed single- and multiple-Doppler techniques for measuring the wind in widespread precipitation, by using the precipitation particles as tracers, or in the boundary layer, by tracking insects, wind-blown debris, or sometimes refractive irregularities in truly clear air. The physicists, having lowered their sights from the ionosphere to the stratosphere and even the troposphere, employed fixed-beam, near-zenith-pointing VHF radars whose limited sensitivity compared with that of the shorter-wavelength microwave radars was compensated for by long dwell times and much signal integration. The theory of clear-air scattering had by then been rounded out by Soviet turbulence theorists, who assumed that the random spatial structure of the refractivity could be explained by the action of homogeneous, isotropic turbulence on an existing background gradient of refractivity (Tatarski 1961). Terminology seems to have settled on Bragg scattering as the name for this mechanism, though Kolmogorov or Tatarski would have been more to the point. In spite of what appear to be restrictive assumptions, the predictions of the theory, including the dependence of scattering cross section on wavelength, seem to be confirmed by observations in the UHF range (wavelengths from about 10 to 100 cm). At the longer wavelengths in the VHF range the theory is satisfactory much of the time, though observations of strongly refractive layers at normal incidence sometimes suggest that specular scattering of the kind considered by Friend may also be important. In short, when Dick Strauch and others at the NOAA laboratories in Boulder proceeded in 1980 with the design of wind profilers for tropospheric and stratospheric wind measurements, they were able to draw on both the experience of meteorologists with microwave measurements in the troposphere and the experimental know-how of the middle-atmosphere physicists. By their early success and quick acceptance, wind profilers have finally brought the two communities together. Important consequences of the NOAA initiative are 1) a network of wind profilers in the tropical Pacific, operated by the Aeronomy Laboratory (Balsley et al. 1991), and 2) the Wind Profiler

Demonstration Network in the central United States (National Weather Service 1994).

Atmospheric structure, waves, and layers

Radar studies of the clear air have revealed more complexity in atmospheric structure than had been imagined. While the buoyant thermals observed by Konrad and Kropfli (1968) provided satisfying evidence for what was already thought to exist, the extremely thin layers with waves and breaking waves, reported by Gossard et al. (1970) and by Bean (1972), were new, unexpected, and a bit disquieting. How could such things have existed without our knowing about them? There was, to be sure, some earlier evidence. Good observers had noted clouds in lee waves that sometimes indicated instability leading to wave breaking, and airplane encounters with clear-air turbulence were evidence for local patches of something like breaking waves.

The layers are explained by thin zones of sharp gradient in temperature or humidity. Gravity waves are commonly observed on the layers, made evident by upward and downward displacements with periods ranging from minutes to hours. Breaking of the waves is understood often to be caused by vertical wind shear and Kelvin–Helmholtz instabilities. But, in addition to the ubiquitous layers, there is often horizontal structure in the clear-air echoes, indicating fronts, convergence zones, lines of shear, gravity currents, roll vortices, and other phenomena not otherwise observable. These observations have given new insight on processes in the boundary layer and pose a challenge for numerical models of atmospheric dynamics and thermodynamics.

Improved forecasts

Planning for the Wind Profiler Demonstration Network began in 1985. The prototype instrument, a 404-MHz (75-cm wavelength), three-beam profiler, was completed by Sperry and installed at Platteville, Colorado, in the summer of 1988. Installation of the network of 30 stations in the central United States was completed in May 1992. A goal of the project was to determine the extent to which this new source of wind data, with high resolution in time and altitude, could improve routine field operations and numerical weather predictions of the National Weather Service. Experience has shown that the profiler information gives a better understanding of the three-dimensional structure of weather systems than other datasets. The usefulness of the data has been proved in thunderstorm forecasting, in locating short-wave troughs and ridges that would otherwise be undetected, and in resolving subsynoptic features such as low-level jets and convergence lines (National Weather Service 1994). Profiler network data are assimilated in a numerical regional prediction model once every three hours. These improve the performance statistics of the model for short-term forecasts in the area of the network. Studies of particular cases show improvements in the forecasting of heavy precipitation. Research applications have included observing fronts and jet streaks, gravity waves, and airflow in relation to mesoscale convective complexes. Further improvements in forecast skill are expected as the frequency of data assimilation is increased.

Retrievals and data assimilation

A tantalizing problem, recognized soon after the first Doppler radar measurements, was to determine the three-dimensional wind vector from the measurable radial component. The first solution was to assume a horizontally uniform wind field and to measure the radial component as a function of azimuth around a low-elevation circle centered on the radar site. Each pointing direction measures a different component of what is assumed to be the same wind vector, so the measurements can be combined to determine the horizontal wind speed and direction. This approach, called the VAD analysis, was generalized by Browning and Wexler (1968) to allow determination of the mean wind speed and direction within the scanned circle, the horizontal divergence, and the stretching and shearing deformations, assuming a linear variation of the wind components with position in space. Integrating the horizontal divergence over altitude could give an estimate of the vertical velocity profile. Peace et al. (1969) first posed the problem of determining the horizontal wind at different locations around the radar site. This requires estimating the speed and direction from Doppler measurements in limited regions defined by azimuth sectors and range intervals. Cal Easterbrook (1975) recast the problem in a more amenable form and gave several examples of two-dimensional circulation patterns derived from Doppler-observed radial velocities.

Separate from the research on wind measurement with a single-Doppler radar, and following the leads of Peace and Brown (1968) and Jay Miller (1972), much progress was made on wind field estimation using two radars separated some distance apart and scanning the same volume in space. Kropfli and Miller (1975) applied the technique to the detailed mapping of the three-dimensional flow field in a thunderstorm, setting the stage for many similar studies over the next decade. So convincing were the boundary-layer wind fields produced by Bob Kropfli and others that Tzvi Gal-Chen (1978) recognized an entirely new application of the data. The wind observations could be combined with the momentum equations of dynamic meteorology to allow unique determination of the fields of pressure and density deviations from their horizontal averages. (The somewhat unfortunate word "retrieval," which suggests getting something back that you once had, was not yet introduced.) This landmark contribution also showed the way wind observations could be incorporated in numerical models to provide regular adjustments to the initial conditions before integrating forward in time—the procedure now called data assimilation. By 1983, Gal-Chen and Kropfli, as well as Roux and Testud, were reporting on pressure and temperature fields determined from dual-Doppler wind observations. An authoritative review by Hane et al. (1988) describes the general problem of determining thermodynamic quantities and even cloud physical quantities from wind information alone.

Doug Lilly saw in Gal-Chen's analysis a way to get at the original problem of determining the three-dimensional wind vector from measurements with one radar. Lilly and Moeng (1984) suggested that the continuity equation and the vorticity equation could evidently be used to constrain the possible solutions to the one that is dynamically possible. Prophetically, they wrote "it seems that the combination of radar winds and numerical simulation, including the data treatment proposed here, could allow true mesoscale prediction of convective storm evolution. . . ." Several methods for determining the three-dimensional wind field have now been applied and compared (Laroche and Zawadzki 1994), showing that the classic problem has essentially been solved. In fact, the Montreal regional weather office employs one of the methods operationally, using data from the Marshall Radar Observatory at McGill.

The subject of data assimilation is currently a topic of intense research stimulated by the availability of Doppler radar data from

NEXRAD radars and wind profiles from the demonstration network. As predicted by Lilly and Moeng, data and numerical models are being used together for mesoscale prediction. The observations provide a recurring check on the model output, and the model provides a dynamically consistent framework for the observations. Among many recent research contributions, Liou et al. (1991) presented a simulation illustrating the feasibility of the approach and assessing sources of error, and Crook and Tuttle (1994) successfully employed real data in predictions of gust-front formation and motion. There are still unsolved problems in the way of incorporating radar winds for numerical forecasting of moist convection, but many researchers are focused on the problem, and early success may be confidently predicted.

The AMS radar conferences

A 20-page set of "minutes" prepared by Alan Bemis, complete with charts and sketches, summarizes the presentations given at the 1947 Weather Radar Conference at MIT, attended by 91 people. Like a good set of class notes, it describes the key points of the papers presented and records the ensuing discussions. In September 1950 a "Symposium on Microwaves, Cloud Physics and Aerosols" was held at McGill University. The report summarizing this meeting consists of roughly 100 pages of notes, sketches, and dialogue. Only about 20 people are recorded as attending, mainly the McGill and GRD groups, though Clarkson, Columbia, and Boston Universities were also represented. A year later, what is now regarded as the Second Radar Meteorology Conference was held as part of a Conference on Water Resources at the Illinois State Water Survey. The papers and discussions are included in the proceedings volume that was published the following year. By the time of the next radar conference, held at McGill in September 1952, the community was confident enough to start numbering the meetings. The proceedings of the third conference were, in fact, preprints, a word not yet much used. Stewart Marshall, as conference chairman, required advance copies of manuscripts from the authors and had the conference volume printed and ready to hand out at the time of registration. An innovation at the time, this idea eventually caught on and is now standard practice at AMS conferences. What was given up were the discussions of the papers, an important part of the earlier conference summaries, which recorded the flavor and dynamics of the meetings more effectively than the papers alone and gave some idea about personalities. At a few of the later meetings, write-ups of discussions were reinstated and published in supplements to the preprints, but that tradition now seems lost and unlikely to return.

Table 4-1 is a list of the radar meteorology conferences to the present time. The first to have official sponsorship by the AMS was the fourth, at the University of Texas, though Ken Spengler had attended the earlier ones, watching after the interests of the Society. To appreciate the development of the subject, the serious student should begin by looking through the proceedings of these meetings. Especially in the early years, the conferences were the main forum for disseminating new ideas and information. Because radar meteorology by its nature was interdisciplinary (before that word was used outside the social sciences), the conferences welcomed contributions from the fringe ar-

TABLE 4-1. AMS RADAR METEOROLOGY CONFERENCES.

1	March 1947	Cambridge, MA
2	October 1951	Urbana, IL
3	September 1952	Montreal, PQ, Canada
4	November 1953	Austin, TX
5	September 1955	Asbury Park, NJ
6	March 1957	Cambridge, MA
7	November 1958	Miami Beach, FL
8	April 1960	San Francisco, CA
9	October 1961	Kansas City, MO
10	April 1963	Washington, DC
11	September 1964	Boulder, CO
12	October 1966	Norman, OK
13	August 1968	Montreal, PQ, Canada
14	November 1970	Tucson, AZ
15	October 1972	Champaign, IL
16	April 1975	Houston, TX
17	October 1976	Seattle, WA
18	March 1978	Atlanta, GA
19	April 1980	Miami Beach, FL
20	November 1981	Boston, MA
21	September 1983	Edmonton, AB, Canada
22	September 1984	Zurich, Switzerland
23	September 1986	Snowmass, CO
24	March 1989	Tallahassee, FL
25	June 1991	Paris, France
26	May 1993	Norman, OK
27	October 1995	Vail, CO

eas that had not yet established their own specialist meetings, such as atmospheric electricity, cloud physics, severe storms, and laser studies. (It was Herb Ligda, incidentally, who coined the acronym lidar. Clever graduate students used to pronounce it "ligdar.")

The proceedings record some of the playfulness that seems to be a trademark of the community. It began with the paper by Bemis and Fleisher at the fourth conference, which reported observational evidence for a mechanism of cyclone development attributed to Heinrich Schimmel, the "hinauf-und-ab" effect, otherwise called seeding from above. Aaron Fleisher outdid himself on homeground at MIT for the sixth conference. For the banquet there was a full-scale musical review, featuring such now famous hits as "Don't Take the Bright Band Away from Me" (for bass solo) and the final chorus and theme song of the conference, "More Data, More Data (from Pole to Equator)." In later years there was a paper on the theory of echo-free regions, including Ray Wexler's poem playing with double negatives, "The Theory of the Void"; a report on YESRAD (Yesterday's Radar) as an economical alternative to NEXRAD; an exposition of the principles of ground clutter detection (GDR); and a lively song poking fun at polarization measurements, "The Differential Reflectivity Blues." Gentle irreverence is a healthy sign. These bits of humor certainly made the conferences more memorable.

Other innovations at radar meteorology conferences include the use of overhead transparencies instead of slides (Montreal 1968), the replacement of individual paper presentations by panels of authors, and the use of reviewers or rapporteurs to present groups of papers in place of the authors. Some ideas worked and others flopped, but the radar conferences have been notable for their attempts to improve communication and have thus contributed to the evolution of all AMS conferences.

Current status and trends

Weather offices have for many years employed radar data as part of standard forecasting procedure. Airplanes and ships are equipped with weather radars to avoid storm hazards and to plot a safer course. No research project concerned with storms, winds, diffusion, air quality, precipitation, or even climate can do without radar support. So complete is the appropriation of radar by meteorology that commercial television stations routinely show the radar weather pattern as part of the evening news. [See chapter 17 by Leep, this volume.]

Because few areas of atmospheric science are untouched by radar, most meteorologists require a working familiarity with radar data. Forecasters have long understood how to use reflectivity patterns of storms; now they are learning to use the additional information provided by Doppler radars and wind profilers. Many researchers, not just the radar specialists, need to have an understanding of velocity ambiguities, ground clutter, the smoothing and sampling problems unique to radar observations, and the uncertainties of radar calibration. Computer aids and special training programs are helping to ease the transfer of knowledge from the research laboratory to weather offices. Experience with the NEXRAD network over the next few years will indicate how successful these efforts have been.

The next important advance for operational meteorology seems likely to be the integration of weather observations with mesoscale numerical forecasting models. This is a subject of vigorous current research aimed at improving the short-term forecasts of heavy rain and severe storms by frequent assimilation of radar data. Farther away seems the employment of polarimetric methods for improved rainfall measurements. Operationally, at least for the next few years, better estimates of areal rainfall are more likely to come from more accurate measurements of reflectivity and better allowance for unresolvable variations in reflectivity.

In research we look forward to the increased use of radar data in validating numerical models, a complement to the operational application just described. Major effort and considerable resources have gone in to the development of airborne Doppler radar. This initiative is already beginning to produce results and should influence the conduct of mesoscale and storm-scale research in the immediate future. Polarimetric measurements hold the promise of soon contributing in a major way to precipitation physics. Radar wind profilers, in addition to stimulating the research on data assimilation in regional and mesoscale models, have revived research in precipitation physics based on measurements of the complete Doppler spectrum. The first satellite-borne precipitation radar, designed to gather data for climatological studies, will soon be launched. Beyond extrapolating these current research trends, it is difficult to speculate on what the future may hold. It seems fair to predict, however, that radar will be more important than ever before as the data become an integral part of the atmospheric observing system.

Acknowledgments. The authors are grateful to Ralph Donaldson for reading an early version of the manuscript and making many constructive suggestions for improvement.

REFERENCES

Armstrong, G.M., and R.J. Donaldson, 1969: Plan-shear indicator for real-time Doppler radar identification of hazardous storm winds. *J. Appl. Meteor.,* **8,** 376–383.

Atlas, D., 1964: Advances in radar meteorology. *Advances in Geophysics,* Vol. 10, Academic Press, 317–478.

_____ , Ed., 1990: *Radar in Meteorology.* Amer. Meteor. Soc., 806 pp.

_____ , and C.W. Ulbrich, 1977: Path- and area-integrated rainfall measurement by microwave attenuation in the 1–3-cm band. *J. Appl. Meteor.,* **16,** 1322–1331.

_____ , and _____ , 1990: Early foundations of the measurement of rainfall by radar. *Radar in Meteorology,* D. Atlas, Ed., Amer. Meteor. Soc., 86–97.

_____ , M. Kerker, and W. Hitschfeld, 1953: Scattering and attenuation by non-spherical atmospheric particles. *J. Atmos. Terr. Phys.,* **3,** 108–119.

_____ , W.G. Harper, F.H. Ludlam, and W.C. Macklin, 1960: Radar scatter by large hail. *Quart. J. Roy. Meteor. Soc.,* **86,** 468–482.

_____ , R. Tatehira, R.C. Srivastava, W. Marker, and R.E. Carbone, 1969: Precipitation-induced mesoscale wind perturbations in the melting layer. *Quart. J. Roy. Meteor. Soc.,* **95,** 544–560.

_____ , T.T. Fujita, S.L. Barnes, A. Pearson, R.A. Anthes, and L.J. Battan, 1976: Severe local storms. *Bull. Amer. Meteor. Soc.,* **57,** 398–435.

Austin, P.M., 1947: Measurement of approximate raindrop size by microwave attenuation. *J. Meteor.,* **4,** 121–124.

_____ , 1987: Relation between measured radar reflectivity and surface rainfall. *Mon. Wea. Rev.,* **115,** 1053–1070.

_____ , and A. Bemis, 1950: A quantitative study of the bright band in radar precipitation echoes. *J. Meteor.,* **7,** 165–171.

_____ , and S.G. Geotis, 1960: The radar equation parameters. Preprints, *Eighth Conf. on Radar Meteorology,* San Francisco, CA, Amer. Meteor. Soc., 15–22.

_____ , and _____ , 1990: Weather radar at MIT. *Radar in Meteorology,* D. Atlas, Ed., Amer. Meteor. Soc., 22–31.

Balsley, B.B., D.A. Carter, A.C. Riddle, W.L. Ecklund, and K.S. Gage, 1991: On the potential of VHF wind profilers for studying convective processes in the tropics. *Bull. Amer. Meteor. Soc.,* **72,** 1355–1360.

Barnston, A.G., 1991: An empirical method of estimating raingage and radar rainfall measurement bias and resolution. *J. Appl. Meteor.,* **30,** 282–296.

Battan, L.J., 1953: Observations of the formation and spread of precipitation in cumulus clouds. *J. Meteor.,* **10,** 311–324.

_____ , 1959: *Radar Meteorology.* University of Chicago Press, 161 pp.

_____ , 1964: Some observations of vertical velocities and precipitation sizes in a thunderstorm. *J. Appl. Meteor.,* **3,** 415–420.

_____ , 1973: *Radar Observation of the Atmosphere.* University of Chicago Press, 324 pp.

Bean, B.R., 1972: Application of FM-CW radar and acoustic echosounder techniques to boundary layer and CAT studies. *Remote Sensing of the Environment,* V.E. Derr, Ed., NOAA/ERL, U.S. Government Printing Office, 20.1–20.28.

Bemis, A.C., 1946: First technical report under Signal Corps project. MIT Weather Radar Research, 94 pp.

_____ , 1951: Census of radar-weather projects. Conference on water resources. Bulletin 41, Illinois State Water Survey, 193–197.

Bent, A.E., 1946: Radar detection of precipitation. *J. Meteor.,* **3,** 78–84.

Bowen, E.G., 1951: Radar observations of rain and their relation to mechanisms of rain formation. *J. Atmos. Terr. Phys.,* **1,** 125–140.

_____ , 1987: *Radar Days.* Adam Hilger, 231 pp.

Bringi, V.N., K. Knupp, and L. Liu, 1995: Microphysical and kinematic study of early-to-mature cloud transition using multiparameter radar, dual-Doppler synthesis, and aircraft penetrations. Preprints, *Conf. on Cloud Physics,* Dallas, TX, Amer. Meteor. Soc., 267–271.

Browne, I.C., and N.P. Robinson, 1952: Cross-polarization of the radar melting band. *Nature,* **170,** 1078–1079.

Browning, K.A., 1964: Airflow and precipitation trajectories within severe local storms which travel to the right of the winds. *J. Atmos. Sci.,* **21,** 634–639.

______ , and R. Wexler, 1968: A determination of kinematic properties of a wind field using Doppler radar. *J. Appl. Meteor.,* **7,** 105–113.

Burgess, D.W., and L.R. Lemon, 1990: Severe thunderstorm detection by radar. *Radar in Meteorology,* D. Atlas, Ed., Amer. Meteor. Soc., 619–656.

Byers, H.R., 1948: The use of radar in determining the amount of rain falling over a small area. *Eos, Trans. Amer. Geophys. Union,* **29,** 187–196.

______ , and R.R. Braham, 1949: *The Thunderstorm.* U.S. Government Printing Office, 287 pp.

Chisholm, A.J., and J.H. Renick, 1972: The kinematics of multicell and supercell Alberta hailstorms. Alberta Hail Studies, Research Council of Alberta Rep. 72-2, 24–31.

Crook, A., and J.D. Tuttle, 1994: Numerical simulations initialized with radar-derived winds. Part II. Forecasts of three gust-front cases. *Mon. Wea. Rev.,* **122,** 1204–1217.

Donaldson, R.J., 1965: Methods for identifying severe thunderstorms by radar: A guide and bibliography. *Bull. Amer. Meteor. Soc.,* **46,** 174–193.

______ , 1970: Vortex signature recognition by a Doppler radar. *J. Appl. Meteor.,* **9,** 661–670.

Doneaud, A.A., S. Ionescu-Niscov, D.L. Priegnitz, and P.L. Smith, 1984: The area-time integral as an indicator for convective rain volumes. *J. Climate Appl. Meteor.,* **23,** 555–561.

Dou, X.K., G. Scialom, and Y. Lemaître, 1991: Three-dimensional analytical mesoscale wind field retrieval from airborne Doppler radar data. Preprints, *25th Conf. on Radar Meteorology,* Paris, France, Amer. Meteor. Soc., 490–492.

Doviak, R.J., and D.S. Zrnić, 1993: *Doppler Radar and Weather Observations.* 2d ed. Academic Press, 562 pp.

Easterbrook, C.C., 1975: Estimating horizontal wind fields by two-dimensional curve fitting of single Doppler radar measurements. Preprints, *16th Conf. on Radar Meteorology,* Houston, TX, Amer. Meteor. Soc., 214–219.

Ecklund, W.L., D.A. Carter, and B.B. Balsley, 1979: Continuous measurement of upper atmospheric winds and turbulence using a VHF Doppler radar. Preliminary results. *J. Atmos. Terr. Phys.,* **41,** 983–994.

Evans, J.V., 1969: Theory and practice of ionospheric study by Thomson scatter radar. *Proc. IEEE,* **57,** 496–530.

Findeisen, W., 1940: Die entstehung der 0°-Isothermie und die fraktocumulus-Bildung unter Nimbostratus. *Meteor. Z.,* **57,** 49–54.

Friend, A.W., 1939: Continuous determination of air-mass boundaries by radio. *Bull. Amer. Meteor. Soc.,* **20,** 202–205.

_____ , and R.C. Colwell, 1937: Measuring the reflecting regions in the troposphere. *Proc. IRE,* **25,** 1531–1541.

Gal-Chen, T., 1978: A method for the initialization of the anelastic equations: Implications for matching models with observations. *Mon. Wea. Rev.,* **106,** 587–606.

_____ , and R.A. Kropfli, 1983: Deduction of thermodynamic properties from dual-Doppler observations of the PBL. Preprints, *21st Conf. on Radar Meteorology,* Edmonton, AB, Canada, Amer. Meteor. Soc., 33–38.

Gish, O.H., and H.G. Booker, 1939: Nonexistence of continuous intense ionization in the troposphere and lower stratosphere. *Proc. IRE,* **27,** 117–125.

Gordon, W.E., 1958: Incoherent scattering of radio waves by free electrons with applications to space exploration by radar. *Proc. IRE,* **46,** 1824–1829.

Gossard, E.E., and R.G. Strauch, 1983: *Radar Observation of Clear Air and Clouds.* Elsevier, 280 pp.

_____ , J.H. Richter, and D. Atlas, 1970: Internal waves in the atmosphere from high-resolution radar measurements. *J. Geophys. Res.,* **75,** 3523–3536.

Hall, M.P.M., S.M. Cherry, J.W.F. Goddard, and G.R. Kennedy, 1980: Raindrop sizes and rainfall rates measured by a dual-polarization radar. *Nature,* **285,** 195–198.

Hane, C.E., C.L. Ziegler, and P.S. Ray, 1988: Use of velocity fields from Doppler radars to retrieve other variables in thunderstorms. *Instruments and Techniques for Thunderstorm Observation and Analysis,* E. Kessler, Ed., University of Oklahoma Press, 215–234.

Hardy, K.R., 1972: Studies of the clear atmosphere using high power radar. *Remote Sensing of the Environment,* V.E. Derr, Ed., NOAA/ERL, U.S. Government Printing Office, 14.1–14.34.

_____ , and H. Ottersten, 1969: Radar investigation of convective patterns in the clear atmosphere. *J. Atmos. Sci.,* **26,** 666–672.

Harrold, T.W., E.J. English, and C.A. Nicholass, 1974: The accuracy of radar-derived rainfall measurements in hilly terrain. *Quart. J. Roy. Meteor. Soc.,* **100,** 331–350.

Herman, B.M., and L.J. Battan, 1961: Calculations of Mie-scattering from melting ice spheres. *J. Meteor.,* **18,** 468–478.

Hitschfeld, W., 1986: The invention of radar meteorology. *Bull. Amer. Meteor. Soc.,* **67,** 33–37.

______ , and J. Bordan, 1954: Errors inherent in the radar measurement of rainfall at attenuating wavelengths. *J. Meteor.,* **11,** 58–67.

Hunter, I.M., 1954: Polarization of radar echoes from meteorological precipitation. *Nature,* **173,** 165–166.

Jones, R.F., 1950: The temperatures at the tops of radar echoes associated with various cloud systems. *Quart. J. Roy. Meteor. Soc.,* **76,** 312–336.

Jorgensen, D.P., and F.D. Marks, 1984: Airborne Doppler radar study of the three-dimensional airflow within a hurricane rainband. Preprints, *22d Conf. on Radar Meteorology,* Zurich, Switzerland, Amer. Meteor. Soc., 572–577.

______ , and J.D. DuGranrut, 1991: A dual-beam technique for deriving wind fields from airborne Doppler radar. Preprints, *25th Conf. on Radar Meteorology,* Paris, France, Amer. Meteor. Soc., 458–461.

Joss, J., and E.G. Gori, 1978: Shapes of raindrop size distributions. *J. Appl. Meteor.,* **17,** 1054–1061.

______ , and A. Waldvogel, 1990: Precipitation measurement and hydrology. *Radar in Meteorology,* D. Atlas, Ed., Amer. Meteor. Soc., 577–606.

______ , and R. Lee, 1995: The application of radar–gauge comparisons to operational precipitation profile corrections. *J. Appl. Meteor.,* **34,** 2612–2630.

______ , J.C. Thams, and A. Waldvogel, 1968: The accuracy of daily rainfall measurements by radar. Preprints, *13th Conf. on Radar Meteorology,* Montreal, PQ, Canada, Amer. Meteor. Soc., 448–451.

Konrad, T.G., and R.A. Kropfli, 1968: Radar observations of clear-air convection over the sea. Preprints, *13th Conf. on Radar Meteorology,* Montreal, PQ, Canada, Amer. Meteor. Soc., 262–269.

Kraus, M.J., 1973: Doppler radar observations of the Brookline, Massachusetts, tornado of 9 August 1972. *Bull. Amer. Meteor. Soc.,* **54,** 519–524.

Kropfli, R.A., and L.J. Miller, 1975: Thunderstorm flow patterns in three dimensions. Preprints, *16th Conf. on Radar Meteorology,* Houston, TX, Amer. Meteor. Soc., 121–127.

Labrum, N.R., 1952: The scattering of radio waves by meteorological particles. *J. Appl. Phys.,* **23,** 1324–1330.

Langille, R.C., and K.L.S. Gunn, 1948: Quantitative analysis of vertical structure in precipitation. *J. Meteor.,* **5,** 301–304.

Laroche, S., and I. Zawadzki, 1994: A variational analysis method for retrieval of three-dimensional wind field from single-Doppler radar data. *J. Atmos. Sci.,* **51,** 2664–2682.

Laws, J.O., and D.A. Parsons, 1943: The relation of raindrop-size to intensity. *Eos, Trans. Amer. Geophys. Union,* **24,** 452–460.

Lhermitte, R.M., 1970: Dual-Doppler radar observation of convective storm circulation. Preprints, *14th Conf. on Radar Meteorology,* Tucson, AZ, Amer. Meteor. Soc., 139–144.

_____ , and D. Atlas, 1963: Doppler fall speeds and particle growth in stratiform precipitation. Preprints, *10th Conf. on Radar Meteorology,* Washington, DC, Amer. Meteor. Soc., 297–302.

Ligda, M.G.H., 1951: Radar storm observation. *Compendium of Meteorology,* T.F. Malone, Ed., Amer. Meteor. Soc., 1265–1282.

Lilly, D.K., and C.H. Moeng, 1984: A technique for recovering the total velocity field from single Doppler measurements. Preprints, *22d Conf. on Radar Meteorology,* Zurich, Switzerland, Amer. Meteor. Soc., 506–508.

Liou, Y.C., T. Gal-Chen, and D.K. Lilly, 1991: Retrieval of wind temperature and pressure from single Doppler radar and a numerical model. Preprints, *25th Conf. on Radar Meteorology,* Paris, France, Amer. Meteor. Soc., 151–154.

Marshall, J.S., and W. McK. Palmer, 1948: The distribution of raindrops with size. *J. Meteor.,* **5,** 165–166.

_____ , and W.E. Gordon, 1957: Radiometeorology. *Meteorological Research Reviews,* Vol. 3, No. 14, Amer. Meteor. Soc., 73–113.

_____ , R.C. Langille, and W. McK. Palmer, 1947: Measurement of rainfall by radar. *J. Meteor.,* **4,** 186–192.

_____ , W. Hitschfeld, and K.L.S. Gunn, 1957: Advances in radar weather. *Advances in Geophysics,* Vol. 2, Academic Press, 1–56.

Maynard, R.H., 1945: Radar and weather. *J. Meteor.,* **2,** 214–226.

McCallister, J.P., J.L. Teague, and C.E. Vicroy, 1966: Operational radar rainfall measurements. Preprints, *12th Conf. on Radar Meteorology,* Norman, OK, Amer. Meteor. Soc., 208–215.

McCormick, G.C., and A. Hendry, 1976: Principles for the radar determination of the polarization properties of precipitation. *Radio Sci.,* **10,** 421–434.

Metcalf, J.I., and K.M. Glover, 1990: A history of weather radar research in the U.S. Air Force. *Radar in Meteorology,* D. Atlas, Ed., Amer. Meteor. Soc., 32–43.

Miller, L.J., 1972: Dual-Doppler radar observations of circulations in snow conditions. Preprints, *15th Conf. on Radar Meteorology,* Champaign–Urbana, IL, Amer. Meteor. Soc., 309–314.

———— , 1975: Internal airflow of a convective storm from dual-Doppler radar measurements. *Pure Appl. Geophys.,* **113,** 765–785.

Nathanson, F.E., and P.L. Smith, 1972: A modified coefficient for the weather radar equation. Preprints, *15th Conf. on Radar Meteorology,* Champaign–Urbana, IL, Amer. Meteor. Soc., 228–230.

National Weather Service, 1994: Wind profiler assessment report. 141 pp. [Available from National Weather Service, Silver Spring, MD 20910–3233.]

Newell, R.E., S.G. Geotis, M.L. Stone, and A. Fleisher, 1955: How round are raindrops? *Proc. Fifth Conf. on Radar Meteorology,* Asbury Park, NJ, Amer. Meteor. Soc., 261–268.

Peace, R.L., and R.A. Brown, 1968: Comparison of single and double Doppler radar velocity measurements in convective storms. Preprints, *13th Conf. on Radar Meteorology,* Montreal, PQ, Canada, Amer. Meteor. Soc., 464–473.

———— , ———— , and H.G. Camnitz, 1969: Horizontal motion field observed with a single pulse Doppler radar. *J. Atmos. Sci.,* **26,** 1096–1103.

Pilié, R.J., J.E. Jiusto, and R.R. Rogers, 1963: Wind velocity measurement with Doppler radar. Preprints, *10th Conf. on Radar Meteorology,* Washington, DC, Amer. Meteor. Soc., 329a–329ℓ.

Probert-Jones, J.R., 1962: The radar equation in meteorology. *Quart. J. Roy. Meteor. Soc.,* **88,** 485–495.

———— , 1984: Resonance component of backscattering by large dielectric spheres. *J. Opt. Soc. Amer., Ser. A,* 1, 822–830.

———— , 1990: A history of radar meteorology in the United Kingdom. *Radar in Meteorology,* D. Atlas, Ed., Amer. Meteor. Soc., 54–60.

Ray, P., 1990: Convective dynamics. *Radar in Meteorology,* D. Atlas, Ed., Amer. Meteor. Soc., 348–390.

Rigby, E.C., J.S. Marshall, and W. Hitschfeld, 1954: The development of the size distribution of raindrops during their fall. *J. Meteor.,* **11,** 362–372.

Rinehart, R.E., 1991: *Radar for Meteorologists.* University of North Dakota Press, 334 pp.

Rogers, R.R., 1984: A review of multiparameter radar observations of precipitation. *Radio Sci.,* **19,** 23–36.

Roux, F., and J. Testud, 1983: Pressure and temperature fields retrieved from dual Doppler radar observations of a West-African squall line. Preprints, *21st Conf. on Radar Meteorology,* Edmonton, AB, Canada, Amer. Meteor. Soc., 27–32.

Ryde, J.W., 1946: The attenuation and radar echoes produced at centimetre wave-lengths by various meteorological phenomena. *Meteorological Factors in Radio-Wave Propagation,* Physical Society of London, 169–189.

Seliga, T.A., and V.N. Bringi, 1976: Potential use of radar differential reflectivity measurements at orthogonal polarizations for measuring precipitation. *J. Appl Meteor.,* **15,** 69–76.

Smith, P.L., 1966: Interpretation of the fluctuating echo from randomly distributed scatterers: Part 3. Preprints, *12th Conf. on Radar Meteorology,* Norman, OK, Amer. Meteor. Soc., 1–6.

_______ , Z. Liu, and J. Joss, 1993: A study of sampling-variability effects in raindrop-size observations. *J. Appl. Meteor.,* **32,** 1259–1269.

Stout, G.E., and J.C. Neill, 1953: Utility of radar in measuring areal rainfall. *Bull. Amer. Meteor. Soc.,* **34,** 21–27.

Tatarski, V.I., 1961: *Wave Propagation in a Turbulent Medium.* Mc-Graw-Hill, 285 pp.

Theon, J.S., and N. Fugono, Eds., 1988: *Tropical Rainfall Measurements.* A. Deepak, 528 pp.

Tuttle, J.D., V.N. Bringi, H.D. Orville, and F.J. Kopp, 1989: Multiparameter radar study of a microburst: Comparison with model relults. *J. Atmos. Sci.,* **46,** 601–620.

Twomey, S., 1953: On the measurement of precipitation by radar. *J. Meteor.,* **10,** 66–67.

Villard, O.G., 1976: The ionospheric sounder and its place in the history of radio science. *Radio Sci.,* **11,** 847–860.

Waldvogel, A., 1974: The N_0 jump of raindrop spectra. *J. Atmos. Sci.,* **31,** 1067–1078.

Watson Watt, R.A., A.F. Wilkens, and E.G. Bowen, 1937: The return of radio waves from the middle atmosphere—I. *Proc. Roy. Soc. London, Ser. A,* **161,** 181–196.

Wexler, R., 1947: Radar detection of a frontal storm 18 June 1946. *J. Meteor.,* **4,** 38–44.

______ , 1951: Theory and observation of radar storm detection. *Compendium of Meteorology,* T.F. Malone, Ed., Amer. Meteor. Soc., 1283–1289.

______ , 1955: An evaluation of the physical effects in the melting layer. *Proc. Fifth Conf. on Radar Meteorology,* Asbury Park, NJ, Amer. Meteor. Soc., 329–334.

______ , and D.M. Swingle, 1947: Radar storm detection. *Bull. Amer. Meteor. Soc.,* **28,** 159–167.

Willis, P.T., and A.J. Heymsfield, 1989: Structure of the melting layer in mesoscale convective system stratiform precipitation. *J. Atmos. Sci.,* **46,** 2008–2025.

Wilson, J.W., and E.A. Brandes, 1979: Radar measurement of rainfall—A summary. *Bull. Amer. Meteor. Soc.,* **60,** 1048–1058.

Woodman, R.F., and A. Guillen, 1974: Radar observations of winds and turbulence in the stratosphere and mesosphere. *J. Atmos. Sci.,* **31,** 493–505.

Zawadzki, I.I., 1984: Factors affecting the precision of radar measurements of rain. Preprints, *22d Conf. on Radar Meteorology,* Zurich, Switzerland, Amer. Meteor. Soc., 251–256.

CHAPTER 5

Evolution of Satellite Observations in the United States and Their Use in Meteorology

JAMES F.W. PURDOM AND W. PAUL MENZEL

Introduction

In his keynote address celebrating the 25th anniversary of the launch of *TIROS-1* at the Second International Satellite Direct Broadcast Services Users' Conference in April 1985, G.O.P. Obasi, secretary-general of the World Meteorological Organization (WMO), remarked:

> We can marvel at the many significant technological developments which have taken place in meteorological satellites since 1 April 1960 . . . . We need to express our gratitude that the countries now operating meteorological satellites recognized TIROS as only the first step and followed up, through generations of improved satellites, to the current global network.
>
> Every day, over 120 countries use information from this network of geostationary and polar-orbiting meteorological satellites. Never before has a technology developed so rapidly to become so essential to so many national weather services. The development of Automatic Picture Transmission (APT) was the prime catalyst in focusing the world's attention on what satellites could do in improving meteorological services.

At that same conference, Richard Hallgren, then National Weather Service (NWS) director and former president of the American Meteorological Society (AMS), remarked on the use of satellite data in research and its phenomenal growth in weather forecasting.

Concerning research, he said: "I know that 20 years ago we were trying to make sure satellites were part of a scientific conference by setting up individual sessions. . . . Now . . . it doesn't matter what session you are attending, even a highly theoretical session, you will find satellites are part of the background information or the most essential information for the particular study." Regarding forecasts and warnings, Hallgren stated: "The use of satellite information simply permeates every aspect of the process and all this in a mere 25 years."

During the decade since these remarks, rapidly growing technology has provided a capability that has allowed the field of satellite meteorology and oceanography to grow dramatically. Countries throughout the world have become dependent on meteorological satellite data for both research and operations, as that data continues to provide new insights into the behavior of the atmosphere and oceans. The scientific literature is rich with articles based on satellite data, covering a wide spectrum of topics ranging from thunderstorms to biomass burning to climate change. Meteorological satellites of today provide more than the early cloud photographs (Fig. 5-1) that caused (and rightly so) so much excitement throughout the meteorological community in the early 1960s. Satellite cloud images are now high-resolution digital renderings of a variety of spectral bands (from the visible to microwave portions of the spectrum), whereby both qualitative and quantitative information is deduced. Thermodynamic soundings from satellites are now operational from both polar-orbiting and geostationary altitudes. New satellites are being designed and built to carry advanced instrumentation such as active radars, advanced multispectral imagers, spectrometers, and advanced microwave sounding units. How did we get to this point in the short 35 years since the launch of *TIROS-1?* Clearly, as pointed out by Kidder and Vonder Haar (1995), the computer revolution played a major role in this development. Indeed, without that revolution many of the science advances that we have realized today would not have come to fruition. But the genesis of the meteorological satellite program did not begin with the computer, nor with the launch of the first meteorological satellite. It began well before satellites and rockets and is a natural part of the evolution of humanity's attempt to improve the ability to observe and understand the weather. Thus, before we proceed from the launch of *TIROS-1,* it is prudent to take a quick glimpse into our past, as we struggled to improve our ability to measure and gain insight into the workings of the atmosphere. Such a glimpse will help us gain a perspective concerning our "roots."

Before satellites and rockets
Sensing the state of the atmosphere

Since colonial times, the interest in today's weather and prediction of tomorrow's has led to attempts at sensing the earth's atmosphere.

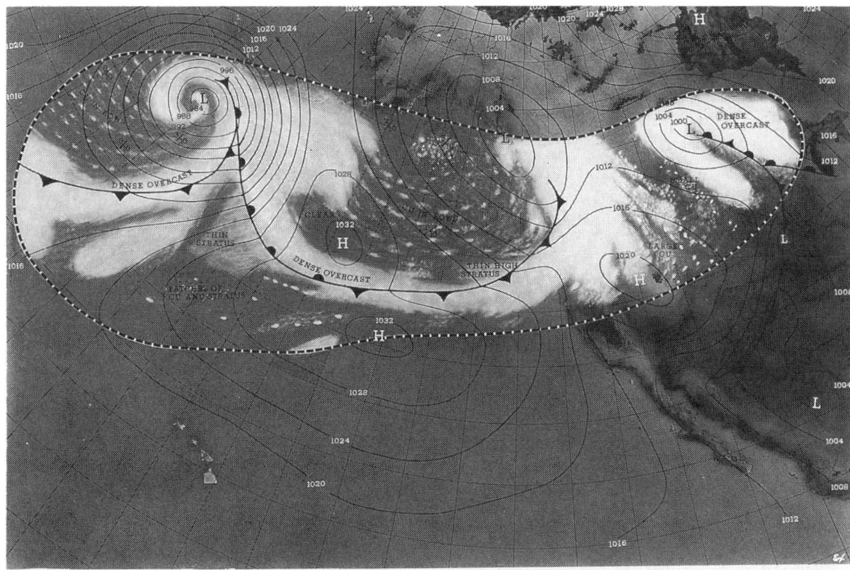

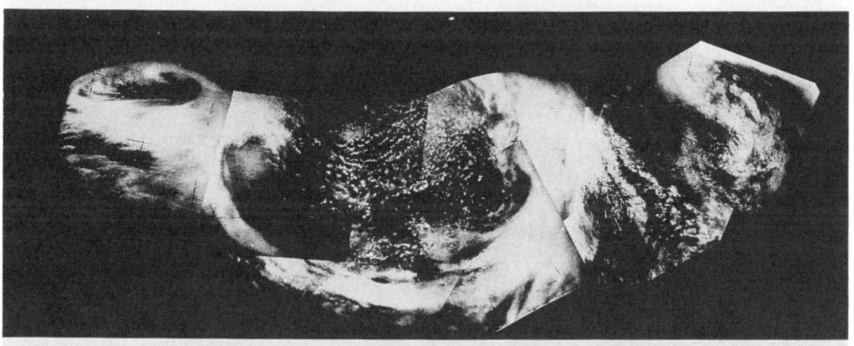

FIG. 5-1. Mosaic of cloud pictures from *TIROS-1* on May 20 1960, (bottom) and artist rendering of those cloud pictures superimposed on a conventional weather map. This picture appeared on the front cover of the October 1960 issue of *Weatherwise,* along with an explanation of the imagery by Vincent J. Oliver. (Photograph courtesy of V.J. Oliver.)

Probing of the atmosphere before rockets and earth-orbiting satellites consisted mostly of balloon and kite flights. Temperature and pressure sensing devices were attached, and some method of recording the data was devised. Benjamin Franklin's kite flights are the most well known of the early meteorological observations. Early in the twentieth century, the Weather Bureau organized a regularly scheduled program of kite observations at as many as seven stations. In this programmed probing of the atmosphere, kites were launched for four or five hours to heights up to three or four miles. The reeling in of the kite was seldom trouble free, and as witnessed in an account by a Weather Bureau kite flyer, "sometimes the operators get knocked down and dragged about by the kite in attempting to land it, while the urchins, at a distance, shout humiliating taunts." Reading the adventures and mishaps of these early kite flights strikes a poignant cord in those of us who watched the birth of the U.S. space program on television, where sometimes rockets would ignite only to collapse upon themselves.

The rapid development of the airplane during World War I stimulated new ways to obtain upper-air observations. In 1925 the Weather Bureau began an experimental program of daily observations from sensors that were attached on the wings of aircraft. The comparative ease of this data gathering soon saw the kite phased out completely. With instrumented-aircraft flights, observations over a relatively large local area were now possible, and early attempts at depicting synoptic flow began. The airplane also gave meteorologists the ability to observe clouds and cloud patterns from above. In 1938, rawinsondes were first used routinely, and the sonde network was greatly expanded to support aviation during World War II (Hubert and Lehr 1967). Since that time, the rawinsonde has remained a mainstay of synoptic analysis and numerical modeling. However, rawinsonde observations are mainly restricted to land, vary greatly in density country to country and continent to continent, and are made by a variety of vendors. Early in the meteorological satellite program, the importance of having one uniformly calibrated instrument making all of the measurements was recognized (Yates 1974), and that capability, plus the ability for routine global observations, led to proposals "as early as 1958 to exploit details of the spectral emission by the earth's atmosphere to infer the vertical structure of its temperature and its content of water vapor" (Smith et al. 1986).

Clouds and cloud patterns

The use of cloud observations to aid in forecasting is not a new concept. From the earliest days of meteorology, the importance of clouds in defining the state of the atmosphere was recognized. Long ago, P.L. Schereschewsky, who served as chief of Weather Services for the French Armies and was an avid fan of satellite cloud imagery, pointed out the importance of clouds and cloud patterns to understanding the weather in his section on "Clouds and States of the Sky" in the *Handbook of Meteorology* (Schereschewsky 1945):

> Clouds have been observed and used for short-period weather forecasting from time immemorial. . . . Clouds are now considered essential and accurate tools for weather forecasting. Every feature of the air masses (discontinuity, subsidence, instability and stability, etc.) is reflected by the shape, amount, and structure of the clouds. Thus close scrutiny of clouds will assist the analyst in identifying and analyzing air masses.

Schereschewsky's observations about the importance of cloud observations are being realized with today's advanced era of geostationary satellites. As Fred Ostby, director of the National Severe Storm Forecast Center (NSSFC), observed at the GOES-J prelaunch press conference in May 1995, "forecasters at NSSFC who routinely use *GOES-8* imagery are able to more closely monitor the development of small-scale features that have severe weather potential."

The rocket era and conception of the meteorological satellite program

In 1929, Robert Goddard launched a payload on a rocket that included a barometer, a thermometer, and a camera. According to Smith et al. (1986), the origins of the meteorological satellite program are often traced back to that effort. Two additional significant technological advances were necessary to make the first meteorological satellite possible: advances in rocket technology and development of television cameras. The advances in rocket technology during World War II led to the first composite cloud photographs from the top of the atmosphere. D. L. Crowson (1949), an air force officer, published the first rocket photography to appear in meteorological literature in 1949 in the *Bulletin of the American Meteorological Society* and commented on its meteorological value. According to Tepper (1982), "the U.S. meteorological satellite program had its earliest beginnings" from that era with a project named Janus II. Kellogg (1966) traced the beginnings of the meteorological satellite era to a classified project at the RAND corporation that was under way in 1947 and a report written by Greenfield and Kellogg in 1951 "on the subject on how one could use satellites for meteorology." Kellogg notes the following:

> This report contained a chapter by J. Bjerknes himself . . . in which he used some of the early V-2 photographic observations recovered at White Sands in 1948. With these photographs he did a magnificent synoptic analysis, using at first just the cloud data to see what he could see. . . . This probably was the first time a serious attempt was ever made to analyze the atmosphere from above, and it did show what a valuable adjunct the cloud pictures could be.

In addressing part of the early history of meteorological satellites, Vaughn (1982) pointed out that Harry Wexler was among the first to discuss the use of and observing requirements for a meteorological satellite, primarily for use as a "storm patrol." Wexler, as head of the Office of Research of the Weather Bureau, was keenly interested in the use of satellites (Wexler 1954) and was a key player in early WMO planning that lead to the World Weather Watch. Before the launch of the first meteorological satellite, under his guidance, Sigmund Fritz became head of the Meteorological Satellite Section in the Office of Research of the Weather Bureau. That small laboratory, with Fritz, Dave Johnson, Jay Winston, Charlie Bristor, and Les Hubert, evolved into what is known today as the National Environmental Satellite and Data Information Service (NESDIS).

Television technology also experienced considerable growth during the late 1940s and 1950s. The sophistication of the newly emerging rocket and television technologies "culminated in the design of the TIROS satellite" (Hubert and Lehr 1967) and the successful launch of *TIROS-1* on April Fools' Day 1960. However, *TIROS-1* was not the first satellite to be launched with meteorological instrumentation. Television equipment was flown unsuccessfully on *Pioneer I* and *II* in late 1958 and *Vanguard II* in February 1959. In August, 1959, television signals were received successfully from *Explorer VI,* and crude images of the earth's cloud cover were constructed. Also flown in 1959 on the army's *Explorer VII* were instruments designed by Verner E. Suomi and Robert Parent of the University of Wisconsin that measured the heat budget of the atmosphere (Kellogg 1966; Hubert and Lehr 1967). The longwave radiation map based on these measurements can be found in Weinstein and Suomi (1961).

Milestones from 1960 to 1994

The polar program

The meteorological satellite era officially began on 1 April 1960 when *TIROS-1* was launched by a Douglas Aircraft–built Thor-Able rocket. The early TIROS satellites provided cloud pictures using shuttered television cameras (photosensitive vidicon tubes) whose image was recorded on tape as an analog signal for later transmission to the ground. While many meteorologists enjoyed the early TIROS years, more than a few commented on the difficulty in earth location of imagery with the original TIROS imagery (van de Boogaard 1966) because the spin axis of those early satellites pointed to a fixed position in space and the images that were directly beneath the satellite subpoint were "few and far between." This can be seen by inspecting the first TV picture from *TIROS-1* (Fig. 5-2). It was perhaps this difficulty in working with early imagery that prompted Kellogg (1966) to say "the real title of this [talk] should be 'Satellite Meteorology Beckons the Academic Community,' or 'Meteorologists, Where Are You?'" Realignment of the camera optic axes with respect to the satellite spin axis, suggested by Tom Haig while working at the Department of Defense (DoD), resolved the imaging difficulties. As seen from Hallgren's remarks in the introduction, the meteorologists came in force.

The civilian and defense sectors have historically formed a strong partnership in the polar satellite arena. In 1958 the DoD established the Advanced Research Project Agency (ARPA) that funded the TIROS program. The first TIROS was built by Radio Corporation of America under a U.S. Army Signal Corps contract.

Later the TIROS program was transferred to the National Aeronautics and Space Administration (NASA) (Johnson 1994). The value of satellite imagery to operational weather services was recognized immediately after pictures began being received from *TIROS-1*. In 1961, a National Operational Meteorological Satellite System (NOMSS) was implemented with NASA to launch and control the satellites in space and the Weather Bureau to be the operator. At that time, Nimbus was planned to be the first operational system, and a single operational meteorological satellite system was planned to meet both civilian and DoD needs. For a number of reasons, including unique sensor payloads and nodal crossing times (Gomberg and McEl-

FIG. 5-2. The first television picture from *TIROS-1*, 1 April 1960. (Photograph courtesy of D.S. Johnson.)

roy 1985), the merging of defense and civilian requirements into a single satellite system proved impractical, and a separate satellite system was developed by the DoD (Fuller 1990). In the years that followed, the DoD meteorological satellite project became known as the Defense Meteorological Satellite Program (DMSP). Until 1973 DMSP was a classified program, and operation prior to 1965 remains classified today. According to Gomberg and McElroy (1985), "the purpose of the DMSP is to provide high-quality weather and other environmental data in a timely fashion to the Armed Forces of the United States for tactical and strategic missions." The DMSP satellite bus evolution paralleled that of the civilian satellite system, and many of the advances with TIROS and National Oceanic and Atmospheric Administration (NOAA) satellites, such as the wheel mode of operation and three-axis, body stabilization, are a result of advances with the DMSP program. It is interesting to note that 35 years after the launch of *TIROS-1* a joint civilian and defense system is being developed by an integrated program office for a national polar-orbiting operational environmental satellite system.

Ten experimental TIROS satellites were launched between 1960 and 1965. These satellites were forerunners to the TIROS Operational System (TOS). Recognizing the importance of meteorological satellites to the nation, a new federal laboratory was formed under the leadership of Sigmund Fritz. This evolved into the current NESDIS.

Important steps in the TIROS program included the following:

(a) the first APT with *TIROS-VIII,* which provided television pictures to ground-receiving sites on a worldwide basis—APT has been recognized as one of the United States greatest "ambassadors" of goodwill (Popham 1985);

(b) the change to a "wheel" mode of operation with *TIROS-IX,* which meant that television pictures were taken when the camera was pointed directly at satellite subpoint—this allowed for ease in navigation of the imagery; and

(c) the introduction of sun-synchronous orbits, which meant that a satellite crossed the equator at the same local solar time each orbit—this allowed for the development of worldwide mosaics of satellite images (Fig. 5-3) and opened the door for a number of important scientific projects, including the World Climate Research Programme.

The experimental TIROS series gave way to nine operational TOS satellites launched between 1966 and 1969, which were named *ESSA-1* through *ESSA-9* for the parent agency at that time, the Environmental Science Services Administration. An operational photomosaic from *ESSA-5* is shown in Fig. 5-4. The year 1965 is noteworthy within the operational satellite community: in that year under the guidance of Bob White, who was then head of ESSA, the National Environmental Satellite Service (NESS) was moved from under the weather service and made a separate line component of ESSA. Under the guidance of Dave Johnson, head of NESS from its inception until

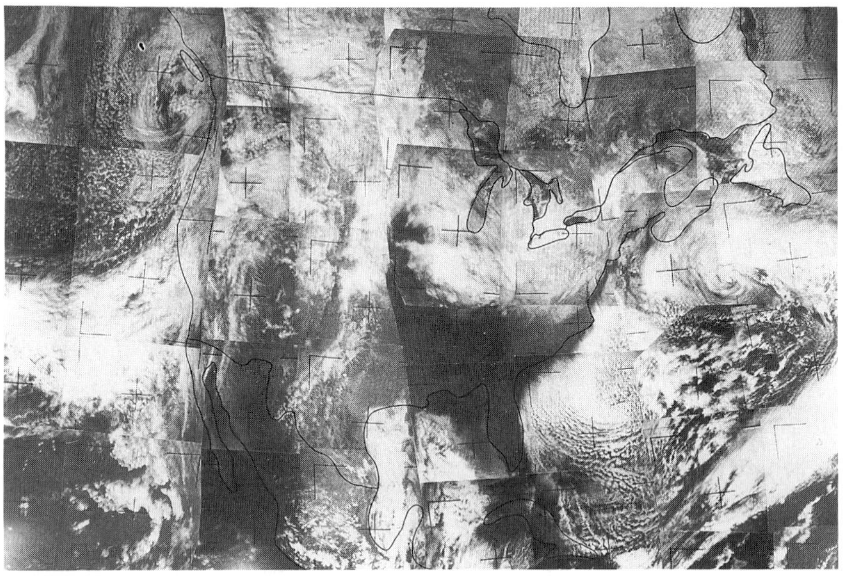

FIG. 5-3. First complete photomosaic from *ESSA-1*, 5 February 1966, showing heavy rain–producing storm along the West Coast, ice in Lake Erie, and snow cover over the northeast United States. (Photograph courtesy of D.S. Johnson.)

he retired in 1982, NESS developed into a world-renowned operational satellite service. In parallel with the operational ESSA series, NASA, under the leadership of Harry Press, Bill Nordberg, and Bill Bandeen, developed and maintained a research series of seven Nimbus satellites that had as one of its major goals to serve as a test bed for future operational polar-orbiting instruments; in that area Nimbus was very fruitful (e.g., three-axis stabilization, advanced vidicon camera systems, infrared imagers, microwave radiometers, and infrared sounders). Additional detail concerning the impact of Nimbus instrumentation on the operational polar-orbiting satellites may be found in Allison et al. (1977). Nimbus provided a number of scientific firsts and laid the foundation for satellite data use in interdisciplinary earth sciences. Hass and Shapiro (1982) provide an excellent overview of Nimbus achievements in the areas of meteorology, oceanography, hydrology, geology, geomorphology, geography, cartography, and agriculture. They point out that the roots of the Landsat program can be traced directly back to Nimbus.

The Nimbus satellites also served as pathfinders for global weather experiments organized by the Committee on Space and Atmo-

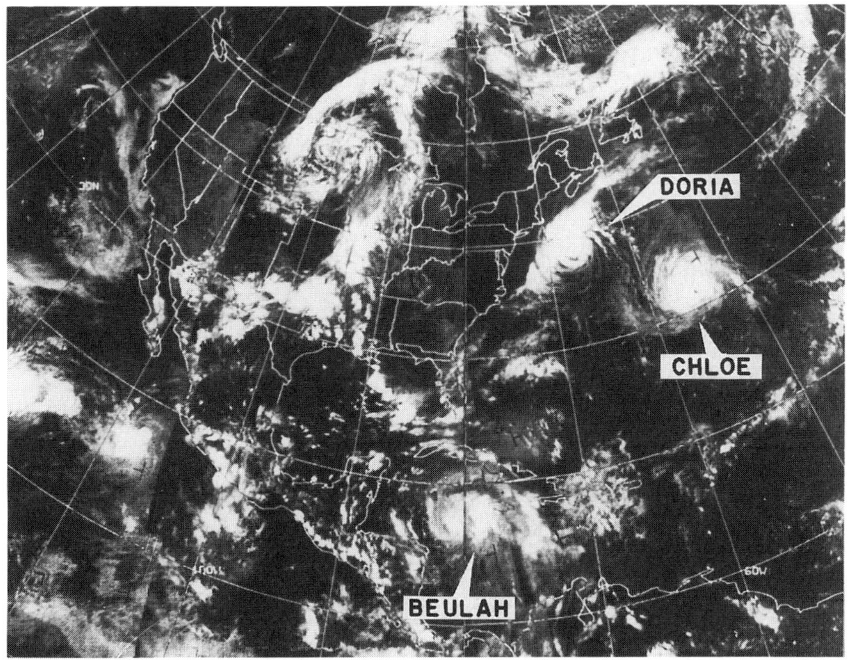

FIG. 5-4. An operational photomosaic from *ESSA-5* showing Hurricanes Beulah, Chloe, and Doria. (Photograph courtesy of D.S. Johnson.)

spheric Research (COSPAR). COSPAR Working Group 6 deliberations and publications served as the foundation for the general nature and detailed composition of the satellite-observing system in the First Global Atmospheric Research Programme Global Experiment and of the international operational system that followed. Main players in this group included Sigmund Fritz, Harry Press, Morris Tepper, Stan Ruttenberg, Bill Nordberg, Dave Johnson, Pierre Morel, Kiryll Kondratyev, Fritz Moeller, and Brian Tucker.

Significant advances with the ESSA series included routine global coverage using onboard recording systems on odd-numbered spacecraft and APT on even-numbered satellites. As forecast by Oliver and Ferguson (1966), the global coverage provided by the onboard storage capability allowed the NESS to provide operational satellite support to forecast centers for the first time. Based on the work of Oliver et al. (1964), jet streams, midtropospheric trough and ridge lines, and vorticity centers were readily located in the ESSA images. While the feasibility of using satellite imagery to locate and track tropical storms was recognized almost immediately in the satellite program (Sadler 1962), it was not until the ESSA series of satellites that routine surveillance was assured and techniques to estimate hurricane intensity from satellites became a regular part of weather forecasting (Dvorak 1972). ESSA was followed by Improved TOS (ITOS), which allowed the global remote coverage and APT services to be combined on one satellite. *ITOS-1,* the first of an operational series of three-axis stabilized satellites, was launched in January 1970. Three-axis stabilization allowed scanning radiometers to be flown on operational meteorological satellites and provided routine infrared window coverage both day and night. The third satellite in the ITOS series was launched on 15 October 1972, called *NOAA-2* for the parent agency, the National Oceanic and Atmospheric Administration. The launch of *NOAA-2* was significant to the operational cloud-imaging program: it marked the end of the vidicon era and the beginning of the era of calibrated multichannel high-resolution radiometers, although only visible and infrared window channel data were available on the Very High Resolution Radiometer (VHRR). The next major step occurred in October of 1978 with the launch of TIROS-N with a four spectral channel Advanced VHRR (AVHRR) imaging system. According to Rao et al. (1990) that step "provided data for not only day and night imaging in the visible and infrared but also for sea surface temperature determination, estimation of heat budget components, and identification of snow and sea ice." AVHRR rapidly moved to a five spectral channel system, with all channels providing imagery at 1.1-km resolution at nadir. Those channels, which have been the backbone of the imaging system, are 0.58–0.68, 0.72–1.1, 3.55–3.93, 10.3–11.3, and 11.5–12.5 µm.

In addition to imaging the earth–atmosphere with visible reflected and thermal emitted radiances, the polar-orbiting satellite series also developed a sounding capability. King (1958) and Kaplan (1959) published works that indicated that it was possible to infer the temperature of the atmosphere or the concentration of the attenuating gas as a function of atmospheric pressure level. King showed that measurements of the atmosphere at several tangential viewing angles could also provide information on temperature changes with altitude. Kaplan suggested this could be done through measurements in several narrow and carefully selected spectral intervals by inverting the process of radiative transfer. Temperature profiles are derived using the emission from CO_2 in the atmosphere, and concentrations of moisture are then inferred with emissions from H_2O in the atmosphere. Surface temperatures are estimated from observations in the spectral regions where the atmosphere is most transparent. Other early work included that of Wark and Fleming (1966), who published one of the first articles on the procedures to extract atmospheric profiles with indirect measurements from satellites.

The first temperature retrievals using satellite data were accomplished with the Satellite Infrared Spectrometer (SIRS), a grating spectrometer aboard *Nimbus-3* in 1969 (Wark et al. 1970). Comparison with radiosonde-observed profiles showed that the satellite-derived temperature profiles were very representative overall, with detailed vertical features smoothed out. The major problems with the early SIRS observations were caused by clouds that usually existed within the instrument's 250-km diameter field of view. Also the SIRS observed only along the suborbital track, and consequently there were large gaps in data between orbits. In spite of these problems, the SIRS data immediately showed promise of benefiting the current weather analysis–forecast operation and was put into operational use on 24 May 1969, barely one month after launch (Smith et al. 1970). An example of SIRS impact on model forecast is shown in Fig. 5-5. *Nimbus-3* also carried the Infrared Interferometer Spectrometer (IRIS), a Michelson interferometer that measured the earth-emitted radiation at high spectral resolution ($5\ \mathrm{cm}^{-1}$). IRIS had 100-km resolution and was nadir viewing only. These interferometer measurements of the terrestrial spectrum revealed details of the carbon dioxide, water vapor, and ozone absorption bands never seen before (Hanel et al. 1971).

In 1972 a scheme was devised to reduce the influence of clouds by employing a higher spatial resolution (30 km) and by taking spatially continuous sounding observations, now possible with cross-track scanning, using the seven-channel Infrared Temperature Profile Radiometer (ITPR) on board *Nimbus-5* (Smith et al. 1974a). With an adjacent field-of-view method, clear radiances could now be inferred by assuming that the variation in radiance between two adjacent fields of view was due to cloud amount only. The ITPR concept was highly successful in alleviating the influence of clouds on the synoptic scale. In fact, satellite soundings to the earth's surface, with an average spacing of 250 n mi, were now possible over 95% of the globe.

Also on board *Nimbus-5* was the first microwave sounding device, the Nimbus Experimental Microwave Spectrometer (NEMS), a nadir-viewing five-channel instrument (Staelin et al. 1973). The NEMS dem-

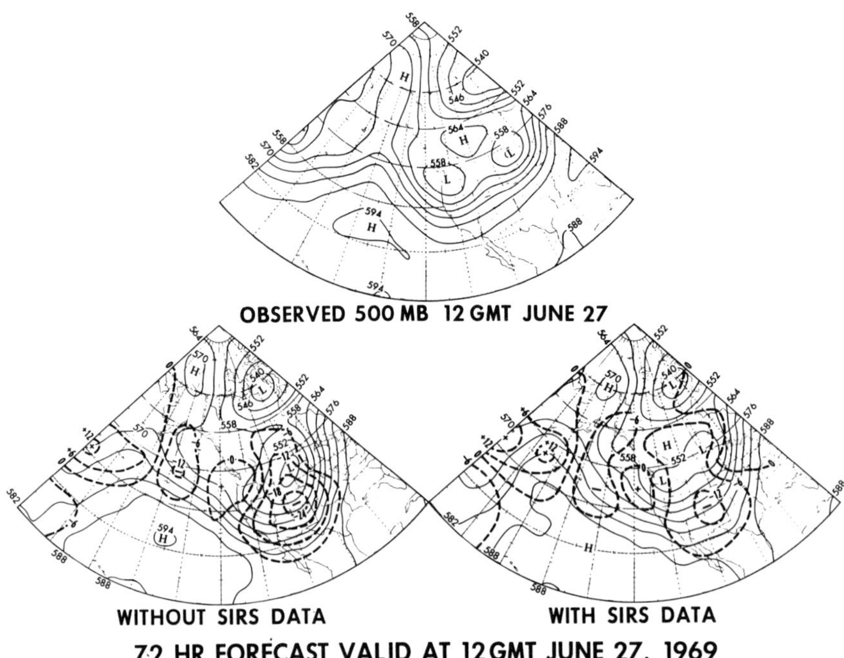

FIG. 5-5. An example of SIRS impact on model analyses.

onstrated the capability to probe through clouds, even dense overcast. Good comparisons of ITPR, NEMS, and radiosonde data were achieved. It was found, however, that best results were achieved from an amalgamation of infrared and microwave radiance data in the temperature profile inversion process, thereby providing the maximum available thermal information, regardless of cloud conditions (Smith et al. 1974b).

NOAA-2, mentioned earlier, also allowed operational thermodynamic soundings with its Vertical Temperature Profile Radiometer (VTPR). While input to numerical forecast models had previously relied on cloud-image interpretation through a program with the catchy acronym of SINAP (Satellite Input to Numerical Analysis and Prediction, Nagel and Hayden 1971), soundings from satellites now allowed for quantitative input (Smith et al. 1986); this topic is discussed in more detail in the next section.

From the available results and studies in the early 1970s, it was recognized that the optimum temperature profile results would be achieved by taking advantage of the unique characteristics offered by the 4.3-μm, 15-μm, and 0.5-cm absorption bands. As a consequence, the *Nimbus-6* High-Resolution Infrared Radiation sounder (HIRS) experiment was designed to accommodate channels in both the 4.3- and 15-μm infrared regions, and these were complemented by the 0.5-cm microwave channels of a Scanning Microwave Spectrometer (SCAMS). The HIRS was designed with passively cooled detectors to allow for complete cross-track scanning; the SCAMS also scanned, but with lower spatial resolution. The HIRS experiment successfully demonstrated an improved sounding capability in the lower troposphere due to the inclusion of the 4.3-μm observations.

The operational implementation of these instruments was achieved on the *TIROS-N* spacecraft in 1978 that carried the HIRS and the Microwave Sounding Unit (MSU). The channels were carefully selected to cover the depth of the atmosphere. Infrared soundings of 30-km resolution horizontally were supplemented with microwave soundings of 150-km resolution horizontally (Smith et al. 1979). The complement of infrared and microwave instruments aboard each of the polar-orbiting spacecraft provided complete global coverage of vertical temperature and moisture profile data every 12 hours at 250-km spacing (Smith et al. 1981a).

The *TIROS-N* series of satellites evolved to the current NOAA advanced TIROS-N satellite series, which carries the following:

(a) a five-channel AVHRR for observing cloud cover and weather systems, deriving sea surface temperature (McClain 1979), detecting urban heat islands (Matson et al. 1978) and fires (Matson and Dozier 1981), and estimating vegetation indexes (Tarpley et al. 1984);

(b) an improved HIRS-2 for deriving global temperature and moisture soundings (Smith et al. 1986);

(c) a low spatial resolution MSU primarily for deriving temperature soundings in cloud-covered regions;

(d) a data-collection system; and

(e) search and rescue (SAR) instruments for aiding in search and rescue operations.

The next generation of NOAA polar satellites will carry improved AVHRR (addition of a channel at 1.6 μm for cloud, ice, and snow discrimination), HIRS instruments that will continue to provide their basic services and, most exciting, two advanced microwave sounding units that will provide temperature sounding information at about

50-km horizontal resolution and moisture sounding information at about 15-km horizontal resolution. With the advent in 1996 of this enhanced microwave sounder (more channels, better spatial resolution) and continuation of the high spatial resolution infrared (good spatial resolution, evolving to higher spectral resolution), an all-weather sounding capability is assured.

The *TIROS-N* instruments are now used routinely to measure atmospheric temperature and moisture, surface temperature, and cloud parameters. A large range of applications have been developed that are pertinent to nowcasting as well as operational short-range forecasting; they include the following:

(a) subsynoptic-scale temperature and moisture analyses for severe weather monitoring and prediction;
(b) subsynoptic-scale analysis of surface temperature;
(c) subsynoptic-scale analysis of atmospheric stability;
(d) subsynoptic-scale updating of aviation gridpoint temperature and wind fields;
(e) estimation of the cloud height and amount;
(f) estimation of tropical cyclone intensity, maximum wind strength, and central position; and
(g) estimation of total ozone amount.

These data have become part of the operational practices of weather services internationally (Chedin 1989).

For details concerning scientific applications of polar-orbiting satellite imagery and sounding data, the reader is referred to Rao et al. (1990). That book contains sections on applications of satellite data in meteorology, applications of satellite data to land and ocean sciences, application for satellites and climate, and the use of weather satellite data in agricultural applications. For details on the use of polar-orbiting imagery, also see Scorer (1990).

As with the civilian polar-orbiting system, DMSP has led to a number of meteorological accomplishments that were made possible by its unique sensor payloads. Sensors of note on DMSP include (a) the Operational Line Scanner (OLS), (b) the Special Sensor Microwave Imager (SSM/I); and (c) the infrared and microwave sounders. The OLS is a two-channel visible and infrared imager with a 0.3-n-mi resolution. The "visible" band of OLS extends well beyond the visible portion of the spectrum, covering from approximately 0.5 to 1.1 µm. The sensor is designed to allow for low-light imaging and can detect features such as city lights at night and the aurora borealis (Meyers 1985). One of the most exciting sensors on DMSP is the SSM/I. SSM/I measures both horizontally and vertically polarized earth–atmosphere radiation at 19.36, 22.23, and 85.5 GHz and horizontally polarized radiation at 37 GHz. SSM/I data are used operationally by the air force to produce precipitation rates (Kidder and Vonder Haar 1995). SSM/I imagery also shows details in convective systems, including hurricane rainbands. For details concerning scientific applications of DMSP data, the reader is referred to Schnapf (1985).

The geostationary program

On 6 December 1966, a stellar day in satellite meteorology, the first Applications Technology Satellite (*ATS-1*) was launched. *ATS-1*'s spin–scan cloud camera (Suomi and Parent 1968) was capable of providing full disk visible images of the earth and its cloud cover every 20 minutes. The inclusion of the spin–scan cloud camera on *ATS-1* occurred because of an extraordinary effort by Verner Suomi, Dave Johnson and Homer Newell, who made it possible to add this new capability to *ATS-1* when the satellite was already well into its fabrication. Meteorologists were astounded by the first global views of clouds and cloud systems in motion. According to Johnson (1982), "as Morris [Tepper of NASA] predicted, geostationary satellite images can be used to track clouds from which winds at cloud altitude can be inferred." Research into tracking clouds and producing winds using image sequences began almost immediately (Hubert and Whitney 1971) and, as will be discussed in section 5, is still an area of intense investigation. By the early 1970s, ATS imagery was being used in operational forecast centers, with the first movie loops being used at the NSSFC in the spring of 1972. Fig. 5-6 shows *ATS-3* imagery received at NSSFC during the most severe outbreak of tornadoes in recorded history.

As with the polar program, NASA research and development led to the operational geostationary satellite program. ATS was mainly a communications system test bed; however, its success in the meteorological arena led to NASA's development of the Synchronous Meteorological Satellite (SMS), an operational prototype dedicated to meteo-

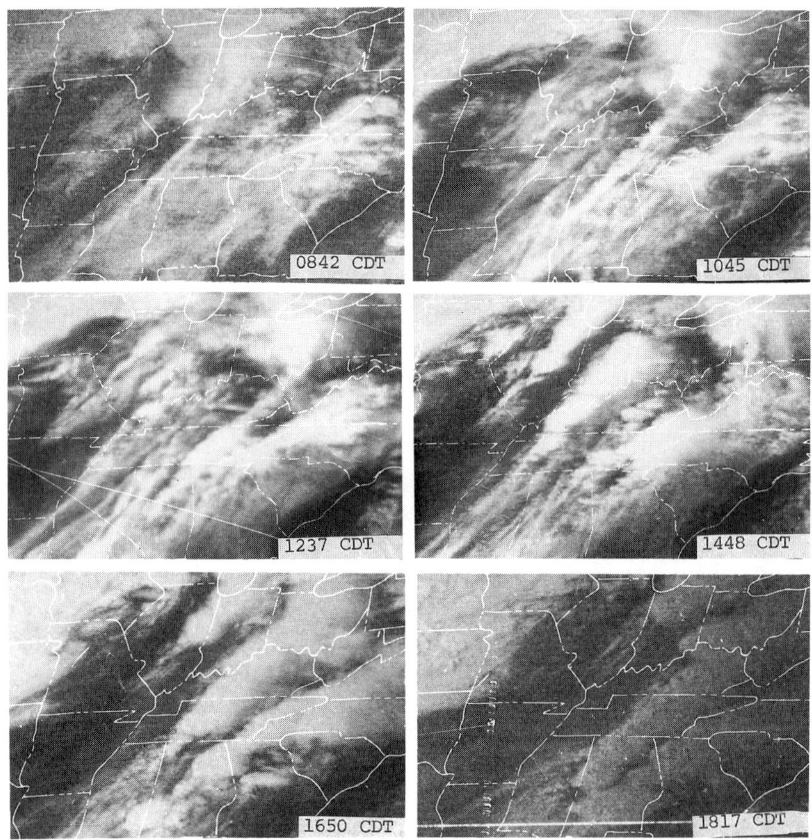

FIG. 5-6(a)–(f). Series of *ATS-3* imagery received at NSSFC during the 3 April 1974 supertornado outbreak. These images show the development of the three major squall lines that produced massive tornado damage from the Mississippi Valley to the Allegheny Mountains.

rology. *SMS-1* was launched May 1974 and *SMS-2* was launched February 1975; those satellites were positioned above the equator at 75° and 135°W, which today remain the nominal positions of the U.S. eastern and western Geostationary Operational Environmental Satellites (GOES). The two NASA prototypes, *SMS-1* and *SMS-2*, and the subsequent NOAA GOES provided three important functions that remain central to today's geostationary satellite program:

1. Multispectral imagery from the Visible and Infrared Spin–Scan Radiometer (VISSR), which provided routine coverage of the earth and its cloud cover in the visible and infrared window

channels. VISSR provided visible imagery at 1-km spatial resolution and infrared window channel imagery at 7-km spatial resolution.

2. Weather Facsimile, which provided the transmission of low-resolution satellite images and conventional weather maps to users with low-cost receiving stations.

3. Data Collection System, which allows for the relay of data from remote data collection platforms through the satellite to a central processing facility.

By 1980, the GOES system evolved to include an atmospheric temperature and moisture sounding capability with the addition of more spectral bands to the spin–scan radiometer; it was called the VISSR Atmospheric Sounder (VAS). The first GOES-VAS, *GOES-4*, was launched in September 1980. The addition of spectral channels represented a major improvement in satellite capabilities (Smith et al. 1981b); however, imaging and sounding could not be done at the same time. Furthermore, a spinning satellite, viewing the earth only 5% of the satellite's duty cycle, made it difficult to attain the instrument signal to noise needed for either high quality soundings or the high spatial resolution infrared views of the earth needed to satisfy the increased sophistication of the data's user community. Recognizing those limitations, NOAA developed its next generation of geostationary satellites, GOES I–M (Menzel and Purdom 1994). The GOES I–M system was introduced with the launch of GOES-I on 13 April 1994, designated *GOES-8* after attaining geostationary orbit; *GOES-9* followed in May 1995. More information on GOES I–M is given in a later section.

As imagery from polar-orbiting satellites helped advance understanding of synoptic-scale phenomena, imagery from geostationary satellites helped advance understanding of the mesoscale. Prior to the geostationary satellite the mesoscale was a "data sparse" region; meteorologists were forced to make inferences about mesoscale phenomena from macroscale observations. Today, geostationary satellite imagery represents the equivalent of a "reporting station" every 1 km with visible data (every 4 km with infrared data); those data reveal mesometeorological features that are infrequently detected by fixed observing sites. The clouds and cloud patterns in a satellite image provide a visualization of mesoscale meteorological processes. When imagery is viewed in animation, the movement, orientation, and development of important mesoscale features can be observed, adding a new dimension to meteorological reasoning. Furthermore, animation provides observations of convective behavior at temporal and spatial resolutions compatible with the scale of the mechanisms responsible for triggering deep and intense convective storms (Purdom 1993). A number of important discoveries using geostationary satellite imagery have had a dramatic impact on mesoscale meteorology and, in turn, short-term forecasts and warnings.

(a) Prior to squall-line formation, organized cumulus development within a surface convergence zone is usually detectable in satellite imagery (Purdom 1976). The ability to use satellite imagery to detect areas of incipient squall-line development has aided in the location and orientation as well as the timing of severe weather watches.

(b) The importance of thunderstorm outflow boundaries, often termed arc cloud lines, in the development and evolution of all types of thunderstorms was first recognized using animated satellite imagery (Purdom 1976). The recognition of one such large-scale boundary resulted in a very accurate, long-lead-time watch box for a major tornado outbreak across North and South Carolina on 28 March 1984 (NOAA 1984). Furthermore, Doppler radar has confirmed the importance of recognizing and tracking arc cloud lines for short-term convective forecasting (COMET 1992).

(c) The size, duration, and high degree of organization of mesoscale convective complexes were not recognized prior to their discovery using infrared satellite imagery (Maddox 1980). These complexes, which are responsible for much of the summer rainfall in the Midwest, have since been the focus of intense investigations.

(d) The location, tracking, and monitoring of hurricanes and tropical storms (Fig. 5-7) has been one of the most successful aspects of the satellite program (Rao et al. 1990), and a technique for using satellite data to estimate hurricane intensity (Dvorak 1972, 1984) is used routinely at the Tropical Prediction Center (formerly the National Hurricane Center). According to former director Robert Sheets (1990), "in the estimation of the author,

FIG. 5-7. One-kilometer resolution visible image of Hurricane Luis taken by *GOES-9* on the 6 September 1995 at 2030 UTC. At the time of this image, Luis is north of Puerto Rico.

and many others, the development of the Dvorak technique (1972, 1984) has been the single greatest achievement in support of operational tropical cyclone forecasting by a researcher to date."

(e) Prior to the polar satellite program, the existence of polar lows (Twitchell et al. 1989) was not widely recognized, and recently, geostationary satellite imagery has been used to study polar low formation near the coast of Labrador (Rasmussen and Purdom 1992).

(f) The influence of early morning cloud cover on the subsequent development of afternoon convection was not realized prior to the geostationary satellite era (Weiss and Purdom 1974; Purdom and Gurka 1974). The importance of this phenomenon has subsequently been simulated using a sophisticated mesoscale model (Segal et al. 1986).

(g) Satellite imagery has had a dramatic impact on our ability to detect and forecast fog behavior. The role of inward mixing in forecasting the dissipation of fog was not well recognized prior to the work of Gurka (1978) with geostationary satellite imagery. Polar-orbiting satellite imagery has been used for a number of years to detect fog, both day and night, using a multi-spectral technique (Eyre 1984), and that technique has been extended with the GOES multispectral imagery (Ellrod 1992) to include monitoring of fog formation at night.

(h) The GOES infrared channel at 6.7 µm is strongly affected by upper-level water vapor and is often called the water vapor channel. The ability of imagery from this channel to depict upper-level flow has provided meteorologists with the ability to view atmospheric systems and their changes in time with a perspective never before possible. As has been shown by Velden (1987) and Weldon and Holmes (1991), this channel may be used for a variety of synoptic and mesoscale applications.

GOES data have become a critical part of NWS operations, with direct reception of the full digital GOES data stream at national centers, while local weather service forecast offices receive limited analog imagery through a land line service known as GOES-TAP. All NWS offices are expected to receive a full complement of digital imagery when a program known as AWIPS (Advanced Weather Interactive Processing System) comes to fruition near the end of this century. Furthermore, quantitative products such as cloud drift winds, thermodynamic soundings and stability parameters, and precipitation amount are routinely produced from GOES data; these are discussed in the next section.

Rapid interval imaging has been an important component of the GOES research program since 1975. Research with SMS and GOES imagery included taking a series of images at 3-min intervals to study severe storm development (Fujita 1982; Shenk and Krein 1975; Purdom 1982). In 1979 during a project known as SESAME (Severe Environmental Storm And Mesoscale Experiment), two GOES satellites were synchronized to produce 3-min interval rapid scan imagery to study storm development. Fujita (1982) and Hasler (1981) used these data to produce very accurate cloud height assignments using stereographic techniques, similar to earlier work by Bristor and Pichel (1974). Other interesting studies with rapid interval GOES imagery include assessing thunderstorm severity (Adler and Fenn 1979; Shenk and Mosher 1987) and tracking cloud motions in the vicinity of hurricanes (Shenk 1985; Shenk et al. 1987). Over the years, results from the research community filtered into satellite operations, and by the mid-

1980s, 5-min interval imagery became a routine part of satellite operations during severe storm outbreaks. Today, with *GOES-8* and *-9*, imagery at 1-min intervals has been taken to study a variety of phenomena including severe storms, hurricanes, and cloud motions (Purdom 1995). Perhaps, based on research using that data and other special datasets, the United States will one day move to a three-satellite configuration with an eastern, western, and central satellite, with the central satellite filling the role of a storm patrol, as envisioned decades ago.

Quantitative and specialized uses of satellite data

Satellite data are unique. No other observational system has the capability to observe atmospheric phenomena and their interactions from the synoptic scale to the mesoscale. Furthermore, those observations are made from one uniformly calibrated sensor that can serve as a baseline for the integration of other datasets. This has proved important in the study of climate and the development of a number of quantitative satellite products; selected phenomena are discussed below.

Thermodynamic information—Satellite sounding data

Use of sounder data in operational weather services is coming of age. Polar-orbiting sounders are filling important data voids at synoptic scales. Applications include temperature and moisture analyses for weather prediction, analysis of atmospheric stability, estimation of tropical cyclone intensity and position, and global analyses of clouds. TOVS includes both infrared and microwave observations, with the latter helping considerably to alleviate the influence of clouds for all-weather soundings. GOES-VAS soundings have been used to derive and study quantitative measurements of atmospheric temperature, moisture, and wind at hourly intervals in the Northern Hemisphere. Temporal and spatial changes in atmospheric moisture and stability are improving severe storm warnings. These satellite soundings have been studied extensively over the past few decades; their quality with respect to radiosonde observations has been documented (Jedlovec 1985; LeMarshall 1988). Typically the global rms differences between VAS or TOVS and radiosonde temperature profiles are about 2.5 K; in the presence of clouds this degrades by about 0.5 K.

Over the past 15 years, the use of TOVS products by the meteorological community has played an important role in improving numerical weather prediction. It has been vital for global analysis, and on the average the impact of TOVS on global forecasts has been significant and positive (Menzel and Chedin 1990). In the Southern Hemisphere, TOVS data are essential for accurate analysis and on average have had a positive impact on the resulting forecasts. In the Northern Hemisphere, the forecast impact has been small, with large regional variations in individual cases. For GOES data, assimilation of time changes in distribution of atmospheric moisture offers great hope for significant positive impact on forecast models.

However, optimal use of the temperature, moisture, and cloud data available from the satellite sounders in numerical analysis and prediction is still to be achieved. Many data assimilation systems use the sounding data in much the same way that radiosonde data (i.e., temperature and moisture profiles or derived thickness and precipitable water) are used without properly accounting for the broad vertical resolution or error characteristics of the satellite sounding data. Also, calculations of radiative transfer through the atmosphere still have some inherent inaccuracies that hinder better assimilation of the measured radiances into the numerical forecast models.

Thermodynamic Information—Derived Product Imagery

When the GOES-VAS program began after the launch of *GOES-4,* a demonstration period was scheduled to determine how well that instrument could be used to provide thermodynamic information from geostationary orbit. The VAS demonstration was a joint NOAA and NASA program and involved a number of scientists from those organizations. The planned six-month demonstration period was cut short when *GOES-4* was placed into operational service. Thus, a detailed assessment never occurred, and scientists were left to make do with datasets that had inadequate signal to noise or insufficient frequency. To compensate, a number of products were developed that used combinations of more frequently available VAS channels. Two derived products that have stood the test of time are atmospheric stability (Smith et al. 1985) and precipitable water vapor (Chesters et al. 1987). The products were made by combining radiances from the VAS "split window" and the water vapor channels and represent the first quantitative use of multiple-channel imager data for the derivation of atmospheric moisture parameters.

These products were displayed in a format where the clear-air product (precipitable water vapor or atmospheric stability) is combined with cloud radiances to make a comprehensive image of the state of the atmosphere. A time sequence of these derived product images allows the forecaster to monitor water vapor convergence and atmospheric stability changes usually antecedent to intense convective weather developments. At the same time, cloud development associated with these atmospheric water vapor and stability tendencies can also be monitored (Smith et al. 1985).

Winds derived from images and soundings

Methods to determine winds from satellite data have existed almost since the beginning of the meteorological satellite era. These winds were, and are, used by forecast centers to help with analyses in data-sparse areas. Some of the earliest techniques used clues in polar-orbiting imagery. Among the more successful were the location of cloud patterns and features related to the axis of the jet stream, identification of sun-glint patterns to infer surface wind conditions over the sea, and the use of anvil blow-off for streamline analysis (Anderson et al. 1974). With the beginning of the geostationary satellite era, the ability to track clouds and relate them to flow patterns in the atmosphere was immediately recognized (Hubert and Whitney 1971). GOES made possible the routine tracking of clouds at high resolution, with infrared imagery being available for cloud-height assignment and nighttime tracking.

Cloud motion vectors (CMVs) derived from geostationary satellite imagery are used in numerical weather prediction (NWP) models. These CMVs represent an important source of meteorological information, especially over the oceans. However, over the past decade, improvements in data assimilation and NWP models have outpaced improvements in satellite-derived CMVs causing both the National Centers for Environmental Prediction (NCEP, formerly the National Meteorological Center) and the European Centre for Medium-Range Weather Forecasts (ECMWF) to restrict the use of upper-level vectors. In the Southern Hemisphere, CMVs continue to show positive impact; however, this has not been the case in the Northern Hemisphere, where more dense in situ data sources demand higher quality CMVs if they are to have a positive impact. This need for improvement is addressed further in the section titled "Requirements for Future Investigations."

Accurate cloud-height assignment is one of the difficult problems associated with better use of CMVs in NWP models. Thin cirrus clouds are the best tracers since they are most likely to be passive tracers of the flow at a single level; however, their height assignments are especially difficult since they are not opaque. Recent improvements in cloud-height assignment have come with the CO_2 slicing approach (Merrill et al. 1991) used by NOAA operationally with GOES-VAS. With GOES I–M, CO_2 slicing is not possible (one of the necessary spectral channels is not available), so a technique developed for Meteosat known as the H_2O intercept technique (Nieman et al. 1993) is used instead.

Water vapor imagery often reveals mid- to upper-tropospheric structure that is not apparent in infrared or visible images (Velden et al. 1992). Holmlund (1993) has shown that winds can be derived by tracking features in water vapor imagery. These water vapor winds, along with thermal gradient winds determined from GOES soundings and high- and low-level cloud drift winds derived from animated visible and infrared window images, can be used to produce a deep-layer mean wind field. That deep-layer mean wind field has been used to help define the steering current in which a hurricane is embedded. In a study of six hurricane seasons, these satellite-derived wind analyses produced a 10%–20% reduction in the mean forecast error in hurricane trajectories (Velden et al. 1992).

Climate

As indicated earlier, in 1959 the first meteorological experiment conducted on a satellite platform measured the radiation budget associated with synoptic weather patterns (Weinstein and Suomi 1961). Those early efforts have been expanded to global scales (Vonder Haar et al. 1981). Estimates of the albedo and outgoing longwave radiation have been made from narrow spectral band observations obtained from the NOAA operational polar-orbiting satellites since 1974 (Gruber 1977). Broadband monitoring of the trends in the global energy balance was accomplished with the Earth Radiation Budget Experiment (ERBE) from 1984 to 1991; the ERBE was devised to measure the balance between the radiation input to the atmosphere from the sun and the radiation exiting from the atmosphere as a result of reflection and emission processes. The spatial distribution of the radiation imbalances between incoming and outgoing radiation (the net radiation) is the primary driving force of the general circulation. Satellites are the only platforms capable of measuring the global distribution of the radiation budget in space and time; geographical and seasonal patterns are revealing interesting features associated with El Niño and Southern Oscillation events.

In addition, the NOAA polar-orbiting AVHRR has been used to estimate global sea surface temperatures (SST). These fields provide good horizontal coverage and detail (McClain et al. 1985), which is useful to monitor global changes in longwave radiation emission.

The operational NOAA satellites have also been used to estimate global cloud cover, which modulates the radiation budget in the earth's climate system. Variations in the amplitude and phase of the seasonal and diurnal cycles of the outgoing longwave radiation can be interpreted as variations in the surface temperature and cloudiness. Estimates of the variation of the earth's outgoing longwave radiation or clouds have been derived primarily from the longwave infrared window (10–12 μm) radiances as observed from polar-orbiting and geostationary satellites (Rossow and Lacis 1990). The frequency of cirrus clouds usually has been underestimated in these cloud population studies because thin cirrus clouds are difficult to detect since they reflect little solar radiation and are transparent to terrestrial radiation in the longwave infrared window. Recently, HIRS multispectral observations in the CO_2 absorption band at 15 μm have been used to compile global cloud cover statistics for the past four years (Wylie et al. 1993); this estimation of upper-level semitransparent clouds complements the International Satellite Cloud Climatology Project determinations of global cloud cover.

Satellite data from the past 35 years are an extremely valuable database for investigating climate and global change. There are many parameters relevant to answering climate questions that can be derived from a careful reprocessing of the available data; the preceding paragraphs mention a few examples.

Aerosols and biomass burning (fires)

The polar-orbiting AVHRR has been extremely useful in monitoring aerosol optical thickness (Rao et al. 1988). The aerosol optical thickness is derived from the reflected solar radiation measured at 0.5 µm, which becomes greater as the optical thickness increases. Recently the circulation of the aerosol emitted by the Mt. Pinatubo eruption in June 1991 was tracked with AVHRR data; analyses of daily composites showed that the volcanic aerosol circled the earth in 21 days with the layer initially confined to the Tropics, but it branched into the midlatitudes of both hemispheres within a year (Stowe et al. 1994). This type of aerosol detection is useful for estimating the net global cooling as well as for providing aviation advisories.

GOES and AVHRR data have been used to monitor fire activity in South America. Both instruments are able to detect subpixel fire activity with their shortwave and longwave infrared windows (Matson and Dozier 1981). The advantage of the geostationary perspective resides in its ability to detect fires as they occur and determine trends in biomass burning (Prins and Menzel 1994). For selva (forest) and cerrado (grassland) regions in South America, the expansion of burning is readily seen in a comparison of GOES data from one year to the next; indeed, the locations of the fires in the Amazon forest have spread to a current area three times as large as in 1983.

Precipitation estimation

Knowledge of precipitation is important throughout meteorology: flash floods, water planning, latent heat release, and climate all come to mind. One of the first attempts to determine rainfall from satellites was developed by Follansbee (1973), who used cloud amount and type, determined from inspection of polar-orbiting imagery, to apportion rainfall over a 24-h period. Since Follansbee's early efforts, a number of different techniques have been developed; all showing different skill depending on rainfall type. The Bristol technique, developed by Barrett et al. (1986), uses either polar-orbiting or geostationary satellite imagery. The Woodley–Griffith technique (Griffith et al. 1978) used hourly GOES visible and infrared imagery brightness thresholds and their changes in time to estimate rainfall—that technique works best in tropical regions. The convective stratiform technique, developed by Adler and Negri (1981) shows considerable skill in estimating both tropical convective and stratiform precipitation under anvils of mature and decaying convective systems using minimum observed infrared brightness temperatures and one-dimensional cloud models. The technique used operationally by NOAA today is an outgrowth of the Scofield–Oliver technique (Scofield and Oliver 1977). That technique uses 30-min interval GOES visible, infrared, and water vapor imagery and an array of "enhancing factors" that either add to or subtract from the estimated rainfall amount. More recent research into rainfall estimation uses microwave measurements from the DMSP SSM/I. A number of interesting papers concerning the use of SSM/I data were presented in a special session on precipitation at the Sixth Conference on Satellite Meteorology and Oceanography held in Atlanta, Georgia, in 1992.

Tropical cyclone intensity estimation

As pointed out earlier, one of the major contributions made by satellites has been in the area of tropical storms and hurricanes. The success of the Dvorak technique (1972, 1984) is without question. Initially, that technique used day-to-day differences in cloud shape and banding shown in visible imagery to assign a T number relating to tropical storm intensity. Also taken into account in forecasting day-to-day development was the environment into which the storm was moving. With the advent of infrared imagery and geostationary observations, the Dvorak technique evolved to incorporate eye temperature and that of the surrounding cloud band to refine the T number (Dvorak and Wright 1977). This basic technique, with refinements, remains at the heart of today's operational satellite hurricane intensity estimates. Research has pointed to the use of microwave data to help determine hurricane intensity (Kidder et al. 1978; Velden and Smith 1983). Those techniques, which relied on poor-resolution microwave data, will be revisited when NOAA-K and its Advanced Microwave Sounding Unit is launched in 1996.

A new geostationary satellite era: GOES I–M

The GOES I–M system is the result of the evolution over the past quarter century of the ability of U.S. satellites to observe weather systems from geostationary altitude. The GOES I–M system, one of the major components of NOAA's extensive modernization program (Friday 1989), is a significant advancement in operational geostationary satellites. Those advancements include the following:

(a) earth-oriented spacecraft for more efficient duty cycles of imager and sounder that yield higher spatial resolution and improved signal to noise;

(b) improved five-channel high-resolution imager for improvement in present services, as well as the development of a number of advanced products;

(c) separate sounding and imaging for avoiding conflicts in scheduling between the two that had previously existed, hence allowing maximal use for each sensor; and

(d) operational sounding from geostationary orbit for the first time.

GOES-8 data provide a considerable enhancement to geostationary remote sensing capabilities. At the time of this writing, *GOES-9* has been launched and is expected to become operational in late 1995. [*GOES-9* is operational.] An extensive assessment and evaluation program is under way using data from both *GOES-8* and *-9* to determine the best use of data from these new resources. Below are listed some of the performance characteristics realized, or expected, from the *GOES-8* and *-9* imagers.

Visible (0.52–0.72 μm): 10-bit imagery with 1024 brightness levels (compared to the previous 6-bit data with 64 brightness levels) and sampling along the line every 0.6 km with 1-km spatial resolution. With proper image enhancement, the visible channel enables improved cloud-edge and cloud-top feature detection for improvements in cloud-drift winds and severe storm identification, extension of the use of visible imagery into low-light situations, detection and potential assessment of pollution and haze, highly accurate cloud-height measurements using both stereo and cloud shadow techniques, and use of imagery well beyond the satellite's 60° zenith angle (Fig. 5-8).

Shortwave infrared window (3.78–4.03 μm): 4-km resolution with sensitivity of 0.23 K for 300-K targets. With proper image enhancement and with information from other channels, these data are useful for identifying fog and cirrus at night (Fig. 5-9), locating water clouds over snow during the daytime, delineating between supercooled and ice cloud during daytime (with longwave IR), detecting hot areas such as fires and volcanoes, locating a hurricane's eye when the eye is covered with thin cirrus, and improving sea surface temperature measurements at night (in part due to decreased diffraction effects in comparison to the longwave IR channels).

FIG. 5-8. *GOES-8* 1-km resolution visible image taken on 31 May 1994 at 1300 UTC. The image extends from about 55°N to near the Arctic Circle. In terms of satellite zenith angle that means from about 63° to 75°. Notice how clearly structure may be seen in the snow and ice field over Hudson Bay, while to the north Coats Island has water around part of it, with ice in much of the Fisher Strait between Coats and Southampton Islands. Notice that the water appears "hazy" around Mansel Island but not Coats Island. Use of 3.9- and 10.7-μm channel imagery revealed that the "hazy" region was thin supercooled fog and stratus.

Water vapor channel (6.47–7.02 μm): 8-km spatial resolution with sensitivity of 0.22 K at 230 K. This enables better winds in cloud-free areas, improved analysis of synoptic-scale features, and ability to identify mesoscale features embedded within synoptic disturbances (Fig. 5-10).

Longwave infrared window (10.2–11.2 μm): 4-km resolution and signal to noise of 0.14 K at 300 K. With proper image enhancement and combined with other channels, this channel is used for better cloud-edge and cloud-top feature detection for improved cloud-drift wind velocities, enhanced capabilities for monitoring thunderstorm development and evolution, severe storm identification and location of storms with heavy rainfall, improved monitoring of land surface heating and cooling, good low-level moisture identification with the 11.5–12.5-μm channel, and better nighttime SST determination with the 11.5–12.5- and 3.78–4.03-μm channels.

Split window (11.5–12.5 μm): 4-km spatial resolution with signal to

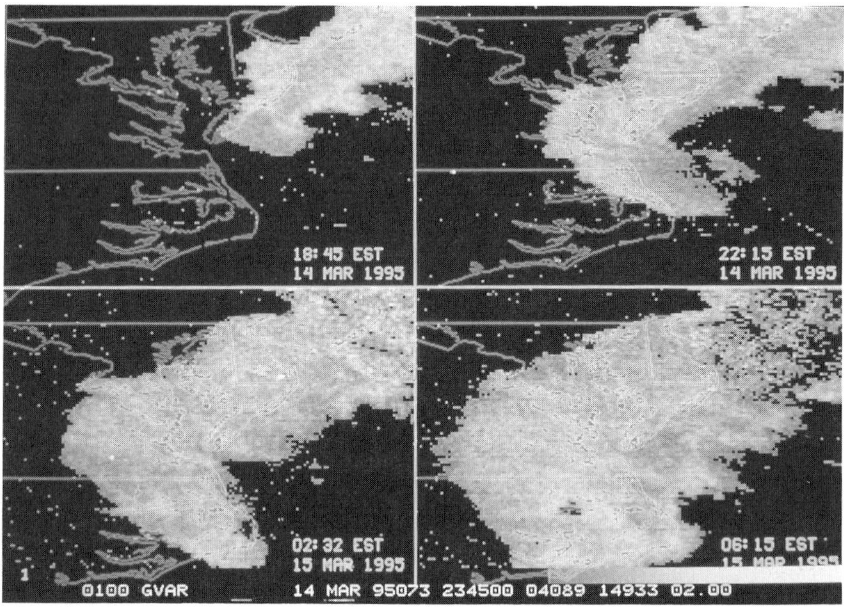

FIG. 5-9. *GOES-8* nighttime fog product, where areas of fog and stratus are shown as white. During the night of 14 March and the early morning of 15 March, coastal fog and stratus spread southward over the Atlantic Ocean and inland reducing visibility to as low as one-tenth of a mile: this had a pronounced effects on ground, air, and sea transportation.

noise of 0.26 K at 300 K. With proper image enhancement and with data from other channels, this channel is used for development of much more accurate low-level moisture products (Fig. 5-11), recognizing areas with a potential for radiation fog development, and improving sea surface temperature products.

A number of special sounder and imager science tests have been performed with *GOES-8* and *-9,* including 1-min interval scan. Examples of 1-min imagery and sounder products may be viewed by acquiring the Cooperative Institute for Research inthe Atmosphere (CIRA) and Cooperative Institute for Meteorological Satellite Studies (CIMSS) home pages on the Internet World Wide Web (CIRA 1994; CIMSS 1995). Tutorials on the imager and sounder, which provide glimpses of 1-min interval imagery, show comparisons of *GOES-7* and *-8* imagery and show improvements realized with the *GOES-8* sounder, are also available on a CD-ROM produced by SS/LORAL (1995), which has been provided (free of charge) at a variety of AMS meetings. A computer-based learning module that focuses on the use of satellite imag-

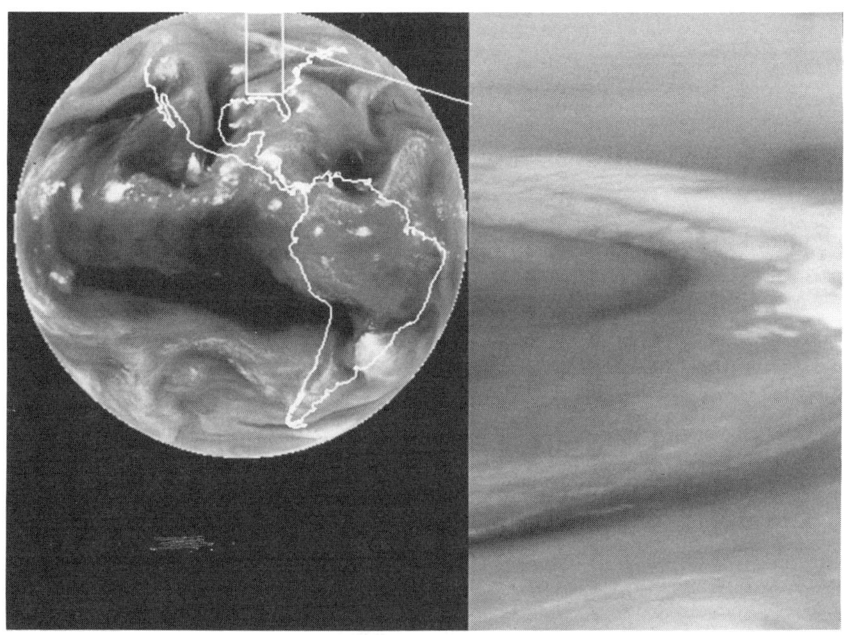

FIG. 5-10. *GOES-8* 6.7-µm infrared channel image on 31 May 1994. Notice how well the structure in the upper-level water vapor field appears in the expanded region.

ery has been prepared by NOAA in conjunction with the Cooperative Program for Operational Meteorology, Education, and Training (COMET 1995).

Requirements for future investigations

The MSU on the TOVS has been used to investigate the strength of tropical cyclones (Velden et al. 1984). Operational use is limited by the poor resolution of the current NOAA polar orbiters. Improved storm intensity estimates should be realized using the higher-resolution microwave data from the advanced microwave sounding unit on NOAA's next generation of polar-orbiting satellites to be launched in 1996. Indeed, this new high vertical and spatial resolution "sounder" should provide very useful image products—they need only to be developed.

The accuracy and density of cloud motion winds, global in coverage and eventually mesoscale in resolution, requires further attention. Large errors in cloud drift wind velocity and height assignment have

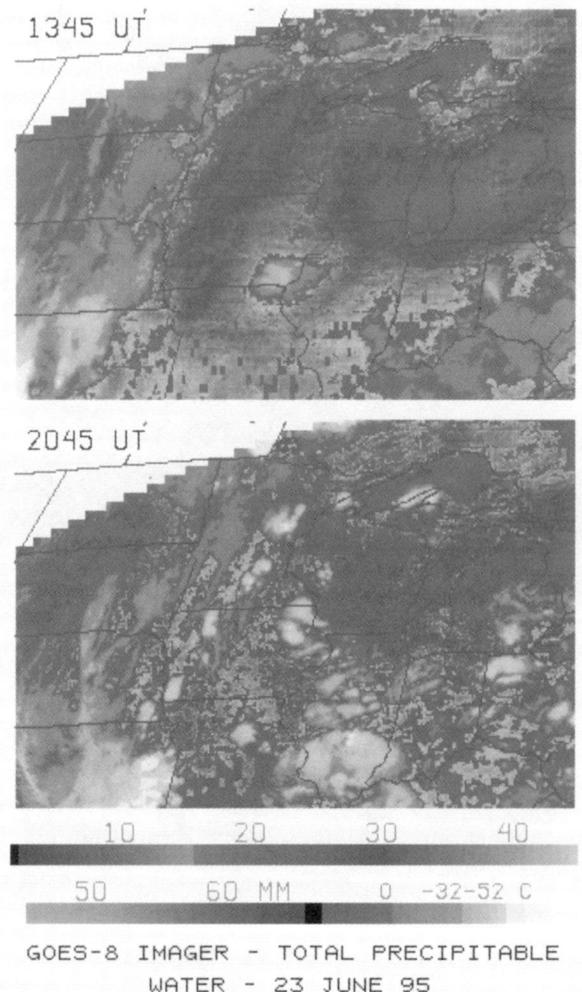

FIG. 5-11. Top: A derived product image of total precipitable water vapor from the *GOES-8* imager is shown for 1345 UTC 23 June 1995 over the north-central United States. Clouds remain in shades of gray, while moisture values are color coded (as shown by the scale on the bottom). Note higher moisture regions (red and bright yellow) over the Ohio Valley, across Missouri, and from central Kansas to eastern North Dakota (along a frontal band). A secondary moisture maximum exists along the upper Mississippi River Valley.

Bottom: Seven hours later the total precipitable water vapor at 2045 UTC 23 June 1995. Convection has developed along the higher moisture regions noted in Fig. 5-11. from the night of 14 March and the early morning of 15 March, and shows coastal fog and stratus spread southward over the Atlantic Ocean and inland to cover parts of Delaware, Maryland, Virginia, and North Carolina, as well as the Delaware and Chesapeake Bays. Visibility as low as one-tenth of a mile had a pronounced effect on ground, air, and sea transportation.

been noted by operational forecast centers. Slightly over 50% of the earth's surface area lies between 30°N and 30°S. That area, and indeed most of the Southern Hemisphere, is virtually devoid of ground-based observations. Also, as one approaches the equatorial region from 30° (north or south), the wind field rather than the mass field takes on increasing importance for meteorological applications. Tropical regions are well known for explosive convection, highly unbalanced flows, and strong vertical wind shears. Furthermore, as the winds becomes stronger, the effect of an error in velocity specification on accelerations within a model becomes greater (Pielke 1987). Measuring a wind with several meters per second velocity error is certainly not desirable; placing that wind at the wrong height can be disastrous. Among the major challenges are to derive winds in complex situations that routinely exist in jet stream regions and around the intertropical convergence zone and to improve the accuracy of height assignment for clouds that are used as tracers. The need exists to develop a generalized method for making the best possible measurements of cloud motions and heights from currently available and future operational satellite platforms and sensors. Since geometry does not rely on cloud properties, it would be preferable to derive cloud heights using geometric techniques. Hasler et al. (1981) showed that stereographic techniques could be used to determine cloud heights very accurately with 1-km resolution visible imagery from two geostationary satellites whose scan patterns were synchronized to within a few seconds. Techniques to overcome the requirement for two-satellite synchronization have been developed (Purdom and Dills 1993) using cloud shadows, cloud relative motion, and time-adjusted stereo; these require manual processing and are available only for research applications at this time, but when used with geostationary imagery, extreme accuracy has been demonstrated. For models with grid scales on the order of tens of kilometers, the need for accurate atmospheric thermodynamic soundings becomes important. While the ability to derive accurate satellite soundings with vertical resolutions required by true mesoscale models may never exist [indeed this is true of most observing systems (Pielke 1987)], techniques need to be developed that can better define the structure of the mass field within a larger-scale field produced by conventional sources. Furthermore, sounding data from geostationary satellites provide the only hope of continually monitoring the development of low-level moisture, the fuel for intense convection, on temporal scales compatible with that of convective phenomena. Knowledge of low-level moisture is important for both modeling applications and applied forecasting. The assessment of moisture includes both monitoring its horizontal advection and measuring its increase due to fluxes from wet ground heated by incoming solar radiation.

A major limitation in the use of satellite soundings in weather analysis and numerical forecast models remains poor vertical resolution in the profile retrievals from measurements with low spectral resolution. Higher spectral resolution is needed to discriminate high-altitude radiance contributions at opaque absorption line centers from low-altitude radiance contributions at transparent regions between absorption lines. A Michelson interferometer or a grating spectrometer using new detector technology offer the opportunity for measurements with spectral resolution ($\lambda/\Delta\lambda$) on the order of 1000 and continuous spectral coverage between 4 and 15 µm; this yields about 2000 spectrally independent measurements and enables much higher vertical resolution. Both such instruments are currently planned; the Atmospheric Infrared Sounder is a grating spectrometer that will be part of the NASA Earth Observing System (EOS), and the Geostationary High Spectral Resolution Interferometer Sounder is an interferometer that will replace the existing filter radiometer GOES sounder. The vertical resolution and accuracy of these instruments are expected to meet or come close to meeting the anticipated requirements (1-km vertical resolution with 1-K rms difference with respect to radiosondes) for global sounding for numerical weather prediction during the next decade (Smith 1991).

Understanding clouds and their development is fundamental to proper analysis and forecasting of the atmosphere on scales ranging from the mesoscale to global climate. Often this understanding is reflected in a qualitative manner. For example, hook echoes, tornado vortex signatures, overhanging vaults and bounded weak echo regions, arc cloud lines, and overshooting cloud-top areas with infrared "V" notches (Fujita 1978; Heymsfield et al. 1983) are basically pattern recognition of phenomena relating to convection that are used to aid forecasters. While such uses are qualitative in nature, they are very important. They have aided meteorologists in developing a deeper understanding of such phenomena as severe thunderstorm structure and thunderstorm triggering through convective-scale interaction. Furthermore, with the new *GOES-8* imagery, forecasters will have the ability precisely to locate and monitor a variety of phenomena not routinely associated with short-term forecasts and warnings. Among those phenomena are haze, dust and pollution, water clouds over snow cover, and nighttime fog and stratus. The multispectral channel capability afforded by the current GOES and NOAA satellites, both alone and in combination, provides a wealth of information for the development of multispectral image products. Indeed, it seems apparent that modern weather forecasting will require the development of multiple data source products. For example, we need a good combined Next Generation Weather Radar and satellite-image product that does not destroy the information in one at the expense of the other.

Challenges of the future

Capitalizing on research advances

The growth of environmental satellite programs in the early days of the science was robust. In the 1960s and 1970s NASA and NOAA participated in an Operational Satellite Improvement Program (OSIP) that saw a number of experimental instruments exercised and their data tested before they entered the operational arena; in addition, the DMSP was vigorous and active. The DoD program contributed to the operational ESSA and NOAA satellite bus configurations, and the OSIP program had as test beds Nimbus, ATS, and SMS, which led to improvements both in NOAA's polar and geostationary programs. The development and growth of next-generation operational polar and geostationary satellite systems has been hindered by the absence of an OSIP-like program. In the area of oceanic observing satellites there are a number of stellar programs (existing or planned), including the European Remote Sensing Satellite (ERS), Japan's Marine Observation Satellite (MOS) and Advanced Earth Observing System (ADEOS), and NASA's Earth Observing System and Ocean Topography Experiment (TOPEX/Poseidon), but there is no operational follow-up based on the expected successes from those programs. Indeed, it appears that there is no overall coordinated effort to combine results from the various research programs and move them into a meaningful, sustained operational system. This is disturbing when it is realized that the development of new satellite systems and sensors may require several years to over a decade. Focused scientific discussions between government, university, and private sector experts are needed to raise critical issues and to foster professional activity regarding new concepts, development, research, operations, and practical application of satellite measurements to meteorological and oceanographic problems.

Environmental monitoring of the planet

Over 70% of this planet is covered by water and is poorly monitored; the remaining 30% is monitored well only over small regions. (The continental United States covers only around 3% of the total area of the earth.) Satellites provide the only system capable of providing measurements from uniformly calibrated instruments over the entire planet on a routine basis. NOAA's modernization and the advanced facilities at the National Center for Atmospheric Research and throughout the university community provide the opportunity for the United States to serve as a highly instrumented laboratory that allows us to learn how to use the variety of available satellite systems to monitor more effectively the remaining 97% of the planet. There needs to be an organized effort to move in that direction.

Training and expanding the user community

Satellites represent one of the primary tools with which we observe our planet and its atmosphere on a variety of scales. The importance of those observations, coupled with the rapidly evolving nature of this specialty, means that many people outside the scope of "satellite meteorology and oceanography" need guidance. However, there are no effective short courses for professionals with little background and knowledge of satellite meteorology and oceanography. An effort is under way at COMET to provide training materials in satellite data use. The COMET materials are directed at operational forecasters and perhaps can meet even broader needs. However, a serious problem exists, as was evident in Eric Smith's review of *Weather Satellites* (1991): there does not exist a definitive text at either the graduate or undergraduate level that covers this topic, and it is questionable whether it might be possible to produce one.

In the past, the number of users of satellite data has been limited because of the cost of acquiring and analyzing satellite data. The computer revolution, coupled with a variety of data services (present and planned), point toward the day when those limitations will be behind us. With the anticipated expanding community of users of satellite data, it is imperative to establish ready access to training materials.

Conclusions

In the past 35 years NOAA, with help from NASA and the DoD has established a remote sensing capability on polar and geostationary platforms that has proven useful in monitoring and predicting severe weather such as tornadic outbreaks, tropical cyclones, and flash floods in the short term, and climate trends indicated by sea surface temperatures, biomass burning, and cloud cover in the longer term. This has become possible first with the visible and infrared window imagery of the 1970s and has been augmented with the temperature and moisture sounding capability of the 1980s. The imagery from the NOAA satellites, especially the time-continuous observations from geostationary instruments, dramatically enhanced our ability to diagnose atmospheric cloud systems and to predict severe thunderstorms. These data were almost immediately incorporated into operational procedures. Use of sounder data in the operational weather systems is more recently coming of age. The polar-orbiting sounders are filling important data voids at synoptic scales. Applications include temperature and moisture analyses for weather prediction, analysis of atmospheric stability, estimation of tropical cyclone intensity and position, and global analyses of clouds. TOVS includes both infrared and microwave observations, with the latter helping considerably to alleviate the influence of clouds for all weather soundings. GOES-VAS has been used to develop procedures for retrieving atmospheric temperature, moisture, and wind at hourly intervals in the Northern Hemisphere. Temporal and spatial changes in atmospheric moisture and stability are improving severe storm warnings. Atmospheric flow fields (deep-layer mean wind field composites from cloud drift, water vapor drift, and thermal gradient winds) are helping to improve hurricane trajectory forecasting. In 1994, the first of NOAA's next generation of geostationary satellites, *GOES-8,* dramatically improved the imaging and sounding capability. Applications of these NOAA data also extend to the climate programs; archives from the last 15 years offer important information about the effects of aerosols and greenhouse gases and possible trends in global temperature.

In the near future, forecasters will be able to use "modernization era" satellite and radar observing systems to investigate the evolution and structure of cloud systems and perform short-term forecasts and warnings. Considering that convective activity ranges from airmass thunderstorms with life cycles well less than an hour to mesoscale convective systems with some aspects of their circulations that last for days, substantial benefits can be realized through correct interpretation of radar and satellite data. For example, there is obvious benefit in a forecaster distinguishing in real time between a supercell storm, associated with prolonged periods of severe weather, and a nonsevere airmass thunderstorm. With the advent of improved mesoscale numerical models, the importance of interpretation of radar and satellite images and soundings is not diminished but rather is magnified as the forecaster now has the added task of using the observations critically to examine the predicted sequence of events.

The quest for improved remote sensing from satellites must continue. There is a need for higher temporal, spatial, and spectral resolution in the future radiometers. Higher temporal resolution is becoming possible with detector array technology; higher spatial resolution may come with active cooling of infrared detectors so that smaller signals can be measured with adequate signal to noise. Higher spectral resolution is being considered through the use of interferometers and grating spectrometers. Advanced microwave radiometers measuring moisture as well as temperature profiles will soon be flying in polar orbit; a geostationary complement is being investigated. Ocean color observations with multispectral narrowband visible measurements are being planned. The challenge of the future is to further the progress realized in the past decades so that environmental remote sensing of the oceans, atmosphere, and earth increases our understanding of the processes affecting our lives and future generations.

Acknowledgments. The authors would like to thank Dave Johnson and Vince Oliver for several of the figures that appear in the text. A number of people provided thoughtful reviews of the manuscript, and their suggestions have been incorporated where possible. In particular, thanks go to Chuck Arnold, Fran Holt, Dave Johnson, Vince Oliver, Bill Smith, Morris Tepper, Paul Try, Paul Twitchell, and Tom Vonder Haar. Finally, we are also grateful for several discussions with Verner Suomi and are saddened by his recent passing.

APPENDIX A

Recognized Pioneers

When writing any history, problems are invariably encountered when singling out individuals for special contributions that they made to the field. However, this was done in 1985 to some extent, when the 25th anniversary of weather satellites occasioned a number of special celebrations. At a National Space Club reception at the Smithsonian Institution's National Air and Space Museum on 1 April 1985, a small group of people received special awards for their contributions that led to the success of the U.S. weather satellite program. For a history not to single out the contributions of this select group would be injustice indeed: their names, where they were working during the period of their contribution, along with excerpts from the citations presented to each are presented below in alphabetical order.

- T. Theodore Fujita, University of Chicago, for creative scientific leadership as an enthusiastic pioneer in the use of satellite imagery to analyze and predict mesoscale weather phenomena and to understand severe thunderstorms, tornadoes, and hurricanes.
- Stanley M. Greenfield and William W. Kellogg, RAND Corporation, for critical scientific leadership as visionary researchers whose studies laid the groundwork for the initial TIROS and subsequent weather satellites. Their research established the ground resolution required to identify major cloud types and a methodology to interpret useful spatial and temporal changes and thus provide useful meteorological information from satellite altitudes.
- Thomas O. Haig, U.S. Air Force Satellite Experimental Program, for his pioneering achievements in engineering design and development while directing the air force program. These achievements proved the "wheel mode" attitude control system that enabled direct photography of clouds over the entire earth from a single satellite.
- Rudolf A. Hanel, NASA/Goddard Space Flight Center, for pioneering achievements in engineering design and development

critical to the evolutionary advancement in instrumentation on civil environmental satellites. These advances added the capability for mapping cloud cover at night, for determining vertical atmospheric profiles of temperature, water vapor, and ozone, and for measuring global sea and land surface temperatures.

- David S. Johnson, NOAA, for exceptional accomplishments in program development while directing the U.S. Civil Operational Environmental Satellite Program. During his tenure, the United States established its preeminent position in the monitoring of the global environment and never suffered a break in operational weather satellite service.

- Dominick J. Juarez, ITT Aerospace/Optical Division, for major contributions in engineering design and development that led to the evolution of cryogenically cooled detector technology into the highly successful imaging and sounding instruments flown on the operational polar-orbiting environmental satellites. These instruments include the HIRS and the AVHRR.

- Lewis D. Kaplan, Massachusetts Institute of Technology, for pioneering contributions to scientific leadership as the creator of the original scientific concept for detecting the structure of the atmosphere using spectral and thermal radiation while at MIT and for subsequent experiments and improvements that made possible atmospheric sounding from environmental satellites.

- Roy Leep of WTVT television, Tampa, Florida (1966–present), for pioneering contributions to public safety and service by being the first television meteorologist to use weather pictures received directly from satellites to enhance the education and warning of the public about the progress of hurricanes and other severe storms. He led the television industry in the progressive addition of new satellite direct readout capabilities as they became available and motivated other television broadcasters to follow his example.

- Vincent J. Oliver, Satellite Data Applications Group, NOAA, for innovative, outstanding scientific leadership while directing research activities that developed many of the techniques used in daily weather forecasting operations in the United States and throughout the world. He developed techniques to determine the jet streams, frontal systems, rainfall estimates, land fog, and other weather-related phenomena from satellite images.

- Abraham Schnapf, RCA AstroElectronics, for outstanding contributions to engineering design and development and to program development at RCA AstroElectronics that led to four successful generations of weather satellites—the TIROS, ESSA, ITOS, and *TIROS-N*/NOAA series, all of which met or exceeded design life expectancy.
- Leonard W. Snellman, National Weather Service, NOAA, for enthusiastic scientific leadership in pioneering the use of satellite imagery in operational weather forecasting by developing applications uniquely suited to the western states, conducting training programs for forecasters, and installing facsimile reception capabilities in even the smallest weather service office in his region.
- Verner E. Suomi, University of Wisconsin, for unparalleled scientific leadership and innovative engineering design and development in conceiving new sensors and applications from the first TIROS satellite through the GOES series. His inventions include the "spin–scan" camera that made geostationary weather satellites possible and McIDAS (Man–computer Interactive Data Access System), the software philosophy that permits interactive access to vast quantities of environmental satellite data.

Also recorded in the program of the National Space Club reception was the following challenge.

> The far-sighted vision of these pioneers created meteorological satellites that examine the earth's environment for the betterment of the world population. The work begun can only be finished when the betterment is available to all people everywhere. A true challenge of the next 25 years of weather satellites will be to fulfill the vision of universal benefits.

In addition to the pioneers recognized above, a number of individuals have distinguished themselves in the eyes of their peers for satellite-related activities. These are reflected in the awards program of the AMS and are listed below (*Bulletin of the American Meteorological Society,* **75,** 1421–1468).

- Verner E. Suomi, in 1968 the Carl-Gustaf Rossby Research

Medal, for helping to transform the satellite as a meteorological probe from a dream to a reality.

- Sean A. Twomey, in 1980 the Carl-Gustaf Rossby Research Medal, for extensive contributions to the development of many areas of atmospheric science, including remote sensing from satellites.
- Rudolph Hanel, Don T. Hilleary, Lewis D. Kaplan, and David Q. Wark, in 1970 the Second Half Century Award, each cited for his pioneering work in the development of the technique of procuring upper-air sounding from earth-orbiting satellites.
- Thomas H. Vonder Haar, in 1981 the Second Half Century Award, for his leadership as a researcher and as a coordinator of national and international programs in atmospheric radiation and satellite meteorology.
- Moustafa T. Chahine, in 1991 the Jule G. Charney Award, for outstanding contributions to satellite sensing through better understanding of the inverse radiative transfer problem and development of its applications.
- M. Patrick McCormick, in 1991 the Jule G. Charney Award, for outstanding contributions to satellite sensing through development of solar occultation instruments and for elucidation of the nature of polar stratospheric clouds.
- Homer E. Newell, in 1972 the Cleveland Abbe Award, for his many services to meteorology exemplified by his perception of problems and promising solutions in making key decisions regarding the meteorological satellite program.
- Verner E. Suomi, in 1961 the Clarence Leroy Meisinger Award, for his pioneering research work on atmospheric radiation problems in which he effectively used both balloon and satellite observations.
- Tetsuya Fujita, in 1967 the Clarence Leroy Meisinger Award, for pioneering research on mesometeorological analysis and broad contributions to the use of meteorological satellites.
- William L. Smith, in 1970 the Clarence Leroy Meisinger Award, for developing a technique for converting satellite spectrometer measurements to vertical temperature profiles, which permitted immediate operational use of the data.
- Lee George Dickinson, in 1969 the Award for Outstanding Ser-

vice by a Weather Forecaster, for imaginative and untiring efforts in the use of satellite meteorological information.

- Vincent Oliver, in 1972 the Award for Outstanding Contribution to the Advance of Applied Meteorology, for his innovative contributions over the past three decades in the application of new meteorological observations and for his creative use of meteorological satellite data.
- William W. Kellogg and Stanley M. Greenfield, in 1961 an AMS special award for their pioneering work in the planning of a meteorological satellite.
- Morris Tepper, in 1978 the AMS recognized him for his extraordinary leadership in persuading the nations of the world to join in providing a global weather observing system based on space technology.
- Bruce A. Wielicki, in 1995 the Houghton Award, for outstanding contributions to the understanding of clouds and earth's radiation budget derived from satellite observations and theoretical radiative transfer models.

When the history of satellite remote sensing is revisited in 2010 at the 50th anniversary of *TIROS-1*, this list will have grown in size but certainly not in eminence.

APPENDIX B

References for further reading

There are a number of resources available for those interested in the history of satellite meteorology. Rather than refer the reader to technical journal articles, listed below are selected books and review papers where both general and more in-depth information may be found along with extensive references on specific subjects.

For the early days of the meteorological program the reader is referred to the following works.

- Schereschewsky (1945) presents an early history of the use of cloud observations by meteorologists.
- H.M.E. van de Boogaard (1966) served as editor for papers pre-

sented at a meeting at NCAR in August 1965. At that meeting a number of papers were presented by scientists that were, and had been, involved with the early days of meteorological satellites. Among the speakers were W.W. Kellogg, J.H. Conover, L.F. Hubert, H. Riehl, W. Kamm, T.T. Fujita, L.D. Sanders, V.J. Oliver, E. Ferguson, F. Sanders, G. Warnecke, S. Fritz, E.R. Reiter, P.D. Thompson, V.E. Suomi, J.S. Winston, W. Nordberg, W.R. Bandeen, S. Teweles, J.I.F. King, D.Q. Wark, J. London, Z. Sekera, J.V. Dave, D.S. Johnson, and M. Tepper. Also included with this interesting group of papers are discussions that accompanied the various presentations.

- William Vaughan (1982) served as editor for several papers that were presented at the American Institute of Aeronautics and Astronautics 20th Aerospace Sciences Meeting. Papers cover the early history of satellites as well as the development of the TIROS, Nimbus, DMSP, and GOES systems.

- The conception, growth, accomplishments, and future of meteorological satellites, NASA Conference Publication 2257, edited by Davis and Cook (1982), covers papers presented in a session of the 62nd Annual Meeting of the AMS in San Antonio, Texas.

- Hubert and Lehr's book *Weather Satellites* (1967) provides a wealth of information on the development and early uses of meteorological satellite data.

- Anderson (1974) covers some of the early uses of meteorological satellite imagery and was the first to do so in detail.

- Allison (1980) provides a nice summary of the early development of the meteorological satellite program and details certain instruments and their uses.

- Smith's chapter in the *Handbook of Applied Meteorology* (1985) summarizes the early satellite instruments and the associated applications. From the net flux radiometer to temperature profile sounders, these very readable pages explain the theories, develop the applications, and interpret the results.

- Schnapf (1985) includes 830 pages of information the interested reader will value. Two chapters deal exclusively with DMSP.

- Smith et al. (1986) provide a concise overview of the first 25 years of meteorological satellite operations. A variety of qualitative and quantitative uses of the data are covered in this most interesting article.

- Isaacs et al. (1986) provide a summary of satellite remote sensing of meteorological parameters and their use in global numerical weather prediction. This article includes an exhaustive list of references.
- Rao et al. (1990) provides details of satellites and applications of their data. It contains a wealth of references.
- The 25 years of weather satellites were celebrated at a direct broadcast users conference in April 1985, and a volume of papers and remarks from that conference was edited by Popham (1985). This interesting publication contains comments from a variety of perspectives including the WMO and local school teachers.
- W. Smith (1991) gives an interesting overview of atmospheric soundings from satellites in his Symons Memorial Lecture given at the Royal Meteorological Society in 1990.
- D. Johnson (1994), in the 1994 Verner E. Suomi Lecture, reflected on the evolution of the U.S. meteorological satellite program. It offers yet another perspective from an early administrator of the NESS.

Our field is vibrant and growing, and there are many other materials of interest. We refer the reader to the following.

- *Satellite Meteorology—An Introduction* by Kidder and Vonder Haar (1995) promises to fill a void that has long existed in the literature.
- *Polar and Arctic Lows* by Twitchell et al. (1989) covers what we knew about polar lows up to its publication. The interesting thing about this publication is that polar lows were largely unrecognized as a phenomenon prior to meteorological satellites.
- *Satellite as Microscope* by Scorer (1990) covers detailed uses of polar-orbiting satellite imagery and is mandatory reading for anyone seriously interested in the applications of meteorological satellite imagery.
- *Nowcasting* by Browning (1982) covers the use of a number of technologies, including satellite image and sounding data, for nowcasting.
- *Remote Sensing of the Lower Atmosphere: An Introduction* by Stephens (1994) is interesting reading for those wishing to un-

derstand more of the physics behind uses of various spectral bands, as well as theory behind remote sensing.

REFERENCES

Adler, R.F., and D.D. Fenn, 1979: Thunderstorm intensity as determined from satellite data. *J. Appl. Meteor.,* **18,** 502–517.

______ , and A.J. Negri, 1981: A satellite technique to estimate tropical convective and stratiform rainfall. *J. Appl. Meteor.,* **27,** 30–51.

Allison, L.J., R. Wexler, C. Laughlin, and W. Bandeen, 1977: Remote sensing of the atmosphere from environmental satellites. X-901-77-132 Preprint, Goddard Space Flight Center, Greenbelt, MD, 111 pp.

______ , A. Schnapf, B.C. Diesen III, P.S. Martin, A. Schwalb, and W.R. Bandeen, Eds., 1980: Meteorological satellites. NASA TM 80704, Goddard Space Flight Center, MD, 71 pp.

Anderson, R.K. 1974: Application of meteorological satellite data in weather analysis and forecasting. Tech. Note 124, WMO No. 333, World Meteorological Organization, Geneva, Switzerland, 275 pp.

Barrett, E.C., G. D'Souza, and C.H. Power, 1986: Bristol techniques for the use of satellite data in rain cloud and rainfall monitoring. *J. Brit. Interplan. Soc.,* **39,** 517–526.

Bristor, C.L., and W. Pichel, 1974: Three-D cloud viewing using overlapping pictures from two geostationary satellites. *Bull. Amer. Meteor. Soc.,* **55,** 1353–1355.

Browning, K.A., 1982: *Nowcasting.* Academic Press, 256 pp.

Chedin, A., Ed., 1989: *Tech. Proc. of the Fifth Int. TOVS Study Conf.,* Toulouse, France, 445 pp.

Chesters, D., W.D. Robinson, and L.W. Uccellini, 1987: Optimized retrievals of precipitable water from the VAS split window. *J. Climate Appl. Meteor.,* **26,** 1059–1066.

CIMSS, 1995: Introducing the GOES-8 sounder. [Available on-line from http://cloud.ssec.wisc.edu/sounder/g8.html.]

CIRA, 1994: Introduction to GOES-8. [Available on-line from http://www.cira.colostate.edu.RAMM.overview.]

COMET, 1992: Boundary Layer Detection and Convective Initiation module. University Corporation for Atmospheric Research, laser disc.

______ , 1995: Introduction to *GOES-8/9:* Computer Based Learning

Module. University Corporation for Atmospheric Research, CD-ROM.

Crowson, D.L., 1949: Cloud observations from rockets. *Bull. Amer. Meteor. Soc.,* **30,** 17–22.

Dvorak, V.F., 1972: A technique for the analysis and forecasting of tropical cyclone intensities from satellite pictures. NOAA TM NESS 36, U.S. Dept. of Commerce, Washington, DC, 15 pp.

______ , 1984: Tropical cyclone intensity analysis using satellite data. NOAA Tech. Rep. NESDIS 11, Washington, DC, 47 pp.

______ , and S. Wright, 1977: Tropical cyclone intensity analysis using enhanced infrared satellite data. *Proc. 11th Technical Conf. on Hurricanes and Tropical Meteorology,* Miami, FL, Amer. Meteor. Soc., 268–273.

Ellrod G., 1992: Potential applications of GOES-I 3.9 µm infrared imagery. *Sixth Conf. on Satellite Meteorology and Oceonography,* Atlanta, GA, Amer. Meteor. Soc., 184–187.

Eyre, J.R., 1984: Detection of fog at night using Advanced Very High Resolution Radiometer Imagery (AVHRR). *Meteor. Mag.,* **113,** 266–271.

Follansbee, W.A., 1973: Estimation of average daily rainfall from satellite cloud photographs. NOAA Tech. Memo. NESS 44, Dept. of Commerce, Washington, DC, 39 pp.

Friday, E.W., Jr., 1989: The National Weather Service forecast and warning program outlook. *GOES I-M Operational Satellite Conf.,* Arlington, VA, NOAA, 110–125.

Fujita, T.T., 1978: Manual of downburst identification for project Nimrod. SMRP 156, University of Chicago, 104 pp.

______ , 1982: Infrared, stereo, cloud motion, and radar-echo analysis of SESAME-day thunderstorms. *12th Conf. on Severe Local Storms,* San Antonio, TX, Amer. Meteor. Soc., 213–216.

Fuller, J.F., 1990: *Thor's Legions.* Amer. Meteor. Soc., 381 pp.

Gomberg, L., and S.M. McElroy, 1985: Remote sensing of the earth with the Defense Meteorological Satellite. *Monitoring Earth's Ocean, Land, and Atmosphere from Space—Sensor, Systems, and Applications,* A. Schnapf, Ed., American Institute of Aeronautics and Astronautics, 96–128.

Greenfield, S.W., and W.W. Kellogg, 1951: Inquiry into the feasibility of weather reconnaissance from a satellite vehicle. USAF Project RAND Rep. R-218, 43 pp. [Unclassified edition, 1960: Rep. N-365.]

Griffith, C.G., W.L. Woodley, P.G. Grube, D.W. Martin, J. Stout, and D.N. Sikdar, 1978: Rain estimates from geosynchronous satellite imagery: visible and infrared studies. *Mon. Wea. Rev.,* **106,** 1153–1171.

Gruber, A., 1977: Determination of the earth–atmosphere radiation budget from NOAA satellite data. NOAA Tech. Rep. NESS 76, Department of Commerce, Washington, DC, 28 pp.

Gurka, J.J., 1978: The role of inward mixing in the dissipation of fog and stratus. *Mon. Wea. Rev.,* **106,** 1633–1635.

Hallgren, R.E., 1985: Welcome from AMS. *Second Int. Satellite Direct Broadcast Services Users' Conf.,* Baltimore, MD, U.S. Dept. of Commerce, 14–15.

Hanel, R.A., B. Schlachman, F.D. Clark, C.H. Prokesh, J.B. Taylor, W.M. Wilson, and L. Chaney, 1970: The NIMBUS-3 Michelson Interferometer. *Appl. Opt.,* **9,** 1767–1773.

Hasler, A.F., 1981: Stereographic observations from geosynchronous satellites: An important new tool for the atmospheric sciences. *Bull. Amer. Meteor. Soc.,* **62,** 194–212.

Hass, I.S., and R. Shapiro, 1982: The NIMBUS satellite system—Remote sensing R&D platform of the 70s. NASA Conf. Publ. 2227, 17–30.

Heymsfield, G.M., G. Szejwach, S. Scholtz, and H. Blackmer, 1983: Upper-tropospheric structure of Oklahoma tornadic storms on 2 May 1979. Part II: Proposed explanation of "V" pattern and internal warm region in infrared observations. *J. Atmos. Sci.,* **40,** 1739–1755.

Holmlund, K., 1993: Operational water vapor wind vectors from Meteosat imagery data. *Second Int. Wind Workshop,* Tokyo, Japan, EUMETSAT, 77–84.

Hubert, L.F., and P.E. Lehr, 1967: *Weather Satellites.* Blaisdell Publishing, 120 pp.

——— , F. and L.F. Whitney Jr., 1971: Wind estimation from geostationary-satellite pictures. *Mon. Wea. Rev.,* **99,** 665–672.

Isaacs, R.G., R.N. Hoffman, and L.D. Kaplan, 1986: Satellite remote sensing of meteorological parameters for global numerical weather prediction. *Rev. Geophys.,* **24,** 701–743.

Jedlovec, G.J., 1985: An evaluation and comparison of vertical profile data from the VISSR Atmospheric Sounder (VAS). *J. Atmos. Oceanic Technol.,* **2,** 559–581.

Johnson, D.S., 1982: Development of the operational program for satellite meteorology. NASA Conf. Publ. 2257, 34–40.

______ , 1994: Evolution of the U.S. meteorological satellite program: 1994 Verner E. Suomi Lecture. *Bull. Amer. Meteor. Soc.*, **75**, 1705–1708.

Kaplan, L.D., 1959: Inferences of atmospheric structures from satellite remote radiation measurements. *J. Opt. Soc. Amer.*, **49**, 1004–1014.

Kellogg, W.W., 1966: Satellite meteorology and the academic community. Satellite data in meteorological research. NCAR-TN-11, National Center For Atmospheric Research, Boulder, CO, 5–14.

Kidder, S.Q., and T.H. Vonder Haar, 1995: *Satellite Meteorology—An Introduction.* Academic Press.

______ , W.M. Gray, and T.H. Vonder Haar, 1978: Estimating tropical cyclone central pressure and outer winds from satellite microwave data. *Mon. Wea. Rev.*, **106**, 1458–1464.

King, J.I.F., 1958: The radiative heat transfer of planet earth. *Scientific Uses of Earth Satellites*, J.A. van Allen, Ed., University of Michigan Press, 316 pp.

LeMarshall, J.F., 1988: An intercomparison of temperature and moisture fields derived from TOVS data by different techniques. Part I: Basic statistics. *J. Appl. Meteor.*, **27**, 1011–1030.

Maddox, R., 1980: Mesoscale convective complexes. *Bull. Amer. Meteor. Soc.*, **61**, 1374–1387.

Matson, M., and J. Dozier, 1981: Identification of subresolution high temperature sources using a thermal IR sensor. *Photogramm. Eng. Remote Sensing*, **47**, 1311–1318.

______ , E.P. McClain, D.F. McGinnis Jr., and J.A. Pritchard, 1978: Satellite detection of urban heat islands. *Mon. Wea. Rev.*, **106**, 1725–1734.

McClain, E.P., W.G. Pichel, and C.C. Walton, 1985: Comparative performance of AVHRR based multichannel sea surface temperatures. *J. Geophys. Res.*, **89** (C6), 11 587–11 601.

Menzel, W.P., and A. Chedin, 1990: Summary of the Fifth International TOVS Study Conference. *Bull. Amer. Meteor. Soc.*, **71**, 691–693.

______ , and J.F.W. Purdom, 1994: Introducing GOES-I: The first of a new generation of Geostationary Operational Environmental Satellites. *Bull. Amer. Meteor. Soc.*, **75**, 757–781.

Merrill, R.T., W.P. Menzel, W. Baker, J. Lynch, and E. Legg, 1991: A report on the recent demonstration of NOAA's upgraded capability to derive satellite cloud motion winds. *Bull. Amer. Meteor. Soc.*, **72**, 372–376.

Meyers, W.D., 1985: The Defense Meteorological Satellite Program: A review of its impact. *Monitoring Earth's Ocean, Land, and Atmosphere from Space—Sensor, Systems, and Applications*, A. Schnapf, Ed., American Institute of Aeronautics and Astronautics, 129–149.

Nagel, R., and C.M. Hayden, 1971: The use of satellite-observed cloud patterns in Northern Hemisphere 500-mb numerical analysis. NOAA Tech. Rep. NESS 55, Dept. of Commerce, Washington, DC, 55 pp.

Nieman, S.A., J. Schmetz, and W.P. Menzel, 1992: A comparison of several techniques to assign heights to cloud tracers. *J. Appl. Meteor.*, **32**, 1559–1568.

NOAA, 1984: The March 28 1984 Carolina tornado outbreak. Disaster survey report to the administrator. NOAA, U.S. Dept. of Commerce, Washington, DC, 48 pp.

Obasi, G.O.P., 1985: Keynote address. *Second Int. Satellite Direct Broadcast Services Users' Conf.*, Baltimore, MD, U.S. Dept. of Commerce, 16–21.

Oliver, V.J., and E.W. Ferguson, 1966: The use of satellite data in weather analysis. *Satellite Data in Meteorological Research*, NCAR-TN-11, National Center For Atmospheric Research, Boulder, CO, 349 pp.

———, R.K. Anderson, and E.W. Ferguson, 1964: Some examples of detection of jet streams from TIROS photographs. *Mon. Wea. Rev.*, **92**, 441–448.

Ostby, F., 1995: Presentation at the GOES-J prelaunch press conference at Kennedy Space Flight Center. NASA, audio/videocassette. [Available from NASA KSC Press Office, Kennedy Space Center, FL 32815.]

Pielke, R.A., 1987: The challenge of using mesoscale data in mesoscale models. *Symp. Mesoscale Analysis and Forecasting*, Vancouver, BC, Canada, ESA, 651–652.

Popham, R., 1985: Setting the stage, the direct broadcast community: who, what, and where. *Second Int. Satellite Direct Broadcast Services Users' Conf.*, Baltimore, MD, U.S. Dept. of Commerce, 22–23.

Prins, E.P., and W.P. Menzel, 1994: Trends in South American biomass burning detected with the GOES-VAS from 1983–1991. *J. Geophys. Rev.,* **99,** 16 719–16 735.

Purdom, J.F.W., 1976: Some uses of high resolution GOES imagery in the mesoscale forecasting of convection and its behavior. *Mon. Wea. Rev.,* **104,** 1474–1483.

_____ , 1982: Integration of research aircraft data and 3 minute interval GOES data to study the genesis and development of deep convective storms. Preprints, *12th Conf. on Severe Local Storms,* San Antonio, TX, Amer. Meteor. Soc., 269–271.

_____ , 1985: The application of satellite sounding and image data to the Carolina tornado outbreak of 28 March 1984. Preprints, *14th Conf. on Severe Local Storms,* Indianapolis, IN, Amer. Meteor. Soc., 276–279.

_____ , 1993: Satellite observations of tornadic thunderstorms. *The Tornado: Its Structure, Dynamics, Prediction, and Hazards, Geophys. Monogr.,* No. 79, Amer. Geophys. Union, 265–274.

_____ , 1995: Observations of thunderstorms and hurricanes using one-minute interval GOES-8 imagery. Abstracts, *Week B, International Union of Geodesy and Geophysics, XXI General Assembly,* Boulder, CO, Amer. Geophys. Union, Washington, DC, B286.

_____ , and J.G. Gurka, 1974: The effect of early morning cloud cover on afternoon thunderstorm development. Preprints, *Fifth Conf. on Weather Forecasts and Analysis,* St. Louis, MO, Amer. Meteor. Soc., 58–60.

_____ , and P.N. Dills, 1993: Cloud motion and height measurements from multiple satellites including cloud heights and motions in polar regions. *Second Int. Wind Workshop,* Tokyo, Japan, EUMETSAT, 245–248.

Rasmussen, E.A., and J.F.W. Purdom, 1992: Investigation of a polar low using geostationary satellite data. Preprints, *Sixth Conf. on Satellite Meteorology and Oceanography,* Atlanta, GA, Amer. Meteor. Soc., 120–122.

Rao, C.R.N., L.L. Stowe, E.P. McClain, and J. Saper, Eds., 1988: Development and application of aerosol remote sensing with AVHRR data from the NOAA satellites. *Aerosols in Climate,* Deepak Publishing, 69–80.

_____ , S.J. Holmes, R.K. Anderson, J.S. Winston, and P.E. Lehr, 1990:

Weather Satellites: Systems, Data, and Environmental Applications. Amer. Meteor. Soc., 503 pp.

Rossow, W.B., and A.A. Lacis, 1990: Global and seasonal cloud variations from satellite radiance measurements. Part II: Cloud properties and radiative effects. *J. Climate,* **3,** 1204–1253.

Sadler, J.C., 1968: *Average Cloudiness in the Tropics from Satellite Observations.* East–West Center Press, 22 pp.

Schereschewsky, P., 1945: Clouds and states of the sky. *Handbook of Meteorology,* F.A. Berry, E. Bollay, and N.R. Beers, Eds., McGraw Hill, 1068 pp.

Schnapf, A., Ed., 1985: *Monitoring Earth's Ocean, Land, and Atmosphere from Space—Sensor, Systems, and Applications.* American Institute of Aeronautics and Astronautics, 830 pp.

Scofield, R., and V.J. Oliver, 1977: A scheme for estimating convective rainfall from satellite imagery. NOAA Tech. Memo NESS 86, Dept. of Commerce, Washington, DC, 47 pp.

Scorer, R.S., 1990: *Satellite as Microscope.* Ellis Horwood, 268 pp.

Segal, M., J.F.W. Purdom, J.L. Song, R.A. Pielke, and Y. Mahrer, 1986: Evaluation of cloud shading effects on the generation and modification of mesoscale circulations. *Mon. Wea. Rev.,* **114,** 1201–1212.

Sheets, R.C., 1990: The National Hurricane Center—Past, present, and future. *Wea. Forecasting,* **5,** 185–232.

Shenk, W.E., 1985: Cloud motion derived winds: their accuracy, coverage, and suggestions for future improvements. *NASA Symp. on Global Wind Measurements,* Columbia, MD, NASA, 123–128.

———, and E.R. Kreins, 1975: The NASA severe storm research program. Preprints, *Ninth Conf. on Severe Local Storms,* Norman, OK, Amer. Meteor. Soc., 468–473.

———, and F. Mosher, 1987: Suggested severe local storm operational scenarios for GOES-I/M. NASA TM 100688, Washington, DC.

———, T.H. Vonder Haar, and W.L. Smith, 1987: An evaluation of observations from satellites for the study and prediction of mesoscale events and cyclone events. *Bull. Amer. Meteor. Soc.,* **68,** 21–35.

Smith, E.A., 1991: Review of *Weather Satellites. Bull. Amer. Meteor. Soc.,* **72,** 1402–1405.

Smith, W.L., 1985: Satellites. *Handbook of Applied Meteorology,* D.D. Houghton, Ed., John Wiley and Sons, 380–472.

_____ , 1991: Atmospheric soundings from satellites—False expectation or the key to improved weather prediction? Royal Meteorological Society, Symons Memorial Lecture, London, UK, May 16, 1990. *Quart. J. Roy. Meteor. Soc.,* **117,** 267–297.

_____ , H.M. Woolf, and W.J. Jacob, 1970: A regression method for obtaining real-time temperature and geopotential height profiles from satellite spectrometer measurements and its application to NIMBUS-3 "SIRS" observations. *Mon. Wea. Rev.,* **98,** 582–603.

_____ , D.T. Hilleary, J.C. Fischer, H.B. Howell, and H.M. Woolf, 1974a: Nimbus ITPR experiment. *Appl. Opt.,* **13,** 499–506.

_____ , D.H. Staelin, and J.T. Houghton, 1974b: Vertical temperature profiles from satellites—Results from second generation instruments aboard Nimbus-5. *Proc. COSPAR Symp. on Approaches to Earth Survey Problems Through the Use of Space Techniques,* Hamburg, Germany, Akademie-Verlag, 123–143.

_____ , H.M. Woolf, C.M. Hayden, D.Q. Wark, and L.M. McMillin, 1979: The TIROS-N operational vertical sounder. *Bull. Amer. Meteor. Soc.,* **60,** 1177–1187.

_____ , F.W. Nagle, C.M. Hayden, and H.M. Woolf, 1981a: Vertical mass and moisture structure from TIROS-N. *Bull. Amer. Meteor. Soc.,* **62,** 388–393.

_____ , V.E. Suomi, W.P. Menzel, H.M. Woolf, L.A. Sromovsky, H.E. Revercomb, C.M. Hayden, D.N. Erickson, and F.R. Mosher, 1981b: First sounding results from VAS-D. *Bull. Amer. Meteor. Soc.,* **62,** 232–236.

_____ , _____ , F.-X. Zhou, and W.P. Menzel, 1982: Nowcasting applications of geostationary satellite atmospheric sounding data. *Nowcasting,* K.A. Browning, Ed., Academic Press, 123–135.

_____ , G.S. Wade, and H.M. Woolf, 1985: Combined atmospheric sounder/cloud imagery—A new forecasting tool. *Bull. Amer. Meteor. Soc.,* **66,** 138–141.

_____ , W.P. Bishop, V.F. Dvorak, C.M. Hayden, J.H. McElroy, F.R. Mosher, V.J. Oliver, J.F. Purdom, and D.Q. Wark, 1986: The meteorological satellite: Overview of 25 years of operation. *Science,* **231,** 455–462.

SS/LORAL, 1995: Performance in space. SS/LORAL, CD-ROM. [Available from Space Systems LORAL, 3825 Fabian Way, Palo Alto, CA 94303.]

Staelin, D.H., A.H. Barrett, J.W. Waters, F.T. Barath, E.J. Johnston,

P.W. Rosenkranz, N.E. Gaut, and W.B. Lenoir, 1973: Microwave spectrometer on the Nimbus-5 satellite: Meteorological and geophysical data. *Science,* **182,** 1339–1341.

Stephens, G., 1994: *Remote Sensing of the Lower Atmosphere: An Introduction.* Oxford University Press, 523 pp.

Stowe, L.L., R.M. Carey, and P.P. Pellegrino, 1992: Monitoring the Mt. Pinatubo aerosol layer with NOAA/11 AVHRR data. *Geophys. Res. Lett.,* **19,** 159–162.

Suomi, V.E., and R. Parent, 1968: A color view of planet earth. *Bull. Amer. Meteor. Soc.,* **49,** 74–75.

Tarpley, J.D., S.R. Schneider, and R.L. Money, 1984: Global vegetation indices from the NOAA-7 meteorological satellite. *J. Climate Appl. Meteor.,* **23,** 491–494.

Tepper, M., 1982: Early program development and implementation. NASA Conf. Publ. 2257, 5–33.

Twitchell, P.F., E.A. Rasmussen, and K.L. Davidson, Eds., 1989: *Polar and Arctic Lows.* A. Deepak Publishing, 420 pp.

van de Boogaard, H.M.E., Ed., 1966: Satellite data in meteorological research. NCAR-TN-11, Boulder, CO, 349 pp.

Vaughn, W.W., 1982: Meteorological satellites—some early history. *Meteorological Satellites—Past, Present and Future.* NASA Conf. Publ. 2227, 1–2.

Velden, C.S., 1987: Satellite observations of Hurricane Elena (1985) using the VAS 6.7 μm "water-vapor" channel. *Bull. Amer. Meteor. Soc.,* **68,** 210–215.

——— , and W.L. Smith, 1983: Monitoring tropical cyclone evolution with NOAA satellite microwave observations. *J. Climate Appl. Meteor.,* **22,** 714–724.

——— , ——— , and M. Mayfield, 1984: Applications of VAS and TOVS to tropical cyclones. *Bull. Amer. Meteor. Soc.,* **65,** 1059–1067.

——— , C.M. Hayden, W.P. Menzel, J.L. Franklin, and J.S. Lynch, 1992: The impact of satellite-derived winds on numerical hurricane track forecasting. *Wea. Forecasting,* **7,** 107–118.

Vonder Haar, T.H., G.G. Campbell, E.A. Smith, A. Arking, K. Coulson, J. Hickey, F. House, A. Ingersoll, H. Jacobowitz, L. Smith, and L. Stowe, 1981: Measurements of the earth radiation budget from satellites during the first GARP global experiment. *Adv. Space Res.,* **1,** 285–297.

Wark, D.Q., and H.E. Fleming, 1966: Indirect measurements of atmospheric temperature profiles from satellites. *Mon. Wea. Rev.,* **94,** 351–362.

______ , D.T. Hilleary, S.P. Anderson, and J.C. Fischer, 1970: Nimbus satellite infrared spectrometer experiments. *IEEE Trans. Geosci. Electron.,* **GE-8,** 264–270.

Weinstein, M., and V.E. Suomi, 1961: Analysis of satellite infrared radiation measurements on a synoptic scale. *Mon. Wea. Rev.,* **89,** 419–428.

Weiss, C.E., and J.F.W. Purdom, 1974: The effect of early morning cloud cover on afternoon thunderstorm activity. *Mon. Wea. Rev.,* **102,** 400–401.

Weldon, R.B., and S.J. Holmes, 1991: Water vapor imagery. NOAA Tech. Rep. NESDIS 57, U.S. Dept. of Commerce, NOAA, NESDIS, Washington, DC, 213 pp.

Wexler, H. 1954: Observing the weather from a satellite vehicle. *Brit. Interplan. Soc.,* **13,** 269–276.

Wylie, D.P., W.P. Menzel, H.M. Woolf, and K.I. Strabala, 1994: Four years of global cirrus cloud statistics using HIRS. *J. Climate,* **7,** 1972–1986.

CHAPTER 6

A History of Calculating Aids in Meteorology

FREDERIK NEBEKER

Though an ancient science, meteorology was almost entirely non-quantitative until the seventeenth century. In the course of that century meteorological observation became largely quantitative as instruments were devised to measure temperature, atmospheric pressure, humidity, precipitation, wind direction, and wind force. Since that time, calculating aids have assumed greater and greater importance in meteorology.

Numerical tables and mechanical calculators

The new instruments gave large quantities of data, but meteorologists needed to do a great deal of calculation in order to make the data useful. One problem was that people in different countries employed different units of measure. For example, in the nineteenth century there were three common temperature scales—centigrade, Fahrenheit, and Reamur. A second problem was that corrections often had to be applied to the observed readings. Barometric readings, for example, were regularly corrected for temperature and capillary action. A third problem was that some quantities were measured indirectly, their values being computed from the observed values of other quantities. For example, humidity was calculated from wet- and dry-bulb temperature readings. A fourth problem was that actual values often needed to be converted to corresponding values under standard conditions, such as actual barometric pressure to corresponding pressure at sea level and at $0°C$.

To solve these problems, meteorologists adopted a computing device long used by astronomers, numerical tables, and in the second half of the nineteenth century there appeared several collections of tables for meteorological data reduction. These tables, unlike earlier tables

Reading of Thermometer, Fahr.		Temp of Dew-Point, Fahr.	Force of Vapor in English Inches.	Weight of Vapor.		Humidity, Saturation = 1.000.	Weight in Grains of a Cubic Foot of Air. Height of the Barometer in English Inches.						
Dry.	Wet.			In a Cubic Foot of Air.	Reqd. for Sat'n. of a Cubic Ft. of Air.		in. 28.0	in. 28.5	in. 29.0	in. 29.5	in. 30.0	in. 30.5	in 31.0
°	°	°	in.	gr.	gr.		gr.	gr.	gr.	gr.	gr.	gr.	gr.
62	62	62.0	0.559	6.25	0.00	1.000	491.2	499.9	508.7	517.5	526.3	535.1	543.9
	61	60.3	0.528	5.91	0.34	0.946	491.4	500.1	508.9	517.7	526.5	535.3	544.1
	60	58.6	0.499	5.58	0.67	0.893	491.5	500.2	509.0	517.8	526.6	535.4	544.2
	59	56.9	0.472	5.27	0.98	0.843	491.7	500.4	509.2	518.0	526.8	535.6	544.4
	58	55.2	0.445	4.99	1.26	0.798	491.9	500.6	509.4	518.2	527.0	535.8	544.6
	57	53.5	0.421	4.70	1.55	0.752	492.0	500.7	509.5	518.3	527.1	535.9	544.7
	56	51.8	0.397	4.44	1.81	0.710	492.1	500.7	509.5	518.4	527.3	536.1	544.9
	55	50.1	0.375	4.19	2.06	0.670	492.2	500.9	509.7	518.5	527.4	536.2	545.0
	54	48.4	0.354	3.95	2.30	0.632	492.4	501.1	509.9	518.7	527.6	536.4	545.2
	53	46.7	0.333	3.72	2.53	0.595	492.5	501.3	510.1	518.9	527.7	536.5	545.3
	52	45.0	0.315	3.52	2.73	0.563	492.7	501.5	510.3	519.1	527.9	536.7	545.5
	51	43.3	0.297	3.31	2.94	0.530	492.8	501.6	510.4	519.2	528.0	536.8	545.6
	50	41.6	0.280	3.13	3.12	0.501	492.9	501.7	510.5	519.3	528.1	536.9	545.7
	49	39.9	0.263	2.95	3.30	0.472	493.0	501.8	510.6	519.4	528.2	537.0	545.8
	48	38.2	0.248	2.77	3.48	0.443	493.1	501.9	510.7	519.5	528.3	537.1	545.9
	47	36.5	0.234	2.61	3.64	0.418	493.2	502.0	510.8	519.6	528.4	537.2	546.0

FIG. 6-1. This is a small part of a 33-page table in the fourth edition of Guyot's *Tables, Meteorological and Physical* (1884). The bottom row tells that the following correspond: 1) dry-bulb temperature 62°F, wet-bulb temperature 47°F, 2) dewpoint 36.5°F, 3) vapor pressure 0.234 in. Hg, 4) water vapor mass 2.61 g ft^{-3}, 5) 3.64 g of water vapor required per cubic foot for saturation, 6) relative humidity 0.418, and 7) total mass of one cubic foot of air, including the water vapor, 493.2 g for barometric pressure of 28.0 in., etc.

for the presentation of meteorological data, were computing devices. Consider, for example, Arnold Guyot's *Tables, Meteorological and Physical,* which appeared in 1852, with subsequent editions in 1857, 1859, and 1884 (Guyot 1884). The fourth edition contains about 700 pages of tables. Roughly one-half of them are for converting units of measure. Of the other tables, some are for making corrections to instrumental readings, some for computing quantities measured indirectly, and some for converting the actual values to the corresponding values under standard conditions (see Fig. 6-1).

Numerical tables remained common computing devices through the first half of the twentieth century. Compilations, such as Guyot's, continued to be published, and new tables were devised. For example, in 1928, G.C. Simpson, as part of a study of the absorption and emission of electromagnetic radiation in the atmosphere, constructed tables for the evaluation of complicated functions (Simpson 1928a, 1b). A reviewer wrote that "one of the most valuable features of the paper consisted in the tables" since they "simplified greatly the application of theoretical results to practical problems" (Simpson 1928b). Simpson used his tables, not for data reduction, but in an early example of numerical experimentation. Others who published new tables for calculation were T.N. Doerr (1921), J.C. Ballard (1931), and S. Hess and S. Fomenko (1945).

Though Blaise Pascal built a mechanical adding machine in 1642, the first commercially successful mechanical calculator was the Arithmometer, patented by the Frenchman Charles Thomas in 1820. Use of the machines spread only gradually. By 1920 mechanical calculators were widely used, in both the commercial and the scientific world (Kidwell and Ceruzzi 1994). Among the most popular machines were Felt's Comptometer (which sold for 50 years following its introduction in 1887), Odhner's Calculating Machine (patented in 1891), the Brunsviga (some 20 000 were sold between 1892 and 1912), Steiger's Millionaire (some 5000 were sold between 1899 and 1935), the Mercedes–Euklid Calculating Machine (first marketed in 1910), and various Monroe calculating machines. Many of these machines were used by meteorologists. The American meteorologist Daniel Draper used various calculating machines, including a Comptometer in about 1890, and the English meteorologist Lewis Fry Richardson (famous for his 1922 book *Weather Prediction by Numerical Process*) used both the Odhner and the Mercedes–Euklid (Ashford 1985, 246; Richardson 1922, 13). One of the reasons meteorologists made greater use of statistical methods in the decades following World War I was that, for the first time, many had the use of mechanical calculators.

Graphical techniques

In 1903 a Norwegian physicist-turned-meteorologist, Vilhelm Bjerknes, began advocating a calculational approach to weather forecasting, believing it possible to bring together the full range of observation and the full range of theory to predict the weather (Friedman 1989). Bjerknes believed that a graphical—rather than an analytical or numerical—solution of the equations was the only feasible method. He believed that a graphical calculus, comparable to Newton's and Leibniz's algebraic calculus, could be developed: "as the observations are presented by means of charts, therefore all mathematical computa-

tions must be recast into graphical operations by means of maps"
(Bjerknes 1914).

Bjerknes and his collaborators pursued the development of the
graphical calculus, and in 1910 and 1911 they published the first two
volumes of *Dynamical Meteorology and Hydrography*. These volumes
present a great many graphical techniques—such as for multiplying
scalar fields, for taking the gradient of a scalar field, and for taking the
divergence or the curl of a vector field—and show their application to
meteorology, as in the calculation of the vertical motion of air from
knowledge of the horizontal motion over a large area.

Bjerknes was not the first to use graphical techniques in meteorology. In 1884 Heinrich Hertz, with whom Bjerknes later studied,
introduced a graphical procedure for studying adiabatic processes in
the atmosphere. Graphical addition and subtraction were described in
Adolf Sprung's widely used *Lehrbuch der Meteorologie* of 1888, and, at
about the same time, Wladimir Köppen and Max Möller used graphical
procedures to construct upper-air weather charts (Kutzbach 1979,
152, 237).

In the first half of the twentieth century a wide range of graphical
techniques were developed. For example, in the first decade of the
century the Austrian meteorologist Felix Exner showed that the rate of
change of pressure is inversely proportional to the area enclosed between consecutive isobars and isotherms (Gold 1930). In 1923, the
Japanese meteorologist S. Fujiwhara proposed that forecasters calculate vorticity by using a weather map and a celluloid scale for gradient
wind (Fujiwhara 1923). Jerome Namias, who was at the Massachusetts Institute of Technology (MIT) in the 1930s, later commented on
the interest then shown in graphical methods of solving equations
(Namias 1986, 6). Other examples of graphical techniques are the
tephigram [a diagram, devised by William Napier Shaw in 1925, used
to predict cloud formation; see Poulter (1938)] and many procedures
used in the search for weather cycles [such as Arthur Schuster's periodogram or F.J.W. Whipple's harmonic dial; see Schuster (1900) and
Whipple (1917)]. Of all the graphical methods, nomography was most
used.

Though the technique was first used almost a century earlier, the
word "nomograph" was coined in 1891 by the French engineer Maurice
d'Ocagne to denote a figure presenting a quantitative law in such a
way that the implication of the law, in any particular case, is readily

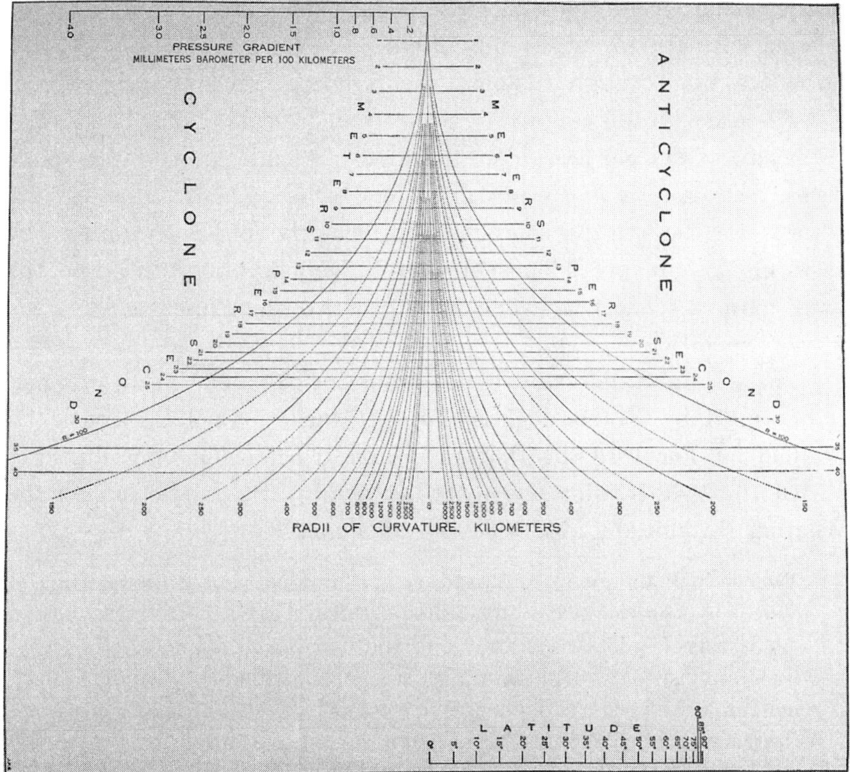

FIG. 6-2. A nomogram, constructed by Herbert Bell of the University of Chicago, for calculating the gradient–wind velocity given the pressure gradient, the latitude, and the radius of curvature of the local isobar (Humphreys 1920, p. 144).

determinable (usually by seeing where a straightedge, placed so that it connects points on two scales, cuts a third scale). Nomography was widely used in engineering and in many of the sciences, including meteorology (Evesham 1986).

Figure 6-2 is an example taken from Humphreys' *Physics of the Air* (1920). It is a graphical representation of the general gradient–velocity equation

$$2 \, \omega \, v \, \sin \phi - (1/\rho) \, (dp/dn) = \pm \, v^2/r \, ,$$

where the upper sign is used for anticyclones and the lower for cyclones. To calculate the gradient–wind velocity, place a straightedge so that it passes through the known value of the pressure gradient (on the scale at the top) and the latitude of the place in question (on the scale at the bottom). Find the intersection of the straightedge and the curve labeled with the known radius of curvature of the local isobar. The horizontal line through this point of intersection gives the wind speed.

The use of graphical techniques by meteorologists after World War I was part of a larger movement to substitute analog procedures—slide rules, nomographs and other graphical techniques, and analog machines—for the (digital) computations formerly done with tables. The resulting loss in precision seemed relatively unimportant. And the great gain in speed was everywhere praised. Slide rules were particularly expeditious.

The standard slide rule, which dates back to about 1620, was regularly used by meteorologists. Nelson Haas, in a 1924 article "A Method for Locating the Decimal Point in Sliderule Computation," stated that slide rules were used extensively in the work of the Weather Bureau and asserted their adequacy:

> Twenty-inch slide rules are used chiefly for such computations. The slide rule is particularly well adapted to this work, for the 20-inch rule yields three figures accurately and the fourth approximately. Four figures represent the maximum accuracy that is readily attainable in meteorological observations, and consequently the 20-inch rule is entirely satisfactory for this work, and it is very expeditious.

Characteristic of meteorology in the interwar period and the succeeding two decades was the development of a great many special-purpose slide rules. Such devices helped forecasters calculate wind direction and speed at various heights. This information could be gained by the use of pilot balloons, but since the raw data were the azimuth and elevation angles measured by a theodolite, much computation was involved. Since using tables was fairly time consuming, special slide rules were constructed for this purpose and were by 1922 commonly employed (Thompson 1922). Another example is the slide rule, devised by Noel Sellick (1937), that allows one to convert barometric pressure at one height to pressure at any other height, for values of the temperature between 0° and 30°C. A sophisticated example is shown in Fig. 6-3. This is only part of an elaborate slide rule system, designed by John C. Bellamy in 1943, for calculating the constant absolute vorticity path of wind (Godske et al. 1957, 716). The part shown consists of three slide rules with four, eight, and six scales, respectively. The slide rule system is based directly on Carl-Gustav Rossby's vorticity equation used to calculate air movement.

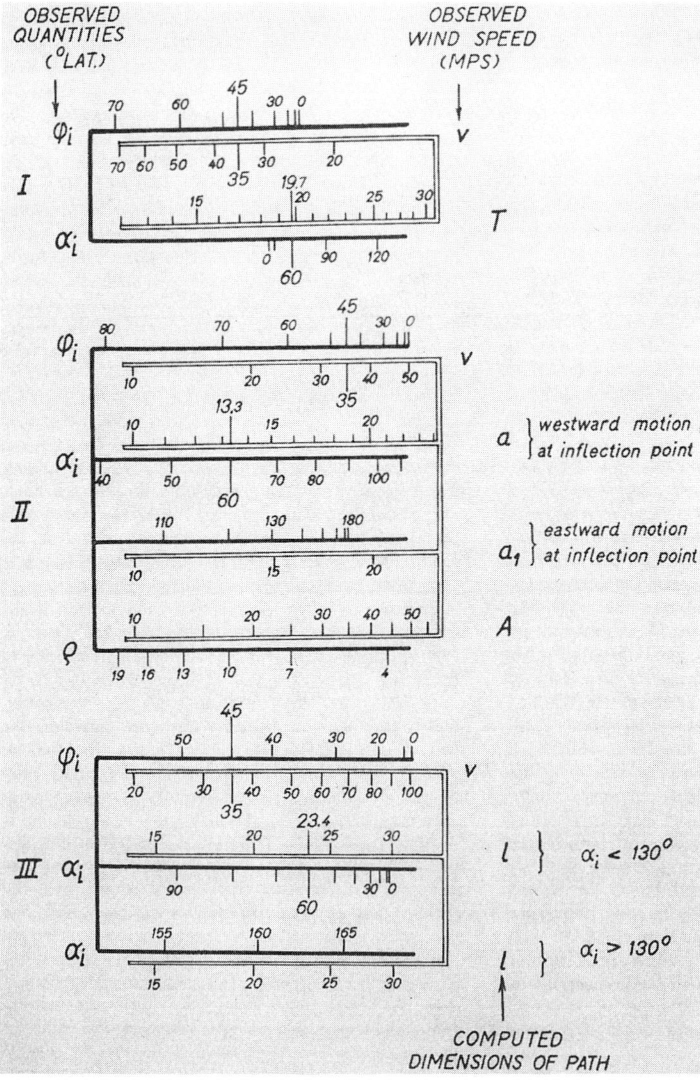

FIG. 6-3. Part of a slide rule system for calculating the constant absolute vorticity path (reproduced in Godske et al. 1957, 716).

Computing forms and punched-card machines

Another calculating aid of the interwar years deserves mention for its anticipation of the computer program. Richardson in his 1922 *Weather Prediction by Numerical Process* showed how to compute future values of atmospheric variables from present values. He presented the complex algorithm, each step of which is either an arith-

COMPUTING FORM P I.　Pressure, Temperature, Density, Water and Continuous Cloud

Longitude = 11° East　　　Latitude = 5400 km North　　　Instant 1910 May 20ᵈ 7ʰ G.M.T.

Height h	δh	p	g	R	$p=\dfrac{R}{\delta h}$	P	$p=\dfrac{P}{\delta h}$	W	$\mu=\dfrac{W}{R}$	θ	Density of saturated vapour w_s	$W_s = w_s\,\delta h$	Precipitated $= W - W_s - 0.4$	Q in continuous cloud	Potential temperature at surface pressure
Ref.:—		p. 185	Helmert	previous		Ch. 8/2/2			p. 185	ρ and p					By Ch. 8/2/6 #13
$10^5\times$	$10^5\times$	$10^2\times$			$10^{-3}\times$	$10^3\times$	$10^2\times$		$10^{-3}\times$	°A	$10^{-6}\times$				
h_0		0													
			9751	210·3		1278		zero	zero	212					
$h_2 = 11·8$		2050													
	4·6		978·2	209·0	0·454	1362	2965	0·0	0·0	227·5		0·0	0·0	0·0	320·8
$h_4 = 7·2$		4090													
	3·0		979·2	203·3	0·677	1501	5003	0·1	0·5	257·7	1·3	0·4	0·0	0·0	312·0
$h_6 = 4·2$		6079													
	2·2		980·0	192·1	0·875	1538	6990	0·4	2·1	279·0	6·0	1·6	0·0	0·0	305·8
$h_8 = 2·0$		7960													
	1·6		980·6	169·9	1·061	*1402	8770	0·9	5·3	287·5	12·3	2·0	0·0	0·0	295·3
$h_{10} = 0·4$		9626													

FIG. 6-4. The first of Richardson's 23 computing forms.

metic operation or the entering of a number from another location, in a set of 23 computing forms, the first of which is shown in Fig. 6-4. The numbers in italics are those that are written in during a computation. The third column of numbers comes from the initial data (printed on page 185 in Richardson's book). The numbers in the next two columns come from tables. The numbers in the fifth column result from dividing each entry in the fourth column by the corresponding entry in the second column, and so forth. Some of the numbers computed here are entered onto other computing forms. Computing forms of this sort were used by many other meteorologists, among them Sachindra Nath Sen (1924), G.C. Simpson (1929), and Samuel B. Solot (1939).

A data-handling device that became important to meteorology in the interwar period was the punched-card machine. In the late 1880s, Herman Hollerith devised a set of card-punching, card-sorting, and card-tabulating machines for use by the U.S. Census Bureau (Goldstine 1972, 65–71). The machines performed well with the 1890 census, and Hollerith's machines were soon in use for census purposes in many countries. The machines were not used for meteorological data, however, until the 1920s.[1]

---

[1]In the mid-1890s, the U.S. Hydrographic Office made a short-lived trial of their use with meteorological observations (Bates 1956).

In about 1920 the Meteorological Office of the British Admiralty began using punched-card methods to compute summary statistics. In 1922 the Dutch Meteorological Institute, having borrowed some British card files, began using punched-card machines. So did Norway, France, and Germany soon thereafter. In the mid-1920s, the Czechoslovakian meteorologist L.W. Pollak designed an inexpensive punch machine and had one placed in every Czechoslovakian weather station, and by 1927 Pollak had used the tabulating machines to produce frequency tables of barometric pressure. By the late 1930s, the weather service of every major European country was analyzing data by means of punched cards (Air Weather Service 1949; Conrad and Pollak 1950, p. 351).

A laggard in this movement was the weather service of the United States, where the Hollerith method had originated. The Weather Bureau had considered acquiring some of Hollerith's punching and tabulating machines in about 1885 and in 1895, but it was not until the mid-1930s that the Weather Bureau actually did so (Austrian 1982, 112; Whitnah 1961, 66). In 1934 a science advisory board recommended a card-punching unit for the central office of the Weather Bureau, and funds were provided for card punching by two of President Franklin Roosevelt's programs to ameliorate the Depression—the Civil Works Administration in 1934 and the Works Progress Administration (WPA) in 1936 (Whitnah 1961, 157).[2] A number of meteorological atlases were produced using these cards, such as *Atlas of Climatic Charts of the Oceans* (1938) and *Airway Meteorological Atlas for the United States* (1941). For the production of the latter, more than 14 million airway observations were transferred to punched cards (Conrad and Pollak 1950, 353).

Tabulating equipment thus made it possible to use many more data than was possible before. Another important result was a higher standard of weather data. Such machines as duplicating punches, reproducing punches, verifiers, and interpreters eliminated most errors from a great many of the routine data processing tasks.[3] More errors

---

[2]Another WPA project, the New York Mathematical Tables Project, which employed about 350 people to carry out computations, did some work for meteorologists.

[3]A duplicating punch was a manually operated key punch with a duplicating facility, so that data common to several cards could be punched automatically, while the data specific to the individual card were punched manually. An interpreter converted punched-card information into printed information.

were eliminated when sorting and tabulating were done by machine. Machines were used also to check for missing data and to identify obviously erroneous data. A third effect of the use of punched-card machines was the facilitation of calculation, particularly for sophisticated statistical analyses. Originally these machines were used only for sorting, searching, and counting. As the tabulating equipment became more sophisticated—IBM introduced in 1931 a multiplying punch and in about 1933 removable control panels—meteorologists were able to do more and more complicated computations mechanically (Bashe et al. 1986, 17).

During World War II punched-card machines were used in new ways in order to improve forecasting. They were used to compute duration frequencies (how often a type of weather continues a given number of hours or days). They were used in correlating surface observations with upper-air information, thus making the latter more valuable, and in correlating between weather elements at different locations. They were used, too, in forecast verification. A wartime technique that continued in use after the war, at least in a few places, was "analog selection." The Air Weather Service placed on punched cards information describing each weather map in a 40-year series of maps. The cards were used to find the map of past weather that most closely resembled the current weather map, and the course of the past weather then served as a guide in forecasting. This method was used by the American forecasters contributing to the D-Day forecasts (Stagg 1971, 30).[4]

Meteorologists used tabulating machines also for more complex calculations. Here they were following the lead of two great pioneers of scientific calculations: L.J. Comrie in England and Wallace Eckert in the United States.[5] In the late 1920s, Comrie, at the National Almanac Office in London, made calculations of lunar motions. In the 1930s, W. Eckert, at the Thomas J. Watson Astronomical Computing Bureau of Columbia University (the head of IBM having provided the funding), worked to make punched-card machines useful to sciences besides as-

---

[4]An account of the analog method developed by Irving Krick, which is probably very similar to the one used at Widewing, is in Krick (1954, 87–106).

[5]The first large calculations on punched-card machines were not scientific calculations. They were done in the early 1920s by American railroad companies that needed to compute car-mile and ton-mile statistics for the Interstate Commerce Commission (Eckert 1940, ix).

tronomy. W. Eckert showed, for example, how to use these machines to solve numerically certain classes of ordinary differential equations. Eckert was effective in proselytizing for his methods by welcoming visitors to his laboratory, by offering the "Watson Laboratory Three-Week Course in Computing" (which was attended, over a number of years, by 1600 people from 20 countries), and by his 1940 book *Punched Card Methods in Scientific Computation.*

Near the end of the war, meteorologists began to use tabulating machines to solve complicated equations. One example is the computation, carried out in 1945 by Gilbert Hunt of the Air Weather Service, of the total amount of water vapor in the atmosphere above a given station; each such computation involved many different operations and hundreds of observations.[6] Another example is a study of the action of winds in the buildup of heavy seas, which involved a lengthy series of operations carried out automatically by a card-programmed calculator (Aerology Branch 1953, 17).

The use of punched cards by the American weather services continued to grow after the war. By the end of 1947 all Weather Bureau stations were entering, as standard practice, current meteorological observations onto punched cards (Whitnah 1961, 226). In 1953 there were almost 200 million punched cards at the national weather records center, up from 80 million at the end of the war. And as electronic computers became available to meteorologists in the mid-1950s, the value of having weather information on punched cards became even greater (Bellamy 1952). Though the principal motivation for all these uses of tabulating machines was to speed up data processing and other computation, a secondary motivation was to eliminate human error in the processes thus automated.

Electrical analog computers

Punched-card machines were also used to search for weather cycles (Pollak 1925; Pollak and Kaiser 1934). In the late 1930s, John Mauchly, an American physicist, sought to show that the sun's 27-day rotational period influenced the weather. Even with the help of a dozen students (paid for by the National Youth Administration, another New Deal program) and several Marchant calculators, the work went

---

[6]Note 14 April 1945 by John Mauchly (Wexler Papers, Library of Congress).

slowly. In 1939 and 1940, Mauchly built an electrical harmonic analyzer, which sped up the calculation of harmonic coefficients by a factor of 5 or 10, and the accuracy was just sufficient—to two significant figures—for the purpose. With this device Mauchly detected periodicities in rainfall of 13.5 days and 27 days.

Mauchly made plans to build a more elaborate harmonic analyzer, but then sometime in 1941 he abandoned these plans to work on a digital electronic device, apparently because of the greater accuracy possible with a digital device. And from this time on Mauchly gave almost all of his time not to the analysis of meteorological data, but to the design of electronic digital computers. His success is well known. He and J. Presper Eckert were principal designers of four famous computers: the ENIAC, the EDVAC, the BINAC, and the UNIVAC.[7] But Mauchly apparently never published, either before 1941 or after, any of the results of his study of meteorological data (Burks and Burks 1988, 103).

Mauchly's harmonic analyzer was an analog computer. A number of other analog computers were employed by meteorologists. The differential analyzer built by Vannevar Bush at MIT and a 10-equation, 10-variable electrical linear equation solver built by RCA, though both designed for other purposes, were applied to meteorological problems. It was especially differential analyzers that meteorologists found useful; they were used to solve a nonlinear system of equations advanced by Edward Lorenz and to find the spectral representation of functions as suggested by George Platzman (Dingle and Young 1965, vi); Maurice Wilkes (1949) modeled atmospheric tides with the differential analyzer at the Mathematical Laboratory of Cambridge University; and Luigi Jacchia and Zdenek Kopal (1952) investigated atmospheric oscillations and temperature profiles with the Rockefeller Differential Analyzer at MIT.

A few meteorologists followed the lead of Mauchly in building an analog computer specifically for meteorological use. R.J. Taylor and E.K. Webb (1955), finding that a study of the vertical turbulent fluxes of heat, water vapor, and momentum involved a great deal of calculation

---

[7]ENIAC was the first large-scale electronic digital computer; EDVAC was the first stored-program computer to be planned; BINAC was the first operational stored-program computer in the United States; and UNIVAC was the first commercially produced electronic digital computer (Stern 1980).

with the raw data, designed a special-purpose differential analyzer for this work. D.P. Brown and R.A. Harvey (1961) built an integrator to record automatically a weighted average daily solar radiation. Seymour Hess (1957) designed an analog computer for finding the Laplacian of any mapped quantity, such as the geostrophic vorticity. And in the late 1950s, the Central Institute of Forecasting in Moscow used a special-purpose electronic analog computer for predicting temperature anomalies (Baum and Thompson 1959).

With all these devices, accuracy was quite limited, usually to one or two significant figures. When it was used in a meteorological study in 1946, the RCA linear equation solver was judged unsatisfactory because of its limited accuracy.[8] In the study made by Jacchia and Kopal (1952), the differential analyzer was abandoned in favor of mechanical desk calculators whenever greater accuracy was required. It was sometimes possible, by reformulating the mathematical task, to increase somewhat the accuracy achievable, but limited accuracy remained a problem that, together with the fact that most analog devices could solve only a narrow range of problems, led to the dominance of the digital computer in the 1960s.

Electronic digital computers

The first large-scale electronic digital computer, the ENIAC, was completed just as World War II ended, and at that time John von Neumann began making plans to build, at the Institute for Advanced Study (IAS) in Princeton, New Jersey, a much more powerful and versatile machine devoted to the advancement of the mathematical sciences. An important objective for von Neumann was to demonstrate, with a particular scientific problem, the revolutionary potential of the computer. He chose for this purpose weather prediction, and in 1946 established the Meteorology Project at the institute. The project had a slow start, and the institute computer took longer to build than was expected, but in 1950 trial forecasts were made using the ENIAC, and in 1952 using the IAS computer. By that year at least three other groups were also working on numerical weather prediction: the Atmospheric Analysis Laboratory of the Air Force Cambridge Research Lab-

---

[8]This machine was accurate to three figures [Travel Report Wexler to Reichelderfer 20 December 1946 (Wexler Papers, Library of Congress)].

oratory, the Napier Shaw Laboratory of the British Meteorological Office, and the International Meteorological Institute of the University of Stockholm. By 1956, when the IAS Project ended, von Neumann's expectations had been fulfilled: it had been shown that a physics-based algorithm could be used to predict large-scale atmospheric motions as accurately as human forecasters could, and it had been shown that computer technology could carry out such algorithms fast enough and reliably enough for the forecasts to be useful (Aspray 1990; Nebeker 1995).

The commercial availability of electronic computers beginning in the early 1950s led to much wider computer use by meteorologists. For example, the IBM 701 was introduced in the spring of 1953; its computing power was comparable to that of the IAS computer. By 1957 numerical forecasts were being produced daily in several countries (Rossby 1959, 32). Meteorologists were soon using computers in many other ways, for research (notably by numerical experimentation) and especially for processing observational data.

The disappearance of other calculating aids

As electronic digital computers proliferated on the meteorological scene, they pushed most other calculating aids off the stage altogether. The oldest of the calculating aids, numerical tables, managed, but just barely, to maintain a presence.[9] For example, in 1949 Thompson published a set of tables for computing the heights of the 850-, 700-, and 500-mb surfaces from the readings of altimeters in aircraft. In each of the early volumes of *Tellus,* the journal Rossby established in 1949, there are several articles that include tables for calculating, and this is true as well of the contemporaneous volumes of the *Bulletin of the American Meteorological Society.* In 1966 the Weather Bureau published a collection of tables to be used in the processing of radiosonde data. But in the 1960s and 1970s, electronic computers and hand-held calculators came to replace tables for routine or complicated tasks,

---

[9]Of course tables are still a standard way of presenting observational data. Tables have served also to present "calculated data," as with Edmond Halley's 1687 calculations of barometric pressure as a function of height, or with Coulson et al.'s 1960 *Tables Related to Radiation Emerging from a Planetary Atmosphere with Rayleigh Scattering* (prepared with the use of an electronic computer). Such tables may be regarded as calculating aids, since they serve to obviate a lengthy calculation rather than to present empirical findings.

though tables continued to be used occasionally.[10] It must be remembered that until quite recently computers were very expensive, so many meteorologists, even in industrial countries, had to make do with tables and other inexpensive calculating aids.

Geometrical methods of calculation, that is, ones that involve the representation of a quantity by a length or an area, have been abandoned almost entirely. It was in the 1940s and 1950s that interest in special-purpose slide rules, nomograms, special instruments and paper for plotting data, and other graphical techniques was at its highest, and interest remained high in the 1960s. In 1949, O.G. Sutton described a slide rule to be used for calculating atmospheric diffusion, and in the 1950s several different circular slide rules were constructed for calculating the winds in the flight path of an aircraft (Buell 1961). In 1959, Frank Hanson and Paul Taft devised a plotting system, involving several movable scales and a two-piece plotting surface, for finding wind speeds by tracking pilot balloons using two theodolites; Hanson and Taft devised the system in order to "eliminate the laborious process of hand computing double-theodolite balloon-measured wind data." In 1961, Robert Weedfall and Walter Jagodzinski presented a different graphical procedure for the same task. Because the calculations to be carried out with the raw data often depended on particularities of the observational instrument, an appropriate slide rule, chart, or plotting form was in many cases provided with the instrument (MacCready 1957).

Calculations involving graphs, either on maps or on more abstract coordinate planes, were for a time much used. In such calculations integrations were sometimes done using planimeters (Robertson and Cameron 1952). In 1950 A.F. Williams showed a way to use a map as a calculating aid: if barometric pressure is plotted on a map for which the scale at each point is proportional to the sine of the latitude, then the geostrophic wind speed at any point is proportional to the contour spacing there. Another example is Harold Crutcher's graphical procedure for estimating the effect of wind on aircraft speed (Crutcher 1956). And because statistical procedures frequently involved a great deal of computation, graphical procedures were sometimes used for their facilitation (Godske et al. 1957, 750–751, 763–765).

---

[10]In 1981 the Board of Certified Consulting Meteorologists of the American Meteorological Society endorsed a list, prepared by Chester Newton, of 50 books useful to a consulting meteorologist; two of the 50 books are compilations of tables (Newton 1981).

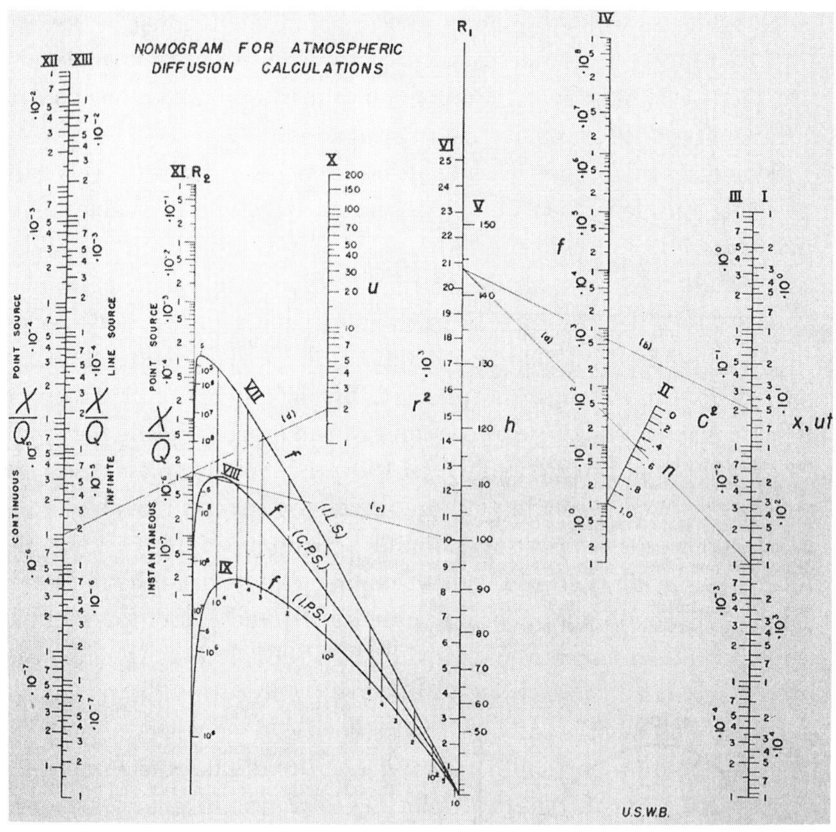

FIG. 6-5. This nomogram, devised by Frank Gifford, allows one to solve quickly nine equations that O.G. Sutton had derived for describing atmospheric diffusion.

In the 1950s and 1960s nomography was perhaps the most commonly used of the graphical methods. Nomograms were constructed for use in predicting the location of the jet stream (Riehl 1952), for estimating the return periods of extreme weather conditions (Weiss 1955), for finding the amount of precipitable water (Peterson 1961), for determining the floating altitude of balloons [*Bull. Amer. Meteor. Soc.,* **45,** 542], and for dozens of other meteorological tasks. A nomogram was generally easy to use, the only physical operation being the alignment of a straightedge, and it could obviate a complicated and time-consuming analytical or numerical procedure. A particularly intricate example, constructed by Frank Gifford in 1952 to calculate atmospheric diffusion, is shown in Fig. 6-5. Whereas most nomograms were made to solve one equation, this nomogram, which comprises 13 scales, solves nine different equations (Gifford 1953).

Nomograms, like other analog devices, had quite limited accuracy. The above nomogram produced only one or two significant figures. According to the inventor, "this is no drawback, since, at present, such accuracy considerably exceeds that of both diffusion theory and experimental concentration measurements" (Gifford 1953). Yet as theory and observation advanced, there came to be fewer and fewer areas where one-place accuracy was satisfactory. The 1960s were the last decade in which an appreciable number of new nomograms were reported in the meteorological literature.

One of the most important graphical techniques of the 1950s was the method devised by Ragnar Fjörtoft (1952) for integrating the barotropic vorticity equation. Fjörtoft was working in the Norwegian tradition of the graphical calculus, mentioned above, and his 1952 paper may be said to have fulfilled Bjerknes's program of calculating the weather graphically. In late 1955 Fjörtoft's method was being used in quite a few places: in Tokyo by both the Central Meteorological Observatory and the Geophysical Institute of Tokyo University, in Dublin by the Irish Meteorological Service, in Wellington by the New Zealand Meteorological Service, and in the United States by seven groups.[11] In about 1951 the Soviet meteorologists N.I. Buleyev and M.I. Yudin devised graphical techniques, similar to Fjörtoft's, for solving the vorticity equation and related equations, and their techniques were used in trial forecasts (Phillips et al. 1960).

Thus, the 1950s and 1960s saw a flourishing of graphical and other analog means of calculation. This reflected the large amount of data meteorologists handled and the increasing mathematization of meteorological theory and practice. However, in the 1970s and 1980s, especially with the general availability of the microcomputer, virtually all other devices for calculation came to be replaced by the electronic digital computer.

---

[11]These were the Air Weather Service of the U.S. Air Force; the Extended Forecasting Section of the Weather Bureau; the Travelers Weather Research Center in Hartford, Connecticut; and the departments of meteorology at MIT, the University of Chicago, The Pennsylvania State University, and the University of Washington [George Platzman Survey (Charles Babbage Institute Archives)]. At the same time or only slightly later, Fjörtoft's method was being used by the Canadian Meteorological Centre (Kwizak 1972).

REFERENCES

Aerology Branch, 1953: *Climatology by Machine Methods for Naval Operations, Plans, and Research.* Department of the Navy.

Air Weather Service, U.S. Department of the Air Force, 1949: *Machine Methods of Weather Statistics.* 4th ed. Department of the Air Force.

Ashford, O.M., 1985: *Prophet—or Professor: The Life and Work of Lewis Fry Richardson.* Adam Hilger, xiv + 304 pp.

Aspray, W., 1990: *John von Neumann and the Origins of Modern Computing.* MIT Press, xvii + 376 pp.

Austrian, G.D., 1982: *Herman Hollerith: Forgotten Giant of Information Processing.* Columbia University Press, xvii + 418 pp.

Ballard, J.C., 1931: Table for facilitating computation of potential temperature. *Mon. Wea. Rev.,* **59,** 199–200.

Bashe, C.J., L.R. Johnson, J.H. Palmer, and E.W. Pugh, 1986: *IBM's Early Computers.* MIT Press, xviii + 717 pp.

Bates, C.C., 1956: Marine meteorology at the U.S. Navy Hydrographic Office—A resume of the past 125 years and the outlook for the future. *Bull. Amer. Meteor. Soc.,* **37,** 519–527.

Baum, W.A., and P.D. Thompson, 1959: Long-range weather forecasting in the Soviet Union. *Bull. Amer. Meteor. Soc.,* **40,** 394–409.

Bellamy, J.C., 1952: Automatic processing of geophysical data. *Advances in Geophysics,* Vol. 1, Academic Press, 1–43.

Bjerknes, V., 1914: Meteorology as an exact science. *Mon. Wea. Rev.,* **42,** 11–14.

———, and J.W. Sandstrom, 1910: *Statics. Dynamic Meteorology and Hydrography,* Vol. 1, Carnegie Institution of Washington, 146 pp. and tables.

———, Th. Hesselberg, and O. Devik, 1911: *Kinematics. Dynamic Meteorology and Hydrography,* Vol. 2, Carnegie Institution of Washington, 175 pp. and 60 plates.

Brown, D.P., and R.A. Harvey, 1961: Solar- and sky-radiation integrator. *Bull. Amer. Meteor. Soc.,* **42,** 325–332.

Buell, C.E., 1961: Determination of the feasibility of a winds-ahead computer. Naval Weather Research Facility, Naval Air Station, Norfolk, VA, Final Report Contract N189(188)50237A.

Burks, A.R., and A.W. Burks, 1988: *The First Electronic Computer, The Atanasoff Story.* University of Michigan Press, xii + 387 pp.

Conrad, V., and L.W. Pollak, 1950: *Methods in Climatology*. 2d ed. Harvard University Press, xxvii + 459 pp.

Crutcher, H.L., 1956: Wind aid from wind roses. *Bull. Amer. Meteor. Soc.*, **37**, 391–402.

Dingle, A.N., and C. Young, 1965: *Computer Applications in the Atmospheric Sciences*. University of Michigan, xvii + 256 pp.

Doerr, T.N., 1921: Tables of 0.288th powers. *Quart. J. Roy. Meteor. Soc.*, **47**, 196.

Eckert, W.J., 1940: *Punched Card Methods in Scientific Computation*. Thomas J. Watson Astronomical Computing Bureau, ix + 136 pp. (reprint, with a new introduction by J.C. McPherson, MIT Press and Thomas Publishers, 1984).

Evesham, H.A., 1986: Origins and development of nomography. *Ann. Hist. Comput.*, **8**, 324–333.

Fjörtoft, R., 1952: On a numerical method of integrating the barotropic vorticity equation. *Tellus*, **4**, 179–194.

Friedman, R.M., 1989: *Appropriating the Weather: Vilhelm Bjerknes and the Construction of a Modern Meteorology*. Cornell University Press, xx + 251 pp.

Fujiwhara, S., 1923: On the mechanism of extratropical cyclones. *Quart. J. Roy. Meteor. Soc.*, **49**, 105–118.

Gifford, F., Jr., 1953: An alignment chart for atmospheric diffusion calculations. *Bull. Amer. Meteor. Soc.*, **34**, 101–105.

Godske, C.L., T. Bergeron, J. Bjerknes, and R.C. Bundgaard, 1957: *Dynamic Meteorology and Weather Forecasting*. Amer. Meteor. Soc. and Carnegie Institution, 800 pp.

Gold, E., 1930: Obituary notice of F.M. Exner. *Quart. J. Roy. Meteor. Soc.*, **56**, 194–196.

Goldstine, H.H., 1972: *The Computer from Pascal to von Neumann*. Princeton University Press, xi + 378 pp.

Guyot, A., 1884: *Tables, Meteorological and Physical*. 4th ed. Smithsonian Institution.

Haas, N.W., 1924: A method for locating the decimal point in slide-rule computation. *Mon. Wea. Rev.*, **52**, 29–30.

Halley, E., 1687: A discourse of the rule of the decrease of the height of the mercury in the barometer. *Philos. Trans. Roy. Soc. London*, **16**, 104–116.

Hanson, F.V., and P.H. Taft, 1959: Plotting systems for the evaluation of double-theodolite balloon-measured winds. *Bull. Amer. Meteor. Soc.,* **40,** 221–224.

Hertz, H., 1884: Graphische Methode zur Bestimmung der adiabatischen Zustandsgleichungen feuchter Luft. *Meteor. Z.,* **1,** 421–431. [An English translation is in C. Abbe, 1891: *The Mechanics of the Earth's Atmosphere.* Smithsonian Institution, 198–211.]

Hess, S.L., 1957: A simple analog computer for determination of the Laplacian of a mapped quantity. *Bull. Amer. Meteor. Soc.,* **38,** 67–73.

Humphreys, W.J., 1920: *Physics of the Air.* Franklin Institute, xi + 665 pp.

Jacchia, L.G., and Z. Kopal, 1952: Atmospheric oscillations and the temperature profile of the upper atmosphere. *J. Meteor.,* **9,** 13–23.

Kidwell, P.A., and P.E. Ceruzzi, 1994: *Landmarks in Digital Computing.* Smithsonian Institution Press, 148 pp.

Krick, I.P., and R. Fleming, 1954: *Sun, Sea and Sky.* J.B. Lippincott, 248 pp.

Kutzbach, G., 1979: *The Thermal Theory of Cyclones: A History of Meteorological Thought in the Nineteenth Century.* Amer. Meteor. Soc., 255 pp.

Kwizak, M., 1972: Computer weather forecasting in Canada. *Bull. Amer. Meteor. Soc.,* **53,** 1154–1157.

MacCready, P.B., Jr., 1957: The Munitalp cloud theodolite. *Bull. Amer. Meteor. Soc.,* **38,** 460–469.

Namias, J., 1986: Autobiography. *Namias Symposium,* John O. Roads, Ed., Scripps Institution of Oceanography, 1–59.

Nebeker, F., 1995: *Calculating the Weather: Meteorology in the 20th Century.* Academic Press, 251 pp.

Newton, C.W., 1981: A fifty-book library and journal selections for a Certified Consulting Meteorologist. *Bull. Amer. Meteor. Soc.,* **62,** 71–77.

Peterson, K.R., 1961: A precipitable water nomogram. *Bull. Amer. Meteor. Soc.,* **42,** 119–121.

Phillips, N.A., W. Blumen, and O. Cote, 1960: Numerical weather prediction in the Soviet Union. *Bull. Amer. Meteor. Soc.,* **41,** 599–617.

Pollak, L.W., 1925: Hilfsmittel zur Aufsuchung versteckter Periodizitäten sowie zur harmonischen Analyse überhaupt. *Ann. Hydrogr. Meteor.,* **53,** 209.

______ , and F. Kaiser, 1934: Neue Anwendungen des Lochkarten-Verfahrens in der Geophysik. *Hollerith Nachrichten,* Heft **44.**

Poulter, R.M., 1938: Cloud forecasting: The daily use of the tephigram. *Quart. J. Roy. Meteor. Soc.,* **64,** 277.

Richardson, L.F., 1922: *Weather Prediction by Numerical Process.* Cambridge University Press, xvi + 236 pp. (reprint, Dover Publications, 1965).

Riehl, H., 1952: A quantitative method for 24-hour jet-stream prognosis. *J. Meteor.,* **9,** 159–166.

Robertson, G.W., and H. Cameron, 1952: A planimetric method for measuring the velocity of the Upper Westerlies. *Bull. Amer. Meteor. Soc.,* **33,** 387–389.

Rossby, C.-G., 1959: Current problems in meteorology. *The Atmosphere and the Sea in Motion,* Bert Bolin, Ed., Rockefeller Institute Press, 9–50.

Schuster, A., 1900: The periodogram of magnetic declination. *Trans. Cambridge Philos. Soc.,* **18,** 107.

Sellick, N.P., 1937: A slide rule for reduction of barometric pressure. *Quart. J. Roy. Meteor. Soc.,* **63,** 439–440.

Sen, S.N., 1924: On the distribution of air density over the globe. *Quart. J. Roy. Meteor. Soc.,* **50,** 29.

Simpson, G.C., 1928a: Some studies in terrestrial radiation. *Roy. Meteor. Soc. Mem.,* **2,** 69–95.

______ , 1928b: Further studies in terrestrial radiation. *Roy. Meteor. Soc. Mem.,* **3,** 1–26.

______ , 1929: The distribution of terrestrial radiation. *Roy. Meteor. Soc. Mem.,* **3,** 53–78.

Solot, S.B., 1939: Computation of depth of precipitable water in a column of air. *Mon. Wea. Rev.,* **67,** 100–103.

Stagg, J.M., 1971: *Forecast for Overlord.* Ian Allan, 128 pp.

Stern, N., 1980: John William Mauchly: 1907–1980. *Ann. Hist. Comp.,* **2,** 100–103.

Sutton, O.G., 1949: *Atmospheric Turbulence.* Methuen, viii + 107 pp.

Taylor, R.J., and E.K. Webb, 1955: A mechanical computer for micrometeorological research. Division of Meteorological Physics Technical Paper 6, Commonwealth Scientific and Industrial Research Organisation, Melbourne, Australia, 16 pp.

Thompson, J.A., 1922: Meteorology. *The Outline of Science,* J.A. Thompson, Ed., G.P. Putnam's Sons, 761–790.

Thompson, J.C., 1949: Tables for computing the height of standard pressure surfaces from aircraft reports. *Bull. Amer. Meteor. Soc.,* **30,** 286–287.

U.S. Weather Bureau, 1966: *Radiosonde Observation Computation Tables.* Government Printing Office, iii + 67 pp.

Weedfall, R.O., and W.M. Jagodzinski, 1961: Comments on double-theodolite evaluations. *Bull. Amer. Meteor. Soc.,* **42,** 322–324.

Weiss, L.L., 1955: A nomogram based on the theory of extreme values for determining values for various return periods. *Mon. Wea. Rev.,* **83,** 69–71.

Whipple, F.J.W., 1917: The significance of the harmonic analysis of diurnal variation of pressure. *Quart. J. Roy. Meteor. Soc.,* **43,** 282–283.

Whitnah, D.R., 1961: *A History of the United States Weather Bureau.* University of Illinois Press, ix + 267 pp.

Wilkes, M.V., 1949: The E.D.S.A.C., Report on a conference on high speed automatic calculating machines, 22–25 June 1949. University Mathematical Laboratory, Cambridge.

Williams, A.F., 1950: A "geostrophic map." *Bull. Amer. Meteor. Soc.,* **31,** 325–317.

Cloud Microphysics and Dynamics

Chapter 7

Formation of Rain:
A Historical Perspective

Roscoe R. Braham Jr.

Every occurrence in nature is preceded by other occurrences which are its causes, and succeeded by others which are its effects. The human mind is not satisfied with observing and studying any natural occurrence alone, but takes pleasure in connecting every natural fact with what has gone before it, and with what is to come after it.

(John Tyndall 1872)

Introduction

On 13 November 1946, Vincent Schaefer scattered dry ice pellets into the top of a supercooled stratiform cloud over Pittsfield, Massachusetts. Within minutes the seeded portion of the cloud was transformed into a mass of ice crystals (*New York Times* 1946; Lampbright 1970). But more than the cloud was transformed! Physicists, chemists, and engineers were attracted by the prospect of benefits to be realized from cloud seeding and by the romance of in situ cloud and aerosol research. Cloud physics quickly emerged as a recognized specialty in meteorology and became much more quantitative and laboratory oriented. Laboratories devoted to the study of clouds sprang up in several nations. Clouds themselves became subjects of experimentation. More than ever before, meteorology became a multidisciplinary science.

Over the same period, with the gradual improvement in numerical models of atmospheric processes, incorporation of cloud effects became critical to their performance. No longer could meteorologists accept the view that clouds were merely embroidery on the face of the winds—fully determined by synoptic-scale air motions and thermodynamics and giving rise to precipitation as a by-product. Clouds are now recognized to be major "players" in the continuous drama of weather and climate through their role in the global heat budget and their interactions with, and feedback into, larger-scale atmospheric processes.

The demonstration that supercooled clouds could be modified by seeding them with dry ice prompted a step function increase in cloud-related research. Although translating this vision into reality proved to be more difficult than first imagined, there can be no doubt that a large fraction of what is known today about the physics of clouds can be traced to research initially motivated by the prospects and promises of weather modification. Before the days of cloud seeding, most cloud studies were driven by a need for better meteorological support for commerce, especially transportation—first, ships at sea, and later, aircraft. Examples include studies of fog and fog dissipation, low ceilings and low visibility, aircraft icing, and storm turbulence. With the buildup to, and the conduct of, World War II, needs of the military establishments prompted a large increase in research in studies of these weather-related problems and added several more to the list, including contrail formation and suppression, vehicle erosion, microwave propagation and scattering, spherics, and lightning.

Weather, with its clouds and storms, is arguably the most important environmental factor faced by humankind. The practical necessity of coping with its vagaries and an intellectual curiosity for understanding and forecasting weather events has always provided strong motivation for conducting cloud studies. Since about 1970, the realization emerged of a need for incorporating cloud effects into models for studying air and water quality, mesoscale weather, and global climate. Most recently these provided a major motivation in cloud physics, although research prompted by intellectual curiosity, support of commerce and industry, and the search for useful weather modification techniques abound.

The roots of cloud physics extend well back into the nineteenth century. As the science of clouds developed, two subspecialties emerged—cloud dynamics and cloud microphysics (Douglass 1934). This paper chronicles some of the most important steps in the development of cloud microphysics. Efforts to understand the development of precipitation, rain, and snow has been a central theme in cloud microphysics and forms the focus of this paper.[1] An account of the development of ideas about the meteorological conditions that lead to

---

[1] The history of weather modification, cloud electrification, and hail formation are not included in this review. The history of weather modification has already been treated in several places (e.g., Havens 1952; Lampbright 1970; Byers 1974; Elliott 1974).

clouds and to theories of rain formation prior to about 1900 were given by Middleton (1965). This paper begins where Middleton ends.[2]

One can identify five overlapping themes in the development of theories of rain formation: 1) early acceptance of condensation, perhaps augmented by collision and coalescence, as a process capable of producing rain; 2) concerns about cloud stability against rain formation and the need for some trigger event; 3) widespread adoption of the view that all significant rains come from melting snowflakes; 4) establishment of a scientific basis for the collision and coalescence (all liquid) rain mechanism; and 5) recognition of interactions between the ice crystal and coalescence mechanisms. We will examine each of these themes along with related issues.

Early views about rain formation

By the late nineteenth century, the dominant role of expansional cooling in the formation of clouds was well established. So too was the role of hygroscopic particles as cloud nuclei. It was thought that most cloud nuclei were either sea salt particles or sulfuric acid droplets formed through oxidation of SO_2 released by combustion. The condensation process was fairly well understood, and it was recognized that the relative humidity inside of clouds was only slightly above 100%. It was widely accepted that rain often developed simply by continuation of the cloud-forming process of condensation or from a combination of condensation plus collision and coalescence between cloud drops. But incomplete understanding of the physics and inadequate data on cloud characteristics precluded quantification of this view.

In 1876, in a paper read before the Manchester Literary and Philosophical Society, Osborne Reynolds[3] (1879, 55) dismissed condensation as a process capable of producing large drops because "there is no way in which the heat developed by condensation can be got rid of." Radiation was the only heat release mechanism considered by Reynolds. This, he correctly concluded, was too slow to be effective. Appar-

---

[2]Süring (1940) gives a historical review of cloud physics up to about 1940. Mason (1971) and Pruppacher and Klett (1978) give considerable amounts of historical information. Findeisen (1938) and Braham and Squires (1974) have given status reports on the subject.

[3]Osborne Reynolds (1842–1912) subsequently gained fame in the scientific world through his study of hydrodynamics, condensation, and heat transfer. The terms "Reynolds stress" and "Reynolds number" are named after him.

ently, he had not yet recognized that the molecular diffusion process responsible for moving vapor to the drop surface would also remove heat from it. But we must recall that this paper was given during the time period in which kinetic theory was being developed (Loeb 1934, p. 5). Reynolds continued (p. 57): "It appears clear, therefore, that the only way in which a falling drop can grow is by the aggregation to itself of the particles of moisture in the air; and the only way in which it can encounter these is by its downward motion through the air." A calculation of the total cloud water contained in the conical volume swept out by a drop growing by collision and coalescence led him to conclude that, "if a particle, after commencing to descend, aggregated to itself all the water suspended in the volume of air through which it swept, . . . there is no difficulty in explaining the size of the drops. The difficulty is rather the other way, in explaining why the drops are not sometimes larger than they are." This latter problem he circumvented by suggesting that 1) only part of the water in the swept volume would collide with the falling drop and 2) when drops reach some maximum size, surface tension would be unable to hold them together against the turbulence of the air. Subsequent research would show the merit of both suggestions.

Reynolds then observed (p. 59) that "a cloud does not always rain; and hence it would seem that in their normal condition the particles of a cloud are all of the same size and have no internal motion, and that the variation in size is due to some irregularity or disturbance in the cloud." In this belief Reynolds was many years before his time. His view that cloud drops are normally all the same size, and without relative motions, foretold the work of Schmauss and Wigand (1929), and others, who believed that clouds have an inherent "stability," which tends to prevent rain development. Fifty years later the idea that some "disturbance" or "irregularity" was required to trigger the precipitation process was expanded by Bergeron (1935) and Findeisen (1938), who argued that ice particles provided just such a trigger in supercooled clouds.

New data and new tools for an emerging science

The last quarter of the nineteenth century and the first quarter of the next saw a rapid increase in efforts to determine cloud- and rain-drop characteristics. Kohler (1921, 1929) captured fog drops on fine greased wires and threads for sizing under a microscope. Another early technique was to measure the angular diameter of corona rings produced by drops when the sun, or moon, was viewed through them. This method gives the predominant, or most frequent, drop size. Kohler (1925a) reported results from several thousands such measurements, showing that the most frequent drop diameter was about 17 μm, larger in fogs and low stratus and smaller in altocumulus and stratocumulus clouds. Coated slides for collecting fog and cloud particles for examination came into use somewhat later; Hagemann (1936) and Deim (1942, 1948) were among the first to report their use.

Usable data on raindrop sizes and concentrations predate those of cloud drops, both because the larger sizes of raindrops made them easier to measure and because good data could be obtained from ground level. Weisner (1895) developed the filter paper method for drop size measurements for a study of the effects of the mechanical impact of intense tropical rains on plant foliage. Lenard (1904) appears to have been the first to sprinkle filter papers with a water-soluble dye to make the drop impact sites more visible. Bentley (1904) came forward with the technique for capturing raindrops in a layer of fine flour. Blanchard (1949) proposed the use of coated screens, which records drop sizes by the holes created when drops pass through the screen. Jones and Dean (1953) developed a ground camera for photographing raindrops. Electro-optical probes for measuring drops were first proposed by Knollenberg in 1970.

Some of the early measurements suggested that raindrop masses tended to group into mass ratios of 1:2:4:8:, etc., thereby suggesting coalescence[4] as the primary growth mode (Defant 1905; Kohler 1925b; Neiderdorfer 1932). Later, however, with improved measurement techniques, the idea of mass grouping of raindrops fell into disfavor.

Among the earliest measurements of drop fall speeds were those made by Lenard (1904) using a vertical wind tunnel. Schmidt (1909) used a device consisting of a pair of rotating plates mounted one above the other. A slit in the top plate allowed drops to pass through. Their

---

[4]Coalescence actually involves two physical processes: 1) the collision between two drops, followed by 2) the permanent union, or coalescence, of these two drops. Meteorologists use the terms "coalescence rain" or "warm rain" to refer to the combined effects of these two processes. Mordy and Eber (1954) may have been the first to use the term "warm rain," though one cannot be sure since recognition of the coalescence process was literally exploding among cloud physicists about this time.

position of arrival on the lower plate, relative to the slit, gave data from which fall speeds, as a function of drop size, could be estimated. More accurate measurements were made by Laws (1941) and Gunn and Kinzer (1949). Stokes' law, the often-used equation for the fall speed of small drops, was first derived in a paper dealing with the friction that air places on a moving pendulum (Stokes 1850).

The development of manned balloons gave meteorologists their first opportunities to explore conditions in and around clouds away from ground effects inherent with mountain observations. According to Rotch (1900), John Jefferies in 1784 made the first manned balloon flight for scientific purposes. In the 1860s James Glaisher[5] was one of the early users of manned balloons to study the atmosphere, in a series of flights between 1862 and 1866. Delightful accounts of some of the early manned balloon flights can be found in Glaisher (1863, 1871) and Camille Flammarion's *L'atmosphere* (1873) as translated and abridged by Glaisher.

With this new vehicle, in just a few years meteorologists learned more about microphysics of clouds than they had gleaned from ground-based observations in the previous century. For the first time they could explore deep layers of supercooled cloud drops.[6] The fact that moderately pure water in small quantities, can be cooled below $0°C$ before freezing has been known at least since the early eighteenth century (Fahrenheit 1724). Supercooling of clouds has been recognized at least since 1783, according to Middleton (1965, 51, 206), who attributes the observation to Horace-Benedict de Saussure (1783). Other early observations of supercooling of water include Mousson (1858), Dufour (1861), Glaisher (1863), and Berson (1900). Theoretical explanations for supercooling, showing that nucleation of drops from vapor was thermodynamically preferred over the direct formation of ice crystals, at temperatures commonly found in natural clouds, came

---

[5]James Glaisher (1809–1903) was an English meteorologist credited with being one of the prime movers in the establishment of the Royal Meteorological Society, serving as its first secretary and later as its president. In 1862 he undertook a series of balloon flights with the aeronaut Henry Coxwell. For a bibliography see Marriott (1930).

[6]Most of the German writers of the 1900–1940 period used the term "undercooled" to denote the condition of liquid water cooled below $0°C$. Most modern writers use the term "supercooled" to denote this condition. Hann (1906) and Weickmann (1947) used the term supercooled.

much later (Becker and Doring 1935; Volmer 1939; Krastanow 1940, 1941).

Airplanes became available as platforms for cloud studies about 1910. The first meteorological flight in Germany was in the fall of 1912. Subsequently, meteorological flights were conducted in England, Norway, and Sweden. During the early 1940s, Weickmann made over 100 meteorological flights, some to heights in excess of 10 000 m, in an open-cockpit Henschel reconnaissance plane. Data from these flights formed the basis for Weickmann's 1947 monograph. Georgii (1930) gives an account of the early use of airplanes for meteorology and of some of the instruments employed. Some of the techniques used to acquire cloud physics data from airplanes in the 1940s are described in Findeisen and Schulz (1944).[7] One of the early uses of airplane data on cloud-top heights and temperatures was to evaluate the Bergeron–Findeisen precipitation mechanism.

Coalescence accepted—For want of anything better

Well into the twentieth century, for want of a better explanation, many meteorologists accepted the view that rain was the end product of condensation or combined condensation and coalescence. Obvious difficulties, such as the time and cloud depth required for drops to grow large enough to reach the ground, or why only a small fraction of the cloud drops grow into raindrops, either were ignored or surmounted by rather novel ideas. Consider a few examples.

In his book *Physical and Dynamical Meteorology* the well-known English meteorologist, David Brunt (1928, 58) had this to say about rain formation:

> If an ascending mass of air is cooled below its dew-point, the water-vapour will be partly condensed into water-drops . . . which will fall as rain if the process of condensation is continued for sufficient time. Practically all rain is formed in this manner.

One of the most widely used texts in meteorology during the period 1910–1940 was Meteorology by Willis Isbister Milham,[8] professor at

---

[7]Reasonably accurate measurements of clear-air temperature and pressure (height) were available almost from the first. However, reliable instruments for making good measurements of cloud particles from rapidly moving airplanes became available only during the past 30–40 years. Temperature measurement inside liquid water clouds still presents difficult technical problems.

[8]Prof. Willis I. Milham (1874–1957) obtained his doctorate degree from the University of

Williams College, Williamstown, Massachusetts. In his 1931 edition, we read the following:

> Raindrops are formed from the cloud particles. These are not of the same size, and the larger ones fall faster than the smaller ones, or, if they are being carried up by ascending currents, it will be the larger particles which are carried up less rapidly. Due to these motions many collisions must occur, and whenever two cloud particles collide they coalesce into a diminutive raindrop. A raindrop thus increases in size, due to collision and condensation, until it reaches the base of the cloud and begins its final fall to the surface of the earth.

When it was recognized that the amount of cloud water was insufficient to allow all cloud drops to grow to raindrop sizes, the search was on for ways to restrict the condensation to a relatively few cloud drops, thereby allowing them to grow much larger. One idea made use of selective filtering of the nuclei. In Gregg[9] (1930, 166) we read the following:

> The small droplets of water in a cloud are formed by condensation, on dust or other nuclei. . . . These droplets . . . fall with reference to the air, and in so doing they filter the air of its nuclei with the result that, as condensation proceeds, the water vapor has less and less nuclei on which to form and are therefore larger individually.

This view was reported as late as 1940 by Humphreys (1940, 277).

Not everyone was happy with the idea that rain could be formed through drop collision and coalescence, even though no one offered an alternative. For example, in his *Manual of Meteorology*, Vol. 3, Sir Napier Shaw (1930, 339) wrote the following:

Strassburg in 1901. He was a charter member of AMS and served as its president from 1924 to 1925. According to one of Milham's necrologies (AMS 1957), his meteorology textbook was "adopted by every American college or university of the time which offered such a course." Thus, it must have had a strong influence in shaping the views about rain formation during the period 1910–1940. An interesting feature of this text is a list of over 300 publications that he considered to be desirable for every meteorologist's library.

[9]Willis Ray Gregg (1880–1938) was president of the AMS in 1938. He was chief of the U.S. Weather Bureau from 1934 to 1938. His textbook titled *Aeronautical Meteorology* was widely considered to be the authoritative reference on the subject prior to, and during, World War II.

There are some reasons that make it seem doubtful whether water-drops will coalesce except in special circumstances. We know of no satisfactory evidence for or against the automatic coalescence of water-drops in the atmosphere. The whole question of coalescence therefore is very imperfectly understood and will not be cleared up until the dust aspect and the electrical aspects are investigated. In the meantime a theory which assumes the coalescence of cloud-drops or water-drops has an opportunity of adding to the stability of our knowledge by underpinning its own foundation.

Clouds were viewed as colloidlike

During the 1920s and 1930s, it was widely held among meteorologists that the initiation of precipitation required some special condition, or trigger, capable of acting quickly and enabling some clouds to develop rain. For example, Hann (1906, 224) states that, "Not infrequently one sees heavy, dark, clouds lingering or remaining stationary in the sky without raining to earth. Then suddenly a release appears immanent and rain formation strikes the entire cloud. . . . What force is it that has prevented the coalescence, or enlargement, of cloud particles, which is suddenly and efficiently released?" Hann continues: "This even reminds us of experiments with colloidal solutions which remain unaltered for years and suddenly undergo a catastrophic change."[10] Hann then suggests electrical effects as a possible "coagulation force," perhaps because of the frequent occurrence of lightning with heavy rain downpours. It would be another 25 years before Bergeron would suggest that ice particles in supercooled clouds would provide such a trigger. But that is getting ahead of our story.

The argument that clouds are inherently stable, requiring a trigger for release of precipitation, was further developed by Schmauss (1919, 1920) and others. A common thread among these authors was the idea that this stability resulted from a mutual repulsion of drops carrying like electrical charge. In 1929, Schmauss and Wigand published a small book entitled *Die Atmosphare als Kolloid,* which attempts to show that the atmosphere, with its ions, small nuclei particles, and cloud drops, should be considered a colloid: inherently stable until disturbed by some external "force."

---

[10]Reference to colloids came about because of their remarkable property of remaining dispersed, unagglomerated, for long periods of time. According to MacDougal (1948, 663), the first detailed study of colloids was by Graham (1861). The word "colloid" comes from Greek, meaning gluelike.

By the 1940s with the expansion of research in cloud physics, it was realized that thermal motions, which are important in maintaining the separation and suspension of particles in colloids, are totally negligible in the behavior of cloud particles and that the analogy between clouds and colloids was not well founded.

Wegener—The father of cloud physics

The foundation of much of cloud physics is the fact that ice particles in the presence of supercooled cloud drops constitute a nonequilibrium physical state that results in growth of the ice and evaporation of the drops, as nature tries to restore thermodynamic equilibrium. The first to clearly enunciate this appears to have been Alfred Lothar Wegener, a German meteorologist and classical physicist. To understand the history of cloud physics, we need to know something about this man.

Wegener was born in Berlin on 1 November 1880. He studied astronomy, geology, and meteorology at the Universities of Heidelberg, Innsbruck, and Berlin. From the latter institution he obtained his doctorate, magna cum laude, in 1905. With his older brother Kurt, also a meteorologist, Wegener engaged in several free balloon flights, one of which set a world's endurance record of 52 hours. Wegener participated in four expeditions to the Greenland icecap; an unfortunate incident on the fourth expedition led to his death in November 1930. He was buried on the icecap[11] (Benndorf 1931; Schwarzbach 1986).

During his first expedition to Greenland, Wegener had many opportunities to observe and explore the growth of hoar frost. This led to his recognition of the meteorological importance of the three-phase system (ice, vapor, and supercooled water). He discusses this topic at some length in his 1910 *Meteorologische Zeitschrift* paper and in his 1911 textbook entitled *Thermodynamik der Atmosphäre*.

Wegener's book on atmospheric thermodynamics is one of the landmark publications in early meteorology. For cloud physicists, the chap-

---

[11]In 1913 Wegener married Else Köppen, daughter of the meteorologist and climatologist, Vladimir Köppen. During the second expedition to Greenland, Else lived in the home of Vilhelm Bjerkness in Oslo. Thus, we know that Wegener and his work was well known among meteorologists at Oslo. Included in this group was Tor Bergeron who, 22 years later, cited Wegener's book on atmospheric thermodynamics as the foundation for his 1933 paper that has become the cornerstone for much of cloud physics. Wegener also wears the mantle of "father of continental drift."

ters on "General Thermodynamics of Real Gases" and "Physics of Clouds" attract attention. Following a phase diagram for water, we find on p. 81 the following statement:

> When the three phases are in contact, the vapor tension will adjust itself to a value in between the saturation values over ice and over water. The effect then must be, that condensation will take place continually on the ice, while at the same time liquid water evaporates, and the process must go on until the liquid phase is entirely consumed.

It appears, however, that the application of this statement to the growth of ice crystals within supercooled clouds was not fully evident to Wegener at this time, perhaps as a result of his training in classical physics, where the emphasis had been on equilibrium thermodynamics. Several more years elapsed before the importance of this process in the development of precipitation was clearly enunciated.

The Bergeron–Findeisen mechanism

One of the most often referenced papers in cloud physics is that presented by Tor Bergeron before the Lisbon meeting of the International Union of Geodesy and Geophysics on 19 September 1933 and published in 1935.[12] Bergeron wrote that it "is clearly proven that cloud masses may be regarded as colloidal suspensions of water in air ... which will remain in this state of colloidal equilibrium, until a sort of 'coagulation' is released within them." He then lists four conditions generally recognized as favoring colloidal stability, namely, uniformity of electrical charge, size, temperature, and motion of the individual cloud elements. After examining each of these conditions for evidence that their nonfulfillment might lead to "colloidal instability," Bergeron then developed the argument that relatively few ice particles in a supercooled cloud could individually grow large enough to provide a release from colloidal stability. As a source for these ice particles, Bergeron suggested that relatively few nuclei would have the special prop-

---

[12]Bergeron was born of Swedish parents in England in 1891. After studying in Stockholm and Leipzig, in 1928 he received his doctorate from the University at Oslo, where he was a student of Bjerknes. He is noted for his studies of air masses and fronts, as well as for cloud physics. The AMS elected him an Honorary Member in 1958 and Fellow in 1968. In 1949 he received the Symons Gold Medal from the Royal Meteorological Society and in 1966 was awarded the WMO Prize. He died in 1977 in Bergen.

erties needed to act as "sublimation" nuclei and that crystallization of "already formed" drops was a more likely source of ice crystals (i.e., drop freezing). After citing Wegener (1911) and his own earlier work (Bergeron 1928) for what he called previous "indications" and "hints" of this theory, he proceeded: "I thus presume that almost every real raindrop (d > 0.5 mm) and all snowflakes originated around an ice crystal in the above mentioned way."

In 1971, Duncan Blanchard queried Bergeron concerning the background of his 1935 paper. Bergeron responded that in February 1922, while visiting a resort near Oslo, he walked along a narrow road in a fir forest that was often immersed in a stratus layer. He noted that when the temperature was above 0°C the stratus invaded the trees and the roadway, but when the temperature was $-10°$C the forest and roadway were essentially free of cloud drops. He reasoned that, in the latter case, hoar frost on the trees was growing by vapor deposition and consuming most of the supercooled drops (see Bergeron 1972; Blanchard 1978).

Bergeron's hypothesis that rain, other than drizzle, has its origin in snowflakes is the cornerstone for much of cloud physics. However, its acceptance and rapid spread through the cloud physics community was due, in large measure, to Theodor Robert Walter Findeisen, whose contributions we consider next.

Findeisen[13] was a German experimental meteorologist trained in laboratory experimentation and in the use of airplanes for the study of clouds. In 1938 he published a very complete summary of the physics of clouds as the subject was then known. He developed Bergeron's hypothesis into a complete conceptual model of precipitation formation (Findeisen 1938).[14] For this reason it is often called the Bergeron–Findeisen precipitation mechanism. Findeisen argued for two kinds of cloud nuclei: condensation and sublimation. He felt that coalescence rain was limited to drizzle and that all important rain originated

---

[13]Findeisen was born in Hamburg, 23 July 1909. He studied in Karlsruhe and Hamburg under A. Wegener, A. Wigand, A. Peppler, and W. Peppler, receiving his doctorate in 1931. His dissertation dealt with the sizes, concentrations, and coagulation of drops in fogs. He served as head of the Aircraft Weather Unit in Munich, and later as head of the Cloud Research Unit of the German Weather Service in Prague. He died in Prague about 9 May 1945.

[14]This 1938 paper by Findeisen is recommended reading for all graduate students in cloud physics. Unfortunately, many students will be unable to read it in its original German, but it has been was translated into English (Stickley 1939; Sharenow 1939).

through the melting of snowflakes. Because of the breadth of his study of clouds and because of his careful application of classical physics to cloud processes, Findeisen fully deserves to be called "the first cloud physicist."

Circumstantial support for the Bergeron–Findeisen mechanism

Beginning about 1930, aerological observations from airplanes became available in Europe and the United States. These new data were examined by Peppler (1936), Mann (1940), Stickley (1940), and Schwerdtfeger (1948) to establish cloud-top temperatures during periods of rain and snow. These authors concluded that, in general, the observations verified the Bergeron–Findeisen hypothesis. The small number of cases of rain from clouds totally warmer than freezing were generally thought to be observational errors or to correspond to drizzle. From our present-day perspective, these studies were seriously flawed in assuming that *any* temperature less than 0°C was sufficient to prove the ice crystal mechanism.

None of these studies provided more than circumstantial support for Bergeron–Findeisen mechanism. But both the argument and its proponents were persuasive, and it gradually came to dominate the thinking of meteorologists as the basis for all significant amounts of rain. Attention had shifted from collision and coalescence of water drops to the growth of ice crystals as the mechanism of rain formation.

Before proceeding with tracing the development of theories of rain, it should be noted that recognition of a possible role of ice particles in rain formation predates the work of Bergeron. McAdie (1895), Bentley (1904), and Guilbert (1921, 1922) reported an apparent association of ice clouds or clouds that visually appeared to contain ice with rain development. Particularly interesting are the views of Wilson Bentley, a Vermont farmer who studied raindrops and snowflakes over a period of 40 years. He wrote that, "We seem to have the best of grounds for assuming that rain often has a dual origin; that there are, in general, two distinct processes concerned in the formation and precipitation of rain, or snow from which rain ultimately results. The processes might not inaptly be termed the fluid and the solid (or crystalline, or snow) processes" (Bentley 1904, 455). Bentley also concluded that the snow process was responsible for the greater part of the total rainfall.

Not everyone convinced

The Bergeron–Findeisen mechanism was widely promoted, but not universally accepted, as the process of formation of *all* rain other than drizzle. In the discussion following Bergeron's presentation, several persons reported having seen rain from clouds that could not have risen above the freezing level. Bergeron accepted these observations but insisted that only light rain or drizzle could fall from warm clouds. Others remained skeptical. Much earlier, Osborne Reynolds had noted that if drops of different temperatures were brought close to one another, the cooler one would grow by the process of vapor diffusion, while the warmer one would evaporate, owing to its warmer temperature and higher surface vapor pressure. Petterssen (1940, 46), noted that this effect could rival the Bergeron–Findeisen mechanism in effectiveness at temperatures slightly below $0°C$. But neither Reynolds nor Petterssen put forth a plausible process whereby drops of different temperature could be brought near one another. In responding to his discussants after the 1933 paper, Bergeron used this possibility to explain the observations of rain from nonfreezing clouds.

In his presidential address before the Royal Meteorological Society, Simpson (1941) reviewed various aspects of rain formation and came to the conclusion (p. 128) that "there can be little doubt that Bergeron has correctly described the formation of rain in certain circumstances. . . . But is he correct in his presumption that almost every real rain drop originated around an ice crystal?" Simpson also stated that (p. 130) "So far no theoretical basis has been found for the statement that water clouds cannot produce large rain drops by coalescence of the water particles." Other early reservations about the Bergeron–Findeisen mechanism as the only process of formation of substantial rains are given by Holzman (1936), Byers (1937, 183 et seq.), Houghton (1938), and Heywood (1940).

Existence of coalescence rain established

World War II brought on a big increase in the number of airplane flights over almost the entire globe. [See chapter 16 by Cartwright and Sprinkle, this volume.] Undoubtedly pilots and weather officers on these flights became aware of rain from clouds that were entirely below the freezing level. After the war, warm-rain observations began to appear in the meteorological literature; encouraged, perhaps, by a request by A. C. Best (1948) for the submission of such observations. Several of the early reports of warm rain referred to sightings made in tropical latitudes during the war period.[15] In his Ph.D. research at the University of Chicago, Battan (1953) used radar measurements from the Thunderstorm Project to show that rain was often initiated by coalescence in midlatitude summer convective clouds, *even though* the cloud tops were above the freezing level. To establish the fraction of clouds containing rain as a function of cloud-top height, studies using aircraft and radar were carried out in the Caribbean, central United States, and Arizona (Braham et al. 1951; Byers 1955; Byers and Hull 1955; Braham et al. 1957). These studies confirmed and extended the findings of Battan. At long last, the reality of warm rain had been established beyond question.

Why did it take so long?

Once it was established that rain, even heavy rain, could form through coalescence in nonfreezing clouds, the obvious question was, Why did it take so long for the coalescence mechanism to be put on a firm scientific footing? A number of contributing factors can be recognized. These are as follows.

1) The Bergeron–Findeisen mechanism was appealing because it was based on correct physics, while its only serious competitor, coalescence, could not be firmly established for lack of good data on cloud drop sizes, collision–coalescence efficiencies, and drop terminal velocities. Petterssen's argument for vapor diffusion from warm to cold drops was also based on solid physical principles, but no one had suggested how drops of substantially different temperatures could come close together.

2) At this time most cloud observations were from midlatitudes where precipitating clouds usually extend above the 0°C level.

---

[15]Among these reports are the following: Wexler (1945), Kotch (1947), Langmuir (1948), Best (1948), Haggard (1948), Leopold and Halstead (1948), Craddock, (1949), Hunt (1949), Virgo (1950), Davies (1950), Foster (1950), Jones (1951), Smith (1951), and Mordy and Eber (1954). The report by Wexler, then a major in the army air force, resulted from a flight into a hurricane to observe turbulence effects on aircraft. He and other members of the flight crew were awarded the U.S. Army Air Force Air Medal for their observations on this flight.

This provided superficial evidence for the ice mechanism and made it difficult to recognize coalescence; moreover, the criterion used for evidence of ice particle involvement was too simplistic.

3) It was not appreciated that only about one in one million cloud drops would have to be somewhat larger in order to initiate coalescence rain.

4) The overarching demands of World War II tended to focus research toward military objectives and disrupted normal channels of publication and exchange of scientific ideas. Although a considerable amount of cloud-related research was conducted during World War II, publication of the results was delayed until after the war.[16]

Several other contributing factors can be recognized: Schmauss and Wigand (1929), and others, had argued that clouds are inherently stable and required some trigger to initiate precipitation. Houghton (1938) had shown that continued condensation should narrow drop spectra and make clouds even more stable. The worldwide eminence of the Bergen School, where Bergeron was located, tended to give prestige to the Bergeron–Findeisen hypothesis. In addition, during this time period large numbers of meteorologists were trained in the United States, but many of their instructors came from Europe and Scandinavia. Most of the textbooks of the period gave prominence to the Bergeron–Findeisen mechanism, suggesting that it explained most, if not all, significant rain. It also seems quite likely that the optimism for early and substantial benefits from glaciogenic seeding delayed recognition of the widespread occurrence of a mechanism that could produce ice particles in clouds at temperatures only slightly below 0°C.

Recognition of the existence of both the ice crystal and warm-rain mechanisms did not constitute complete scientific understanding. A great deal of research has been required to bring us to our current (1995) level of understanding of precipitation mechanisms. Some of the highlights of this research is covered in the following sections. Selecting these highlights is more difficult than tracing the development of

---

[16]Some wartime reports have become available as unpublished reports; probably some very good work is still unrecognized, buried in archives or lost forever.

the theories of rain formation because the discovery process is still in progress.

Scientific underpinning of coalescence growth

Central to the development of a scientific understanding of clouds and precipitation was the need for accurate calculations of particle growth, both by vapor diffusion and by collision and coalescence. For most purposes in meteorology the growth (evaporation) of particles by diffusion of water molecules to (or from) its surface is assumed to be in steady state and governed by Fick's law of diffusion. The mathematical description of this process uses an analogy between the vapor density around the particle and the electrostatic potential around a charged body of the same size and shape. According to McDonald (1963), William Thompson (later Lord Kelvin) in 1814 may have been the first to note this analogy. It was subsequently used by Maxwell in 1875 in discussing vapor exchange at the surface of a wet-bulb thermometer. It then was used by Stefan (1881) and Jeffreys (1918) before Houghton (1933) used it in a study of drop evaporation.[17] The fact that the boundary conditions are not strictly identical in the two problems and that accurate calculations require the use of gas kinetics rather than gas diffusion to describe the vapor flux through a thin layer of air close to the drop surface was recognized somewhat later by Langmuir (1944) and Rooth (1957).

Kohler (1926) appears to have been the first to combine the effects of drop size and its soluble nucleus with the drop growth equation and the thermodynamic equation to describe the initial phase of conden-

---

[17]Houghton was born in 1905 and died in 1987. He was a member of the MIT Department of Meteorology from its founding in 1941 until his retirement in 1970. He was president of AMS (1946–1947), elected Honorary Member in 1975, and received its Charles Franklin Brooks Award in 1958 and its Cleveland Abbe Award in 1982.

From a historical perspective it is interesting that in using his equation for the evaporation of drops, Houghton (1933) found substantial differences between predicted evaporation rates and those he measured in a small closed chamber of known temperature and relative humidity. Houghton recognized that this could be due to the evaporating drops being colder than the air. He had no way to measure the drop temperatures and he felt that exact calculation of the equilibrium drop temperature was impractical. However, when he assumed that the drop temperatures were the same as those measured by a small wet-bulb thermometer, placed within the chamber, the rates agreed within experimental error. Apparently it did not occur to Houghton to apply the equation that he developed for vapor flux *to* the drops to the problem of heat flux *from* the drops.

sation of cloud drops. Numerical integration of these equations to describe the growth of drops in the base of a cloud was first carried out by Krauss and Smith (1949) and Howell (1949). These were followed by Squires (1952), Mordy (1959), and Neiburger and Chien (1960). Mason (1953) simplified the calculations by making use of the Clausius–Clapeyron equation, and McDonald (1953a) called attention to an erroneous use of Raoult's law of solution effects that had entered some of the earlier calculations.

In parallel, but apparently independent efforts, Frossling (1938), Findeisen (1939), and Langmuir (1944)[18] developed semiempirical equations virtually identical to that of Houghton for the growth and evaporation of water drops. Findeisen used his equation to calculate the time and fall distance required for drops to evaporate under typical cloud-base conditions. On the basis of these calculations, he suggested that the lower size limit for raindrops be taken as a 200-μm diameter, a value commonly used today in cloud physics. Langmuir developed his growth equation in connection with aircraft icing studies on Mt. Washington, New Hampshire (Langmuir 1944). Apparently Houghton and Langmuir first became acquainted with each other's work in April 1944 when Langmuir visited the Massachusetts Institute of Technology (MIT) to discuss the icing studies under way at Mt. Washington (Langmuir 1944, 216–219).

An adequate theory for the growth of drops by collision and coalescence rests upon having adequate data on three factors, namely, drop fall speeds, the probability that a drop will collide with other drops, and the likelihood that a collision between drops will result in permanent union. Early measurements of drop fall speed were considered in the section entitled "New Data and New Tools for an Emerging Science." Experiments by Rayleigh (1879b) had suggested that coalescence was limited to drops of opposite electrical charge. As a result of these and other similar experiments, through the last quarter of the nineteenth century there was considerable skepticism as to whether colliding drops would coalescence. In his doctoral dissertation, Findeisen (1932) pointed out that the prevalent idea that cloud drops were all of a similar size was not based on observations and had led to erroneous conclusions about the probability of coalescence. Although

---

[18]In applying his drop growth equation, it appears that Langmuir overlooked the solution effect that is of crucial importance in the initial growth of cloud drops.

he confessed not to understand how coalescence came about, he had no doubt that it occurred. More recently a number of experiments have shown that colliding drops do indeed coalesce under a wide range of, but not all, conditions.

Langmuir and Blodgett (1945) appear to have been first to calculate drop collision efficiencies starting from first principles of hydrodynamics as part of their studies of screening smokes and aircraft icing for the military. Because of their complexity, the governing equations had to be solved on an analog computer. After the war, when these results became available in the open literature, they were used by several scientists in making calculations of drop growth by collection under several different meteorological conditions. These include Langmuir (1948), Bowen (1950), Houghton (1950), and Ludlam (1951). These studies demonstrated a theoretical basis for the formation of rain, even heavy rain, without the involvement of ice, *assuming* that a source could be found for the initial, or starter, drops.

Understanding the origin for a spectrum of cloud drop sizes, including the "starter" drops for coalescence growth, came from the merging of two different lines of inquiry: 1) discovery of regional differences in cloud drop spectra and clarification of the role of large soluble particles as cloud nuclei, and 2) recognition of the role of random collisions in drop growth.

Early investigators of cloud drops used different equipment and were generally limited to clouds of a single region. This obscured one of the most fundamental properties of clouds, namely, regional differences in cloud-drop concentrations and size spectra. The first systematic measurements using the same equipment to study similar clouds in different regions were conducted by two groups: the Commonwealth Scientific and Industrial Research Organisation (CSIRO) Cloud Physics Group of Australia,[19] who made measurements in Hawaii and New South Wales, and the University of Chicago Cloud Physics Group, who made measurements in the Caribbean, the central United States, and Arizona. The resulting data clearly illustrated major differences in the cloud-drop spectra between cumuli of marine and continental regions. Battan and Reitan (1955) and Squires (1956) published the first ac-

---

[19]CSIRO, in Sydney, Australia, was one of the first to organize a cloud physics effort after Langmuir and Schaefer had demonstrated the possibilities of weather modification through glaciogenic seeding.

counts of these findings. Through these, and numerous other measurements that quickly followed, cloud physicists came to recognize that drop concentrations typically become larger, and drop sizes smaller, as one goes from maritime regions to moist continental, to dry continental, and then to urban settings. It also was realized that these differences in cloud-drop populations would lead to differences in the subsequent development of precipitation. The search was on for an explanation.

Since Kraus and Smith (1949) and Howell (1949) had already shown the relationship between cloud condensation nuclei (CCN) and cloud drop spectra, it was natural to suspect that regional differences in cloud drop spectra were caused by regional differences in CCN populations. But before this idea could be investigated, some means for measuring CCN concentrations at very low levels of supersaturations had to be found. Sean Twomey and Patrick Squires of CSIRO were the first to pursue this goal vigorously. At first, Twomey used a chemical diffusion chamber similar to that developed by Vollrath (1936) but he found it more convenient to use a thermal diffusion chamber similar to one developed by Langsdorf (1936). By combining measurements on cloud-drop concentrations *and* CCN spectra from the same region, Squires and Twomey (1960) were the first to explain satisfactorily the observed regional differences in drop spectra. Then in 1972, Fitzgerald, in his doctoral studies at the University of Chicago, conducted the first measurements in which all parameters governing drop growth were measured on several individual clouds. These studies clearly showed why it was no accident that the earliest observations of warm rain came from the Tropics.

The second important element in developing the scientific underpinning of coalescence rain was recognition of the importance of random collisions and coalescence between drops of similar sizes. Smoluchowsky (1918) and Müller (1928) gave the foundation for the classic theory of random coalescence of particles, but it remained for Schumann (1940) to apply the foundation to the growth of fog drops. Melzak and Hitschfeld (1953) and Telford (1955) applied it to the growth of cloud drops. These studies showed how random collisions between cloud drops, followed by coalescence, can broaden drop size spectra and initiate coalescence. Edward Berry, in his doctoral dissertation at the University of Nevada, further explored the consequences of the stochastic collection process (Berry 1967; Berry and Reinhardt 1974a–c).

Scientific underpinning of ice crystal mechanism

It seems likely that ice crystals and snow have attracted attention since the dawn of civilization. Interesting accounts of some of the early observations on snow and ice crystals can be found in Needham and Lu Gwei-Djen (1961), Mason (1966), and Hobbs (1974). But prior to about 1800, relatively little was contributed to the scientific underpinning of the role of ice particles in precipitation.

One of the early concerns was to account for the seemingly endless variety of crystal shapes. Scoresby (1820) and Fritsch (1853) were among the first to note that crystal shapes appeared to be related to temperature. However, Dobrowolski (1903)[20] is usually regarded as the first to make a serious effort to link crystal shape and temperature. But the evidence was blurred, because crystals collected at the ground show growth characteristics of all of the different temperatures through which they had fallen. Growth of ice crystals under controlled laboratory conditions was undertaken in the mid-1930s by Ukichiro Nakaya and his colleagues and graduate students at Hokkaido University. They found that the basic crystal habit (i.e., plate or column growth) was indeed controlled by temperature (Nakaya 1938, 1951, 1954; Nakaya and Sato 1935; Hanajima 1944, 1949).[21] These findings were verified and extended by a number of other scientists, including aufm Kampe et al. (1951) and Kobayashi (1957). In 1958 Hallett and Mason contrived a beautiful experiment that allowed growing ice crystals to be moved from one temperature to another, showing that crystal growth at the new temperature also changed in the manner found by Nakaya. Subsequently, other investigators showed that the supply of vapor molecules controlled the rate of growth and the degree of crystal compactness and secondary growth (Marshall and Langleben 1954; Kobayashi 1961). Magono and Lee (1966) provided the ice crystal classification system used today in cloud physics. Schaefer (1941) introduced the replica technique for studying ice crystals.

---

[20]Dobrowolski was a scientist on the *Belgica,* a 33-m-long, three-masted sailing ship (with 150-hp auxiliary engine), which overwintered in the Antarctic pack ice during 1888/1889. Interesting details are given by Arctowski (1901a,b).

[21]A major Japanese publication summarizing Nakaya's studies was in press in 1945 when the publishing house was destroyed by bombing. Fortunately a proof copy of the text and originals of almost 3000 photomicrographs survived. From these a new book was prepared in English and published by Harvard University Press long after the war (Nakaya 1954).

The consequences of ice particle formation in supercooled clouds came into focus as a result of the writings of Bergeron and Findeisen, but the nature of the nuclei responsible for ice initiation was unknown. Wegener and Findeisen thought that ice crystals were initiated from the vapor by "sublimation nuclei."[22] Wegener (1911) proposed that these nuclei may be quartz grains that he thought would have crystal properties favorable for ice initiation. Findeisen (1938) accepted the idea of sublimation nuclei but felt that quartz would be of limited significance. A series of laboratory and flight experiments led Weickmann (1942, 1947) and Wall (1943) to conclude that few if any true sublimation nuclei existed because ice appeared only at or near water saturation.

In an effort to identify the nuclei responsible for snow crystal formation, Kumai (1951), aufm Kampe et al. (1952), and Isono (1955) used the electron microscope to examine snowflakes for particles that could have served as their nuclei. And indeed, they found particles about 1 μm in diameter at the crystallographic centers of most (but not all) snowflakes, where one would expect the nuclei to be located. In addition, they found many smaller particles scattered over the crystal surface. Using electron beam diffraction, Isono (1955), Kumai (1957, 1961), and Kumai and Francis (1962) identified the majority of the central particles as soil particles, clay mineral grains being most common, though a minor fraction were identified as sea salt and other substances. (The smaller particles were thought to be nuclei of cloud drops collected by the snowflakes.) Snowflakes from several locations scattered around the world gave essentially similar results. These findings were subsequently confirmed by several other investigators using a variety of different techniques. However, Mossop (1963) criticized the conclusion that most ice nuclei are clay minerals on the basis that these particles were not found in every snowflake and that any one of the smaller particles found on a snowflake might have been the nucleus.

Bergeron (1935) thought that the requirements for sublimation nuclei would be so stringent as to make it unlikely that many such particles could exist and that most ice particles originated from freez-

---

[22]During the period 1910–1940, several authors used the term "sublimation nuclei" when referring to frozen drops that continued to grow by vapor deposition, rather than nuclei on which ice first formed.

ing of cloud drops. While examining natural snowflakes, Bentley (1924) and other early snowflake observers noted that the centers of many of them appeared to be frozen droplets. Subsequently, using improved techniques, several investigators verified this observation (Weickmann 1947; Auer 1971, 1972). Hallett (1964) showed that upon freezing a drop might become either a single crystal or a cluster of several, depending upon certain circumstances. This discovery paved the way for understanding the origin of the three-dimensional dendrites and bullet crystal rosettes as polycrystalline frozen drops. Although it may not have been obvious at the time, proponents of drop freezing as the source of ice in clouds were sidestepping the more fundamental question of the nature of nuclei responsible for the freezing.

The first comprehensive laboratory study of drop freezing appears to be that of J. Heverly (1948, 1949) at The Pennsylvania State College. This was followed by Keith Bigg (1953a–c) at Imperial College.[23] Using freshly distilled water, Bigg reached the conclusion that drop freezing temperatures were controlled mainly by a stochastic process. Subsequent experiments by Vali and Stansbury (1966) showed that the temperature at which a drop freezes was determined by several factors: the nature of the nuclei contained within the water, the rate of cooling, and the length of time the drop remained at any given degree of supercooling. All three of these investigations were carried out as doctoral research.

The temperature at which drops of pure water freeze without the help of nuclei, the so-called homogeneous freezing temperature, was determined experimentally to be about $-40°C$ by Schaefer (1946), Mason (1952), Pound et al. (1953), and Jacobi (1955). Theoretical calculations of this quantity have encountered numerous difficulties; for discussions see McDonald (1953b), Eadie (1971), and Pruppacher and Klett (1978).

Measurements of ice nuclei have centered around two different techniques: cooling saturated samples of air and filtering atmospheric aerosols, including nuclei, then subjecting the filter to cold saturated conditions. Findeisen (1942) and Findeisen and Schulz (1944) used large, slowly cooled, expansion chambers to observe the number of crystals formed as a function of temperature. Other early users of

---

[23]An extensive list of references to studies of water freezing, going back to Fahrenheit (1724), can be found in Bigg (1953a).

expansion chambers for counting ice nuclei include Cwilong (1947), Bigg (1956, 1957a), and Warner (1957). Aufm Kampe et al. (1951) used a small refrigerated room in which air samples were allowed to mix into a supercooled cloud for counting ice nuclei. Smith and Heffernan (1954) developed a temperature-controlled mixing chamber to count ice nuclei. More recently, a variety of other techniques have been used. Because different techniques emphasize different physical principles, it is not too surprising that the results have shown wide scatter. Fletcher (1962, 241) combined measurements from several sources to derive an average curve of ice nucleus concentration as a function of temperature. This came to be known as the Fletcher curve. Mossop (1985), Cooper (1986), and others found that the concentration of ice crystals in stratified clouds, or early in the lifetime of cold-based cumulus clouds, matched the Fletcher curve reasonably well. But during the 1960s, cloud physicists found that concentrations of ice particles in warm-based cumulus clouds often exceeded values expected from the Fletcher curve by several orders of magnitude. The events that led to this discovery, and its consequences on our understanding of precipitation growth, are covered in a later section.

The question of natural ice nuclei became even more complex when Vali (1968), and then Schnell (1974) in his doctoral research at the University of Wyoming, showed that residues of certain bacteria are capable of initiating ice nucleation at temperatures near 0°C. The nature of nuclei responsible for initiation of ice in clouds remains a subject of research. In 1995 many cloud physicists are of the view that drop freezing is the major source of ice particles in clouds; however, the issue is far from settled.[24]

A brief side trip in cloud physics occurred from about 1953 to 1963 when Bowen (1953) suggested that meteoritic dust particles might have a pronounced effect upon the occurrence of rain, perhaps acting as ice nuclei. Subsequently, however, it was shown that the rainfall patterns that Bowen had linked to meteor showers were more likely explained by a lunar contribution to atmospheric tides. For further details see Bowen (1953), Fletcher (1962), Adderley and Bowen (1962), and Bradley et al. (1962).

---

[24]Readers may consult Mason (1971), Hobbs (1974, 1990), Vali (1985), Rangno and Hobbs (1988), Beard (1992), and Young (1993) for other references and greater detail about ice nucleation in the atmosphere.

Accurate calculations of ice crystal growth require accurate data on the saturation vapor pressure of water over ice and over supercooled drops. Useful values of the saturated vapor pressure over water and ice were available in the late 1800s. Hann (1906), in his *Lehrbuch für Meteorologie,* gives a table of saturation vapor pressures over ice and over supercooled water as a function of temperature. As late as 1924, the best values of saturation vapor pressure over supercooled water were obtained from calculations using the Clausius–Clapeyron equation (Washburn 1924, 489). Houghton (1950) was the first to calculate the growth ice crystals using the electrostatic capacitance of shapes similar to those of simple ice crystals. McDonald (1963) clarified the use of the capacitance factor for such calculations.

Barkow (1908) may have been the first to recognize that the formation of graupel involved ice crystals colliding with cloud drops—a process called coagulation by the early German cloud physicists but now called riming. Schumann (1938) recognized that riming growth would increase the temperature of the riming particle and that such growth would be limited by a maximum particle surface temperature of $0°C$. This concept was further developed by Ludlam (1958) to recognize two different riming growth regimes: the dry growth regime, where the surface temperature remains below freezing, and the wet growth regime, where the surface temperature is equal to $0°C$. Conditions that give rise to a surface temperature of exactly $0°C$ is known as the Schumann–Ludlam limit. Alternating periods of wet and dry growth give rise to the banded appearance of hailstones.

It appears that Zaytsev (1950) was the first to publish measurements of the height distributions of drop size, concentration, and liquid water content in cumulus clouds. This was followed by Weickmann and aufm Kampe (1953). In a series of papers, beginning in 1955, Jack Warner critically reviewed available data on cumulus cloud microstructure and pointed the way for future research. For references to these papers see Warner (1955, 1973).

The microphysics of midlatitude stratified clouds has received less attention than that of convective clouds. Cunningham (1951, 1952) was the first to report on the growth and distribution of precipitation in a large winter storm based upon combined aircraft and radar data. This study was carried out in conjunction with his Ph.D. studies at MIT. This was followed by Browning and Harrold (1969). Detailed calculations of snow crystal growth in stratified clouds were carried out by

Wexler (1952). Pioneers in the use of radar to study cirrus-level "generating cells" and their role as a source of ice particles for growth in lower supercooled layers include Bowen (1951), Lhermitte (1952), and Douglas et al. (1957). In 1967, Braham and Spyers-Duran found that cirrus crystals can survive several thousand meters of fall in clear air before evaporating completely. The Stormy Weather Group at McGill University may have been the first to launch an all-out study of clouds using radar. [See chapter 4 by Rogers and Smith, this volume.] In 1973, Professor Peter Hobbs and his associates at the University of Washington began a series of comprehensive studies on winter cyclonic storms of the Pacific Northwest using aircraft, radar, and surface measurements. For references to these extensive studies see Hobbs et al. (1978) and Hertzman and Hobbs (1988).

Interaction between coalescence and ice crystal mechanisms established

It is now recognized that precipitation often involves both the coalescence and ice crystal mechanisms. This discovery came about in 1960 when scientists at the University of Chicago began a study of the response of summertime cumulus clouds in Missouri to silver iodide seeding. It was found that a large fraction of the cumulus clouds developed warm rain before reaching the freezing level. It was also observed that when the raindrops were carried up to levels where the temperature was $-5°C$ or colder, they froze and quickly began to rime to form graupel. In this way the clouds developed ice particle concentrations that were several orders of magnitude larger than expected from local ice nucleus measurements. Thus, the target clouds for the seeding experiment developed ice particle concentrations equal to or greater than necessary for efficient precipitation development at temperatures warmer than the threshold of silver iodide aerosols used for the seeding.[25] These observations were first reported at the 1961 cloud physics conference in Australia[26] (Braham 1961) and subsequently

---

[25]Observations of ice particles at fairly warm temperatures in convective clouds had previously been reported by Coons et al. (1949) and Murgatroyd and Garrod (1960). However, the implications of their findings were not apparent to the cloud physics community. The University of Chicago project was the first to study these particles in detail and discover their role in the formation of precipitation in warm-based clouds.

[26]I recall this meeting very well. After I told the assembled group that we regularly found ice particle concentrations in nonseeded clouds exceeding those needed for efficient

verified by many other investigators (Koenig 1962, 1963; Mossop 1968; Hallett et al. 1978).

In 1962 Hoffer and Braham collected graupel from tops of cumulus clouds as they rose through the $-10°C$ level. The graupel were returned to the laboratory, melted, and then cooled to measure their freezing temperature. Every melted graupel particle froze at a temperature 10°C or more degrees *colder* than the coldest temperature in the cloud from which it had been collected. This led to the conclusion that few, if any, of the nuclei responsible for the initial freezing of the raindrops could have been mineral grains, such as those found by Kumai (1961) in the centers of snowflakes, because those should have survived the melting and freezing process.[27]

Hoffer and Braham (1962) concluded that the coalescence-grown raindrops may have been frozen by contact with fragments of ice produced by some yet unknown process during a prior drop freezing. Koenig (1962, 1963), in his doctoral research, developed this idea into a model of a chain reaction process in which ice fragments produced by freezing of a large drop are encountered by other large drops causing them to freeze and release ice fragments. This model made use of the findings of Bigg (1957b) and Mason and Maybank (1960), who had shown that freezing drops sometimes shed small ice fragments. Subsequently, Hallett and Mossop (1974) and others found that riming, in certain combinations of temperature, updraft speed, and cloud-drop spectra, could be a major source of ice fragments in supercooled clouds.[28]

Recognition of the fact that warm raindrops could freeze at relatively warm temperatures, then rime to form graupel earlier in the life of clouds and at much warmer temperatures than ice can be initiated by the Bergeron–Findeisen mechanism, led to a reexamination of the role of ice in natural precipitation. Braham (1964, 1986a,b) and Johnson (1987) showed that, for warm-based clouds, the growth of large drops by coalescence, followed by freezing and riming into graupel, was a more

---

rain formation, and at temperatures warmer than the threshold of efficient nucleation through silver iodide, there was very little discussion. Afterward, I realized that they simply did not believe me, perhaps in part because we had not yet developed a rational explanation for the findings.

[27]In fact, many studies have shown that ice nuclei often are more effective once they have served to initiate ice.

[28]Ice particles formed by any process involving preexisting ice particles, in contrast to nucleation on ice nuclei, have been called "secondary ice particles" and have been the subject of considerable research since their importance was first recognized.

efficient mechanism than either the Bergeron–Findeisen or the coalescence mechanisms acting alone. This mechanism was called the "coalescence–freezing" mechanism. The full importance of this mechanism for precipitation formation is not yet known, though it could be enormous. For example, investigators of the 1976 storm that caused disastrous flooding in the Big Thompson Canyon of Colorado[29] found indications that coalescence rain may have played a major role (McCain et al. 1979).

A look ahead

If "the past is prologue,"[30] what can one say about the course of cloud physics based upon current trends? As better techniques are developed for bridging the space–timescale differences between cloud physics and cloud dynamics, numerical models should allow researchers to explore more fully the interactions and feedbacks between cloud microphysics and cloud dynamics. Use of radar depolarization data may allow detailed study of the coalescence–freezing mechanism in deep convective clouds. Better understanding of hydrometeor growth in deep convection should allow better prediction of severe storms and better measurements of precipitation from them using remote sensors. This would allow more realistic calculations of the global hydrologic cycle.

It seems likely that mesoscale meteorology will remain attractive; it may tend to become more site specific, as new tools such as Next Generation Weather Radar (NEXRAD), profilers, and other remote sounding devices illuminate local weather peculiarities resulting from topographic and anthropogenic influences. Topics that interface with other scientific disciplines and that address socioeconomic issues, both local and global, seem likely to be favored for productive research in the next decade. These include air and water quality, aerosols and atmospheric chemistry, hydrometeorology, and quantitative precipitation forecasts. Hopefully within the next decade we will understand

[29]Over 10 inches of rain in about 5 hours, in the mountains of north-central Colorado on 31 July and 1 August 1976, produced disastrous flooding in many mountain canyons, resulting in the loss of about 140 lives and substantial property damage. The storm will be remembered by many cloud physicists and meteorologists who were attending the International Conference on Cloud Physics and the World Meteorological Organization Scientific Conference on Weather Modification in Boulder, Colorado, at the time of the storm.

[30]Shakespeare's *The Tempest,* 2.1.261.

the formation of ice in the atmosphere and clarify the role of biological components as ice nuclei and condensation nuclei.

Predicting trends (a practice not entirely new to meteorologists) is a high-risk occupation. Experience shows that the best way to "predict" the future is to take an active role in shaping it. Strong backgrounds mixing meteorology with classical physics, chemistry, biology, oceanography, and hydrology, added to a willingness to work hard, will help students prepare for a role in shaping the future of cloud physics.

Acknowledgments. The author freely acknowledges that many excellent papers may have been overlooked, undervalued, or passed over because of space restrictions in this brief summary of the development of cloud microphysics. To those authors, I apologize. Several libraries and librarians assisted me in finding some of the earlier materials. The Interlibrary Loan and Reference Departments of Hill Library, North Carolina State University; Ms. Carol Watts, director, NOAA Central Library; Ms. Maureen Dungey and Ms. Kathleen Zar, University of Chicago; and Dr. David Kristovich, Illinois State Water Survey, were especially helpful.

REFERENCES

Adderley, E.E., and E.G. Bowen, 1962: Lunar component in precipitation data. *Science,* **137,** 749–750.

AMS, 1957: Willis Isbister Milham. *Bull. Amer. Meteor. Soc.,* **38,** 301.

Arctowski, H., 1901a: Exploration of Antarctic lands. *Geogr. J.,* **17,** 150–180.

_____ , 1901b: The Antarctic voyage of the Belgica during the years of 1897, 1898, and 1899. Smithsonian Institution, Annual Rep. 377–388.

Auer, A.H., 1971: Observations of ice crystal nucleation by drop freezing in natural clouds. *J. Atmos. Sci.,* **28,** 285–290.

_____ , 1972: Inferences about ice nucleation from ice crystal observations. *J. Atmos. Sci.,* **29,** 311–317.

aufm Kampe, H.J., H.K. Weickmann, and J.J. Kelly, 1951: The influence of temperature on the shape of ice crystals growing at water saturation. *J. Meteor.,* **8,** 168–174.

_____ , _____ , and H.H. Kedesby, 1952: Remarks on "Electron-microscope study of snow crystal nuclei." *J. Meteor.,* **9,** 374–375.

Barkow, E., 1908: Zur entstehung der graupeln. *Meteor. Z.* **25,** 456–458.

Battan, L.J., 1953: Observations on the formation and spread of precipitation in convective clouds. *J. Meteor.,* **10,** 311–324.

_______ , and C. Reitan, 1957: Droplet size measurements in convective clouds. *Artificial Stimulation of Rain,* H. Weickmann and W. Smith, Eds., Pergamon, 184–191.

Beard, K.V., 1992: Ice initiation in warm-base convective clouds: An assessment of microphysical mechanisms. *Atmos. Res.,* **28,** 125–152.

Becker, R., and W. Doring, 1935: Kinetische behandlung der keimbildung in ubersattigten Dampfen. *Ann. Phys.,* Leipzig, Ser. 6, **24,** 719–752. (Translated by J. Vanier in U.S. NACA Tech. Memo. 1374, 43 pp.)

Benndorf, H., 1931: Alfred Wegener. *Gerl. Beitr. Geophys.,* **31,** 337–377.

Bentley, W.A., 1904: Studies of raindrops and raindrop phenomena *Mon. Wea. Rev.,* **32,** 450–456.

_______ , 1924: Forty years of study of snow crystals. *Mon. Wea. Rev.,* **52,** 530–532.

Bergeron, T., 1928: *Über die Dreidimensional Verknüpfende Wetteranalyse.* Geofysiske Publikosjoner, 111 pp.

_______ , 1935: On the physics of cloud and precipitation. *Proc. Verbaux Assoc. Météor. Int. Union Geodesy Geophys., Fifth General Assembly,* Lisbon, Portugal, Dupont, Paris, 3–19, 173–178.

_______ , 1972: L'origine de la théorie des noyaux de glace comme declencheurs de précipitation un cinquantenaire. *J. Rech. Atmos.,* **6,** 49–53.

Berry, E.X., 1967: Cloud droplet growth by collection. *J. Atmos. Sci.,* **24,** 688–701.

_______ , and R.L. Reinhardt, 1974a: An analysis of cloud drop growth by collection. Part I: Double distributions. *J. Atmos. Sci.,* **31,** 1814–1824.

_______ , and _______ , 1974b: An analysis of cloud drop growth by collection. Part II: Single initial distributions. *J. Atmos. Sci.,* **31,** 1825–1831.

_______ , and _______ , 1974c: An analysis of cloud drop growth by collection. Part III: Accretion and self-collection. *J. Atmos. Sci.,* **31,** 2127–2135.

Berson, A., 1900: Fahrt des Ballons, "Phonix," von 19 October 1893.

Wissenschaftliche Luftfahrten, II, R. Assmann and A. Berson, Eds., Braunschweig, 183–194.

Best, A.C., 1948: Formation of rain. *Meteor. Mag.,* **77,** 231–232.

Bigg, E.J., 1953a: The supercooling of water. Ph.D. thesis, Imperial College, University of London, 170 pp.

______ , 1953b: The supercooling of water. *Proc. Phys. Soc. London, Ser. B,* **66,** 688–694.

______ , 1953c: The formation of atmospheric ice crystals by the freezing of droplets. *Quart. J. Roy. Meteor. Soc.,* **79,** 510–519.

______ , 1956: Counts of atmospheric freezing nuclei at Carnarvon, Western Australia, January 1956. *Aust. J. Phys.,* **9,** 561–565.

______ , 1957a: A new technique for counting ice-forming nuclei in aerosols. *Tellus,* **9,** 394–400.

______ , 1957b: The fragmentation of freezing water drops *Bull. Obs. Puy-de-Dome,* **3,** 65–67.

Blanchard, D.C., 1949: The use of sooted screens for determining raindrop size and distribution. Project Cirrus Occasional Rep. 16, General Electric Research Laboratory, Schenectady, NY, 11 pp.

______ , 1978: Tor Bergeron, 1891–1977. *Bull. Amer. Meteor. Soc.,* **59,** 387–392.

Bowen, E.G., 1950: The formation of rain by coalescence. *Aust. J. Sci. Res., Ser. A,* **3,** 193–213.

______ , 1951: Radar observations on rain and their relation to the mechanisms of rain formation. *J. Atmos. Terr. Phys.,* **1,** 125–140.

______ , 1953: The influence of meteoritic dust on rainfall. *Aust. J. Phys.,* **4,** 490–497.

Bradley, D.A., M.A. Woodbury, and G.W. Brier, 1962: Lunar synodical period and widespread precipitation. *Science,* **137,** 748–749.

Braham, R.R., Jr., 1961: Project Whitetop. Abstracts, *Int. Conf. on Cloud Physics,* Canberra and Sydney, Australia, Australian Academy of Science and CSIRO, 9.1.

______ , 1964: What is the role of ice in summer rain-showers? *J. Atmos. Sci.,* **21,** 640–645.

______ , 1986a: The cloud physics of weather modification. *WMO Bull.,* **55,** 215–221, 308–315.

______ , 1986b: Coalescence-freezing precipitation mechanism. Preprints, Preprints, *10th Conf. on Planned and Inadvertent Weather Modification.,* Arlington, VA, Amer. Meteor. Soc., 142–145.

______ , and P. Spyers-Duran, 1967: Survival of cirrus crystals in clear air. *J. Appl. Meteor.*, **6**, 1053–1061.

______ , and P. Squires, 1974: Cloud physics—1974. *Bull. Amer. Meteor. Soc.*, **55**, 543–586.

______ , S.E. Reynolds, and J.H. Harrell, 1951: Possibilities for cloud seeding as determined by a study of cloud height versus precipitation. *J. Meteor.*, **8**, 416–418.

______ , L.J. Battan, and H.R. Byers, 1957: Artificial nucleation of clouds. *Cloud and Weather Modification, Meteor. Monogr.*, No. 11, Amer Meteor Soc., 47–85.

Browning, K.A., and T.W. Harrold, 1969: Air motion and precipitation growth in a wave depression. *Quart. J. Roy. Meteor. Soc.*, **95**, 288–309.

Brunt, D., 1928: *Physical and Dynamical Meteorology*. Cambridge University Press, 428 pp.

Byers, H.R., 1937: *Synoptic and Aeronautical Meteorology*. McGraw-Hill, 279 pp.

______ , 1955: Geographical differences in cloud populations. *Arch. Meteor. Geophys. Bioklimatol.*, **8**, 180–184.

______ , 1974: History of weather modification. *Weather and Climate Modification*, W. Hess, Ed., Wiley, 3–44.

______ , and R.K., Hull, 1955: A census of cumulus cloud height versus precipitation in the vicinity of Puerto Rico during the winter and spring of 1953–54. *J. Meteor.*, **12**, 176–178.

Coons, R.D., E.L. Jones, and R. Gunn., 1949: Artificial production of precipitation in cumulus clouds. Research Paper 33, U.S. Weather Bureau, Washington, DC, 46 pp.

Cooper, W.A., 1986: Ice initiation in natural clouds. *Precipitation Enhancement—A Scientific Challenge*, R. Braham Jr., Ed., Amer. Meteor. Soc., 29–41.

Craddock, J.M., 1949: The development of cumulus cloud: Results of observations in Malaya. *Quart. J. Roy. Meteor. Soc.*, **75**, 147–153.

Cunningham, R.M., 1951: Some observations of natural precipitation processes. *Bull. Amer. Meteor. Soc.*, **32**, 334–343.

______ , 1952: Distribution and growth of hydrometeors around a deep cyclone. Massachusetts Institute of Technology Research Rep. 18, 59 pp.

Cwilong, B.M., 1947: Sublimation in a Wilson chamber. *Proc. Roy. Soc. London, Ser. A,* **190,** 137–143.

Davies, D.A., 1950: Tropical rainfall from cloud which did not extend to the freezing level. *Meteor. Mag.,* **79,** 354.

Defant, A., 1905: Gesetzmässigkeiten in der verteilung der verschiedenen tropfengrössen bei regenfällen. *Akademie der Wissenschaften, Vienna Sitzungsberichte, Mathematisch-Wissenschaftliche Klasse, 2a,* **114,** 585–646.

de Saussure, H.B., 1783: *Essais sur l'hygrométrie.* S. Fauche Pere et Fils, 524 pp.

Diem, M., 1942: Messungen der grösse von wolkenelementen, I. *Ann. Hydrogr.,* **70,** 142–150.

_____ , 1948: Messungen der Grösse von Wolkenelementen, II. *Meteor. Rundsch.,* **1,** 261–273.

Dobrowolski, A., 1903: La neige et le givre. Résultats du voyage du s.y. Belgica en 1897–1898–1899. Rapports Scientifiques. Météorologie, Commision de la Belgica, 78 pp.

Douglass, C.K.M., 1934: The physical processes of cloud formation. *Quart. J. Roy. Meteor. Soc.,* **60,** 333–341.

Douglass, R.H., K.L.S. Gunn, and J.S. Marshall, 1957: Pattern in the vertical of snow generation. *J. Meteor.,* **14,** 95–114.

Dufour, L., 1861: Über das gefrieren des wasser und über die bildund des hagels. *Ann. Phys.,* **114,** 530–554.

Eadie, W.J., 1971: A molecular theory of the homogeneous nucleation of ice from supercooled water. Ph.D. dissertation, University of Chicago, 177 pp.

Elliott, R.D., 1974: Experience of the private sector. *Weather and Climate Modification,* W. Hess, Ed., Wiley, 45–89.

Fahrenheit, D.G., 1724: Experimenta e observationes de congelatione aquae in vacuo factae. *Philos. Trans. Roy. Soc. London,* **33,** 78–84.

Findeisen, W., 1932: Messungen der grosse und anzahl der nebeltropfen zum studium der koagulation inhomogenen nebels. *Gerl. Beit. Geophys.,* **35,** 295–340.

_____ , 1938: Die kolloidmeteorologischen vorgänge bei der niederschlagsbildung. *Meteor. Z.,* **55,** 121–133.

_____ , 1939: Das verdampfen der wolken und regentropfen. *Meteor. Z.,* **56,** 453–460.

_____ , 1942: Experimentelle untersuchungen über die atmosphärishen eisteilchenbildung. *Meteor. Z.,* **59,** 349–353.

______ , and G. Schulz, 1944: Experimentelle untersuchugen zur atmosphärischen eisteilchenbildung, I. *Forschungs Erf. Reichswett., Ser. A,* **27,** 38–48.

Fitzgerald, J.W., 1972: A study of the initial phase of cloud droplet growth by condensation. Ph.D. dissertation, University of Chicago, 144 pp.

Flammarion, C., 1873: *The Atmosphere.* Translated by J. Glaisher. Harper, 453 pp.

Fletcher, N., 1962: *The Physics of Rainclouds.* Cambridge University Press, 386 pp.

Frössling, N., 1938: Über die verdunstung fallender tropfen. *Gerl. Beitr. Geophys,* **52,** 170–216.

Foster, H., 1950: An unusual observation of lightning. *Bull. Amer. Meteor. Soc.,* **31,** 140–141.

Fritsch, K., 1853: Uber schneefiguren. *Sitzungsber. Akad. Wiss. Wein (Math.-Nature Kl.),* **11,** 492–499.

Georgii, W., 1930: Das flugzeug als aerologisches forschungmittel. *Beitr. Phys. Atmos.,* **16,** 199–223.

Glaisher, J., 1863: An account of meteorological and physical observations in five balloon ascents in the year 1863. Rep. on the 33d Meeting of the British Association for Advancement of Science, 426–516.

______ , 1871: *Travels in the Air.* Bentley, 398 pp.

Graham, T., 1861: Liquid diffusion applied to analysis. *Philos. Trans. Roy. Soc. London,* **151,** 183–224.

Gregg, W.R., 1930: *Aeronautical Meteorology.* 2d ed. Ronald Press, 403 pp.

Guilbert, G., 1921: On the formation of rain and the origin of cirrus. *Compt. Rend. Acad. Sci.,* **173,** 999–1001.

______ , 1922: *La prévision scientifique du temps, traité pratique.* Augustin Challemel, 439 pp.

Gunn, R., and G.D. Kinzer, 1949: The terminal velocity of fall for water droplets in stagnant air. *J. Meteor.,* **6,** 243–248.

Hagemann, V., 1936: Eine methode zur bestimmung der grösse der nebel und wolkenelemente. *Gerl. Beitr. Phys. freien Atmos.* **46,** 261–282.

Haggard, W.H., 1948: Colloidal instability of tropical clouds. *Bull. Amer. Meteor. Soc.,* **29,** 139–140.

Hallett, J., 1964: Experimental studies of the crystallization of super-cooled water. *J. Atmos. Sci.*, **21**, 671–682.

———— , and B.J. Mason, 1958: The influence of temperature and super-saturation on the habit of ice crystals grown from the vapor. *Proc. Roy. Soc. London, Ser. A*, **247**, 440–453.

———— and S.C. Mossop, 1974: Production of secondary ice particles during the riming process. *Nature*, **249**, 26.

———— , R.I. Sax, D. Lamb, and A.S.R. Murty, 1978: Aircraft measure-ments of ice in Florida cumuli. *Quart. J. Roy. Meteor. Soc.*, **104**, 631–651.

Hanajima, M., 1944: On the conditions of growth of snow crystals. *Low Temp. Sci.*, **A1**, 53–65

———— , 1949: On the growth conditions of man-made snow. *Low Temp. Sci.*, **A2**, 23–29

Hann, J.V., 1906: *Lehrbuch der Meteorologie.* 2d ed. Leipzig. 642 pp.

Havens, B.S., 1952: History of Project Cirrus. General Electric Re-search Lab. Rep. RL-756, 102 pp.

Hertzman, O., and P.V. Hobbs, 1988: The mesoscale and microscale structure and organizations of clouds in midlatitude cyclones. XIV: Three-dimensional airflow and vorticity budget of rainbands in a warm occlusion. *J. Atmos. Sci.*, **45**, 893–914.

Heverly, J.R., 1948: A study in atmospheric condensation. Ph.D. dis-sertation, The Pennsylvania State College, 141 pp.

———— , 1949: Supercooling and crystallization. *Eos, Trans. Amer. Geo-phys. Union*, **30**, 205–210.

Heywood, G.S.P., 1940: Rain formation in the tropics. *Quart. J. Roy. Meteor. Soc.*, **66**, 46.

Hobbs, P.V., 1974: *Ice Physics.* Clarendon, 837 pp.

———— , 1978: Organization and structure of clouds and precipitation on the mesoscale and microscale in cyclonic storms. *Rev. Geophys. Space Phys.*, **16**, 741–755.

———— , 1990: Ice in clouds. Preprints, *Cloud Physics Conf.*, San Fran-cisco, CA, Amer. Meteor. Soc., 600–606.

Hoffer, T.E., and R.R. Braham, 1962: A laboratory study of atmo-spheric ice particles. *J. Atmos. Sci.*, **19**, 232–235.

Holzman, B., 1936: A note on Bergeron's ice-nuclei hypothesis for the formation of rain. *Bull. Amer. Meteor. Soc.*, **17**, 331–333.

Houghton, H.G., 1933: A study of the evaporation of small water drops. *Phys.*, **4**, 419–424.

———, 1938: Problems connected with the condensation and precipitation processes in the atmosphere. *Bull. Amer. Meteor. Soc.*, **19**, 152–159.

———, 1950: A preliminary quantitative analysis of precipitation mechanisms. *J. Meteor.*, **7**, 363–369.

Howell, W., 1949: The growth of cloud drops in uniformly cooled air. *J. Meteor.*, **6**, 134–149.

Humphreys, W.J., 1940: *Physics of the Air.* 3d ed. McGraw-Hill, 676 pp.

Hunt, T.L., 1949: Formation of rain. *Meteor. Mag.*, **78**, 26.

Isono, K., 1955: On ice crystal nuclei and other substances found in snow crystals. *J. Meteor.*, **12**, 456–462.

Jacobi, W., 1955: Homogeneous nucleation in supercooled water. *J. Meteor.*, **12**, 408–409.

Jeffreys, H., 1918: Some problems of evaporation. *Philos. Mag.*, **35**, 270–280.

Johnson, D.B., 1987: On the relative efficiency of coalescence and riming. *J. Atmos. Sci.*, **44**, 1671–1680.

Jones, D.M.A., and L.A. Dean, 1953: A raindrop camera. Illinois State Water Survey Research Rep. 31, 50–64.

Jones, R.F., 1951: Rain from non-freezing clouds. *Meteor. Mag.*, **80**, 273–274.

Kepler, J., [1611] 1966: On the shapes of snow crystals. *The Six-Cornered Snowflake.* Translated by B.J. Mason. Clarendon, Reprint, 47–56.

Knollenberg, R.G., 1970: The optical array: An alternative to scattering or extinction for airborne particle size determination. *J. Appl. Meteor.*, **9**, 86–103.

Kobayashi, T., 1957: Experimental researches on the snow crystal habit and growth by means of a diffusion cloud chamber. *J. Meteor. Soc. Japan*, **35**, 38–44.

———, 1961: The growth of snow-crystals at low supersaturations. *Philos. Mag.*, **6**, 1363–1370.

Koenig, L.R., 1962: Ice in the summer atmosphere; an inquiry into its structure, genesis and metamorphosis. Ph.D. dissertation, University of Chicago, 176 pp.

———, 1963: The glaciating behavior of small cumulonimbus clouds. *J. Atmos., Sci.*, **20**, 29–47.

Köhler, H., 1921: Über die tropfengrössen der wolken und die kondensation. *Meteor. Z.*, **38**, 359–365.

______ , 1925a: Untersuchengen über die elemente des nebels und der wolken. *Medd. Meteor.-hydr. Anst. Stockholm*, **2**, 73 pp.

______ , 1925b: Über tropfengruppen in wolken. *Meteor. Z.*, **42**, 137–143, 463–467.

______ , 1926: Zur thermodynamik der kondensation an hygroskopischen kernen and bemerkungen über das zusammenfliessen der tropfen. *Medd. Statens Meteor.-Hydrograf.*, **3**, 16 pp.

______ , 1929. Wolkenuntersuchungen dem sonnblick im herbst 1928. *Meteor. Z.*, **46**, 409–420.

Kotch, W.J., 1947: An example of colloidal instability of clouds in tropical latitudes. *Bull. Amer. Meteor. Soc.*, **28**, 87–89.

Krastanow, L., 1940: Über die bildung der unterkuhlten wassertropfen und der eiskristalle in der freien atmosphäre. *Meteor. Z.*, **57**, 357–371.

______ , 1941: Beitrag zur theorie der tropfen und kristallbildung in der atmosphäre. *Meteor. Z.*, **58**, 37–45.

Kraus, E.B., and B. Smith, 1949: Theoretical aspects of cloud drop distributions. *Aust. J. Sci. Res., Ser. A.*, **2**, 376–388.

Kumai, M., 1951: Electron-microscope study of snow crystal nuclei. *J. Meteor.*, **8**, 151–156.

______ , 1957: Electron-micoscope study of snow crystal nuclei, II. *Geopfys. Pura Appl.*, **36**, 169–181.

______ , 1961: Snow crystals and the identification of the nuclei in northern United States of America. *J. Meteor.*, **18**, 139–150.

______ , and K.E. Francis, 1962: Nuclei in snow and ice crystals on the Greenland ice cap under natural and artificially simulated conditions. *J. Atmos. Sci.*, **19**, 474–481.

Lampbright, H., 1970: *The Politics of an Emergent Technology*. Inter-University Case Prog., 506 pp.

Langmuir, I., 1944: Supercooled water droplets in rising currents of saturated air. *Collected Works of Irving Langmuir*, C.G. Suits, Ed., Pergamon, 199–334.

______ , 1948: The production of rain by a chain reaction in cumulus clouds at temperatures above freezing. *J. Meteor.*, **5**, 175–192.

______ , and K.B. Blodgett, 1945. A mathematical investigation of water-drop trajectories. *Collected Works of Irving Langmuir*, C.G. Suits, Ed., Pergamon, 348–393.

Langsdorf, A., 1936: A continuously sensitive diffusion cloud chamber. *Rev. Sci. Insts.,* **10,** 91–103.

Laws, J.O., 1941: Measurements of the fall-velocity of water-drops and rain-drops. *Eos, Trans. Amer. Geophys. Union,* **22,** 709–721.

Lenard, P., 1904: Ueber regen. *Meteor. Zeit.,* **21,** 248–262. Translated by R.H. Scott in *Quart. J. Roy. Meteor. Soc.,* **31,** 62–73.

Leopold, L.B., and M.H. Halstead, 1948: First trials of the Schaefer–Langmuir dry-ice cloud-seeding technique in Hawaii. *Bull. Amer. Meteor. Soc.,* **29,** 525–534.

Lhermitte, R.M., 1952: Les "bandes supérieures" dans la structure verticale des echos de pluie. *Compt. Rend. Acad. Sci.,* **235,** 1414–1416.

Loeb, L.B., 1934: *The Kinetic Theory of Gases.* McGraw-Hill, 687 pp.

Ludlam, F.H., 1951: The production of showers by the coalescence of cloud droplets. *Quart. J. Roy. Meteor. Soc.,* **77,** 402–417.

———— , 1958: The hail problem. *Nubila,* **1,** 12–94.

MacDougal, F.H., 1948: *Physical Chemistry.* Macmillan, 722 pp.

Magono, C., and C.W. Lee, 1966: Meteorological classification of natural snow crystals. *J. Fac. Sci., Hokkaido Univ., Ser VII (Geophysics),* **2,** 321–335.

Mann, G., 1940: Untersuchen über die aerologischen bedingungen für die niederschlagsbildung in der atmosphäre an hand des aufstiegsmaterials der wetterflugstelle zu königsberg/Pr. *Beit. Phys. Atmos.,* **26,** 121–151.

Marriott, W., 1904: Some account of the meteorological work of the late James Glaisher, F.R.S. *Quart. J. Roy. Meteor. Soc.,* **30,** 1–28.

Marshall, J.S., and M.P. Langleben, 1954: A theory of snow-crystal habit and growth. *J. Meteor.,* **11,** 104–120.

Mason, B.J., 1952: The spontaneous crystallization of supercooled water. *Quart. J. Roy. Meteor. Soc.,* **78,** 22–27.

———— , 1953: The growth of ice crystals in a supercooled water cloud. *Quart. J. Roy. Meteor. Soc.,* **79,** 104–111.

———— , 1971: *The Physics of Clouds.* Clarendon Press, 671 pp.

———— , and J. Maybank, 1960: The fragmentation and electrification of freezing water drops. *Quart. J. Roy. Meteor. Soc.,* **86,** 176–186.

McAdie, A., 1895: Natural rain-makers. *Pop. Sci. Mon.,* **47,** 642–648.

McCain, J.F., L.R. Hoxit, R.A. Maddox, C.F. Chappell, and F. Caracena, 1979: Meteorology and hydrology in Big Thompson River and Cache la Poudre Basins. Storm of July 31–August 1, 1976, in the Big Thompson River and Cache la Poudre River Basins, Larimer and Weld Counties, Colorado. U.S. Geological Survey Paper 1115, 152 pp.

McDonald, J.E., 1953a: Erroneous applications of Raoult's Law. *J. Meteor.,* **10,** 68–70.

_____ , 1953b: The physics of cloud modification. *Advances in Geophysics,* Vol. 5, Academic Press, 223–303.

_____ , 1963: Use of the electrostatic analogy in studies of ice crystal growth. *J. Appl. Math. Phys.,* **14,** 610–620.

Melzak, A.F., and W. Hitschfeld, 1953: A mathematical treatment of random coalescence. Stormy Weather Group Science Rep. MW-11, McGill University, 28 pp.

Middleton, W.E.K., 1965: *A History of the Theories of Rain and Other Forms of Precipitation.* Watts, 206 pp.

Milham, W.I., 1931: *Meteorology.* Macmillan, 549 pp.

Mordy, W.A., 1959: Computations of the growth by condensation of a population of cloud droplets. *Tellus,* **11,** 16–44.

_____ , and L.E. Eber, 1954: Observations of rainfall from warm clouds. *Quart. J. Roy. Meteor. Soc.,* **80,** 48–57.

Mossop, S.C., 1963: Atmospheric ice nuclei. *Z. Angew. Math. Phys.,* **14,** 456–486.

_____ , 1968: Comparison between concentration of ice crystals in cloud and the concentration of ice nuclei. *J. Rech. Atmos.,* **3,** 119–124.

_____ , 1985: The origin and concentration of ice crystals in clouds. *Bull. Amer. Meteor. Soc.,* **66,** 264–273.

Mousson, A., 1858: Eine tatsachen betreffend das schmelzen und gefrieren des wassers. *Ann. Phys.,* **105,** 161–174.

Müller, H., 1928: Zur allgemeinen theorie der raschen koagulation. *Kolloidchem,* **27,** 223–250.

Murgatroyd, R.J., and M.P. Garrod, 1960: Observations of precipitation elements in cumulus clouds. *Quart. J. Roy. Meteor. Soc.,* **86,** 167–175.

Nakaya, U., 1938: Artificial snow. *Quart. J. Roy. Meteor. Soc.,* **64,** 619–624.

_____ , 1951: The formation of ice crystals. *Compendium of Meteorology,* T. Malone, Ed., Amer. Meteor. Soc., 207–220.

_____ , 1954: *Snow Crystals: Natural and Artificial.* Harvard University Press, 510 pp.

_____ , and I. Sato, 1935: On the artificial production of frost crystals with reference to the mechanism of formation of snow crystals. *J. Fac. Sci., Hokkaido Univ., Ser. 2*, **1**, 206–214.

Needham, J., and Lu Gwen-Djen, 1961: The earliest snow crystal observations. *Weather*, **12**, 319–327.

Neiburger, M., and C.W. Chien, 1960: Computations of the growth of cloud drops by condensation using an electronic digital computer. *Physics of Precipitation, Geophys. Monogr.*, No. 5, Amer. Geophys. Union, 191–210.

New York Times, 1946: Three mile cloud made into snow by dry ice dropped from plane. 14 November, p. 25.

Niederdorfer, E., 1932: Messungen der grösse der regentropfen. *Meteor. Z.*, **49**, 1–14.

Peppler, W., 1936: Zur aerologie der wolken, besonders des nimbus. *Beit. Phys. Atmos.*, **23**, 275–288.

Petterssen, S., 1940: *Weather Analysis and Forecasting*. McGraw-Hill, 503 pp.

Pound, G.M., L.A. Madonna, and S.L. Peake, 1953: Critical supercooling of pure water droplets by a new microscopic technique. *J. Colloid Sci.*, **8**, 187–193.

Pruppacher, H.R., and J.D. Klett, 1978: *Microphysics of Clouds and Precipitation*. Reidel, 714 pp.

Rangno, A.R., and P.V. Hobbs, 1988: Criteria for the onset of significant concentrations of ice particles in cumulus clouds. *Atmos. Res.*, **22**, 1–13.

Reynolds, O., 1879a: On the manner in which raindrops and hailstones are formed. *Mem. Manchester Lit. Philos. Soc.*, **6**, 48–60.

———, 1879b: The influence of electricity on colliding water drops. *Proc. Roy. Soc. London*, **28**, 406–409.

Rooth, C., 1957: On a special aspect of the condensation process and its importance in the treatment of cloud particle growth. *Tellus*, **9**, 372–337.

Rotch, L., 1900: *Sounding the Ocean of Air*. Society for Promoting Christian Knowledge, 184 pp.

Schaefer, V.J., 1941: A method of making snowflake replicas. *Science*, **93**, 239–240.

———, 1946: The production of ice crystals in a cloud of supercooled water droplets. *Science*, **104**, 457–459.

Schmauss, A., 1919: Randbemerkungen II. *Meteor. Z.*, **36**, 11–16.

———, 1920: Kolloidchemie und meteorologie. *Meteor. Z.*, **37**, 1–8.

_____, and A. Wigand, 1929: *Die Atmospháre als Kolloid*. Braunschweig, 74 pp.

Schmidt, W., 1909: Eine unmittelbare bestimmung der fallgeschwindigkeit von regentropfen. *Akademie der Wissenschaften, Vienna, Sitzungsberichte, Mathematisch-Wissenschaftliche Klasse, 2a,* **118,** 71–84

Schnell, R.C., 1974: Biogenic sources of atmospheric ice nuclei. Ph.D. dissertation, University of Wyoming, 113 pp.

Schumann, T.E.W., 1938: The theory of hail formation. *Quart. J. Roy. Meteor. Soc.,* **64,** 3–21.

_____, 1940: Theoretical aspects of the size distribution of fog particles. *Quart. J. Roy. Meteor. Soc.,* **66,** 195–207.

Schwarzbach, M., 1986: *Alfred Wegener, The Father of Continental Drift*. Science Tech., Inc., 241 pp.

Schwerdtfeger, W., 1948: Über die bildung von reganschauern uber see. *Meteor. Rundsch.,* **1,** 453–456.

Scoresby, W., 1820: *An Account of the Arctic Regions with a History and Description of the Northern Whale-Fishery*. Vol. 1. Archibald Constable, 419–445.

Sharenow, M., 1939: *Translation of W. Findeisen's Colloidal–Meteorological Processes in the Formation of Precipitation*. Evans Signal Laboratory.

Shaw, N., 1930: *Manual of Meteorology*. Vol. III, *The Physical Processes of Weather,* Cambridge, 340 pp.

Simpson, G.C., 1941: On the formation of cloud and rain. *Quart. J. Roy. Meteor. Soc.,* **67,** 99–133.

Smith, E.J., 1951: Observations of rain from non-freezing clouds. *Quart. J. Roy. Meteor. Soc.,* **77,** 33–43.

_____, and K.J. Heffernan, 1954: Airborne measurements of the concentration of natural and artificial freezing nuclei. *Quart. J. Roy. Meteor. Soc.,* **80,** 182–197.

Smoluchowsky, M.V., 1918: Versuch einer mathematischen theorie der koagulationskinetic kolloidal lösunger. *Z. Physik. Chem., Frankfort,* **92,** 129–168.

Squires, P., 1952: The growth of cloud drops by condensation. *Aust. J. Sci. Res., Ser. A—Phys. Sci.,* **5,** 59–86.

_____, 1956: The microstructure of cumuli in maritime and continental air. *Tellus,* **8,** 443–444.

_____, and S. Twomey, 1960: The relation between cloud droplet spectra and the spectrum of cloud nuclei. *Physics of Precipitation, Geophys. Monogr.*, No. 5, Amer. Geophys. Union, 211–219.

Stefan, J., 1881: Theorie des psychrometers. *Meteor. Z., Wien,* **16,** 180–182.

Stickley, A.R., 1939: *Selected Papers on the Physics of Condensation and Precipitation.* U.S. Weather Bureau, 1–24.

______ , 1940: An evaluation of the Bergeron–Findeisen precipitation theory. *Mon. Wea. Rev.,* **68,** 272–279.

Stokes, G.G., 1850: On the effect of the internal friction of fluids on the motion of pendulums. *Trans. Cambridge Philos. Soc.,* **9,** 8–106. Reprint, 1966: *Mathematical and Physical Papers of G.G. Stokes.* Vol. 3. Sources of Science, No. 33, Johnson Reprint Corp., 1–141.

Süring, R., 1940: Wolkenbildung, Wolkenstruktur und Niederschlag. *Z. Geophys., Meteor., Geod., Berlin,* **5,** 1–10.

Telford, J., 1955: A new aspect of coalescence theory. *J. Meteor.,* **12,** 436–444.

Tyndall, J., 1872: *The Forms of Water, in Clouds and Rivers, Ice and Glaciers.* Appleton, 196 pp.

Vali, G., 1968: Ice nucleation relevant to the formation of hail. Stormy Weather Group Science Rep. MW-58, McGill University, 51 pp.

______ , 1985: Atmospheric ice nucleation—A review. *J. Rech. Atmos.,* **19,** 105–115.

______ , and E.J. Stansbury, 1966: Time-dependent characteristics of the heterogeneous nucleation of ice. *Can. J. Phys.* **44,** 477–502.

Virgo, S.E., 1950: Tropical rainfall from cloud which did not extend to the freezing level. *Meteor. Mag.,* **79,** 237–238.

Vollrath, R.E., 1936: Continuously active cloud chamber. *Rev. Sci. Inst.,* **7,** 409–410.

Volmer, M., 1939: *Kinetik der Phasenbildung.* Steinkoff, 283 pp.

Wall, E., 1943: Die eiskeimbildung in losungskernen. *Meteor. Z.,* **60,** 94–104.

Warner, J., 1955: The water content of cumuliform cloud. *Tellus,* **7,** 449–457.

______ , 1957: An instrument for the measurement of freezing nucleus concentrations. *Bull. Obs. Puy de Dome* **2,** 33–46.

______ , 1973: The microstructure of cumulus cloud. Part V: changes in droplet size distribution with cloud age. *J. Atmos. Sci.,* **30,** 1724–1726.

Washburn, E.W., 1924: The vapor pressure of ice and water below the freezing point. *Mon. Wea. Rev.,* **52,** 488–490.

Wegener, A., 1910: Über die eisphase des wasserdampfes in der atmosphäre. *Meteor. Z.,* **27,** 451–459.

______ , 1911: *Thermodynamik der Atmosphäre.* Leipzig, 331 pp.

Weickmann, H. 1942: Experimetelle Utersuchungen zur Bildung von Eis und Wasser an Keimen bi tiefen Temperature. Forschung Bericht No. 1730, Zentral für Wissenschaftliche der Luftfahrtforschung, Berlin-Adlershof. Reprint, 1948: translated by V. Conrad, Air Materiel Command Tech. Rep. 5676.

______ , 1947: Die Eisphase in der Atmosphäre. Reports and Translations, Volkenrode, No. 716. Translation by M.G. Sutton. Ministry of Supply, Royal Aircraft Establishment, 95 pp.

______ , and H.J. aufm Kampe, 1953: Physical properties of clouds. *J. Meteor.,* **10,** 204–211.

Weisener, J., 1895: Beitrage zür kenntnis des tropischen regens. *Sitzungsber. Akad. Wiss. Wien,* **104,** 1397–1434.

Wexler, H., 1945: The structure of the September 1944 hurricane when off Cape Henry, Virginia. *Bull. Amer. Meteor. Soc.,* **26,** 156–159.

Wexler, R., 1952: Precipitation growth in stratiform clouds. *Quart. J. Roy. Meteor. Soc.,* **78,** 363–371.

Young, K.C., 1993: *Microphysical Processes in Clouds.* Oxford, 421 pp.

Zaytsev, V.A., 1950: Liquid water content and distribution of drops in cumulus cloud (in Russian). *Tr. Glav. Geofiz. Obs.,* **19,** 122–132. Reprint, 1953: Translated by G. Belko, Tech. Trans. TT-395. National Research Council, 21 pp.

CHAPTER 8

A History of Research in Cloud Dynamics and Microphysics

HAROLD D. ORVILLE

Introduction

Most cloud physicists break into two groups: those who concentrate on the microphysics of clouds and revel in making observations of clouds and precipitation, and those who concentrate on the dynamics of clouds and revel in conceptual and numerical models of clouds, concentrating on the mathematical description of the subject. This dichotomy may have existed in the early days of cloud studies, too, as scientists strove to define what a cloud was and how it worked.

"What it was," at least with respect to the type of cloud, was classified by Luke Howard in 1803 with contributions from Lamarck's (1802) work. The names of the clouds, based on the Latin for certain characteristics of the clouds, can be attributed to Howard, while the three height stages (low, middle, and high) are due to Lamarck (Houze 1993). The World Meteorological Organization's (WMO) *International Cloud Atlas* describes the 10 cloud genera, 14 species, and nine varieties of clouds (WMO 1969). Of relevance to cloud dynamics is the fact that the cumuliform clouds are the more dynamic and bubbly, of small horizontal extent, with vertical motions of one to several meters per second (the most severe reaching 50 m s^{-1} or more), while the stratiform clouds are more quiescent and layered, covering extensive horizontal areas, with very weak vertical motions (normally 1 m s^{-1} or less).

The patterns and type of clouds indicate the principal dynamics of the clouds. For example, the formation of cloud streets signifies specific atmospheric stability conditions and wind shear. Another example is the open and closed cells that occur over the oceans most often, the open ones over the warmer waters east of the continents and the closed cells over the cooler waters west of the continents. The open cells with large cumulonimbus at the vertices of the hexagonal pattern are characterized by large heat fluxes at the surface. The closed cells are distinguished by cloud-top cooling and infrared destabilization. Agee and Asai (1982) caution that these are statistically favored regions and that the two types of cells can even coexist. Emanuel (1994) gives a thorough treatment and review of the theory.

Clouds can also be classified according to their formation via shallow or deep convection. The shallow ones form within the boundary layer and are limited to within about 3 km of the surface. Their primary importance is in the radiative balance of the earth because of their extensive horizontal breadth. Their vertical growth is normally capped by an inversion. Deep clouds may extend from near the surface of the earth to the tropopause. Some relatively high base clouds are also deep if they extend to the upper regions of the troposphere. Clouds developed in deep convection are important for their production of precipitation, hail, strong winds, tornadoes, blizzards, lightning, and other weather phenomena.

This historical review of cloud dynamics research is closely coupled with microphysics and emphasizes those research developments that show the effects of cloud and precipitation particles on cloud dynamics. The review will use the third equation of motion to provide a framework for the discussion that follows. Attention will center on the forcing of vertical motions, which produce clouds. The review will then focus on the early theories and concepts relative to cloud dynamics before about 1960. Discussion of the period after 1960 will focus on cloud observations and the development of numerical cloud models. The interactions of cloud microphysics and cloud dynamics will be emphasized. The text by Emanuel (1994) treats the more purely dynamical aspects of clouds.

Theoretical framework: The third equation of motion

Some of the discussion in the following sections of the text will refer to forcing terms in the third equation of motion. This equation relates to the vertical component of airflow, usually denoted by w. One form of the equation can be written as follows:

$$dw/dt = (1/\rho_o)\partial p'/\partial z + g\,(\theta'/\Theta + Eq_v' - \ell) + Ent + F_e, \qquad (1)$$

where ρ_o is the air density of the base state, p' is the pressure deviation from the base state, θ' is a potential temperature deviation (equal to temperature in shallow convection), g is gravity, Θ is the potential temperature of the base state, q_v' is the water vapor perturbation, ℓ is the total condensate, E is 0.608, Ent is a symbol for the effects of turbulent mixing and entrainment, F_e is a symbol representing the electrical effects, and z and t are independent variables for height and time, respectively.

This equation indicates several effects on the vertical acceleration of cloudy (or clear) air. The first term on the right is the pressure perturbation gradient; the second is the temperature, water vapor, and hydrometeor effects on buoyancy; the third is the influence of entrainment and mixing; and the fourth is the possible effects of electrical processes on vertical motion in a cloud. It is important to realize that turbulent mixing also affects the temperature and water constituencies through equations for energy and water mass conservation, equations that help form a complete set for cloud development. These indirect effects of mixing are usually more important to the vertical motion than the direct effect in term three of (1), as will be seen below.

Add to these forces the destabilizing effect of mesoscale convergence, the moisture and stability conditions of the environment, the topography and the moisture conditions of the underlying surface, the effects of atmospheric and solar radiation, the possibility of cloud shadow effects, and certain conditions of the vertical shear of the horizontal wind to provide the ingredients for cloud formation and evolution. It is little wonder then that the computer has played a large role in helping scientists to understand clouds.

Equation (1) is written in Lagrangian form, and the terms on the right of (1) represent physical effects. More often in numerical cloud models (1) is displayed in Eulerian form, which explicitly breaks up the total derivative into a local derivative and an advection term. The advection term is moved to the right side of the equation with all of the physical effects. Advection is inherently a nonlinear term and leads to many numerical difficulties in solving the equations for cloud development and dissipation.

Before moving on to the early years in the history of cloud dynamics research, I would like to make mention of the work of Saunders (1957), which relates to (1). In this paper, Saunders reviews the various pseudoadiabatic and saturated adiabatic processes for liquid and ice processes. In addition, he calculates the dimunition of buoyancy due to the condensate load in the saturated adiabatic processes. Saunders uses WPA for the temperature attained in a parcel undergoing a water pseudoadiabatic process, WSA for the temperature in a water-saturated adiabatic process, and similar meanings regarding ice for IPA and ISA. A summary of Saunders's calculations show that

> IPA > WPA (by 1° to more than 2°C over a broad range of pressure or height change),
>
> WSA > WPA (by 1° to 2°C after a pressure decrease of 500–600 mb), and
>
> ISA > WPA (by 1° to 5°C over a broad range).

He also shows that if virtual temperatures and condensate loading are considered, then

> WSA (T_{cv}) < WPA (T_v) (by 1° to 2°C over a drop of 200–400 mb),

where T_{cv} is the cloud's virtual temperature diminished by the loading effect of the condensate, and T_v is the virtual temperature. Emanuel (1994) refers to T_{cv} as the "density" temperature. The points demonstrated here are that the load of the condensate can completely negate the positive buoyancy built up in a saturated adiabatic process and that ice processes can have an important influence on the buoyancy of clouds.

The values for temperature deviation substituted into (1) indicate that the second term on the right of the equation is the dominant term in producing vertical motion in a cloud and can be of order 0.1 m s^{-2}. The temperature term can be matched (but is of opposite sign) by the loading term, particularly in heavy rain- and hail-producing clouds where the water- and ice-mixing ratios may approach 10–20 g kg^{-1}. The pressure perturbation term is important in large clouds and in certain regions of the updraft. Values of 1 mb per 1000 m have been inferred for the pressure perturbation change. The entrainment term is crucial for many moderate strength clouds and at the periphery of many clouds. It may be more important in decreasing the temperature by mixing in cool, dry air to the parcel (Hess 1959). This decrease in thermal buoyancy then has a strong effect on the vertical motion through (1). The electrical force term is normally neglected but may have influence in thunderstorms in regions of lightning. [See chapter 11 by Krider, this volume.] The significance of the pressure perturbation term was clarified in some of Marwitz's observations (Marwitz 1972); it was large enough and in an opposite sense to overcome negative buoyancy near cloud base and to support an updraft. Entrainment and the temperature terms were of concern early in the study of clouds, as we shall see next.

The early years

Concepts

In addition to the classification of clouds mentioned above, the development of atmospheric thermodynamics began in the ninteenth century, leading eventually to the concepts of stability and the parcel model. The pseudoadiabatic process was developed in the late 1880s by Peslin (1868) and was applied by von Bezold (1888) and others (see a review by McDonald 1963). The process led to the calculations of temperature within a rising parcel of cloudy air. The parcel model of stability is based on comparisons of the temperature in the parcel and in the environment, resulting in vertical motion via the second term on the right of (1) above, if there is a temperature difference. Modifications of this, taking into account the effects of subsiding environmental air, were discussed by Bjerknes (1938) and were generally called the slice model of convection. The effect of the subsiding air is to warm the environment and decrease the stability of the cloudy parcel (if conditional stability originally existed) compared with the parcel model results. The theory also indicated that the smaller the ratio of updraft air to downdraft air, the greater the instability [as shown by Hess (1959) and Emanuel (1994)].

The original parcel model also ignored the effects of entrainment, a topic that continues to confound the cloud dynamicist to this day, although progress is certainly being made. Stommel (1947, 1951) and Schmidt (1947) applied entraining jet concepts to the growth of cumulus clouds. Lateral entrainment of dry outside air into the cloud was postulated to account for the depletion of cloud water below its adiabatic value and a cooling of the parcel below its pseudoadiabatic temperature value. Stommel implied a tripling of the mass flux over the clouds' depth (which was quite limited, ranging from 500 to 1400 m) in order to account for the observations of temperature inside the clouds (no measurements of water content were made). For a steady-state cloud, this requires either a continually increasing updraft or an increasing cross section with height, conditions that are not very realistic for trade wind cumuli, the subject of Stommel's work. Further development of the entrainment calculations were accomplished by Austin and Fleisher (1948). Houghton and Cramer (1951) examined Stommel's assumption of conservation of vertical momentum and found it necessary to assert lateral entrainment to satisfy continuity conditions. Homogeneous, instantaneous mixing across the width of the cloud or parcel was normally assumed in these studies.

To account for the variable state of most clouds, the bubble theory of convection was postulated by Ludlam and Scorer (1953) and refined by Malkus and Scorer (1955). The theory, as envisioned by Ludlam (1958), pictured a cloud to form from a succession of thermals, the initial ones eroding away and leaving a more favorable path for subsequent bubbles to reach greater heights. This would account for the pyramidal shape of some clouds. The entrainment rate in such clouds was equal to $0.6/R$, where R is the radius of the thermal or bubble.

Plumes or jets were also thought to approximate the shape of rising cumulus clouds. Similarity theory, an application of dimensional analysis, was applied to the concept by Morton et al. (1956). [See the development of the theory and application in Emanuel (1994).] The entrainment rate for plumes was determined to be $0.2/R$. The combination of the bubble and plume models resulted in the starting plume model (Turner 1962) with the same entrainment rate as a plume. The concept was one of a plume capped with a bubble or thermal. The advantage of this for the thermal is that the air mixing into the underside of the thermal is a mixture of the plume and outside air.

Much of the discussion and theoretical work focused on the form of entrainment, whether it was dynamic entrainment or gross entrainment [Squires, in Fletcher (1962)], how much detrainment was there, and whether there was form drag or "skin friction" affecting the cloud (Malkus and Scorer 1955; see an excellent discussion by Houze 1993). One of many useful results was the fact that the updraft velocity in the center of the thermal was greater than the rise rate of the leading edge of the cloud, perhaps by a factor of 2 or more (Woodward 1959). Laboratory work (Scorer 1957) and observations made by soaring in thermals were important for many of these results.

The concepts of lateral, homogeneous, instantaneous mixing (Houze 1993) when applied in 1D, steady-state models (discussed below) failed, in general, to predict the proper dilution of the cloud

(Warner 1955; Warner and Squires 1958) while doing an adequate job of predicting the growth of the cloud top, given the proper radius. Squires (1958) proposed the entrainment of air from the top of the cloud to model a cloud more realistically, an idea that was to resurface nearly 20 years later (Paluch 1979).

Observational programs

Observational studies led to many of the discoveries about clouds in these early years, although the instrumentation to measure characteristics of clouds was not that well developed. Studies by Bunker and associates of trade wind cumuli (Bunker et al. 1949) provided much of the data for the entrainment analyses. Malkus continued the observational programs to develop theories regarding the effects of wind shear on the slopes of clouds and other cloud characteristics (Malkus 1952).

The largest field program in the United States during this period was the Thunderstorm Project (Byers and Braham 1949). Five aircraft sampled the storms simultaneously measuring air motions and temperatures. At the ground, 10 radars were used to track balloons and measure the horizontal flow into the clouds so that calculations of divergence could be made. [See chapter 7 by Braham, this volume.] Three stages of the thunderstorm were depicted: the first, the cumulus stage, with updrafts throughout the cell; the second, the mature stage, with both updrafts and downdrafts (but with the downdrafts in the lower half of the cloud) and with precipitation first reaching the ground; and the third, the dissipating stage, with weak downdrafts throughout the cell. The downdraft and cool outflow were associated with midlevel lateral inflow of dry air as the initiating mechanism. Lapse rates in the clouds were considerably steeper than the appropriate pseudoadiabatic ones. Cells lasted for 20 minutes or more and the storm for 1 hour or more.

It is obvious from material in translated Russian texts that the Russians had also used aircraft to sample the interior of clouds. Zaitsev (1950) reported on the distribution of cloud liquid water in cumulus and cumulus–congestus clouds. His picture of the typical cloud water distribution was one of a maximum value slightly above the center of the cloud and decreasing in all directions from the maximum. Warner's work (1955) in Australia presented a much more complicated picture, with regions of dryness intermixed with moist volumes. Vul'fson and his Russian colleagues (Vul'fson 1954) measured the vertical motions and other characteristics of cumulus clouds. Gutman (1957) and Shishkin (1958) constructed mathematical models of cumulus and cumulonimbus clouds. Their theoretical calculations indicated that updrafts of 50 m s^{-1} were possible.

Other observational programs were being pursued in other countries. The Australian scientists Squires and Warner were very active. The observation by Kraus and Squires (1947) of dynamic growth of a cumulus cloud after seeding with dry ice sparked interest in the possible modification of the dynamics of clouds to influence their production of rain. In the United Kingdom, Yates reported on measurements in the subcloud layer (Yates 1953). Earlier, Diem (1942, 1948) had reported on measurements made in clouds in Germany. The study of clouds had taken on a vigorous status in the 1940s and 1950s, which was only to grow in the later decades.

The use of computers and the development of cloud models distinguishes the next period in the history of cloud dynamics and its relationship with microphysics and will be taken up next.

The advent of computers and numerical cloud models

Numerical cloud models

The first numerical models of thermals and clouds were developed in the late 1950s. The models made it possible to test the various physical processes and their effects on cloud development. Malkus and Witt (1959) developed a numerical model of a dry thermal in two dimensions. Figure 8-1 shows some results. The integration was not carried on very long, but the results showed that the methods being developed for numerical weather prediction could also be applied in cloud dynamics. More will be said about this class of models shortly.

We classify the models as zero-, one-, two-, and three-dimensional (0D, 1D, 2D, 3D), and time dependent or steady state (TD or SS). The models can be fully coupled among the many physical processes, such as the microphysics and the dynamics, or uncoupled, such as occur in kinematic models where the airflow may be prescribed and the microphysics are calculated.

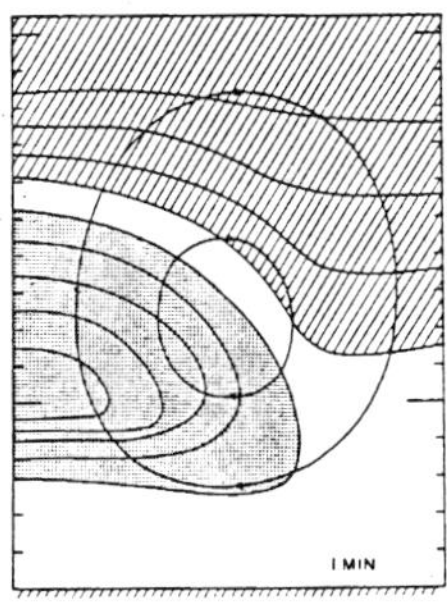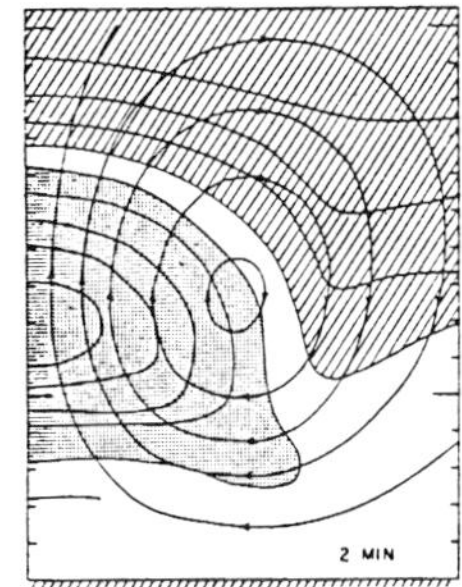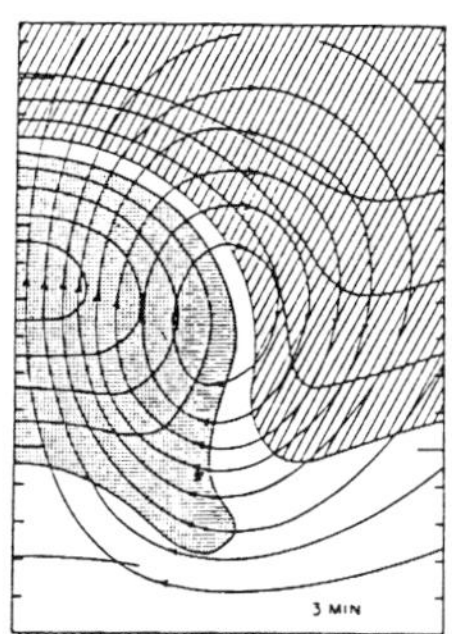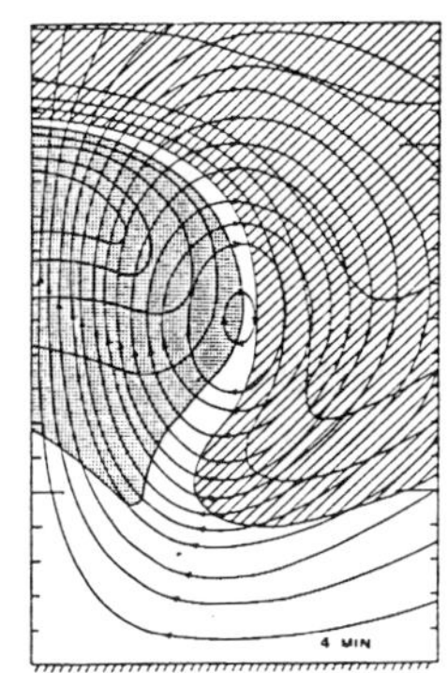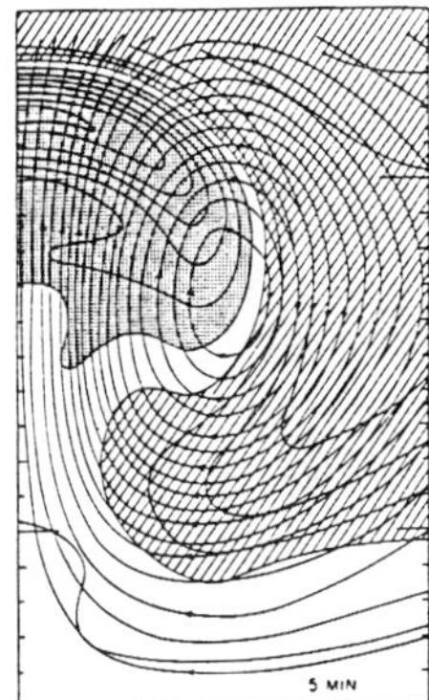

FIG. 8-1. Growth of "up in the air" bubble at 1-min intervals for the first 5 min. The ticks are vertically spaced 20 m apart, with large ticks at z = 100 m (initial perturbation base) and z = 300 m (beginning of upper stable layer). Potential temperature isopleths (intervals of 0.1°C) are solid lines without arrows; the dotted region includes all potential temperatures higher than 289 K, which is characteristic of the lower neutral layer, while the hatched region includes all values lower than 288.9 K. Lines with arrows are streamlines (from Malkus and Witt 1959).

0D

These are often called box models and normally are Lagrangian in nature. Some examples are the models by Elliott (1981), Juisto (1971), Johnson (1979, 1980, 1982), Klazura and Todd (1978), Koenig (1971, 1972), Pastre and Rosset (1977), Rokicki and Young (1978), Strapp et al. (1979), and Heymsfield (1986). These models concentrate on the development of individual particles or groups of particles in a cloud parcel. They simulate the formation of precipitation in a cloud. The models have little to do with cloud dynamics and will not be discussed further.

1DSS

These are sometimes called stick models because the models provide solutions for the dependent variables along the vertical axis. A set of ordinary differential equations constitutes the model. Some examples are the models by Squires and Turner (1962), Simpson et al. (1965), Weinstein and Davis (1968), Simpson and Wiggert (1971), Cotton (1972), Hirsch (1972), Kachurin et al. (1973), Cotton and Boulanger (1975), Orville et al. (1980), Matthews (1981), Wiggert et al. (1982), and Ackerman and Sun (1985). These all have coupled microphysics and dynamics. A few others use a prescribed airflow and are thus uncoupled. These are models by Kessler (1969), Nelson (1971), Young (1977), and Vali et al. (1988).

The principal use of the coupled models has been to provide a sophisticated analysis of an atmospheric sounding. The models have also illustrated some of the interactions between the microphysics and the updraft in a cloud, as indicated in (1). Predictions of cloud-top height and maximum vertical velocities are most often produced in the coupled 1DSS models, although more detailed predictions of precipitation development have been made (Cotton 1972). Although the models have been used to predict precipitation, they should not be in general.[1] An acceptable practice would be to use the predicted cloud-top height from a 1DSS cloud model with a good radar climatology of cloud-top height versus rainfall to estimate rainfall potential, but this has not been done yet to my knowledge.

These models were used extensively in the mid-1960s and early 1970s. Warner (1970) and Simpson (1971) engaged in lively debate about the ability of the 1DSS model to predict successfully both the cloud-top height and the liquid water content. The lateral entrainment assumption of most of these models precluded a more realistic treatment of the mixing of outside air into a cloud.

1DTD

Time is added as an independent variable to the 1DSS models. Consequently, the models involve the solution of nonlinear, partial

---

[1] These models lack a realistic simulation of the precipitation process, primarily the interaction of one parcel with another and the fallout of the particles to lower levels.

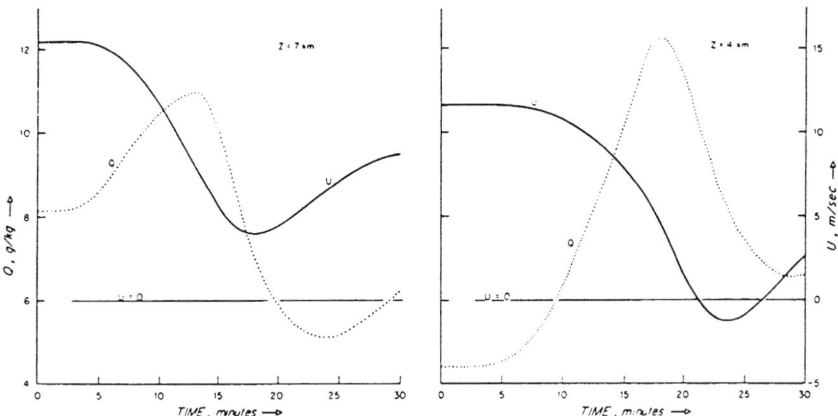

FIG. 8-2. Time variations of updraft U and total water content Q = (R + W) at z = 7 km (left) and at z = 4 km (right); R equals rainwater, and W equals cloud water mixing ratios (from Srivastava 1967).

differential equations, a much more difficult task than solving ordinary differential equations.[2]

An example from this type of model is shown in Fig. 8-2 and comes from Srivastava (1967). This figure shows clearly the strong interactions between the development of precipitation and vertical motion. The maximum loading of the updraft lagged the maximum updraft; a periodic updraft–downdraft ensued, coupled with the development and fallout of the precipitation.

The effect of the microphysics on the dynamics was also clearly demonstrated in Wisner et al. (1972). Mixing was proportional to the radius of the updraft and the slope of the cloud surface. The production of cloud, rain, and hail was simulated. If the formation of hail was inhibited, the updraft would reach a steady state (Orville et al. 1975), probably an artifact of the dynamic framework. The inclusion of hail created a strong downdraft. The melting of the hail and the resultant cooling of the atmosphere was the crucial element in forming the downdraft. The authors concluded that,

> Hail plays a very important role in the chain of events leading to precipitation at the ground. It provides a mechanism for transporting

---

[2]Some examples of 1DTD models include Srivastava (1967), Asai and Kasahara (1967), Weinstein (1970), Ogura and Takahashi (1971), Wisner et al. (1972), Silverman and Glass (1973), Farley and Chen (1975), Farley et al. (1976), Hindman et al. (1977), Nelson (1979), Srivastava (1985, 1987), Hu and He (1989), and Ferrier and Houze (1989).

water down through strong updrafts (> 10 m s^{-1}) and into the lower levels where its weight and the cooling due to its melting are important factors in the initiation of a downdraft. The evaporation of the rain formed by the melted hail then influences strongly the characteristics of the downdraft.

These comments presage the work on microbursts, which was to occur many years later.

2DSS

A second space dimension is added as an independent variable. These models also have uncoupled microphysics and dynamics.[3]

Young's model is particularly detailed in the treatment of the ice processes. No interaction of the microphysics with the airflow is allowed, which seems reasonable in many orographic cases but should be checked for relatively moist soundings (i.e., those with relatively large amounts of supercooled liquid water). It may be possible in some situations for the development of ice precipitation to cause embedded convection to occur, but this effect cannot be simulated in this type of model.

2DTD

These models can be either axisymmetrical or slab symmetric and are fully coupled models. A vorticity form of the equations is normally used for the solutions.[4]

These models have been successful in simulating many features of cumulus and storm dynamics. These features include a multitude of

---

[3]Some examples of 2DSS models include Elliott (1969), Hobbs et al. (1973), Young (1974, 1993), and Plooster and Fukuta (1975).

[4]Some examples of 2DTD models include Lilly (1962), Ogura (1963), Orville (1965), Liu and Orville (1969), Orville and Sloan (1970), Takeda (1971), Wilhelmson and Ogura (1972), Murray and Koenig (1972), Soong and Ogura (1973), Johnson et al. (1975), Orville and Kopp (1977, 1978), Tag (1979), Helsdon (1980), Hsie et al. (1980), Hsie et al. (1980), Hane (1978), Chui (1978), Thorpe et al. (1980, 1982), Chen and Orville (1980), Hall (1980), Orville and Chen (1982), Koenig and Murray (1975, 1983), Kopp et al. (1983), Lin et al. (1983), Orville et al. (1984, 1987), Klaassen and Clark (1985), Clark et al. (1986), Farley and Orville (1986), Helsdon and Farley (1987a,b), Farley (1987a,b), Droegemeier and Wilhelmson (1987), Kubesh et al. (1988), Kopp (1988), Proctor (1988, 1989), Hjelmfelt et al. (1989), Aleksić et al. (1989), Fovell and Ogura (1988, 1989), Tuttle et al. (1989), Orville et al. (1989a,b), Orville and Kopp (1990), Scala et al. (1990), Tao et al. (1993), Ferrier (1994), Kopp and Orville (1994), Tzivion et al. (1994), Randell et al. (1994), and Murakami et al. (1994)

microphysical and dynamical interactions; hail embryo and hailstone circulations in hailstorms; cloud mergers; convergence effects on convection; power park (excess) heat and vapor effects on severe storms; the physics and dynamics of microbursts; some of the causes of long-lasting storms; aerosol transport from the boundary layer to the tropopause; the formation of accumulation zones; cloud-seeding effects; cloud electricity and lightning; the formation and evolution of gust fronts, heavy rains, and flash floods; recycling of precipitation in strong updrafts; small-scale mixing; lower surface and gravity wave interactions with cloud and storm formation and evolution; and radiation effects. The models have also been used for precipitation forecasts and simulations.

An example from Koenig and Murray (1983) is shown in Fig. 8-3 for an axisymmetrical cloud model. The various forces that affect the vertical motion were quantitatively compared. The net force at a given location within the cloud was often quite small compared to several individual components of the force as represented in (1). Changes in the ice and liquid ratios and the attendant thermal effects influenced the production of updrafts and downdrafts in the cloud, depending on the type of cloud modeled. A warm-base cloud was much more affected than a cold base cloud.

3DTD

Three-dimensional, time-dependent cloud models have been applied to most of the above topics, plus those features that are uniquely 3D, such as tornadoes and storm splitting. The models have been used to clarify conditions necessary for particular types of storms to form and the physics of tornado formation.[5]

The 3D models alleviate the shortcomings of the geometrical simplifications of 1D and 2D models and are thus able to address a number

---

[5]Some examples of 3DTD models and/or studies include Steiner (1973), Wilhelmson (1974), Schlesinger (1975, 1978, 1984), Sommeria (1976), Lipps (1977), Klemp and Wilhelmson (1978a,b), Wilhelmson and Klemp (1978, 1981), Cotton and Tripoli (1978), Clark (1979), Clark and Gall (1982), Weisman and Klemp (1982, 1984), Rotunno and Klemp (1982, 1985), Wilhelmson and Chen (1982), Clark and Farley (1984), Levy and Cotton (1984), Knupp (1985), Droegemeier and Wilhelmson (1985), Smolarkiewicz and Clark (1985), Proctor (1987a,b), Weisman et al. (1988), Redelsperger and Lafore (1988), Balaji and Clark (1988), Dudhia and Moncrieff (1987), Murakami (1990), Hall et al. (1990), Farley et al. (1992, 1994), Tao and Simpson (1993), Skamarock et al. (1994), and Bruintjes et al. (1994a–c).

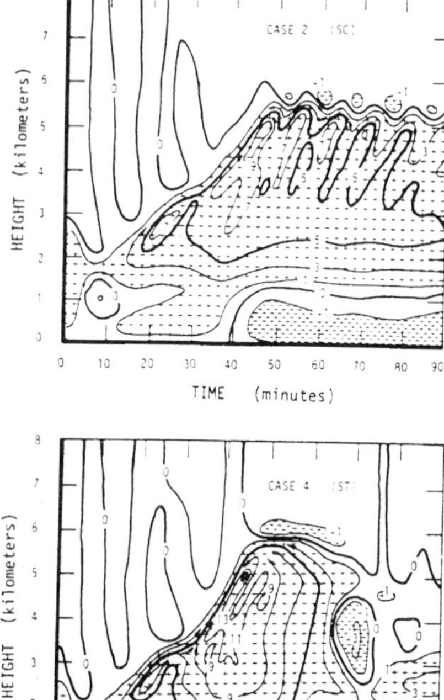

FIG. 8-3. Time–height sections showing the evolution of the vertical draft velocity on the axis of cases 2(SC) and 4(ST). Regions of greater than ±1 m s^{-1} drafts are shaded, contours for drafts of 0 and 5 m s^{-1} are emphasized; contour of +15 m s^{-1} in case 4(ST) is not labeled; C is clean air; T is turbid air (from Koenig and Murray 1983).

of topics, primarily of a dynamic character. Among the important areas addressed by 3D modeling to date have been the role of vertical wind shears, the physics of storm splitting, the divergent motion of individual cells within multicellular storms, storm and cloud rotation, etc. The 3D modeling requires sacrifices, however, in terms of grid resolution and microphysical detail. These models also require large and fast computer systems. Analysis is also time consuming. Consequently, their use for entrainment, merger, and other cloud cell studies has not been as extensive as in the 2D models.

The example from Proctor (1987b) in Fig. 8-4 illustrates a realistic simulation of a microburst. Srivastava (1985, 1987), Proctor (1988, 1989), Hjelmfelt et al. (1989), and Orville et al. (1989b) provide strong arguments that cloud microphysical properties are the primary forcing effects of the microburst. The drag of the precipitation, melting of the graupel and hail, and evaporation of rain are key elements for the formation of a microburst.

The sample result from Farley et al. (1992) in Fig. 8-5 shows the growth of a single cell thunderstorm from the Convective Cooperative Precipitation Experiment (CCOPE) in 1981. The growth of this storm lasted just long enough for a small shower to develop and for one lightning flash to occur. Development of the precipitation through the ice phase led to the loading of the updraft and the formation of the

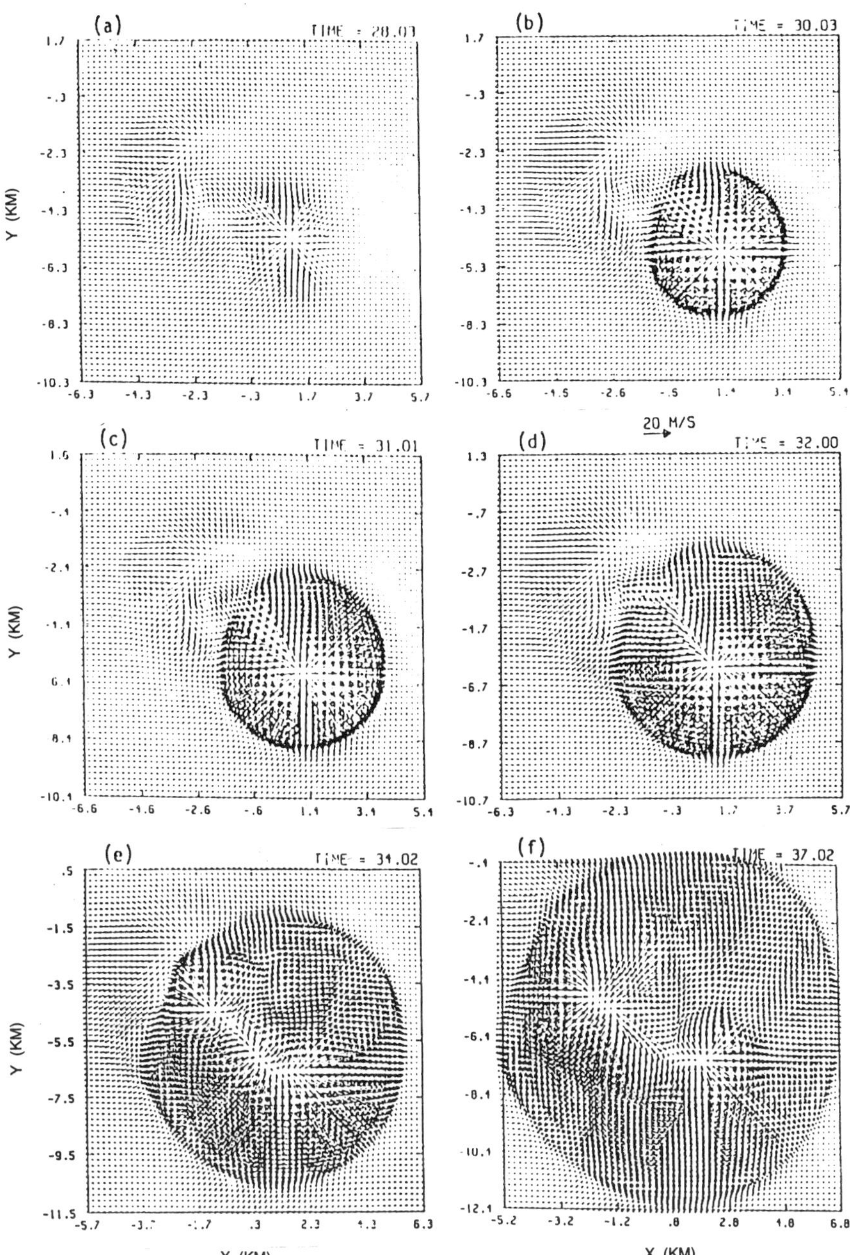

FIG. 8-4. Simulated low-level wind vectors for the 2 August 1985 Dallas–Ft. Worth, Texas, microburst case. All panels show results at an altitude of 100 m. The model time is shown in the upper-right corner of each panel, and the vector scale length is shown between (c) and (d) (adapted from Proctor 1987b).

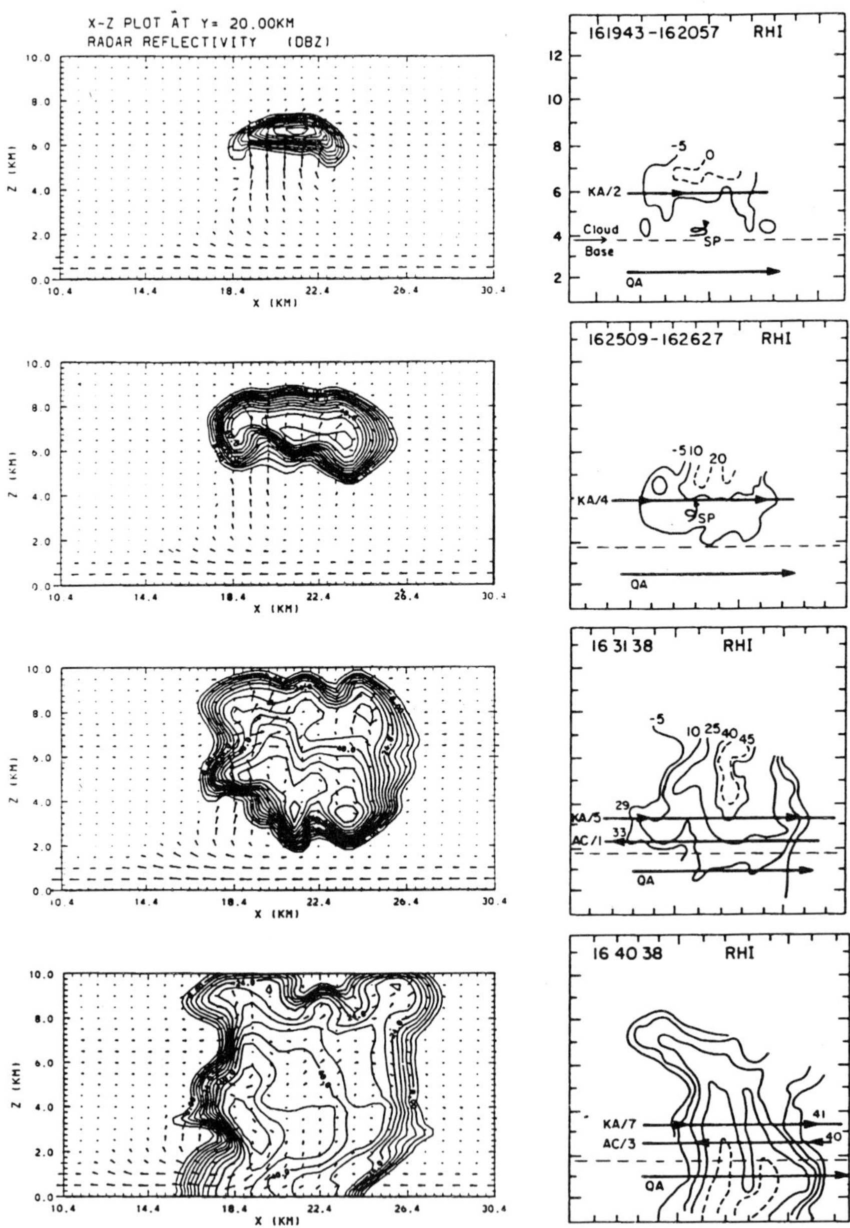

FIG. 8-5. Comparative plots of model-generated west–east vertical cross section of radar reflectivity at Y = 20 km (left panel) and actual CP-2 RHIs (right panel) for case MB. Model plots have 4-dB contour intervals. CP-2 RHIs have 15-dB contour intervals except for the dashed lines, which are labeled. The positions of aircraft are indicated by bold arrows. Tick marks are at every 1 km. The CP-2 RHIs are extracted from Fig. 4 of Dye et al. (1986) (from Farley et al. 1992).

downdraft. A strong microburst developed in the dry subcloud layer of the environment.

Much excellent work has been done using 3DTD models to characterize the dynamics of severe storms. Those studies are reviewed in the texts by Cotton and Anthes (1989), Houze (1993), and Emanuel (1994).

MODELING WORKSHOPS

Cloud modeling workshops have been used to test the models on similar soundings (Silverman et al. 1976; World Meteorological Organization 1985, 1988, 1992a). Comparisons of results from model to model as well as with the appropriate observations from aircraft, radar, and other surface equipment were made. Some excellent datasets were developed, and many publications resulted from the workshops. The 1976 workshop illustrated the weaknesses in 1DTD models in predicting precipitation and the possibilities of successful predictions with 2DTD models. No 3D models were presented.

The 1985 and 1988 workshops showed a strong trend to more 3D simulations. In 1985, questions arose as to the proper path to precipitation and the appropriateness of models of lesser than three dimensions. The observations and modeling of atmospheric electricity effects and their good agreement were reported in 1985. Also, at this workshop, the three-dimensional simulation of a hailstorm proved a worthy test for the 3DTD models. The proper location of the hailfall depended on the microphysical representation of the hail.

The 1988 workshop provided a format for more complex and complete models. The proper path to precipitation was simulated more frequently in the test cases. The first use of a multidimensional model to predict (*not* simulate) cloud and precipitation evolution was demonstrated. The difficulties in comparing model and observational results were noted, and some data uniformity and standard formats were discussed. This 1988 workshop also focused on the advisability of using surface heat and vapor fluxes in place of thermal bubbles for the initiation of convection in the cloud models, at least for those models concerned with the prediction (as opposed to simulation) of cloud and precipitation development.

The 1985 workshop attracted 66 scientists from 12 countries. The numbers rose to 86 scientists from 18 countries in 1988. In both work-

1992 WMO Workshop on Cloud Microphysics & Applications to
Global Change: Data Summaries

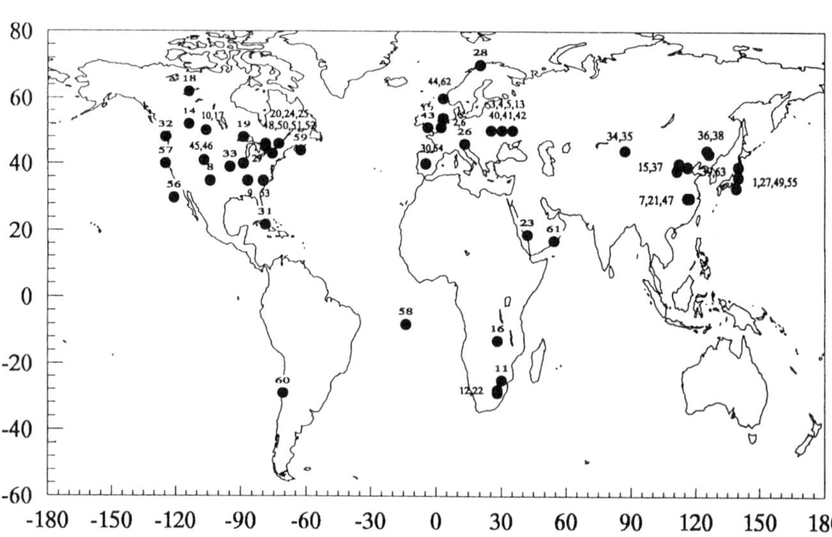

FIG. 8-6. Locations of data collection (from WMO 1992b).

shops, key observationalists for all of the cases were at the workshop
and presented the observers' viewpoints and data.

The workshop in 1992 in Toronto, Canada, was held under different conditions than the other three. A collocated workshop on cloud
microphysics and applications to global change was held at the same
time. Some diffusion of effort was noted, but nearly 200 scientists
attended one or both of the workshops. Multidimensional, time-dependent cloud models continued to give good comparisons with many features of the observations. Problems continued with respect to uniform
initiation methods. The first marine boundary-layer clouds were included as a test case. The survey of microphysical datasets revealed
how extensive the studies of clouds had become over the past 20 years
(Fig. 8-6), much of the research having to do with the interaction of the
microphysics and dynamics of clouds.

Conceptual models

In addition to numerical models, conceptual models of convection,
clouds, and storms have progressed rapidly since the 1960s. Kessler
(1969, 1975) used simple kinematic models of airflow to calculate the
development of precipitation and the distribution of cloud water and

rain in the vertical. He showed the accumulation of rain above a strong updraft but not of the magnitude envisioned by Sulakvelidze et al. (1967) in his studies of the formation of hail.

Conceptual models of entrainment were further developed during this period. The concept of continuous lateral entrainment with instantaneous and homogeneous mixing across the cloud, mentioned above, was replaced with discontinuous, inhomogeneous entrainment. The key studies contributing to the latest theory were those by Paluch (1979), Baker et al. (1980), and Raymond and Blyth (1986). Houze (1993) summarizes the work very nicely, showing how the cloud top is determined by the rise of the few lucky parcels that suffer no entrainment. The main mass of the cloud is made up of parcels that have risen from different levels and have encountered various amounts of entrainment. Comparisons of in-flight measurements of cloud droplet spectra were important for establishing the theory (Baker et al. 1980).

In the 1970s, scientists at the Imperial College produced realistic models of severe storm circulation. Moncrieff and Green (1972), Moncrieff and Miller (1976), and Thorpe et al. (1982) produced models of steady, convective, overturning storms (Cotton and Anthes 1989). Work by Rotunno (1981) showed how the loading of an updraft by the precipitation content (aided perhaps by entrainment) could result in the splitting of a thunderstorm into two cells. The directional shear of the wind was found to be crucial for determining which of the cells would predominate. [See Klemp (1987) for an excellent review of this and related subjects.]

Hailstorms and thunderstorms were the objects of study for Browning and Ludlam (1962), Newton (1963), Fankhauser (1976), Browning and Foote (1976), Knight and Squires (1982), Foote and Wade (1982), and many others. An excellent monograph regarding hail is still pertinent (Foote and Knight 1977). Concepts of multicell and supercell storms were developed from observations of a supercell storm in England and multicell storms during the National Hail Research Experiment (Fankhauser 1976).

Observational efforts

Numerous field experiments have been conducted in the United States and other countries concentrating on the dynamics of clouds or providing data concerning clouds. The Thunderstorm Project and other projects were referred to above. Cloud-seeding experiments provided many opportunities to measure the internal properties of clouds. Notable among them were Project STORMFURY, Florida Area Cumulus Experiment, National Hail Research Experiment, Skywater, High Plains Experiment, Sierra Cooperative Pilot Project, Precipitation Enhancement Project, Israeli I and II field experiments, Alberta Research Council Hail Experiment, Project Hailswath of the South Dakota School of Mines and Technology, and the National Oceanic and Atmospheric Administration's federal–state cooperative programs: Precipitation Augmentation for Crops Experiment, North Dakota Thunderstorm Project, North Dakota Tracer Experiment, plus field experiments in Utah, Arizona, Texas, and Nevada. Numerous cloud-seeding projects in China, Southeast Asia, South Africa, the former Soviet Union, and Morocco also provided information on cloud dynamics and the microphysics of clouds.

Field projects concentrating on other aspects of cloud microphysics and dynamics were the Global Atmospheric Research Program Atlantic Tropical Experiment, Severe Environmental Storm and Mesoscale Experiment, PRE-STORM, Cooperative Huntsville Meteorological Experiment, Northern Illinois Meteorological Research on Downburst, Joint Airport Wind Shear Experiment, Convective Cooperative Precipitation Experiment, National Severe Storm Laboratories projects, field studies of the New Mexico Institute of Mining and Technology, the Thunderstorm Research International Project, Cumulus and Precipitation Experiment, the U.K. studies on clouds and mesoscale systems, and the French cloud studies in boundary layers and frontal situations.

The growth of the number and extent of field studies since the 1940s and 1950s concerning clouds is astounding. Most of the studies in the 1960s and 1970s were associated with cloud seeding and weather modification. The number of these projects peaked in the early 1970s. Although studies concerning weather modification have declined, field studies of the radiative properties of clouds, characteristics needed for climate models used in global change studies, are increasing. Also, interest in severe storms continues, which assures a steady progression of field programs.

Closing remarks

There have been a number of questions concerning the dynamics and microphysics of clouds that have endured throughout the years.

Some of the earliest studies concerned entrainment and mixing in clouds, topics that are still being investigated. Some of the more important problems concerned the interactions between the microphysics and dynamics of clouds, the place and timing of ice initiation, its effect on precipitation formation and the dynamics of the cloud, the production of microbursts, the effects of ice phase seeding on cloud dynamics, the effects of surface heat and moisture sources, the effects of gravity waves, and the influence of atmospheric electricity on cloudy air motions.

Of great importance for the advancement of the understanding of clouds and microphysical interactions has been the significant strides made in the development of new or improved equipment, the use of more powerful and convenient computer power, and the development of realistic numerical models. Aircraft instrumentation, dual-channel microwave radiometers, Doppler and multiparameter radars, satellite remote sensing, wind profilers, automated rain gauge networks, and mesoscale network stations have all contributed significantly to the studies of clouds. Numerical cloud models of various complexities and dimensions have been successfully applied to many of the societal problems caused by severe storms and their products. Much further insight into cloud and precipitation processes and their influence on human beings will occur in the future.

Acknowledgments. Support for this research was provided by the National Science Foundation, Division of Atmospheric Sciences, under Grant ATM-9215188. Acknowledgment is made to the National Center for Atmospheric Research, which is sponsored by the National Science Foundation, for use of its computing facility.

I thank Ms. Joie Robinson, Ms. Carol Hirsch, and Ms. Mary Holter for their help in preparing this manuscript for publication.

REFERENCES

Ackerman, B., and R.Y. Sun, 1985: Predictions by two one-dimensional cloud models: A comparison. *J. Climate Appl. Meteor.,* **24,** 617–628.

Agee, E.M., and T. Asai, 1982: *Cloud Dyamics.* Reidel, 423 pp.

Aleksić, N.M., R.D. Farley, and H.D. Orville, 1989: A numerical cloud model study of the Hallett–Mossop ice multiplication process in strong convection. *Atmos. Res.,* **23,** 1–30.

Asai, T., and A. Kasahara, 1967: A theoretical study of the compensating downward motions associated with cumulus clouds. *J. Atmos. Sci.,* **24,** 487–496.

Austin, J.M., and A. Fleisher, 1948: A thermodynamic analysis of cumulus convection. *J. Meteor.,* **5,** 240.

Baker, M.B., R.J. Corbin, and J. Latham, 1980: The influence of entrainment on the evolution of cloud droplet spectra. 1. A model of inhomogeneous mixing. *Quart. J. Roy. Meteor. Soc.,* **106,** 581–598.

Balaji, V., and T.L. Clark, 1988: Scale selection in locally forced convective fields and the initiation of deep convection. *J. Atmos. Sci.,* **45,** 3188–3211.

Bjerknes, J., 1938: Saturated-adiabatic ascent of air through dry-adiabatically descending environment. *Quart. J. Roy. Meteor. Soc.,* **64,** 325–330.

Browning K.A., and F.H. Ludlam, 1962: Airflow in convective storms. *Quart. J. Roy. Meteor. Soc.,* **88,** 117–135.

_____ and G.B. Foote, 1976: Airflow and hail growth in supercell storms and some implications for hail suppression. *Quart. J. Roy. Meteor. Soc.,* **102,** 499–533.

Bruintjes, R.T., T.L. Clark, and W.D. Hall, 1994a: The effects of gravity waves on the production of supercooled liquid water (SLW) in complex terrain. *Proc. Sixth Scientific Conf. on Weather Modification,* Paestum, Italy, WMP, 287–290.

_____ , _____ , and _____ , 1994b: The dispersion and transport of tracer plumes in complex terrain and the implications for cloud seeding experiments. *Proc. Sixth Scientific Conf. on Weather Modification,* Paestum, Italy, WMP, 291–294.

_____ , _____ , and _____ , 1994c: Interactions between topographic airflow and cloud/precipitation development during the passage of a winter storm in Arizona. *J. Atmos. Sci.,* **51,** 48–67.

Bunker, A.F., B. Haurwitz, J.S. Malkus, and H. Stommel, 1949: Vertical distribution of temperature and humidity over the Caribbean Sea. Woods Hole Oceanographic Histitution and Massachusetts Institute of Technology Papers in Physical Oceanography and Meteorology, Vol. 18, No. 1, 82 pp.

Byers, H.R., and R.R. Braham, 1949: The thunderstorm: Report of the Thunderstorm Project. U.S. Govt. Printing Office, Washington, DC, 287 pp.

Chen, C.H., and H.D. Orville, 1980: Effects of mesoscale convergence on cloud convection. *J. Appl. Meteor.,* **19,** 256–274.

Chiu, C.S., 1978: Numerical study of cloud electrification in an axisymmetric, time-dependent cloud model. *J. Geophys. Res.,* **83** (C10), 5025–5049.

Clark, T., 1979: Numerical simulations with a three-dimensional cloud model: Lateral boundary condition experiments and multicellular severe storm simulations. *J. Atmos. Sci.,* **36,** 2191–2215.

_____, and R. Gall, 1982: Three-dimensional numerical model simulations of airflow over mountainous terrain: A comparison with observations. *Mon. Wea. Rev.,* **110,** 766–791.

_____, and R.D. Farley, 1984: Severe downslope windstorm calculations in two and three spatial dimensions using anelastic interactive grid nesting: A possible mechanism for gustiness. *J. Atmos. Sci.,* **41,** 329–350.

_____, T. Hauf, and J.P. Kuettner, 1986: Convectively forced internal gravity waves: Results from two-dimensional numerical experiments. *Quart. J. Roy. Meteor. Soc.,* **112,** 899–925.

Cotton, W.R., 1972: Numerical simulation of precipitation development in a supercooled cumuli. Part II. *Mon. Wea. Rev.,* **100,** 764–784.

_____, and A. Boulanger, 1975: On the variability of "dynamic seedability" as a function of time and location over south Florida: Part I. Spatial variability. *J. Appl. Meteor.,* **14,** 710–717.

_____, and G. Tripoli, 1978: Cumulus convection in shear flow—Three-dimensional numerical experiment. *J. Atmos. Sci.,* **35,** 1503–1521.

_____, and R.A. Anthes, 1989: *Storm and Cloud Dynamics.* Academic Press, 883 pp.

Diem, M., 1942: Messungen der Grösse von Wolkenelementen. I. *Ann. Hydrogr.,* **70,** 142.

_____, 1948: Messungen der Grösse von Wolkenelementen. II. *Met. Rdsch.* **1,** 261.

Drogemeier, K.K., and R.B. Wilhelmson, 1985: Three-dimensional numerical modeling of convection produced by interacting outflows. Part I: Control simulation and low-level moisture variations. *J. Atmos. Sci.,* **42,** 2381–2403.

_____, and _____, 1987: Numerical simulation of thunderstorm outflow dynamics. Part I: Outflow sensitivity experiments and turbulence dynamics. *J. Atmos. Sci.,* **44,** 1180–1210.

Dudhia, J., and M.W. Moncrieff, 1987: Numerical simulation of quasi-stationary tropical convective bands. *Quart. J. Roy. Meteor. Soc.,* **113,** 929–967.

Dye, J.E., J.J. Jones, W.P. Winn, T.A. Cerni, B. Gardiner, D. Lamb, R.L. Pitter, J. Hallet, and C.P.R. Saunders, 1986: Early electrification and precipitation development in a small, isolated Montana cumulonimbus. *J. Geophys. Res.,* **91,** 1231–1247.

Elliott, R.D., 1969: Cloud seeding area of effect numerical model. Aerometric Research, Inc. Rep., Santa Barbara, CA.

———, 1981: A seeding effect targeting model. Preprints, *Eighth Conf. on Inadvertent and Planned Weather Modification,* Reno, NV, Amer. Meteor. Soc., 28–29.

Emanuel, K.A., 1994: *Atmospheric Convection.* Oxford University Press, 580 pp.

Fankhauser, J.C., 1976: Structure of evolving hailstorm. II. Thermodynamic structure and airflow in the near environment. *Mon. Wea. Rev.,* **104,** 576–587.

Farley, R.D., 1987a: Numerical modeling of hailstorms and hailstone growth: Part II. The role of low density riming growth in hail production. *J. Climate Appl. Meteor.,* **26,** 234–254.

———, 1987b: Numerical modeling of hailstorms and hailstone growth: Part III. Simulation of an Alberta hailstorm—Natural and seeded cases. *J. Climate Appl. Meteor.,* **26,** 789–812.

———, and C.S. Chen, 1975: A detailed microphysical simulation of hygroscopic seeding on the warm rain process. *J. Appl. Meteor.,* **14,** 718–733.

———, and H.D. Orville, 1986: Numerical modeling of hailstorms and hailstone growth: Part I. Preliminary model verification and sensitivity tests. *J. Climate Appl. Meteor.,* **25,** 2014–2035.

———, F.J. Kopp, C.S. Chen, and H.D. Orville, 1976: Use of cloud models to test hail suppression concepts. Preprints, *Int. Conf. on Weather Modification,* Boulder, CO, Amer. Meteor. Soc., 349–356.

———, S. Wang, and H.D. Orville, 1992: A comparison of 3D model results with observations for an isolated CCOPE thunderstorm. *J. Meteor. Atmos. Phys.,* **49,** 187–207.

———, P. Nguyen, and H.D. Orville, 1994: Numerical simulation of cloud seeding using a three-dimensional cloud model. *J. Wea. Modif.,* **26,** 113–124.

Ferrier, B.S., 1994: A double-moment multiple-phase four-class bulk ice scheme. Part I: Description. *J. Atmos. Sci.,* **51,** 249–280.

_____ , and R.A. Houze Jr., 1989: One-dimensional time-dependent modeling of GATE cumulonimbus convection. *J. Atmos. Sci.,* **46,** 330–352.

Fletcher, N.H., 1962: *The Physics of Rain Clouds.* Cambridge University Press, 390 pp.

Foote, G.B., and C.A. Knight, 1977: *Hail: A Review of Hail Science and Hail Suppression. Meteor. Monogr.* No. 16, Amer. Meteor. Soc., 277 pp.

_____ , and C.G. Wade, 1982: Case study of a hailstorm in Colorado. Part I: Radar echo structure and evolution. *J. Atmos. Sci.,* **39,** 2828–2846.

Fovell, R.G., and Y. Ogura, 1988: Numerical simulation of a midlatitude squall line in two dimensions. *J. Atmos. Sci.,* **45,** 3846–3879.

_____ , and _____ , 1989: Effect of vertical wind shear on numerically simulated multicell storm structure. *J. Atmos. Sci.,* **46,** 3144–3176.

Gutman, L.N., 1957: Teoreticheskaya model' kuchevogo oblaka (Theoretical model of a cumulus cloud). *Dokl. Akad. Nauk SSSR,* **112.** (6).

Hall, W.D., 1980: A detailed microphysical model within a two-dimensional dynamic framework: Model description and preliminary results. *J. Atmos. Sci.,* **37,** 2486–2507.

_____ , D. Matthews, and A. Super, 1990: Orographic production of supercooled liquid water and precipitation over the Mogollon Rim of Arizona. Preprints, *Conf. on Cloud Physics,* San Francisco, CA, Amer. Meteor. Soc., 754–757.

Hane, C.E., 1978: The application of a two-dimensional convective cloud model to waste heat release from proposed nuclear energy centers. *Atmos. Environ.,* **12,** 1839–1848.

Helsdon, J.H., Jr., 1980: Chaff seeding effects in a dynamical-electrical cloud model. *J. Appl. Meteor.,* **19,** 1101–1125.

_____ , and R.D. Farley, 1987a: A numerical modeling study of a Montana thunderstorm: Part I. Model results versus observations involving non-electric aspects. *J. Geophys. Res.,* **92,** 5645–5659.

_____ , and _____ , 1987b: A numerical modeling study of a Montana thunderstorm: Part II. Model results versus observations involving electrical aspects. *J. Geophys. Res.,* **92,** 5661–5675.

Hess, S., 1959: *Introduction to Theoretical Meteorology.* Holt, Rinehart, Winston, 362 pp.

Heymsfield, A.J., 1986: Ice particle evolution in the anvil of a severe thunderstorm during CCOPE. *J. Atmos. Sci.*, **43**, 2463–2478.

Hindman, E., P.M. Tag, B.A. Silverman, and P.V. Hobbs, 1977: Cloud condensation nuclei from a paper mill: Part II. Calculated effects on rainfall. *J. Appl. Meteor.*, **16**, 753–755.

Hirsch, J.H., 1972: A numerical cloud model—Its use during Project Cloud Catcher. Preprints, *Third Nat'l. Conf. Wea. Modif.*, Rapid City, SD, Amer. Meteor. Soc., 192–185.

Hjelmfelt, M.R., H.D. Orville, R.D. Roberts, J.P. Chen, and F.J. Kopp, 1989: Observational and numerical study of a microburst line-producing storm. *J. Atmos. Sci.*, **46**, 2731–2743.

Hobbs, P.V., R.C. Easter, and A.B. Fraser, 1973: A theoretical study of the flow of air and fallout of solid precipitation over mountainous terrain. Part II: Microphysics. *J. Atmos. Sci.*, **30**, 813–823.

Houghton, H.G., and H.E. Cramer, 1951: A theory of entrainment in convective currents. *J. Meteor.*, **8**, 95–102.

Houze, R.A., Jr., 1993: *Cloud Physics*. Academic Press, 573 pp.

Howard, L., 1803: *On the Modifications of Clouds*. London, 32 pp.

Hsie, E.-Y., R.D. Farley, and H.D. Orville, 1980: Numerical simulation of ice-phase convective cloud seeding. *J. Appl. Meteor.*, **19**, 950–977.

Hu, Z., and G. He, 1989: Numerical experiments of seeding in cumulonimbus clouds. *Proc. Fifth Scientific Conf. Weather Modification and Applied Cloud Physics*, Beijing, China, WMO, 289–292.

Johnson, D.B., 1979: The role of coalescence nuclei in warm rain initiation. Ph.D. dissertation, University of Chicago, 119 pp.

———, 1980: Hygroscopic seeding to initiate precipitation. *Proc. Third Scientific Conf. Weather Modification*, Clermont-Ferrand, France, WMO, 41–45.

———, 1982: The role of giant and ultra-giant aerosol particles in warm rain initiation. *J. Atmos. Sci.*, **39**, 448–460.

———, E.H. Barker, and R.R. Lowe, 1975: A numerical study of fog clearing by helicopter downwash. *J. Appl. Meteor.*, **14**, 1284–1292.

Juisto, J.E., 1971: Crystal development and glaciation of a supercooled cloud. *J. Rech. Atmos.*, **5**, 69–85.

Kachurin, L.G., N.D. Artemyeva, A.T. Kartsivadze, S. Stoyanov, and M. Tekle, 1973: Simulation of the natural process of hail formation and its transformation under the influence of artificial crystallization. *Proc. Scientific Conf. Weather Modification*, WMO/IAMAP, 231–237.

Kessler, E., 1969: *On the Distribution and Continuity of Water Substance in Atmospheric Circulation. Meteor. Monogr.*, No. 10, Amer. Meteor. Soc., 84 pp.

______ , 1975: On the condensed water mass in rising air. *Pure Appl. Geophys.*, **113,** 971–982.

Klaassen, G.P., and T.L. Clark, 1985: Dynamics of the cloud-environment interface and entrainment in small cumuli: Two-dimensional simulations in the absence of ambient shear. *J. Atmos. Sci.*, **42,** 2621–2642.

Klazura, G.E., and C.J. Todd, 1978: A model of hygroscopic seeding in cumulus clouds. *J. Appl. Meteor.*, **17,** 1758–1768.

Klemp, J.B., 1987: Dynamics of tornadic thunderstorms. *Annu. Rev. Fluid. Mech.*, **19,** 369–402.

______ , and R.B. Wilhelmson, 1978a: The simulation of three-dimensional convective storm dynamics. *J. Atmos. Sci.*, **35,** 1070–1096.

______ , and ______ , 1978b: Simulations of right- and left-moving storms produced through storm splitting. *J. Atmos. Sci.*, **35,** 1097–1110.

Knight, C.A., and P. Squires, 1982: *Hailstorms of the Central High Plains.* Vol. 1, *The National Hail Research Experiment,* Colorado University Press, 282 pp.

Knupp, K.R., 1985: Precipitation convective downdraft structure: A synthesis of observations and modeling. Ph.D. dissertation, Colorado State University, 296 pp.

Koenig, L.R., 1971: Numerical experiments pertaining to warm-fog clearing. *Mon. Wea. Rev.*, **99,** 227–241.

______ , 1972: Parameterization of ice growth for numerical calculations of cloud dynamics. *Mon. Wea. Rev.*, **100,** 417–423.

______ , and F.W. Murray, 1976: Ice-bearing cumulus cloud evolution: Numerical simulation and general comparison against observations. *J. Appl. Meteor.*, **15,** 747–762.

______ , and ______ , 1983: Theoretical experiments on cumulus dynamics. *J. Atmos. Sci.*, **40,** 1241–1256.

Kopp, F.J., 1988: A simulation of Alberta cumulus. *J. Appl. Meteor.*, **27,** 626–641.

______ , and H.D. Orville, 1994: The use of a two-dimensional, time-dependent cloud model to predict convective and stratiform clouds and precipitation. *Wea. Forecasting,* **9,** 62–77.

______ , ______ , R.D. Farley, and J.H. Hirsch, 1983: Numerical simulation of dry ice cloud seeding experiments. *J. Climate Appl. Meteor.*, **22**, 1542–1556.

Kraus, E.B., and P. Squires, 1947: Experiments on the stimulation of clouds to produce rain. *Nature,* **159**, 489–491.

Kubesh, R.J., D.J. Musil, R.D. Farley, and H.D. Orville, 1988: The 1 August 1981 CCOPE storm: Observations and modeling results. *J. Appl. Meteor.*, **27**, 216–243.

Lamark, J.B., 1802: Sur la forme des nuages. *Annuaire Meteor.*, **3**, 151–153, 155–161, 166.

Levy, G., and W.R. Cotton, 1984: A numerical investigation of mechanisms linking glaciation of the ice-phase to the boundary layer. *J. Climate Appl. Meteor.*, **23**, 1505–1519.

Lilly, D.K., 1962: On the numerical simulation of buoyant convection. *Tellus,* **14**, 148–172.

Lin, Y.-L., R.D. Farley, and H.D. Orville, 1983: Bulk parameterization of the snow field in a cloud model. *J. Climate Appl. Meteor.*, **22**, 1065–1092.

Lipps, F.B., 1977: A study of turbulent parameterization in a cloud model. *J. Atmos. Sci.*, **34**, 1751–1772.

Liu, J.Y., and H.D. Orville, 1969: Numerical modeling of precipitation and cloud shadow effects on mountain-induced cumuli. *J. Atmos. Sci.*, **26**, 1283–1298.

Ludlam, F.H., 1958: The hail problem. *Nubila,* **1**, 12–95.

———, and R.S. Scorer, 1953: Convection in the atmosphere. *Quart. J. Roy. Meteor. Soc.*, **79**, 317.

Malkus, J.S., 1952: The slopes of cumulus clouds in relation to external wind shear. *Quart. J. Roy. Meteor. Soc.*, **78**, 520.

———, and R.A. Scorer, 1955: The erosion of cumulus towers. *J. Meteor.*, **12**, 423.

———, and G. Witt, 1959: The evolution of a convective element. A numerical calculation. *The Atmosphere and Sea in Motion,* B. Bolin, Ed., Rockefeller Institute Press and Oxford University Press, 425–439.

Marwitz, J.D., 1972: The structure and motion of severe hailstorms. Part I: Supercell storms. *J. Appl. Meteor.*, **11**, 166–179.

Matthews, D.A., 1981: Natural variability of thermodynamic features affecting convective cloud growth and dynamic seeding: A comparative summary of three High Plains sites from 1975 to 1977. *J. Appl. Meteor.*, **20**, 971–996.

McDonald, J.E., 1963: Early developments in the theory of the saturated adiabatic process. *Bull. Amer. Meteor. Soc.,* **44,** 203–211.

Moncrieff, M.W., and J.S.A. Green, 1972: The propagation and transfer properties of steady convective overturning in shear. *Quart. J. Roy. Meteor. Soc.,* **98,** 336–352.

______ , and M.J. Miller, 1976: The dynamics and simulation of tropical cumulonimbus and squall lines. *Quart. J. Roy. Meteor. Soc.,* **102,** 373–394.

Morton, B.R., G.I. Taylor, and J.S. Turner, 1956: Turbulent gravitational convection from maintained and instantaneous sources. *Proc. Roy. Soc. A,* **234,** 1–23.

Murakami, M., 1990: Numerical modeling of dynamical and microphysical evolution of an isolated convective cloud: The 19 July 1981 CCOPE cloud. *J. Meteor. Soc. Japan,* **68,** 107–128.

______ , T.L. Clark, and W.D. Hall, 1994: Numerical simulations of convective snow clouds over the sea of Japan: Two-dimensional simulations of mixed layer development and convective snow cloud formation. *J. Meteor. Soc. Japan,* **72,** 1–20.

Murray, F.W., and L.R. Koenig, 1972: Numerical experiment on the relation between microphysics and dynamics in cumulus convection. *Mon. Wea. Rev.,* **100,** 717–732.

Nelson, L.D., 1971: A numerical study on the initiation of warm rain. *J. Atmos. Sci.,* **28,** 752–762.

______ , 1979: Observations and numerical simulations of precipitation mechanisms in natural and seeded convective clouds. Ph.D. dissertation, University of Chicago, 188 pp.

Newton, C.W., 1963: Dynamics of severe convective storms. *Severe Local Storms, Meteor. Monogr.,* No. 5, Amer. Meteor. Soc., 33–58.

Ogura, Y., 1963: The evaluation of a moist convection element in a shallow, conditionally unstable atmosphere. A numerical calculation. *J. Atmos. Sci.,* **20,** 407–424.

______ , and T. Takahashi, 1971: Numerical simulation of the life cycle of a thunderstorm cell. *Mon. Wea. Rev.,* **99,** 895–911.

Orville, H.D., 1965: A numerical study of the initiation of cumulus clouds over mountainous terrain. *J. Atmos. Sci.,* **22,** 700–709.

______ , and L.J. Sloan, 1970: A numerical simulation of the life history of a rainstorm. *J. Atmos. Sci.,* **28,** 1148–1159.

______ , and F.J. Kopp, 1977: Numerical simulation of the history of a hailstorm. *J. Atmos. Sci.,* **34,** 1596–1618.

_____ , and _____ , 1978: Reply. *J. Atmos. Sci.*, **35,** 1554–1555.

_____ , and J.-M. Chen, 1982: Effects of cloud seeding, latent heat of fusion, and condensate loading on cloud dynamics and precipitation evolution: A numerical study. *J. Atmos. Sci.*, **39,** 2807–2827.

_____ , and F.J. Kopp, 1990: A numerical simulation of the 22 July 1979 HIPLEX-1 case. *J. Appl. Meteor.*, **29,** 539–550.

_____ , _____ , and C.G. Myers, 1975: The dynamics and thermodynamics of precipitation loading. *Pure Appl. Geophys.*, **113,** 983–1004.

_____ , J.H. Hirsch and L.E. May, 1980: Application of a cloud model to cooling tower plumes and clouds. *J. Appl. Meteor.*, **19,** 1260–1272.

_____ , R.D. Farley, and J.H. Hirsch, 1984: Some surprising results from simulated seeding of stratiform-type clouds. *J. Climate Appl. Meteor.*, **23,** 1585–1600.

_____ , J.H. Hirsch, and R.D. Farley, 1987: Further results on numerical cloud seeding simulations of stratiform-type clouds. *J. Wea. Modif.*, **19,** 57–61.

_____ , F.J. Kopp, R.D. Farley, and R.B. Hoffman, 1989a: The numerical modeling of ice-phase cloud seeding effects in a warm-base cloud: Preliminary results. *J. Wea. Modif.*, **21,** 4–8.

_____ , R.D. Farley, Y.-C. Chi, and F.J. Kopp, 1989b: The primary cloud physics mechanisms of microburst formation. *Atmos. Res.*, **24,** 343–357.

Paluch, I.R., 1979: The entrainment mechanism in Colorado cumuli. *J. Atmos. Sci.*, **36,** 2467–2478.

Pastre, J., and R. Rosset, 1977: Interaction entre une panache humide et une nuage: Étude préliminaire. *J. Rech. Atmos.*, **11,** 213–225.

Peslin, H., 1868: Sur les mouvements généraux de l'atmosphère. *Bull. Assoc. Sci. France, **3,** 299–319.

Plooster, M.N., and N. Fukuta, 1975: A numerical model of precipitation from seeded and unseeded cold orographic clouds. *J. Appl. Meteor.*, **14,** 859–867.

Proctor, F.H., 1987a: The terminal area simulation system. Vol. I: Theoretical formulation. NASA Rep. CR-4046, 123 pp.

_____ , 1987b: The terminal area simulation system. Vol. II: Verification cases. NASA Rep. CR-4047, 97 pp.

_____ , 1988: Numerical simulation of an isolated microburst. Part I: Dynamics and structure. *J. Atmos. Sci.*, **45,** 3137–3160.

_____ , 1989: Numerical simulations of an isolated microburst. Part II: Sensitivity experiments. *J. Atmos. Sci.*, **46,** 2143–2165.

Randell, S.C., S.A. Rutledge, R.D. Farley, and J.H. Helsdon Jr., 1994: A modeling study on the early electrical development of tropical convection: Continental and oceanic (monsoon) storms. *Mon. Wea. Rev.*, **122**, 1852–1877.

Raymond, D.J., and A.M. Blythe, 1986: A stochastic mixing model for nonprecipitating cumulus clouds. *J. Atmos. Sci.*, **43**, 2708–2718.

Redelsperger, J.-L., and J.-P. Lafore, 1988: A three-dimensional simulation of a tropical squall line: Convective organization and thermodynamic vertical transport. *J. Atmos. Sci.*, **45**, 1334–1356.

Rokicki, M.L., and K.C. Young, 1978: The initiation of precipitation in updrafts. *J. Appl. Meteor.*, **17**, 745–754.

Rotunno, R., 1981: On the evolution of thunderstorm rotation. *Mon. Wea. Rev.*, **109**, 171–180.

———, and J.B. Klemp, 1982: The influence of shear-induced pressure gradient on thunderstorm motion. *Mon. Wea. Rev.*, **110**, 136–151.

———, and ———, 1985: On the rotation and propagation of simulated supercell thunderstorms. *J. Atmos. Sci.*, **42**, 271–292.

Saunders, P.M., 1957: The thermodynamics of saturated air: A contribution to the classical theory. *Quart. J. Roy. Meteor. Soc.*, **83**, 342–350.

Scala, J., M. Garstang, W.-K. Tao, K. Pickering, A. Thompson, J. Simpson, V. Hirchhoff, E. Browell, G. Sachse, A. Tores, G. Gregory, R. Rasmussen, and M. Khalil, 1990: Cloud draft structure and trace gas transport. *J. Geophys. Res.*, **95**, 17 015–17 030.

Schlesinger, R.E., 1975: A three-dimensional numerical model of an isolated deep convective cloud: Preliminary result. *J. Atmos. Sci.*, **32**, 934–957.

———, 1978: A three-dimensional numerical model of an isolated thunderstorm: Part I. Comparative experiments for variable ambient wind shear. *J. Atmos. Sci.*, **35**, 690–713.

———, 1984: Effects of the pressure perturbation field in numerical models of unidirectionally sheared thunderstorm convection: Two versus three dimensions. *J. Atmos. Sci.*, **41**, 1517–1587.

Schmidt, F.H., 1947: Some speculations on the resistance to the motion of cumuliform clouds. Koninklijk Nederlands Meteorologisch Instituut De Bilt, No. 125, Meded. Verhand Serie B, I, (8).

Scorer, R.S., 1957: Experiments on convection of isolated masses of buoyant fluid. *J. Fluid Mech.*, **2**, 583–594.

Shishkin, N.S., 1958: K voprosu rascheta vertikal'nykh skorostei v oblake (On the problem of the calculation of vertical velocities in a cloud). *Trudy GGO,* **104.**

Silverman, B., and M. Glass, 1973: A numerical simulation of warm cumulus clouds: Part I. Parameterized versus nonparameterized microphysics. *J. Atmos. Sci.,* **30,** 1620–1637.

_____ , D.A. Matthews, L.D. Nelson, H.D. Orville, F.J. Kopp, and R.D. Farley, 1976: Comparisons of cloud model predictions: A case study analysis of one- and two-dimensional models. Preprints, *Int. Conf. on Cloud Physics,* Boulder, CO, Amer. Meteor, Soc., 343–348.

Simpson, J., 1971: On cumulus entrainment and one-dimensional models. *J. Atmos. Sci.,* **28,** 449–455.

_____ , and V. Wiggert, 1971: 1968 Florida cumulus seeding experiment: Numerical model results. *Mon. Wea. Rev.,* **99,** 87–118.

_____ , R.H. Simpson, D.A. Andrews, and M.A. Eaton, 1965: Experimental cumulus dynamics. *Rev. Geophys.,* **3,** 387–396.

Skamarock, W.C., M.L. Weisman, and J.B. Klemp, 1994: Three-dimensional evolution of simulations of long-lived squall lines. *J. Atmos. Sci.,* **51,** 2563–2584.

Smolarkiewicz, P.K., and T.L. Clark, 1985: Numerical simulations of the evolution of a three-dimensional field of cumulus clouds. Part I: Model description, comparison with observations and sensitivity studies. *J. Atmos. Sci.,* **42,** 502–522.

Sommeria, G., 1976: Three-dimensional simulation of turbulent processes in an undisturbed trade wind boundary layer. *J. Atmos. Sci.,* **33,** 216–241.

Soong, S.T., and Y. Ogura, 1973: A comparison between axisymmetric and slab-symmetric cumulus cloud models. *J. Atmos. Sci.,* **30,** 879–893.

Squires, P., 1958: The spatial variation of liquid water and droplet concentration in cumuli. *Tellus,* **10,** 372–380.

_____ , and J.S. Turner, 1962: An entraining jet model for cumulonimbus updraughts. *Tellus,* **14,** 422–434.

Srivastava, R.C., 1967: A study of the effects of precipitation on cumulus dynamics. *J. Atmos. Sci.,* **24,** 36–45.

_____ , 1985: A simple model of evaporatively driven downdraft: Application to microburst downdraft. *J. Atmos. Sci.,* **42,** 1004–1023.

_____ , 1987: A model of intense downdrafts driven by the melting and evaporation of precipitation. *J. Atmos. Sci.,* **44,** 1752–1773.

Steiner, J.T., 1973: A three-dimensional model of cumulus cloud development. *J. Atmos. Sci.,* **30,** 414–435.

Stommel, H., 1947: Entrainment of air into a cumulus cloud. *J. Meteor.,* **4,** 91–94.

______ , 1951: Entrainment of air into a cumulus cloud. *J. Meteor.,* **8,** 127–129.

Strapp, J.W., H.G. Leighton, and G.A. Isaac, 1979: A comparison of model calculations of ice crystal growth with observations following silver iodide seeding. *Atmos.–Ocean,* **17,** 234–252.

Sulakvelidze, G.K., N. Sh. Bibilashvili and V.F. Lapcheva, 1967: *Formation of Precipitation and Modification of Hail Processes.* Israel Program for Scientific Translations, 208 pp.

Tag, P.M., 1979: A numerical simulation of fog dissipation using passive burner lines. Part I: Model development and comparison with observations. *J. Appl. Meteor.,* **18,** 1442–1454.

Takeda, T., 1971: Numerical simulation of a precipitating convective cloud: The formation of a long lasting cloud. *J. Atmos. Sci.,* **28,** 350–376.

Tao, W.-K., and J. Simpson, 1993: Goddard cumulus ensemble model. Part I: Model description. *Terres., Atmos., Oceanic Sci.,* **4,** 35–72.

______ , ______ , C.-H. Sui, S. Lang, J. Scala, B. Ferrier, M.-D. Chou, and K. Pickering, 1993: Heating, moisture, and water budgets of tropical and midlatitude squall lines: Comparisons and sensitivity to longwave radiation. *J. Atmos. Sci.,* **50,** 673–690.

Thorpe, A.J., M.J. Miller, and M.W. Moncrieff, 1980: Dynamical models of two-dimensional downdrafts. *Quart. J. Roy. Meteor. Soc.,* **106,** 463–484.

______ , ______ , and ______ , 1982: Two-dimensional convection in nonconstant shear: A model of mid-latitude squall lines. *Quart. J. Roy. Meteor. Soc.,* **108,** 739–762.

Turner, J.S., 1962: The starting plume in neutral surroundings. *J. Fluid Mech.,* **13,** 356–368.

Tuttle, J.D., V.N. Bringi, H.D. Orville, and F.J. Kopp, 1989: Multiparameter radar study of a microburst: Comparison with model results. *J. Atmos. Sci.,* **46,** 601–620.

Tzivion, S., T. Reisin, and Z. Levin, 1994: Numerical simulation of hygroscopic seeding in a convective cloud. *J. Appl. Meteor.,* **33,** 252–267.

Vali, G., L.R. Koenig, and T.C. Yoksas, 1988: Estimate of precipitation enhancement potential for the Deuro Basin of Spain. *J. Appl. Meteor.*, **27,** 829–850.

von Bezold, W., 1888: Zur thermodynamik der Atmosphäre. *Sitzungsber. Preuss. Akad. Wiss.,* 485–522.

Vul'fson, N.I., 1954: Konvektivnye dvizheniya v kuchevykh oblakakh (Convective motions in cumulus clouds). *DAN SSSR,* **97.**

Warner, J., 1955: The water content of cumuliform cloud. *Tellus,* **7,** 449.

______ , 1970: On steady-state, one-dimensional models of cumulus convection. *J. Atmos. Sci.,* **27,** 1035–1040.

______ , and P. Squires, 1958: Liquid water content and the adiabatic model of cumulus development. *Tellus,* **10,** 390.

Weinstein, A.I., 1970: A numerical model of cumulus dynamics and microphysics. *J. Atmos. Sci.,* **27,** 246–255.

______ , and L.G. Davis, 1968: A parameterized numerical model of cumulus convection. Dept. of Meteorology Rep. 11, The Pennsylvania State University.

Weisman, M.L., and J.B. Klemp, 1982: The dependence of numerically simulated convective storms on vertical wind shear and buoyancy. *Mon. Wea. Rev.,* **110,** 2479–2498.

______ , and ______ , 1984: The structure and classification of numerically simulated convective storms in directionally varying wind shear. *Mon. Wea. Rev.,* **112,** 2479–2498.

______ , ______ , and R. Rotunno, 1988: The structure and evolution of numerically simulated squall lines. *J. Atmos. Sci.,* **45,** 1990–2013.

Wiggert, V., R.I. Sax, and R.L. Holle, 1982: On the modification potential of Illinois summertime convective clouds with comparisons to Florida and FACE observations. *J. Appl. Meteor.,* **21,** 1293–1322.

Wilhelmson, R.B., 1974: The life cycle of a thunderstorm in three dimensions. *J. Atmos. Sci.,* **31,** 1629–1651.

______ , and Y. Ogura, 1972: The pressure perturbation and numerical modeling of a cloud. *J. Atmos. Sci.,* **29,** 1295–1307.

______ , and J.B. Klemp, 1978: A numerical study of storm splitting that leads to a long-lived storm. *J. Atmos. Sci.,* **35,** 1974–1986.

______ , and ______ , 1981: A three-dimensional numerical simulation of splitting severe storms on 3 April 1963. *J. Atmos. Sci.,* **38,** 1581–1600.

______ , and C.-S. Chen, 1982: A simulation of the development of successive cells along a cold outflow boundary. *J. Atmos. Sci.,* **39,** 1466–1483.

Wisner, C., H.D. Orville, and C. Myers, 1972: A numerical model of a hail-bearing cloud. *J. Atmos. Sci.,* **29,** 1160–1181.

Woodward, B., 1959: The motion around isolated thermals. *Quart. J. Roy. Meteor. Soc.,* **85,** 144.

World Meteorological Organization, 1969: *International Cloud Atlas.* Vol. 1, Geneva, Switzerland, 138 pp.

_____ , 1985: Report of the International Cloud Modelling Workshop/ Conference, Irsee, Federal Republic of Germany. WMP Rep. 8, WMO/TD No. 139, Geneva, Switzerland, 460 pp.

_____ , 1988: Report of the International Cloud Modelling Workshop, Toulouse, France. WMP Rep. 11, WMO/TD No. 268, Geneva, Switzerland, 330 pp.

_____ , 1992a: Report of the International Cloud Modelling Workshop, Toronto, ON, Canada. WMP Rep. 20, WMO/TD No. 565, Geneva, Switzerland, 228 pp.

_____ , 1992b: Programme on physics and chemistry of clouds and weather modification research. WMP Rep. 19, Proc. WMO Workshop on Cloud Microphysics and Applications to Global Change, WMO/TD No. 537, Toronto, ON, Canada, 406 pp.

Yates, A.H., 1953: Atmospheric convection: The structure of the thermals below cloud base. *Quart. J. Roy. Meteor. Soc.,* **79.**

Young, K.C., 1974: A numerical simulation of wintertime, orographic precipitation. Part I: Description of model microphysics and numerical techniques. *J. Atmos. Sci.,* **31,** 1735–1748.

_____ , 1977: A numerical examination of some hail suppression concepts. *Hail: A Review of Hail Science and Hail Suppression, Meteor. Monogr.,* No. 16, Amer. Meteor. Soc., 195–214.

_____ , 1993: *Microphysical Processes in Clouds.* Oxford University Press, 427 pp.

Zaitsev, V.A., 1950: Liquid water content and distribution of drops in cumulus clouds. *Trudy GGO,* **19** (81), 12.

Hurricanes, Convective Storms, and Lightning

A History of Hurricane Forecasting for the Atlantic Basin, 1920–1995

MARK DEMARIA

Introduction

Whether measured in terms of loss of life or property damage, tropical cyclones are among the world's costliest natural hazards (White 1994). Hebert et al. (1993) have shown that the loss of life in the United States from tropical cyclones has dramatically decreased in the last century, even though the number of people at risk and the economic losses have increased. However, they also stress that large loss of life in the United States from tropical cyclones is still possible. The decrease in fatalities is largely due to improvements in hurricane forecasts and warnings. In this paper the history of hurricane forecasting from the time of the establishment of the American Meteorological Society (1920) to the present will be reviewed. Although tropical cyclones are natural hazards in many parts of the world, the focus of this review will be on forecasting of tropical cyclones in the Atlantic basin, which includes the North Atlantic, Caribbean, and Gulf of Mexico.

At the present time, the National Hurricane Center (NHC, renamed the Tropical Prediction Center in 1995) in Miami, Florida, has responsibility for tropical cyclone forecasts and warnings out to 72 hours for the Atlantic and eastern North Pacific basins, and the Central Pacific Hurricane Center (CPHC) in Honolulu, Hawaii, has the responsibility for the North Pacific from 140°W to the date line. Prior to 1988, the Weather Service Forecast Office in San Francisco, produced the forecasts for eastern North Pacific tropical cyclones. The forecast and warning process requires a coordinated effort between several agencies and involves many meteorological and nonmeteorological factors (Sheets 1990). A distinction is made between the estimation of the current and future storm track and intensity (forecasting) and the communication of this information to emergency managers and the public so that appropriate action will be taken (warning). In this paper, the emphasis will be on the history of forecasting, although some discussion of the warning process will also be included.

In the Atlantic basin, tropical cyclones with 1-min maximum sustained surface winds >33 m s^{-1} are referred to as hurricanes. Historically, much of the loss of life and property damage has resulted from hurricanes of category 3 or greater (maximum winds >50 m s^{-1}) on the Saffir–Simpson scale (Hebert et al. 1993). Occasionally, however, extensive damage occurs from less intense tropical cyclones, due to inland flooding (e.g., Hurricane Diane in 1955 and Tropical Storm Alberto in 1994). The warning and emergency response procedures vary greatly depending on the storm intensity at landfall. However, the forecasting methods are very similar for hurricanes and weaker tropical cyclones. For simplicity, the forecasts for all tropical cyclones will be referred to as hurricane forecasts.

The methods for hurricane forecasting have undergone substantial changes during the past 75 years. Many of these changes can be attributed to technological advances such as the implementation of the upper-air network in the late 1930s, the establishment of routine aircraft reconnaissance in the 1940s, the advent of numerical weather prediction beginning in the 1950s, and the availability of satellite observations starting in the 1960s. The agencies responsible for producing the forecasts have also undergone considerable evolution since 1920. For organizational purposes, the 75-year history is divided into several smaller time periods as shown in Table 9-1. The boundaries of these periods loosely correspond to times when significant changes in hurricane forecasting procedures occurred.

Table 9-1. Organization of the history of hurricane forecasting.

Time period	Description
1920–1934	Hurricane diagnosis and extrapolation
1935–1942	The use of upper-air data and forecast office reorganization
1943–1955	Technological advancements: aircraft reconnaissance and radar
1956–1965	Increased research efforts and the development of objective forecast methods
1966–1973	The use of satellite observations and improved forecast models
1974–1987	Geostationary satellites and three-dimensional forecast models
1988–1994	Increased interaction with the research community and the use of global model output
1995–2005	Present status and future outlook

1920–1934, Hurricane diagnosis and extrapolation

At the beginning of this period (1920) hurricane forecasts and warnings for the United States and surrounding areas were the responsibility of the U.S. Weather Bureau in Washington, D.C. This responsibility had been transferred from Havana, Cuba, in 1902 (Calvert 1935). Considerable effort was required to diagnose the current position and intensity of a storm. Calvert (1920) made an analogy between the work of a forecaster and that of a physician. He stated that,

> The means of detecting the beginning of hurricanes . . . can be compared to a physician who diagnoses a case in which there is menace and death to his patient by symptoms that he recognizes because of his technical skill and experience. . . . The forecaster has less than the physician on which to base his diagnosis, oftentimes a single wireless report of weather conditions from a vessel at sea or a land station a hundred miles or more from the storm center.

Despite this difficulty, Calvert (1920) stated that hurricanes rarely occurred without being detected. The primary tools for diagnosing hurricanes were reports from ships at sea and from surface stations in the tropical and subtropical regions. The tides and sea swells at coastal stations were also closely monitored for signs of approaching storms (Cline 1926). Except for an occasional pilot balloon, there was very little direct information about the upper-air conditions (Bowie 1922). However, based upon the work of Father Benito Viñes in the late nineteenth century, indirect information was obtained from the motion of upper- and lower-level clouds (Mitchell 1924).

Knowledge of tropical cyclone structure was fragmentary during this period. The structure of the low-level circulation, including the approximate scale of the outer wind field and the existence of the hurricane eye, was fairly well known (Bowie 1922). Until the mid-1930s, however, there was some uncertainty concerning the vertical structure of storms (Haurwitz 1935). Hurricanes were often described as "flat" systems, suggesting they were confined to the low levels.

The climatology of the formation regions of Atlantic hurricanes was well documented by Mitchell (1924), but considerable disagreement existed concerning the mechanisms responsible for storm formation (Tannehill 1938a). In a review of recent scientific studies, Bowie (1922) summarized two theories of tropical cyclone formation. One theory, referred to as the "convectional hypothesis," suggested that tropical cyclones form when air becomes warm and moist on a large scale. This air rises, which forces horizontal convergence and rotation due to conservation of angular momentum on the rotating earth. As part of this theory, it was suggested that tropical cyclone formation is a relatively rare event because uplifting usually occurs on smaller scales in the Tropics, leading to more isolated thunderstorms. A second theory, referred to as the "countercurrent hypothesis," suggested that atmospheric vortices form along discontinuities between different air masses. In the case of Atlantic tropical cyclones, the discontinuity was the doldrum trough [also referred to as the intertropical convergence zone (ITCZ)] between the northeast and southeast (or sometimes southwest) trades found off the coast of Africa during the peak of the hurricane season (August and September) and in the western Caribbean early and late in the hurricane season. In the western North Pacific, tropical cyclones typically form in the monsoon trough between the northeast trades and the southwest flow at lower latitudes. In a sense, both of these theories were partially correct. The convectional theory is consistent with the results of Pfeffer and Challa (1981) and Montgomery and Farrell (1993), which suggest that tropical cyclone genesis occurs when low-level circulation features interact with upper-level synoptic features in a way that enhances the vertical motion over a large area. However, much of the vertical motion occurs in smaller-scale convective elements rather than as one large-scale region of overturning as suggested by the convectional hypothesis. The countercurrent hypothesis is correct in the sense that the monsoon trough is the source of the low-level rotation for many west Pacific tropical cyclones. The doldrum trough (ITCZ) and frontal discontinuities are the source of the rotation for some Atlantic storms, although easterly waves are more commonly involved in the formation of Atlantic tropical cyclones (Avila and Clark 1989).

The climatology of the motion of Atlantic storms was also well documented by Mitchell (1924), but factors affecting storm motion were not completely understood, primarily due to the lack of upper-air data. A large emphasis was placed on the movement of anticyclones to the north of Atlantic hurricanes. It was generally recognized that hurricanes tend to move out of the tropical and subtropical regions unless a large anticyclone prevented the storm from moving north. For example, Mitchell (1924) stated that hurricanes "seek to move northward at the first available opportunity." For this reason, the movement of anticyclones in relation to the hurricane track was closely monitored as part of the forecasting process (Bowie 1922). Although it was recognized that the flow around an anticyclone must extend above the surface in order to influence the motion of a storm, the extent to which the upper-level flow affected storm motion was not fully appreciated.

Hurricane forecasts during this time were limited to short-range extrapolations (12–24 hours) of current storm positions and trends in motion, with some modifications to take into account large-scale pressure patterns and limited pilot balloon and upper-level cloud observations. As an example of the level of forecast skill during this period, consider the hurricane that struck Miami, Florida, on the morning of 18 September 1926. According to the best-track data from a poststorm analysis of all available information maintained by NHC (Jarvinen et al. 1988), this storm made landfall in Miami just before 1200 UTC (0600 EST) with maximum sustained winds of ~62 m s^{-1} (120 kt). The landfall of the 1926 storm had many similarities with Hurricane Andrew, which made landfall just south of Miami in the early morning of 24 August 1992 with maximum sustained winds of ~64 m s^{-1} (125 kt). The tracks of the 1926 Hurricane and Hurricane Andrew are shown in Fig. 9-1.

According to Mitchell (1926), the 1926 hurricane was first detected

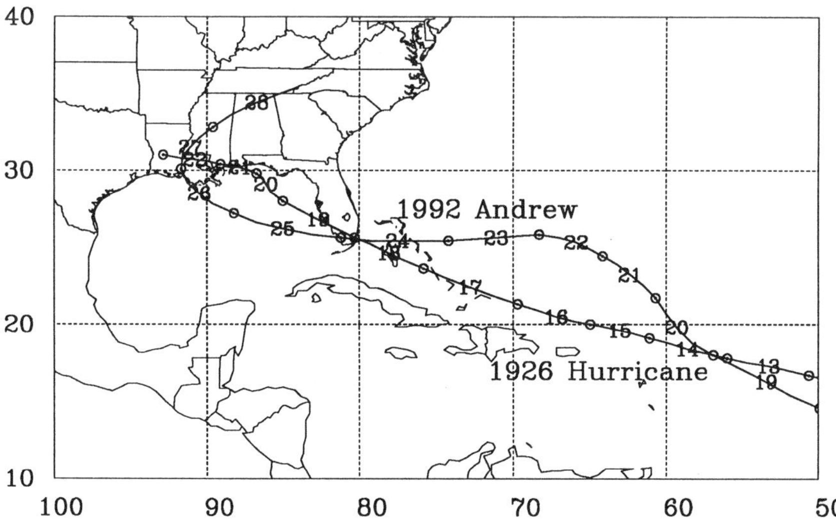

FIG. 9-1. The tracks of the 1926 south Florida Hurricane and Hurricane Andrew (1992). The open circles are the positions at 1200 UTC, and the days of the month (September 1926 or August 1992) are indicated at the 0000 UTC positions.

on 14 September, east-northeast of Puerto Rico. At this time, there was also a hurricane just southwest of Bermuda and a tropical storm moving north over central Cuba. The storm near Bermuda moved to the north in the next few days, but the storm near Cuba moved north over the Bahamas and then turned to the southwest just off the south Florida coast. The latter storm dissipated in the Florida Straits by 17 September (just one day before the 1926 hurricane made landfall). As the 1926 hurricane approached south Florida, the Weather Bureau issued northeast storm warnings (gale warnings) at noon on 17 September (18 hours prior to landfall) from Jupiter Inlet to Key West. The warning message also stated that "this is a very severe storm" and that "precautions should be taken for strong winds Saturday morning" (18 September). However, hurricane warnings were not issued until midnight on 17 September (6 hours prior to landfall).

Historical descriptions by Parks (1986) and Doehring et al. (1994) indicate that the 1926 hurricane took many residents of south Florida by surprise. Contributing factors were that the hurricane warnings were issued after many people had gone to sleep and that many of the residents at that time had not experienced a major storm. The weak tropical system that passed just to the south of Miami the previous day also added to the confusion. The morning edition of the Miami newspaper on 17 September 1926 included a 4-in. story on page one under the heading "Hurricane Reported." However, the editors added that it was not expected to hit Florida. The afternoon paper placed a greater emphasis on the hurricane and included a front-page headline "Miami Warned of Tropical Storm" but did not mention hurricane warnings since none had been issued at the time of publication.

In contrast to the 6-h advance warning for the 1926 storm, a hurricane warning for Andrew was issued at 0800 EDT on 23 August 1992 (21 hours prior to landfall) from Vero Beach to the upper Florida Keys (U.S. Department of Commerce 1993a). A hurricane watch had been issued at 1700 EDT on 22 August (36 hours prior to landfall) from Titusville through the Florida Keys. On the morning of 23 August, the local Miami paper had a front-page headline that read "Bigger, Stronger, Closer." Based upon the author's experience on 23 August, it would have been very difficult to find a person in the south Florida area who was unaware that a hurricane was approaching. However, similar to the 1926 hurricane, a large fraction of the south Florida residents had never experienced a major storm. Thus, despite the effectiveness of the warnings, much of the population was still surprised by the devastating effects of the storm.

In addition to the 1926 hurricane, there were several other strong storms that hit with little advance warning during this period (U.S. Department of Commerce 1993b). A severe hurricane struck Florida in 1928, and about 1800 people drowned when Lake Okeechobee flooded. This storm was the second deadliest in U.S. history (Hebert et al. 1993). Dunn (1971) described another example of a difficult forecast situation. In August 1934 a tropical storm was approaching the upper Texas coast and the Weather Bureau in Washington, D.C., had issued a hurricane warning on a Sunday morning. When the Chamber of Commerce in Galveston wired the Washington office for an update that Sunday afternoon, the forecaster had left his shift (as was the usual procedure) and was scheduled to return in the early evening when more observations were going to be available. The map plotter on duty at the time wired back the comment, "Forecaster on golf course—unable to contact." Although the response was accurate, the Chamber of Commerce in Galveston was somewhat upset. As it turned out, the storm in question did not actually make landfall in Texas.

1935–1942, The use of upper-air data and forecast office reorganization

The growing dissatisfaction with the performance of the Weather Bureau in Washington, D.C., described in the previous section led Congress and the president to decentralize and improve the hurricane warning service (Calvert 1935; Sheets 1990). The forecast service in Washington, D.C., was continued; new centers were established in Jacksonville, Florida, and New Orleans, Louisiana; and a center was reestablished in San Juan, Puerto Rico. A center was also established in Boston, Massachusetts, in 1940. These centers were responsible for issuing warnings every 6 hours, seven days a week, from June to November. The Jacksonville office was the most complete center and had the largest area of responsibility (from Cape Hatteras, North Carolina, to Apalachicola, Florida, and most of the Atlantic). In addition to the new centers, a 24-h hurricane teletype network was set up from Wilmington, North Carolina, to Brownsville, Texas, to improve the distribution of observations and facilitate communications between forecast offices and field locations. In addition, arrangements were made for special ship observations during storm conditions. Because of these improvements, storms were no longer "lost" for up to several days at a time, as had occasionally occurred previously (Dunn 1940a).

Some important conceptual advances were made during this period. Using a theoretical argument based upon hydrostatic considerations, Haurwitz (1935) presented evidence that well-developed tropical cyclones extend through the depth of the troposphere. It was generally accepted that once a hurricane developed, it was maintained by the release of latent heat due to the convection near the storm center (Scofield 1938). McDonald (1942) presented a fairly accurate description of the life cycle of tropical cyclones that included a preliminary phase when a system first becomes organized, a deepening stage when most of the intensification occurs, an expanding stage where the storm increases in size but the maximum winds do not increase, and a declining stage when the storm dissipates. He also suggested that the maximum intensity of hurricanes is limited to a minimum pressure of about 880 mb (26 in.) and that thermodynamic factors are probably responsible for this limitation. To date, the minimum observed sea level pressure for an Atlantic hurricane is 888 mb, which occurred in Hurricane Gilbert (Willoughby et al. 1989a).

Although some aspects of hurricane evolution were understood, there were still some misconceptions about hurricane genesis. By the late 1930s the frontal theory of tropical cyclones had gained general acceptance (Scofield 1938), perhaps on the "coat tails" of the polar front theory of midlatitude cyclones that was making rapid advances during this time (Palmen and Newton 1969). However, this theory came into question as more observations became available. For example, Dunn (1940b) showed that Atlantic hurricane genesis commonly results from the amplification of "isallobaric" (easterly) waves that originate near the African coast.

Another important advancement was the use of upper-air observations. According to Dunn (1940a), the number of pilot balloon observations available in the Caribbean region increased dramatically during the late 1930s, primarily due to efforts by Pan American Airways. In addition, a few radiosonde sites had been established. By the late 1930s, upper-air winds were routinely plotted at 2000-ft intervals up to 14000 ft for hurricane forecasting purposes. It soon became apparent that knowledge of the upper-level flow was extremely useful for determining storm motion, and less emphasis was placed on surface pressure patterns in areas where upper-air data were available (Byers 1935; Dunn 1940a).

Despite the above advances, hurricane forecasting was far from perfect during this period. The upper-air data had a very limited area of coverage, and forecasters still relied heavily on extrapolation of intensity and position estimates based upon surface observations. These extrapolations were often hampered by incomplete information. For example, the hurricane that struck the Florida Keys in September 1935 with a minimum sea level pressure of 892 mb still holds the record for the most intense U.S. landfalling storm (Hebert et al. 1993). However, the only surface observation available to the forecasters just prior to landfall was from the southern Bahamas, which indicated a maximum surface wind of only $\sim$18 m s^{-1} (Burpee 1988). The warning message issued from the Jacksonville office just before landfall indicated that hurricane force winds *probably* existed in a small area near the center (McDonald 1935).

The lack of information near the storm center and the reliance on extrapolation also made it difficult to anticipate rapid changes in the storm speed or direction of motion. An extreme example of this problem is illustrated by the hurricane that struck New England in September 1938. As shown in Fig. 9-2, this storm moved from a location east of Jacksonville, Florida, to the coast of Long Island, New York, in less than 24 hours. According to Tannehill (1938b), this hurricane made landfall on the Long Island coast at about 1900 UTC 21 September and on the Connecticut coast 2 hours later. The Washington, D.C., forecast issued gale warnings at 1500 UTC for the coast from Virginia to New Jersey when the storm was off the Virginia coast but only 4 hours before its landfall on Long Island. The final warning was issued at 1900 UTC for the Long Island coast and Connecticut as the storm was making landfall on Long Island.

To put the forecasts for the 1938 storm in perspective, Fig. 9-2 also shows the track of Hurricane Bob, which made landfall on the Rhode Island coast at about 1800 UTC 19 August 1991 (Pasch and Avila 1992). Hurricane Bob also accelerated rapidly toward the north, although not quite as dramatically as the 1938 storm. At the time of landfall in Rhode Island, Bob was moving at $\sim$14 m s^{-1}, compared with a forward speed of $\sim$24 m s^{-1} when the 1938 hurricane made landfall in New England. These speeds were estimated from the 6-h best-track

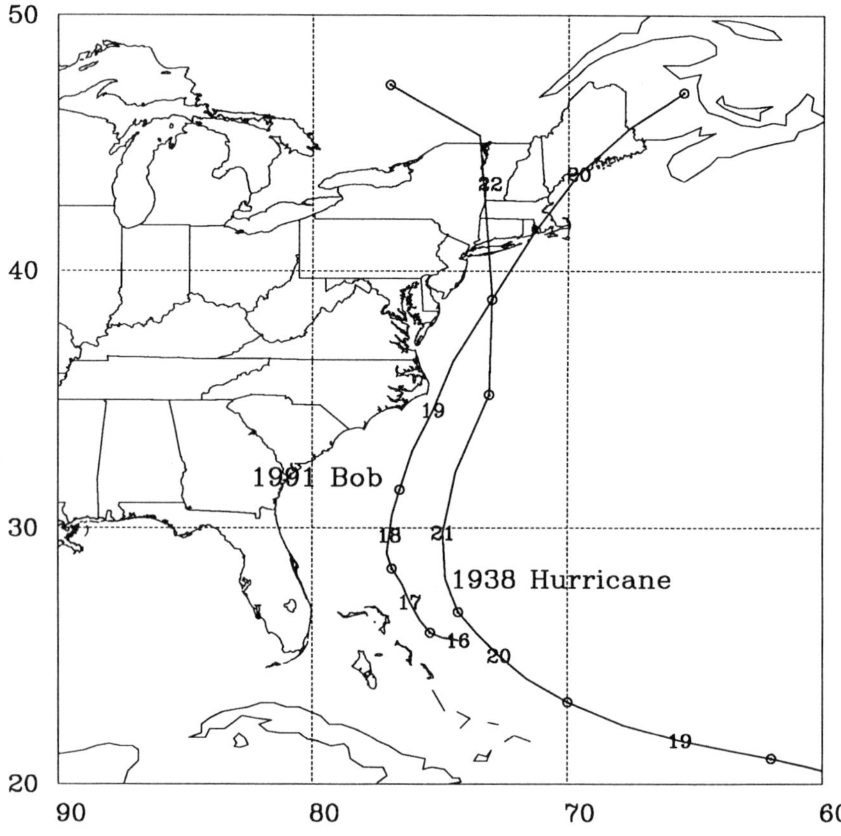

FIG. 9-2. The tracks of the 1938 New England Hurricane and Hurricane Bob (1991). The open circles are the positions at 1200 UTC, and the days of the month (September 1938 or August 1991) are indicated at the 0000 UTC positions.

positions. As described above, warnings for the 1938 hurricane were not issued for New England until just a few hours before landfall. In contrast, The NHC issued hurricane warnings for the region that included Rhode Island 20 hours prior to the landfall of Bob, when the storm was just south of Cape Hatteras, North Carolina (Pasch and Avila 1992).

1943–1955, Technological advancements: Aircraft reconnaissance and radar

At the beginning of this time period, the primary hurricane forecast office in Jacksonville was moved to Miami. The meteorologist in charge (MIC) of this office (which became the NHC in 1955) was Grady

Norton, who was previously in charge of hurricane forecasting at the Jacksonville office. Although Norton died before the Miami office officially became the NHC, he is generally regarded as the first NHC director (Sheets 1990). Norton was recognized for his excellent forecasting and communication skills (Burpee 1988). Considerable insight into the state of hurricane forecasting during this period can be gained from his unpublished soliloquy (Norton 1947). Gordon Dunn became the first official director of the NHC in 1955 (Burpee 1989).

A period of major advancement in hurricane forecasting started when Col. Joseph P. Duckworth completed the first intentional flight into the eye of a hurricane in July of 1943 (Markus et al. 1987). The U.S. Air Force and Navy began regular missions in 1944. Although hurricane forecasting in the Atlantic basin has greatly benefited from aircraft reconnaissance, the initial motivation for these flights came from the massive damage incurred by the U.S. Navy during World War II due to typhoons in the Pacific. The instrumentation and aircraft used for hurricane reconnaissance have varied considerably since 1943 (see Sheets 1990 for a thorough review), but the primary objective of documenting the current location and intensity of a storm has remained the same. As described by Norton (1947), "No forecaster can do much at hurricane forecasting unless he knows where his hurricane is and how intense it is, and there is no quicker way to find out than with aircraft." It is interesting that Norton also stated, "This (aircraft) is an expensive way to obtain reports." The debate over the cost–benefit ratio of aircraft reconnaissance continues today (Gray et al. 1991).

Another objective of the reconnaissance aircraft was to search for tropical disturbances via routine "synoptic track" flights when other observations such as ship reports suggested their existence. However, the advent of weather satellites in the 1960s and 1970s allowed more efficient use of the aircraft and, therefore, reduced the number required to collect needed data. For this reason it became more difficult to justify the cost of reconnaissance, and the U.S. Navy discontinued hurricane missions after 1974. At the present time, aircraft reconnaissance for the Atlantic basin is the responsibility of the U.S. Air Force Reserve at Keesler Air Force Base near Biloxi, Mississippi, and the National Oceanic and Atmospheric Administration's (NOAA) Aircraft Operations Center at MacDill Air Force Base in Tampa, Florida.

The development of meteorological radar also had an impact on hurricane forecasting. Although the radar echoes from precipitation were considered a nuisance in the development of military applications during World War II (Bigler 1981), it soon became apparent that radar could be a valuable meteorological tool. Early radar observations revealed the complicated structure of the inner core of hurricanes and documented the existence of spiral rainbands (Maynard 1945; Wexler 1947). In combination with early aircraft observations, it was discovered that the vertical motion in hurricanes occurs in a narrow band (the eyewall) near the storm center (Wexler 1945; Deppermann 1946). In 1955, a new radar was installed at Hatteras, North Carolina, and all three storms that made landfall in the United States that year passed within its range (Dunn and Miller 1960). Experience during this year demonstrated the usefulness of radar for tracking storms near the U.S. coast. [See chapter 4 by Rogers and Smith, this volume.]

As more accurate storm position and intensity estimates with higher time resolution became available, it was discovered that the intensity and track of the storm can vary more rapidly than was believed previously. For example, prior to 1943, it was generally accepted that once a storm attained hurricane intensity, it remained at hurricane intensity provided that it did not move out of tropical regions (Simpson 1966). In addition, small-amplitude track oscillations were observed by careful examination of aircraft (e.g., Horn 1951) and radar (Dunn and Miller 1960) observations. Modeling studies by Yeh (1950) and later by Kuo (1969) and others indicated that these small-amplitude track variations (often referred to as trochoidal oscillations) are due to internal processes near the storm center and are superimposed on the more smooth path due to large-scale "steering." Forecasting methods have been developed to help separate these trochoidal oscillations from the larger-scale motion of the storm (Sheets 1986).

Another important development during this period was the wartime expansion of the upper-air network through the implementation of radiosonde and rawinsonde equipment (Thompson 1961). Although the coverage in the Tropics was still somewhat meager, numerous synoptic analyses for storms near the U.S. coast and island stations were performed. Riehl and Shafer (1944) used upper-air observations to show that the recurvature of hurricanes is very sensitive to the movement and structure of upper-level troughs in the westerlies. These ideas led to the steering principle for track forecasting, where the winds at the "top" of the hurricane determine the future direction of storm motion (Norton 1947; Gentry 1951). Further studies (Riehl and Burgner 1950; Jordan 1952) deemphasized the use of a single level and advocated mass-weighted layer-mean winds as indicators of future storm motion. The observational study by Jordan (1952) confirmed the theoretical results of Haurwitz (1935), which showed that the hurricane circulation extends through a large fraction of the troposphere.

Synoptic studies also provided new insight into the formation and intensification of hurricanes. Palmen (1948) demonstrated that warm ocean temperatures (>26°C) are required to maintain tropical cyclones, and Riehl and Shafer (1944) indicated that storms do not intensify if the vertical shear of the horizontal wind becomes too large. In an analysis of the genesis of west Pacific storms, Riehl (1948) presented additional observational evidence that the frontal theory of tropical cyclone formation (the countercurrent theory described in the previous section) was not valid and stressed the importance of the interaction between low-level rotation and upper-level troughs.

Hurricane Hazel (1954) provides an example of the improvement of hurricane forecasting techniques during this period. The track of Hazel is shown in Fig. 9-3. As described by Dunn and Miller (1960), aircraft monitoring of Hazel began on 5 October 1954 and continued until the storm made landfall near the North Carolina–South Carolina border at ~1500 UTC 15 October. The aircraft observations documented the intensity and positions of Hazel, including the turn to the north on 10 October and the acceleration toward the north-northwest that began on 13 October. The 500-mb analyses were also used to forecast the turn toward the Carolina coast. Hurricane warnings for the North Carolina outer banks were issued at 1600 UTC on October 14 (23 hours prior to landfall), and gale warnings were issued from Charleston, South Carolina, to Wilmington, North Carolina. The hurricane warnings were extended southward to Charleston at 0700 UTC on 15 October (8 hours prior to landfall). The posting of hurricane warnings near the observed landfall point 23 hours in advance for Hazel is similar to the lead time of the warnings for Hurricane Bob in 1991 and Hurricane Andrew in 1992, described previously, and is a considerable improvement over the very short range warnings issued for the 1938 New England hurricane.

1956–1965, Increased research efforts and the development of objective forecast methods

In addition to Hurricane Hazel, five other significant hurricanes made landfall along the northeast U.S. coast in 1954 and 1955 (Carol

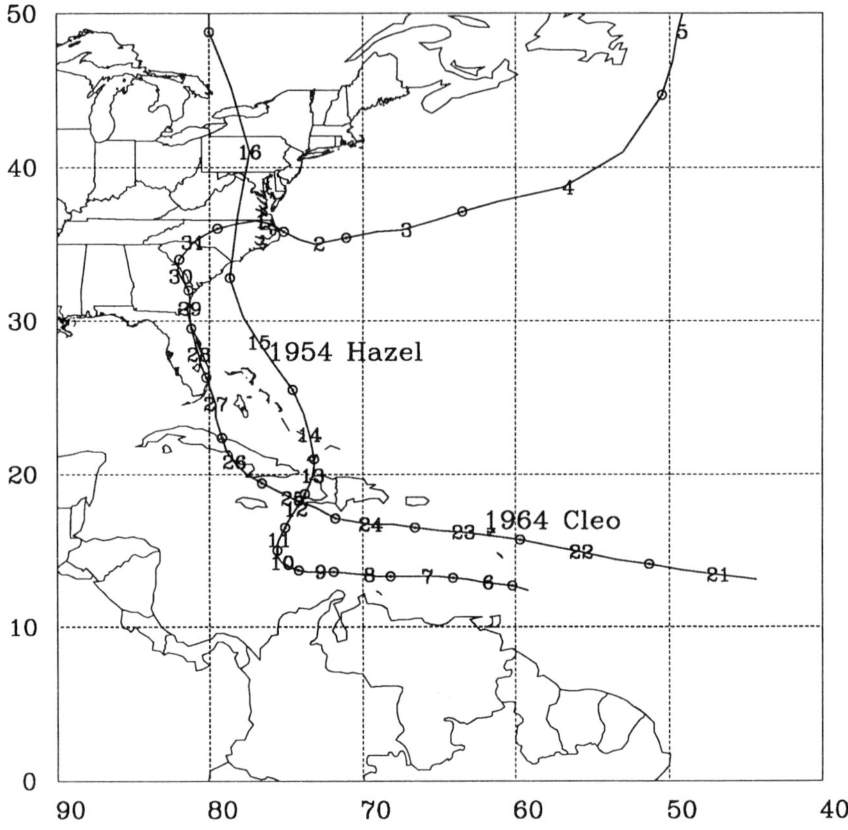

FIG. 9-3. The tracks of Hurricane Hazel (1954) and Hurricane Cleo (1964). The open circles are the positions at 1200 UTC, and the days of the month (October 1958 or August–September 1964) are indicated at the 0000 UTC positions.

and Edna in 1954; Connie, Diane, and Ione in 1955). Although the forecasts were reasonably good (Bigler 1981), the massive damage from these storms (Hebert et al. 1993) heightened the awareness of Congress and led to the formation of the National Hurricane Research Project in 1956 (NHRP Staff 1956; Simpson 1980). The initial goals of this project were to examine the structure of hurricanes and prehurricane disturbances and to determine important parameters for hurricane forecasting using data from instrumented aircraft and conventional sources. After organizational meetings in Washington, D.C., the operations base for NHRP was established in West Palm Beach, Florida. Three aircraft (two WB-50s and one B-47) were furnished by the U.S. Air Force and instrumented by Weather Bureau contractors. In

1959, NHRP was moved to Miami where it was collocated with the NHC. This project has undergone several name changes and reorganizations since 1956 and is currently known as the Hurricane Research Division (HRD) of NOAA's Atlantic Oceanographic and Meteorological Laboratory (AOML). Directors of HRD and its predecessor organizations include Robert H. Simpson (1956–1958), R. Cecil Gentry (1959–1974), Noel E. LaSeur (1975–1977), Stanley L. Rosenthal (1977–1992), Robert W. Burpee (1993–1995), and Hugh E. Willoughby (1995–present). The instrumented aircraft have also undergone considerable evolution since 1956 (Gentry 1980). In 1961, the Research Flight Facility (RFF), which maintained and operated the aircraft was separated from NHRP. HRD now uses NOAA's two WP-3D aircraft with sophisticated instrumentation systems including Doppler radar (Griffin et al. 1992). The WP-3D aircraft are operated by NOAA's Aircraft Operations Center. The NHRP and its offspring organizations have made numerous contributions to the hurricane forecasting problem and the general understanding of tropical cyclones. Gentry (1980), Marks (1990), and Burpee et al. (1994) summarize some of these contributions.

Another consequence of the landfalling storms in 1954–1955 was the expansion of the coastal radar network (Wells 1961). Following these storms, Congress appropriated funds for the installation of WSR-57 radars, which provided nearly continuous coverage for the coastal areas from Texas to Maine by the early 1960s (Sheets 1990).

Another important development was the introduction of objective hurricane forecast models. Riehl et al. (1956) described the first of these models, a statistical regression technique relating 24-h storm displacements to 500-mb geostrophic steering currents. By 1965, five other statistical prediction models were being used by NHC for operational track forecasting (Tracy 1966), many of which had been developed in cooperation with NHRP (e.g., Miller and Moore 1960). Statistical model development was also aided by interactions with the Travelers Weather Research Center (Neumann 1979) and the Spaceflight Meteorology Group (SMG), which provided meteorological support for the Apollo launches from Cape Canaveral. The SMG was collocated with NHC from the early 1960s to the early 1970s. Predictors of future storm positions included past motion of the storm (persistence), sea level pressure, 700- and 500-mb heights and tendencies, and 1000–700- and 700–500-mb thicknesses. Forecasts were made for periods up to 72 hours. Dunn et al. (1968) showed that the 24-h forecast track forecast errors were reduced by 10%–12% during the period 1959–1966. Dunn et al. (1968) also indicated that the use of objective forecast methods and the cooperation between researchers and forecasters were the primary reasons for the error reduction.

Parallel with the advancements being made in numerical weather prediction for the synoptic-scale flow (e.g., Schuman 1989), the Joint Numerical Weather Prediction Unit at the National Meteorological Center (renamed the National Centers for Environmental Prediction in 1995) was developing dynamical models for hurricane track forecasting. The first of these models was run operationally from 1956 to 1958 and tracked the streamfunction minimum in a barotropic prediction initialized with a conventional 500-mb height analysis (Hubert 1957). Due to inadequate resolution and the lack of observations in the Tropics, this model was replaced by a model where the storm circulation was removed from the initial condition (Vanderman 1961). In this case, the model was used to obtain the steering flow from which the storm track was determined. This model was later generalized to include the effect of the storm circulation on the steering flow (Vandermann 1962).

There was also a considerable research effort at the University of Chicago supported by the Weather Bureau that focused on numerical prediction of hurricane movement and intensification. Birchfield (1961) presented results from a relatively high resolution barotropic model that demonstrated the feasibility of hurricane prediction without removal of the vortex circulation. These results also indicated that the initialization of the model with vertically averaged fields is preferable to the use of a single level. Jones (1961) presented preliminary results from forecasts with a two-level model, and Kasahara and Platzmann (1963) described an elegant method for partitioning the vortex and large-scale circulation. Kasahara and Platzmann also explained the reason for the failure of earlier models that separated the vortex and larger-scale flows.

Despite the efforts described above, track forecasts from numerical models were generally inferior to those obtained using statistical methods (Tracy 1966; Miller and Chase 1966). For this reason, numerical prediction of hurricane tracks received little attention until the late 1960s (Sanders and Burpee 1968).

The importance of the ocean as an energy source for tropical cyclones became increasingly clear during this period. Fisher (1958) estimated the surface energy fluxes from the ocean and concluded that "the hurricane is a mechanism which massively siphons energy from the sea." Malkus and Riehl (1960) demonstrated the importance of the ocean for increasing the energy of low-level air spiraling into a storm. Miller (1958) introduced the concept that the ocean temperature controls the maximum possible intensity of tropical cyclones. He also showed that hurricanes rarely reach this theoretical upper bound and suggested that interactions with the large-scale environment probably limit the intensity of most storms. It was also becoming clearer that interaction with the large-scale environment, particularly in the upper levels, is important for hurricane genesis (Ramage 1959). Advances were also made in the understanding of the hurricane inner core, including the importance of the eyewall in the maintenance of the entire storm circulation (LeSeur and Hawkins 1963).

A storm that impacted forecasting methods during this period was Hurricane Cleo (1964), whose track is shown in Fig. 9-3. This storm formed from an easterly wave on 20 August 1964 and developed into a small but intense hurricane by 24 August. On 26 August the storm weakened as it crossed Cuba and then headed toward south Florida. Cleo was closely monitored by reconnaissance aircraft prior to reaching Florida and was under constant radar surveillance after it crossed Cuba (Dunn et al. 1965). The last aircraft left Cleo about 3 hours prior to landfall in Miami, having found a central pressure of 984 mb and no hurricane force winds on the western side of the storm. However, during the next 3 hours, the storm intensified rapidly and reached Miami with a pressure of ~967 mb. Sustained surface winds well above hurricane force were observed over land. Some of the forecasts prior to landfall in south Florida suggested that the center of the storm would remain just offshore, although, as described by Dunn et al. (1965), the storm was following a "mildly zigzag course." This "zigzag course" was probably a trochoidal oscillation that caused the storm to move more to the west than was predicted. This relatively small change in course, combined with the rapid intensification, resulted in much stronger effects from the storm than were anticipated. In hurricane warnings after Cleo, less emphasis was placed on the location of the exact center of the storm.

1966–1973, The use of satellite observations and improved forecast models

The potential of meteorological satellites for hurricane surveillance (e.g., Hubert 1961) and tropical analysis (e.g., Frank 1963) was demonstrated shortly after the launch of the first U.S. weather satellite (*TIROS-1*) in April of 1960. The beginning of the era where satellites were heavily relied upon for operational tropical analysis and forecasting followed the launch of *ESSA-1* and *ESSA-2* in February of 1966 (Sheets 1990). [See Chapter 5 by Purdom and Menzel, this volume.]

Another important advance was the continued improvement of statistical track forecast models. Hope and Neumann (1977) categorized these models as analog, simulated analog, statistical–synoptic, and statistical–dynamical. The analog technique, which uses information from previous storms to determine the future storm track, was applied in the HURRAN (hurricane analog) model (Hope and Neumann 1970). HURRAN became operational in 1968 and was run until the early 1990s. In some cases, it is not possible to find an historical analog. For this reason, Neumann (1972) developed the CLIPER (climatology and persistence) model, which uses climatological and persistence predictors to forecast the storm track. Because CLIPER includes information from previous storms but does not find specific analogs, it was categorized as a simulated analog model. CLIPER is still run at NHC and is often considered a benchmark for the evaluation of other models. If a model has smaller average errors than CLIPER, it is considered skillful (Neumann and Pelissier 1981a).

The statistical–synoptic and statistical–dynamical models are similar to the early regression models described in the previous section. Statistical–synoptic models combine current storm information and synoptic predictors. By the end of this period (1974), two models of this type were run operationally for storms in the Atlantic basin: NHC-67 (Miller et al. 1968) and NHC-72 (Neumann et al. 1972). Statistical–dynamical models include variables from numerical forecasts in addition to current storm information and synoptic predictors. The early attempts to use predictors from numerical forecasts were unsuccessful (Veigas 1966). This lack of success was attributed to the poor quality of the barotropic forecasts in low latitudes. The six-layer primitive equation model became operational at NMC in 1966, which led to improved forecasts of the 500-mb flow (Schuman 1989). In 1973, the first successful statistical–dynamical model became operational at NHC (Neumann and Lawrence 1975).

In the late 1960s, there was renewed interest in the development of numerical models for track forecasting. Under support from the National Hurricane Research Laboratory (formerly NHRP), a new barotropic model was developed at the Massachusetts Institute of Technology (Sanders and Burpee 1968). Special emphasis was placed on the analysis in the vicinity of the storm, and the large-scale flow was determined from a vertically averaged wind field. This model was nicknamed SANBAR (Sanders barotropic) and became operational in 1970 (Neumann and Pelissier 1981a). A number of modifications were made to SANBAR in later years (Sanders et al. 1975, 1980), and the model was used operationally until 1989.

During this time period, there was an apparent divergence of interests between the forecasting and research communities. Ooyama (1964) and Charney and Eliassen (1964) presented theoretical results, which suggested that the development of tropical cyclones could be explained by a cooperative interaction between the cumulus convection near the storm center and the larger-scale tropical cyclone circulation. This cooperative interaction is often referred to as conditional instability of the second kind (CISK). Ooyama (1969) and Yamasaki (1968) described numerical simulations that could simulate the life cycle of tropical cyclones including the mature stage. Following the publication of these papers, a considerable research effort was aimed at the numerical prediction of tropical cyclones (e.g., Rosenthal 1970; Anthes 1972; Kurihara 1975). Although these numerical modeling studies provided an increased understanding of tropical cyclones, they did not have direct forecasting applications. These models were either axisymmetric (due to computer limitations) or used idealized three-dimensional initial conditions, and the focus was often on the specific representation of cumulus convection (e.g., Rosenthal 1978; Anthes 1977). For a more detailed description of the early development of numerical hurricane models, see Anthes (1982).

Another problem that provided a research focus was Project STORMFURY, which was conducted from 1962 to 1983 (Willoughby et al. 1985). The purpose of this project was to determine if hurricanes could be modified by seeding the convective clouds in the eyewall or those surrounding the eyewall with silver iodide. This was a joint project by the Department of Commerce and the U.S. Navy and fulfilled one of the original goals of the NHRP (NHRP Staff 1956). As part of this project, four storms were seeded on a total of eight days during the period 1961–1971. Project STORMFURY ended as it became apparent that the microphysical requirements for seeding were usually not satisfied in hurricane convection. In addition, it was discovered that the natural evolution of the inner core of many intense hurricanes is very similar to behavior expected from seeding (Jordan and Schatzle 1961; Willoughby et al. 1982). This similarity made it difficult to distinguish between responses to seeding and the natural evolution of a storm. Despite the fact that STORMFURY did not lead to an operational method for hurricane modification, many worthwhile investigations were performed as part of this experiment, and STORMFURY funds were used to obtain the NOAA WP-3D aircraft (Willoughby et al. 1985). Also, many of the early modeling studies were motivated by the need to improve understanding of seeding responses (e.g., Rosenthal 1971; Jones 1976). However, the attention devoted to this project may have reduced the emphasis on studies that would have had more direct forecast applications.

In 1968, Gordon Dunn retired and Robert H. Simpson became the director of NHC (Sheets 1990). Simpson established a small research and development unit at NHC, where many of the statistical track forecast models described previously in this section were developed. This unit was also briefly involved in numerical model development and attempted one of the first three-dimensional hurricane simulations initialized with real data (Miller 1969). Simpson trained with Dunn in 1967 and created the "hurricane specialist" positions, with the responsibilities of tropical cyclone forecasting during the hurricane season and research and public service in the remainder of the year (G. Clark 1995, personal communication). Arnold Sugg became the first official hurricane specialist in 1968.

1974–1987, Geostationary satellites and three-dimensional forecast models

The first Geostationary Operational Environmental Satellite (GOES) became available at the beginning of GARP (Global Atmospheric Research Program) Atlantic Tropical Experiment in June 1974 (Kuettner and Parker 1976). The importance of these satellites for tropical analysis and forecasting was stressed by Sheets (1990), who stated: "If there was a choice of only one observing tool for meeting the responsibilities of the NHC, the author would clearly choose the geostationary satellite." GOES satellite observations are extremely useful for determining storm positions and for estimating storm intensities using the method first described by Dvorak (1973). In addition, satellite data are used to determine "cloud track" winds, primarily at upper and lower levels. These winds contribute significantly to the operational database in tropical regions where conventional upper-air observations are not available (Dey 1989).

Another technological advance that had an impact on hurricane forecasting was NOAA's acquisition of two WP-3D aircraft for hurricane research and reconnaissance, which were put into operation in 1976 and 1977 (Sheets 1990). Although these aircraft were obtained in connection with Project STORMFURY (Willoughby et al. 1985), they were never actually used in a hurricane-seeding experiment. However, the observations from these aircraft have contributed greatly to knowledge of hurricane structure (e.g., Marks 1990). In 1977, the Aircraft Satellite Data Link (ASDL) was implemented, which allows observations from the NOAA aircraft to be transmitted directly to NHC in real time (Parrish et al. 1984). The ASDL system greatly increased the utility of the research aircraft for operational forecasting. A similar data observation and transmission system (the Improved Weather Reconnaissance System) was developed in the late 1980s for the WC-130 reconnaissance aircraft operated by the air force reserve.

Another important development was the implementation of a three-dimensional hurricane model for operational track forecasting in the Atlantic basin (Hovermale and Livezey 1977). This model was developed at NMC and included 10 vertical levels with a horizontal resolution of 60 km. Due to computer limitations, the model had a fairly small horizontal domain (3000 km $\times$ 3000 km), but the domain moved so that it remained approximately centered on the storm. For this reason, the model was referred to as the Moveable Fine Mesh (MFM). The MFM became operational in 1976 and was run until the end of the 1988 hurricane season. Due to computer limitations, the MFM was run only to 48 hours during the period 1976–1983. After 1983, it was run for 72 hours. Due to initialization difficulties, the MFM had relatively large forecast errors at 12–24 hours. However, the longer-range forecasts had considerable skill, especially for higher-latitude storms (DeMaria et al. 1990). As described by Hovermale and Livezey (1977), the

MFM was developed because the feasibility of three-dimensional tropical cyclone simulations had been demonstrated in the research environment. Thus, some of the modeling efforts in the late 1960s and early 1970s began to have forecast applications.

Statistical track forecast models also continued to advance during this period. Neumann (1988) developed a new statistical–dynamical model (NHC83) that used output from NMC's global spectral model. On average, this model outperformed all other guidance models during the period 1983–1987, including the MFM (DeMaria et al. 1990). Because of the success of NHC83, the older statistical models (NHC67, NHC72, and NHC73) were no longer run at NHC after 1988 (Sheets 1990). NHC90 (Neumann and McAdie 1991) is a modified version of NHC83 that is still used at NHC.

When the number of guidance models increases, it becomes more likely that there is disagreement between the track predictions. For example, Fig. 9-4 shows some of the model forecasts that were available for Hurricane Allen (1980). The forecaster must use subjective judgement of the model guidance and other available information to produce the official NHC forecast. Partially because of the proliferation of models, greater emphasis was placed on model verification during this period. Detailed descriptions of the official NHC forecasts for 1970–1979 and the guidance models for the period 1973–1979 were presented by Neumann and Pelissier (1981a,b). DeMaria et al. (1990) provided a verification of the Atlantic track guidance models for the period 1983–1988. Verification of the official and model forecasts for the past several years have also been published in the minutes of the Annual NOAA Interdepartmental Hurricane Conference (e.g., Lawrence and Gross 1994).

Some minor progress was also made in intensity forecasting during this period. Jarvinen and Neumann (1979) developed a Statistical Huricane Intensity Forcast (SHIFOR) model that uses climatological and persistence predictors in combination with current storm characteristics. This model is analogous to the CLIPER track prediction model. Because SHIFOR does not include any direct synoptic information, it is useful only for predicting average intensity changes. Hebert (1977, 1978) modified a decision ladder approach that was used to forecast whether or not a tropical cyclone would intensify. This approach was based upon the earlier work of Simpson (1971), modified to include some of the factors such as vertical wind shear suggested by Gray

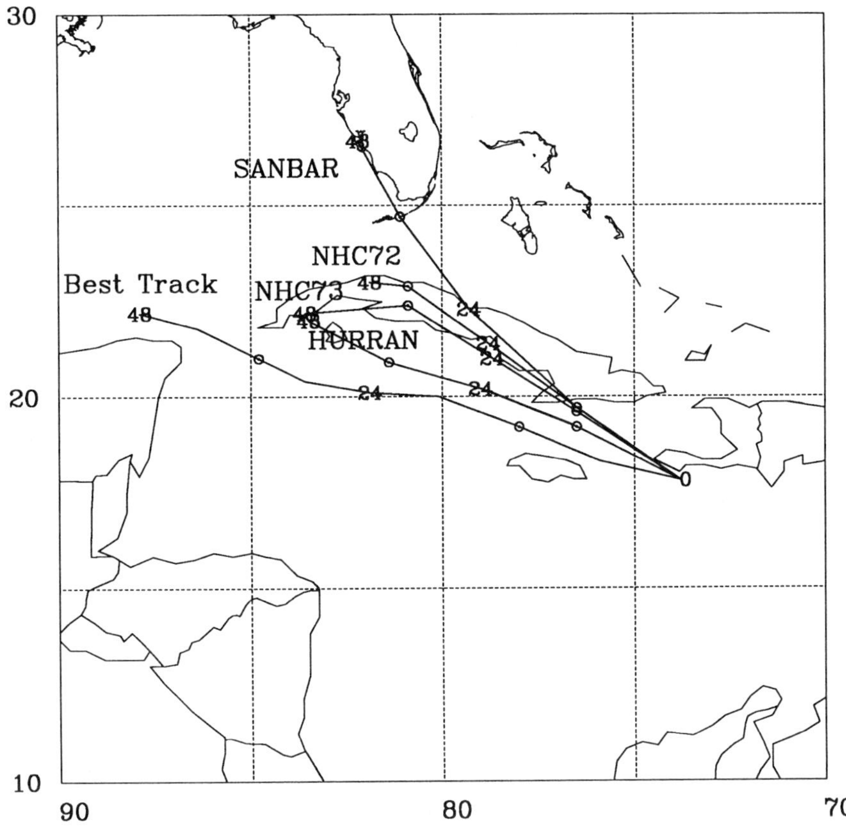

FIG. 9-4. The 48-h best track for Hurricane Allen (1980) beginning at 0000 UTC 6 August 1980. Also shown are the 48-h forecasts from the HURRAN, NHC72, NHC73, and SANBAR track forecast models.

(1968). Although the revised decision ladder approach includes synoptic information, it does not provide a quantitative intensity estimate. Pike (1985) attempted to include geopotential heights as additional predictors in a model similar to SHIFOR. However, this synoptic information resulted in only a marginal improvement relative to SHIFOR.

Another change that affected hurricane forecasting during this period was the appointment of Neil L. Frank as NHC director in 1974 following the retirement of Simpson (Sheets 1990). Frank placed great emphasis on the hurricane warning and preparedness aspect of the hurricane forecasting problem (Frank 1984), and considerable improvements were made in this area during his 13-year tenure. Also,

Frank effectively used electronic media as a means to provide information to people in areas threatened by storms.

Another administrative change was that all tropical cyclone forecasts for the Atlantic basin became the responsibility of the NHC in 1980. As described previously, this responsibility had been shared by the Miami, New Orleans, San Juan, Washington, and Boston offices following the reorganization of the hurricane warning service in 1935.

1988–1994, Increased interaction with the research community and the use of global model output

In 1988 Robert C. Sheets became the director of NHC. He was the acting director of NHC in 1987 after the retirement of Frank. Sheets had previously been one of the directors of Project STORMFURY (Gentry 1980).

Considerable progress was made during this time period in the numerical prediction of hurricane tracks. In 1988 the MFM was replaced by the Quasi-Lagrangian Model (QLM). The QLM had higher vertical and horizontal resolution than the MFM and used a larger model domain (Mathur 1991). In 1988, there were a large number of low-latitude storms, and the QLM did not perform well relative to other guidance models (DeMaria et al. 1990). However, these early difficulties were related to the vortex initialization and model modifications corrected this problem (Mathur 1991). The QLM produced skillful forecasts for the period 1989–1993 (Aberson and DeMaria 1994).

In 1992, the research model developed at the Geophysical Fluid Dynamics Laboratory (GFDL) was adapted for real-time forecasting. This model includes moving nested grids and a sophisticated vortex initialization scheme (Bender et al. 1993). The horizontal grid spacing on the inner mesh is ~20 km, which is one-half that of the QLM. Near-real-time forecasts were made during the 1992–1993 hurricane seasons, and in 1994 the model was transferred to NMC. Although the sample size is fairly small, the GFDL model outperformed all other models (in terms of average track error) during the 1992–1993 seasons (Aberson and DeMaria 1994). The GFDL model replaced the QLM for the 1995 season (Kalnay et al. 1994). During 1995 the GFDL model had smaller average position errors than any of the other track models.

In 1989, the SANBAR model was retired after 19 operational seasons. Beginning that year, an experimental barotropic track prediction model (Vic Ooyama barotropic model, VICBAR) was run in near real time (DeMaria et al. 1992). Since that time, VICBAR has been run in an experimental mode in real time. The main advantages of VICBAR relative to SANBAR are that VICBAR obtains lateral boundary conditions from the NMC global forecast model and uses a much more accurate numerical method.

The Beta and Advection Model (BAM) also uses the barotropic steering concept. The track forecast is obtained by following a trajectory in the vertically averaged horizontal wind from the NMC aviation global forecast model, with a correction that accounts for vortex drift (Marks 1992). The BAM model that uses a deep-layer mean horizontal wind has been run since 1989. Starting in 1990, two additional versions of BAM (with medium- and shallow-layer mean winds) have been run. The medium and deep BAMs have considerable track forecast skill (Aberson and DeMaria 1994), and the differences between the three versions of the model provide useful diagnostic information.

The large-scale regional and global prediction models from NMC have long been used as an aid in hurricane forecasting by providing forecasts of the hurricane environment. The skill of the global forecast models has improved considerably during the 1980s and 1990s (Kalnay 1994; Bonner 1989). This improvement can be attributed to improved model physics, increased horizontal and vertical resolution, and improved data assimilation techniques. Bengtsson et al. (1982) first presented examples of warm-core vortices produced by a global forecast model that were similar to tropical cyclones. In the following several years, the feasibility of forecasting tropical cyclones with research global models was demonstrated (e.g., Krishnamurti and Oosterhof 1989). Beginning in 1992, a tropical cyclone bogussing system was implemented in the NMC operational global model (the aviation model) for track forecasting (Lord 1991). Preliminary results from the 1992–1993 Atlantic seasons are encouraging (Aberson and DeMaria 1994), although the results from the 1994 and 1995 Atlantic hurricane seasons suggest that some modifications to the model or initialization procedures might be necessary (N. Surgi 1995, personal communication).

The current suite of models for track forecasting in the Atlantic includes the simple analog model CLIPER, the statistical–dynamical model NHC90, the barotropic model VICBAR, the three versions of the trajectory model BAM, the GFDL regional baroclinic model, and the aviation model. All of the these models except CLIPER use information from the aviation model for calculation of predictors, for trajectories, or as boundary conditions. One other exception is a version of NHC90 (referred to as UK90) that uses output from the U.K. Meteorological Office global forecast model.

Some improvements have also been made in the models for intensity prediction during this period. Research results have emphasized the importance of the influence of the storm environment (especially in the upper levels) on hurricane intensity changes (e.g., Merrill 1988; Molinari and Vollaro 1989). In addition, Emanuel (1988) developed a theory showing that the maximum intensity of tropical cyclones is determined by the sea surface temperature and the temperature near the tropopause, in qualitative agreement with the earlier study of Miller (1958). These ideas were incorporated into a Statistical Hurricane Intensity Prediction Scheme (SHIPS) for the Atlantic basin (DeMaria and Kaplan 1994). The SHIPS model has been run in an experimental mode, similar to that described above for VICBAR, since 1990. Preliminary results suggest that the average intensity errors for SHIPS are 10%–15% less than those from the simple climatology and persistence intensity model (SHIFOR), although this improvement has not been demonstrated in an operational setting.

In addition to the SHIPS model, intensity prediction has also been attempted in the GFDL model, beginning in 1992 (Kurihara et al. 1993). Although, at the present time, the horizontal resolution (20 km) is marginal for the representation of the inner core of some intense hurricanes, this effort represents the beginning of the era in which three-dimensional models predict intensity as well as track.

The operational application of the GFDL model and the implementation of VICBAR and SHIPS are examples of the increased interaction between the research and forecasting communities that occurred during this period. The development of the BAM model described above was also based upon research results (Holland 1983). The implementation of the ASDL system on the NOAA research aircraft (described previously) also made a number of research applications available for operational use (Burpee et al 1994). During Hurricane Emily (1993), a series of radar reflectivities were transmitted from the aircraft to NHC in real time. In addition, omega dropwindsonde (ODW) observations collected in the storm environment were included in the operational database. Similar ODW experiments have been performed in several storms beginning in 1982, and research results have demonstrated that this information improves the track forecasts in barotropic (Franklin and DeMaria 1992) and baroclinic (Lord 1993) models. In addition, methods for analyzing and displaying the NOAA flight-level data (developed at HRD) have been used for operational forecasting at NHC since the late 1980s (Willoughby et al. 1989b; Sheets 1990). Methods are also being developed at HRD for providing real-time surface wind analyses by combining observations from a number of different platforms (Powell et al. 1996).

1995–2005, Present status and future outlook

Before attempting to project into the future, it is useful to consider what has happened in the recent past and to summarize current hurricane forecasting problems. McAdie and Lawrence (1993) have examined the NHC official track forecast errors for the period 1970–1991, where the errors were corrected to account for interannual variability in the forecast difficulty (Neumann 1981). These results indicate that the 24-, 48-, and 72-h average track forecast errors have decreased by 0.7%, 1.0%, and 1.2% per year during this period. This error reduction is undoubtedly due to the advances in the tools available for track prediction, as described earlier in this chapter. The greater rate of improvement for the longer forecast periods is consistent with the improvements in the NMC global forecast model and the statistical–dynamical and baroclinic track prediction models that provide the most accurate longer-range forecasts.

Although many advancements have been made during the past 75 years, the problem of hurricane forecasting has not been solved. The population at risk from tropical cyclones along the U.S. coastline has increased at a much faster rate than the improvement in track forecast errors. For example, in recent years, the coastal populations of Florida and Texas have increased at about 3% per year (Sheets 1990), while the average 24-h track forecast error (probably the most important for coastal evacuation) has decreased by only about 0.7% per year. In addition, it is currently not possible to provide hurricane warnings long enough in advance, relative to the time it takes to evacuate some of the most hurricane-prone areas (e.g., New Orleans or the Florida Keys) unless the size of the warning area is significantly increased. However, expanding the warning area is very costly and increases the areas that are overwarned (Sheets 1990).

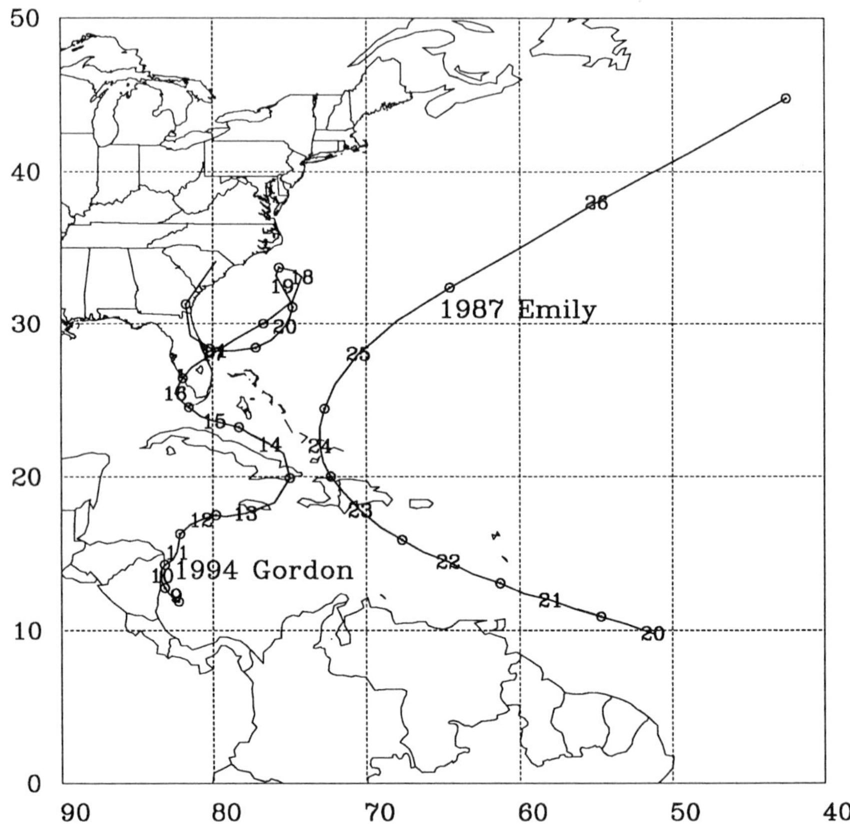

FIG. 9-5. The tracks of Hurricane Emily (1987) and Hurricane Gordon (1994).
The open circles are the positions at 1200 UTC, and the days of the month (September 1987 or November 1994) are indicated at the 0000 UTC positions.

Although the possibility of a tropical cyclone striking the U.S. coast
without being detected (as occasionally occurred in the first half of this
century) is extremely unlikely, some meteorological surprises are still
possible. For example, Fig. 9-5 shows the track of Hurricane Gordon,
which was the last storm of the 1994 Atlantic season. This storm had
a rather erratic track, and the forecast errors from the guidance models were generally larger than average. However, the most difficult
forecasts were for later stages of the storm. After crossing over Florida
on 16 November, Gordon intensified to hurricane strength and appeared to be accelerating toward the northeast. Nearly all of the guidance models initiated at 1200 and 1800 UTC 17 November predicted a
rapid recurvature, and the official NHC forecast was consistent with

the models. However, Gordon interacted with an upper-level trough to its west, rapidly decelerated, and turned toward the south by 19 November as it weakend to below hurricane strength. Although Gordon did not make landfall in the Carolinas, the lead time for the forecast in that region would have been quite short had the loop in the track of Gordon on 18–19 November been a little larger.

Another example of a difficult forecast was for Hurricane Emily (1987), whose track is also shown in Fig. 9-5. Emily formed from a disturbance on the ITCZ and intensified rapidly to a category-3 hurricane on 22 September (Case and Gerrish 1988). Emily was a small storm and weakend rapidly as it crossed Hispaniola on 23 September. Emily was expected to regain hurricane intensity after emerging into the Atlantic. However, the maximum winds remained less than $\sim$30 m s^{-1} for the next 24 hours. By 0000 UTC 25 September, the storm began to accelerate toward the northeast and remained weak. In addition, there was a frontal system to the northwest of the system, so Emily was forecast to weaken. However, in the next 12 hours the storm intensified rapidly and reached Bermuda by 1145 UTC 25 September with sustained winds of 39 m s^{-1} and gusts to 52 m s^{-1}. As described by Case and Gerrish (1988), this intensification was not anticipated, and the rapid acceleration was underestimated. Fortunately, no deaths were reported in Bermuda, although there was considerable damage from the storm.

The occurrence of large, unanticipated track changes as described above are not very common. More typically, the track guidance models in combination with satellite observations (especially water vapor imagery) illustrate the range of possible storm tracks. However, rapid intensity changes are much more difficult to forecast, and underestimates for rapidly intensifying storms are fairly common. For example, the intensities of Hurricanes Hugo in 1989 and Andrew in 1992 were both underestimated prior to landfall in South Carolina and in South Florida, respectively (U.S. Department of Commerce 1993a; National Research Council 1994). Despite the advances described in the previous section, there is considerable room for improvement of hurricane intensity forecasts.

The next several years should be another era of rapid change in hurricane forecasting. Robert W. Burpee became the director of NHC in 1995, following the retirement of Sheets, and the forecast office has moved to the Florida International University campus in Miami. In addition, as part of the restructuring of NMC, NHC is now the Tropical Prediction Center/National Hurricane Center in recognition of its past and expanded role of providing general weather guidance in the Tropics, in addition to tropical cyclone forecasts (McPherson 1994).

There are also several technological changes that will have an impact on hurricane forecasting. The WSR-88D Doppler radar network is in its implementation phase. Although these radars will probably not impact longer-range forecasts due to their limited range, they have the potential to improve forecasting as storms approach the coast (McAdie and Sandrik 1994). Perhaps the most improvement from these radars will be for short-term forecasts and site-specific warnings for inland winds associated with hurricanes. The recent landfalls of Hurricanes Hugo (1989) and Andrew (1992) have emphasized that hurricane winds can be a threat to inland locations as well as along the coast. Another technological change is the implementation of the next generation of geostationary satellites (GOES-I). These satellites have improved imaging systems, which will provide much higher spatial resolution than the *GOES-7* systems (Menzel and Purdom 1994). Technology is also being developed for improving estimates of surface winds from aircraft and satellites (e.g., Tanner et al. 1987; Rappaport and Black 1989). More accurate determination of the extent of high winds could lead to refinements in the areas warned for landfalling storms. [See chapter 5 by Purdom and Menzel, this volume.]

In the longer term, a Gulfstream IV jet aircraft, obtained by NOAA, is scheduled to perform hurricane surveillance in July 1996 and research missions by 1998. The main advantage of the jet relative to the aircraft currently used for hurricane reconnaissance (the Air Force Reserve WC-130s and the NOAA WP-3Ds) is that it will be possible to examine the upper levels of tropical cyclones. Theoretical results suggest that interactions with the environment in the upper levels can have significant impacts on intensity changes (e.g., Holland and Merrill 1984; Montgomery and Farrell 1993).

Even with the new jet aircraft, Doppler radar and GOES-I satellites, the lack of observations over the tropical oceans (especially in the mid levels that are important for storm motion) will still be a problem in many forecast cases. This problem could be even worse if the conventional upper-air network continues to degrade (e.g., Bosart 1990). A technological development that has the potential to make a major impact in this area is the unmanned aircraft. There are currently two types of these aircraft under development: a very small "aerosonde" that can provide in situ data (Holland et al. 1992) and a somewhat larger aircraft capable of releasing sondes to obtain vertical soundings (Langford et al. 1993). The feasibility of these types of systems will be evaluated within the next five years. [See Chapter 3 by Serafin, this chapter.]

Perhaps a fitting way to end this review is to speculate about the state of hurricane forecasting at the end of the next period of change. In Table 9-1, the 75-year history was divided into seven sections of roughly 10 years each. In the next 10 years, the general skill of hurricane track forecasts should continue to improve. Although there is probably an upper limit to the forecast accuracy, it is likely that this limit has not yet been reached. For example, Neumann (1987) indicated that the minimum possible average 24-h track error from statistical models is ~120 km. This minimum is 35% smaller than the average 24-h error of the official NHC forecast (185 km) for the period 1983–1992 (Lawrence and Gross 1994). There has been a dramatic increase in research on tropical cyclone motion that began in the mid-1980s, partially in connection with field experiments in the western Pacific sponsored by the Office of Naval Research (TCM-90, TCM-92, and TCM-93; Harr 1994). The impact of this research should continue to help guide hurricane forecast model development. In addition, the NMC global forecast model, upon which nearly all of the track models depend, should continue to improve.

The above discussion suggests that the average track errors will decrease slowly. However, more significant improvements should be attained in a few individual cases where strategically placed dropwindsondes are released from the NOAA Gulfstream jet and WP-3D aircraft. Preliminary estimates by Burpee et al. (1995) suggest that error reductions of 20%–30% are possible for the 24–60-h forecasts. Track forecasts will also be significantly improved if the unmanned aircraft technology comes to fruition.

Ten years from now, the method for making track forecasts will probably include ensemble forecasting techniques. In the last few years, there has been a considerable emphasis at NMC on this technique (Toth and Kalnay 1994). The method is well suited to the track forecasting problem, where the range of possible tracks could be estimated.

There will probably be some improvement in intensity forecasting, although it is not obvious that this will be accomplished by direct numerical simulation with three-dimensional models. The increased understanding from upper-level observations of hurricanes using the Gulfstream jet, in combination with new theoretical tools being developed (e.g., Shapiro and Montgomery 1993), has the potential to improve intensity forecasts. The ability to transmit aircraft radar information to NHC in real time might also have an impact on short-term intensity forecasts (12–36 hours) and the WSR-88D radars will be useful for short-range forecasting of storms near the coast. It may also be necessary to include the effects of the ocean response when predicting intensity changes, especially for intense and slow-moving storms (e.g., Khain and Ginis 1991). Ensemble forecasting techniques might also be feasible for evaluating the ranges of intensity that are possible, especially if a three-dimensional atmospheric model is successfully coupled with an ocean model.

Acknowledgments. The author would like to thank Peter Black, Robert Burpee, Gilbert Clark, James Fleming, Paul Hebert, Charles Neumann, Stanley Rosenthal, and Robert Sheets for their many helpful comments and suggestions during the preparation of this chapter.

REFERENCES

Aberson, S.D., and M. DeMaria, 1994: Verification of a nested barotropic hurricane track forecast model (VICBAR). *Mon. Wea. Rev.,* **122,** 2804–2815.

Anthes. R.A., 1972: The development of asymmetries in a three-dimensional model of the tropical cyclone. *Mon. Wea. Rev.,* **100,** 461–476.

_____ , 1977: Hurricane model experiments with a new cumulus parameterization scheme. *Mon. Wea. Rev.,* **105,** 287–300.

_____ , 1982: *Tropical Cyclones: Their Evolution, Structure and Effects. Meteor. Monogr.,* No. 41, Amer. Meteor. Soc., 208 pp.

Avila, L.A., and G.B. Clark, 1989: Atlantic tropical systems of 1988. *Mon. Wea. Rev.,* **117,** 2260–2265.

Bender, M.A., R.J. Ross, R.E. Tuleya, and Y. Kurihara, 1993: Improvements in tropical cyclone track and intensity forecasts using the GFDL initialization system. *Mon. Wea. Rev.,* **121,** 2046–2061.

Bengtsson, L., H. Bottger, and M. Kanamitsu, 1982: Simulation of hurricane type vortices in a general circulation model. *Tellus,* **34,** 440–457.

Bigler, S.G., 1981: Radar: A short history. *Weatherwise,* **34,** 158–163.

Birchfield, G.E., 1961: Numerical prediction of hurricane movement with the equivalent-barotropic model. *J. Meteor.,* **18,** 402–409.

Bonner, W.D., 1989: NMC overview: Recent progress and future plans. *Wea. Forecasting,* **4,** 275–285.

Bosart, L.F., 1990: Degradation of the North American radiosonde network. *Wea. Forecasting,* **5,** 527–528.

Bowie, E.H., 1922: Formation and movement of West Indian hurricanes. *Mon. Wea. Rev.,* **50,** 173–179.

Burpee, R.W., 1988: Grady Norton: Hurricane forecaster and communicator extraordinaire. *Wea. Forecasting,* **3,** 247–254.

_____ , 1989: Gordon E. Dunn: Preeminent forecaster of midlatitude storms and tropical cyclones. *Wea. Forecasting,* **4,** 573–584.

_____ , S.D. Aberson, P.G. Black, M. DeMaria, J.L. Franklin, J.S. Griffin, S.H. Houston, J. Kaplan, S.J. Lord, F.D. Marks Jr., M.D. Powell, and H.E. Willoughby, 1994: Real-time guidance provided by NOAA's Hurricane Research Division to forecasters during Emily of 1993. *Bull. Amer. Meteor. Soc.,* **75,** 1765–1783.

_____ , J.L. Franklin, S.J. Lord, and R.E. Tuleya, 1995: The performance of hurricane track guidance models with and without Omega dropwindsondes. Preprints, *21th Conf. Hurricanes and Tropical Meteorology,* Miami, FL, Amer. Meteor. Soc.

Byers, H.R., 1935: On the meteorological history of the hurricane of November 1935. *Mon. Wea. Rev.,* **63,** 318–322.

Calvert, E.B., 1920: Diagnosing a hurricane. *Bull. Amer. Meteor. Soc.,* **1,** 113–114.

_____ , 1935: The hurricane warning service and its reorganization. *Mon. Wea. Rev.,* **63,** 85–88.

Case, R.A., and H.P. Gerrish, 1988: Atlantic hurricane season of 1987. *Mon. Wea. Rev.,* **116,** 939–949.

Charney, J.G., and A. Eliassen, 1964: On the growth of the hurricane depression. *J. Atmos. Sci.,* **21,** 68–74.

Cline, I.M., 1926: *Tropical Cyclones.* MacMillan Company, 301 pp.

De Maria, M., and J. Kaplan, 1994: A statistical hurricane intensity prediction scheme (SHIPS) for the Atlantic basin. *Wea. Forecasting,* **9,** 209–220.

_____ , M.B. Lawrence, and J.T. Kroll, 1990: An error analysis of Atlantic tropical cyclone track guidance models. *Wea. Forecasting,* **5,** 47–61.

———, S.D. Aberson, K.V. Ooyama, and S.J. Lord, 1992: A nested spectral model for hurricane track forecasting. *Mon. Wea. Rev.,* **120,** 1628–1643.

Deppermann, C.E., 1946: Is there a ring of violent upward convection in hurricanes and typhoons? *Bull. Amer. Meteor. Soc.,* **27,** 6–8.

Dey, C.H., 1989: The evolution of objective analysis methodology at the National Meteorological Center. *Wea. Forecasting,* **4,** 297–312.

Doehring, F., I.W. Duedall, and I.M. Williams, 1994: Florida hurricanes and tropical storms 1871–1993: An historical survey. Tech. Paper 71, Florida Sea Grant College Program, Gainesville, FL, 118 pp.

Dunn, G.E., 1940a: Aerology in the hurricane warning service. *Mon. Wea. Rev.,* **68,** 303–315.

———, 1940b: Cyclogenesis in the tropical Atlantic. *Bull. Amer. Meteor. Soc.,* **21,** 215–229.

———, 1971: A brief history of the United States hurricane warning service. *Muse News,* **3,** 140–143. [Available from the Museum of Science, 3280 South Miami Avenue, Miami, FL 33133.]

———, and B.I. Miller, 1960: *Atlantic Hurricanes.* Louisiana State University Press, 326 pp.

———, and Coauthors, 1965: The hurricane season of 1964. *Mon. Wea. Rev.,* **93,** 175–187.

———, R.C. Gentry, and B.M. Lewis, 1968: An eight-year experiment in improving forecasts of hurricane motion. *Mon. Wea. Rev.,* **96,** 708–713.

Dvorak, V.F., 1973: A technique for the analysis and forecasting of tropical cyclone intensities from satellite pictures. NOAA Tech. Memo. NESS 45, 19 pp.

Emanuel, K.A., 1988: The maximum intensity of hurricanes. *J. Atmos. Sci.,* **45,** 1143–1155.

Fisher, E.L., 1958: The exchange of energy between the sea and the atmosphere in relation to hurricane behavior. *J. Meteor.,* **15,** 164–171.

Frank, N.L., 1963: Synoptic case study of tropical cyclogenesis utilizing TIROS data. *Mon. Wea. Rev.,* **91,** 355–366.

———, 1984: The meteorological disservice in the United States. Postprints, *15th Conf. on Hurricanes and Tropical Meteorology,* Miami, FL, Amer. Meteor. Soc., J3–J4.

Franklin, J.L., and M. DeMaria, 1992: The impact of Omega dropwindsonde observations on barotropic hurricane track forecasts. *Mon. Wea. Rev.,* **120,** 381–391.

Gentry, R.C., 1951: Forecasting the formation and movement of the Cedar Keys hurricane, 1–7 September 1950. *Mon. Wea. Rev.,* **79,** 107–115.

_____ , 1980: History of hurricane research in the United States with special emphasis on the National Hurricane Research Laboratory and associated groups. Selected Papers, *13th Tech. Conf. on Hurricanes and Tropical Meteorology,* Miami Beach, FL, Amer. Meteor. Soc., 6–16.

Gray, W.M., 1968: Global view of the origin of tropical disturbances and storms. *Mon. Wea. Rev.,* **96,** 669–700.

_____ , C.J. Neumann, and T.L. Tsui, 1991: Assessment of the role of aircraft reconnaissance on tropical cyclone analysis and forecasting. *Bull. Amer. Meteor. Soc.,* **72,** 1867–1883.

Griffin, J.S., R.W. Burpee, F.D. Marks Jr., and J.L. Franklin, 1992: Real-time airborne analysis of aircraft data supporting operational hurricane forecasting. *Wea. Forecasting,* **7,** 480–490.

Harr, P.A., 1994: The TCM-93 western Pacific mini-field experiment. *Minutes of the 48th Interdepartmental Hurricane Conf.,* Office of the Federal Coordinator, Washington, DC, A3–A6.

Haurwitz, B., 1935: The height of tropical cyclones and of the "eye" of the storm. *Mon. Wea. Rev.,* **63,** 45–49.

Hebert, P.J., 1977: Intensification criteria for tropical depressions in the western North Atlantic. NOAA Tech. Memo. NWS NHC-3, 22 pp.

_____ , 1978: Intensification criteria for tropical depressions of the western North Atlantic. *Mon. Wea. Rev.,* **106,** 831–840.

_____ , J.D. Jarrell, and M. Mayfield, 1993: The deadliest, costliest, and most intense United States hurricanes of this century (and other frequently requested hurricane facts). NOAA Tech. Memo. NWS NHC-31, 40 pp.

Holland, G.J., 1983: Tropical cyclone motion: Environmental interaction plus a beta effect. *J. Atmos. Sci.,* **40,** 328–342.

_____ , and R.T. Merrill, 1984: On the dynamics of tropical cyclone structure changes. *Quart. J. Roy. Meteor. Soc.,* **110,** 723–745.

_____ , T. McGeer, and H. Youngren, 1992: Autonomous aerosondes for economical atmospheric soundings anywhere on the globe. *Bull. Amer. Meteor. Soc.,* **73,** 1987–1988.

Hope, J.R., and C.J. Neumann, 1970: An operational technique for relating the movement of existing tropical cyclones to past tracks. *Mon. Wea. Rev.,* **98,** 925–933.

______, and ______, 1977: A survey of world wide tropical cyclone prediction models. Preprints, *11th Tech. Conf. Hurricanes and Tropical Meteorology,* Miami Beach, FL, Amer. Meteor. Soc., 367–374.

Horn, J.D., 1951: On irregular movements of tropical cyclones in the Pacific. *Bull. Amer. Meteor. Soc.,* **32,** 344–345.

Hovermale, J.B., and R.E. Livezey, 1977: Three-year performance characteristics of the NMC hurricane model. Preprints, *11th Tech. Conf. Hurricanes and Tropical Meteorology,* Miami Beach, FL, Amer. Meteor. Soc., 122–125.

Hubert, W.E., 1957: Hurricane trajectory forecasts from a nondivergent, nongeostrophic, barotropic model. *Mon. Wea. Rev.,* **85,** 83–87.

______, 1961: The potential of meteorological satellites for hurricane surveillance. National Hurricane Research Project Rep. 50, Proc. Second Tech. Conf. on Hurricanes, Miami Beach, FL, Dept. of Commerce, 156–164.

Jarvinen, B.R., and C.J. Neumann, 1979: Statistical forecasts of tropical cyclone intensity. NOAA Tech. Memo. NWS NHC-10, 22 pp.

______, ______, and M.A.S. Davis, 1988: A tropical cyclone data tape for the North Atlantic Basin, 1886–1983: Contents, limitations, and uses. NOAA Tech. Memo. NWS NHC-22, Miami, FL, 21 pp.

Jones, R.W., 1961: The tracking of hurricane Audrey 1957 by numerical prediction. *J. Meteor.,* **18,** 127–138.

______, 1976: A preliminary estimate of a tropical cyclone track change caused by artificial enhancement of convective heat release. NOAA Tech. Memo. ERL WMPO-27, 23 pp.

Jordan, C.L., and F.J. Schatzle, 1961: The "double eye" of Hurricane Donna. *Mon. Wea. Rev.,* **89,** 354–356.

Jordan, E.S., 1952: An observational study of the upper wind-circulation around tropical storms. *J. Meteor.,* **9,** 340–346.

Kalnay, E., G. Dimego, S. Lord, M. Kanamitu, A. Leetmaa, and D.B. Rao, 1994: NMC modeling and data assimilation plans for 1994–1998. Preprints, *10th Conf. on Numerical Weather Prediction,* Portland, OR, Amer. Meteor. Soc., 143–148.

Kasahara, A., and G.W. Platzman, 1963: Interaction of a hurricane with the steering flow and its effect upon the hurricane trajectory. *Tellus,* **15,** 321–335.

Khain, A., and I. Ginis, 1991: The mutual response of a moving tropical cyclone and the ocean. *Contrib. Atmos. Phys.*, **64**, 125–142.

Krishnamurti, T.N., and D. Oosterhof, 1989: Prediction of the life cycle of a super typhoon with a high resolution global model. *Bull. Amer. Meteor. Soc.*, **70**, 1218–1230.

Kuettner, J.P., and D.E. Parker, 1976: GATE: Report on the field phase. *Bull. Amer. Meteor. Soc.*, **57**, 11–27.

Kuo, H.L., 1969: Motions of vortices and circulating cylinder in shear flow with friction. *J. Atmos. Sci.*, **26**, 390–398.

Kurihara, Y., 1975: Budget analysis of a tropical cyclone simulated in an axisymmetric numerical model. *J. Atmos. Sci.*, **32**, 25–59.

_____ , M.A. Bender, R.E. Tuleya, and R.J. Ross, 1993: Hurricane forecasting with the GFDL automated prediction system. Preprints, *20th Conf. on Hurricanes and Tropical Meteorology,* San Antonio, TX, Amer. Meteor. Soc., 323–326.

Langford, J.S., G. Zarlengo, and C. Tracy, 1993: A global monitoring system for hurricanes and tropical meteorology. Preprints, *20th Conf. on Hurricanes and Tropical Meteorology,* San Antonio, TX, Amer. Meteor. Soc., 300–302.

Lawrence, M.B., and J.M. Gross, 1994: 1993 National Hurricane Center forecast verification. *Minutes of the 48th Interdepartmental Hurricane Conf.,* Office of the Federal Coordinator, Washington, DC, B15–B24.

LeSeur, N.E., and H.F. Hawkins, 1963: An analysis of Hurricane Cleo (1958) based on data from research reconnaissance aircraft. *Mon. Wea. Rev.*, **91**, 694–709.

Lord, S.J., 1991: A bogussing system for vortex circulations in the National Meteorological Center global forecast model season. Preprints, *19th Conf. on Hurricanes and Tropical Meteorology,* Miami, FL, Amer. Meteor. Soc., 328–330.

_____ , 1993: Recent developments in tropical cyclone track forecasting with the NMC global model analysis and forecast system. Preprints, *20th Conf. on Hurricanes and Tropical Meteorology,* San Antonio, TX, Amer. Meteor. Soc., 290–291.

Malkus, J.S., and H. Riehl, 1960: On the dynamics of energy transformations in steady-state hurricanes. *Tellus,* **12**, 1–20.

Marks, D.G., 1992: The beta and advection model for hurricane track forecasting. NOAA Tech. Memo. NWS NMC 70, National Meteorological Center, Camp Springs, MD, 89 pp.

Marks, F.D., Jr., 1990: Tropical cyclones. *Radar in Meteorology: Battan Memorial and 40th Anniversary Radar Meteorology Conference,* David Atlas, Ed., Amer. Meteor. Soc., 412–425.

Markus, R.M., N.F. Halbeisen, and J.F. Fuller, 1987: Air Weather Service; our heritage, 1937–1987. Military Airlift Command U.S. Air Force, Scott AFB, IL, 167 pp.

Mathur, M.B., 1991: The National Meteorological Center's quasi-Lagrangian model for hurricane prediction. *Mon. Wea. Rev.,* **119,** 1419–1447.

Maynard, R.H., 1945: Radar and weather. *J. Meteor.,* **2,** 214–226.

McAdie, C.J., and M.B. Lawrence, 1993: Long-term trends in National Hurricane Center track forecast errors in the Atlantic basin. Preprints, *20th Conf. on Hurricanes and Tropical Meteorology,* San Antonio, TX, Amer. Meteor. Soc., 281–284.

______ , and A. Sandrik Jr., 1994: Operational use of the WSR-88D during hurricane Emily, 1 September 1993. *Minutes of the 48th Interdepartmental Hurricane Conf.,* Office of the Federal Coordinator, Washington, DC, A44.

McDonald, W.F., 1935: The hurricane of August 31 to September 6, 1935. *Mon. Wea. Rev.,* **63,** 269–271.

______ , 1942: On a hypothesis concerning the normal development and disintegration of tropical hurricanes. *Mon. Wea. Rev.,* **70,** 1–7.

McPherson, R.D., 1994: The National Centers for Environmental Prediction: Operational climate, ocean, and weather prediction for the 21st century. *Bull. Amer. Meteor. Soc.,* **75,** 363–373.

Menzel, W.P., and J.F.W. Purdom, 1994: Introducing GOES-I: The first of a new generation of geostationary operational environmental satellites. *Bull. Amer. Meteor. Soc.,* **75,** 757–781.

Merrill, R.T., 1988: Environmental influences on hurricane intensification. *J. Atmos. Sci.,* **45,** 1678–1687.

Miller, B.I., 1958: On the maximum intensity of hurricanes. *J. Meteor.,* **15,** 184–195.

______ , 1969: Experiment in forecasting hurricane development with real data. NOAA Tech. Memo. ERLTM-NHRL 85, 28 pp.

______ , and P.L. Moore, 1960: A comparison of hurricane steering levels. *Bull. Amer. Meteor. Soc.,* **41,** 59–63.

______ , and P.P. Chase, 1966: Predictions of hurricane motion by statistical methods. *Mon. Wea. Rev.,* **94,** 399–405.

______ , E.C. Hill, and P.P. Chase, 1968: Revised technique for forecasting hurricane motion by statistical methods. *Mon. Wea. Rev.,* **96,** 540–548.

Mitchell, C.L., 1924: West Indian hurricanes and other tropical cyclones of the north Atlantic ocean. *Mon. Wea. Rev.,* **52** (Suppl. 24), 47 pp.

_____ , 1926: The West Indian hurricane of 14–22 September 1926. *Mon. Wea. Rev.,* **54,** 409–417.

Molinari, J., and D. Vollaro, 1989: External influences on hurricane intensity. Part I: Outflow layer eddy angular momentum fluxes. *J. Atmos. Sci.,* **46,** 1093–1105.

Montgomery, M.T., and B.F. Farrell, 1993: Tropical cyclone formation. *J. Atmos. Sci.,* **50,** 285–310.

National Research Council, 1994: *Hurricane Hugo: Puerto Rico, The U.S. Virgin Islands, and South Carolina, September 17–22, 1989.* National Academy Press, 276 pp.

Neumann, C.J., 1972: An alternate to the HURRAN tropical cyclone forecast system. NOAA Tech. Memo. NWS SR-62, 22 pp.

_____ , 1979: *Operational Techniques for Forecasting Tropical Cyclone Intensity and Movement.* World Meteorological Organization Publ. No. 528.

_____ , 1981: Trends in forecasting the tracks of Atlantic tropical cyclones. *Bull. Amer. Meteor. Soc.,* **62,** 1473–1485.

_____ , 1987: Prediction of tropical cyclone motion: Some practical aspects. Preprints, *17th Conf. Hurricanes and Tropical Meteorology,* Miami, FL, Amer. Meteor. Soc., 266–269.

_____ , 1988: The National Hurricane Center NHC83 model. NOAA Tech. Memo. NWS NHC-41, 44 pp.

_____ , and M.B. Lawrence, 1975: An operational experiment in the statistical-dynamical prediction of tropical cyclone motion. *Mon. Wea. Rev.,* **103,** 665–673.

_____ , and J.M. Pelissier, 1981a: Models for the prediction of tropical cyclone motion over the north Atlantic: An operational evaluation. *Mon. Wea. Rev.,* **109,** 522–538.

_____ , and _____ , 1981b: An analysis of Atlantic tropical cyclone forecast errors, 1970–1979. *Mon. Wea. Rev.,* **109,** 1248–1266.

_____ , and C.J. McAdie, 1991: A revised National Hurricane Center NHC83 model (NHC90). NOAA Tech. Memo. NWS NHC-44, 35 pp.

_____ , J.R. Hope, and B.I. Miller, 1972: A statistical method of combining synoptic and empirical tropical cyclone predictions systems. NOAA Tech. Memo. NWS SR-63, 32 pp.

NHRP Staff, 1956: Objectives and basic design of the National Hurricane Research Project. National Hurricane Research Project Rep. 1, U.S. Dept. of Commerce, Washington, DC, 6 pp.

Ooyama, K.V., 1964: A dynamical model for the study of tropical cyclone development. *Geofis. Int.,* **4,** 187–198.

______ , 1969: Numerical simulation of the life cycle of tropical cyclones. *J. Atmos. Sci.,* **26,** 3–40.

Palmen, E., 1948: On the formation and structure of tropical cyclones. *Geophys.,* **3,** 26–38.

______ , and C.W. Newton, 1969: *Atmospheric Circulation Systems.* Academic Press, 603 pp.

Parks, A.R., 1986: *The Florida Hurricane and Disaster 1926.* Arva Parks and Company, 112 pp.

Parrish, J.R., E.R. Darby, J.D. DuGranrut, and A.S. Goldstein, 1984: The NOAA aircraft satellite data link (ASDL). NOAA Tech. Memo. OAO-3, 34 pp.

Pasch, R.J., and L.A. Avila, 1992: Atlantic hurricane season of 1991. *Mon. Wea. Rev.,* **120,** 2671–2687.

Pfeffer, R.L., and M. Challa, 1981: A numerical study of the role of eddy fluxes of momentum in the development of Atlantic hurricanes. *J. Atmos. Sci.,* **38,** 2393–2398.

Pike, A.C., 1985: Geopotential heights and thicknesses as predictors of Atlantic tropical cyclone motion and intensity. *Mon. Wea. Rev.,* **113,** 931–939.

Powell, M.D., S.H. Houston, and N.M. Dorst, 1996: Hurricane Andrew's landfall in South Florida. Part II: Surface wind fields and potential real-time applications. *Wea. Forecasting,* in press.

Ramage, C.S., 1959: Hurricane development. *J. Meteor.,* **16,** 227–237.

Rappaport, E.N., and P.G. Black, 1989: The utility of Special Sensor Microwave/Imager data in the operational analysis of tropical cyclones. Preprints, *18th Conf. on Hurricanes and Tropical Meteorology,* San Diego, CA, Amer. Meteor. Soc., J21–J24.

Riehl, H., 1948: On the formation of typhoons. *J. Meteor.,* **5,** 247–264.

______ , and R.J. Shafer, 1944: The recurvature of tropical storms. *J. Meteor.,* **1,** 42–54.

______ , and N.M. Burgner, 1950: Further studies of the movement and formation of hurricanes and their forecasting. *Bull. Amer. Meteor. Soc.,* **31,** 244–253.

______ , W.H. Haggard, and R.W. Sanborn, 1956: On the prediction of 24-hour hurricane motion. *J. Meteor.,* **13,** 415–420.

Rosenthal, S.L., 1970: A circularly symmetric primitive equation model of tropical cyclone development containing an explicit water vapor cycle. *Mon. Wea. Rev.,* **98,** 643–663.

______ , 1971: A circularly symmetric primitive equation model of tropical cyclones and its response to artificial enhancement of heating functions. *Mon. Wea. Rev.,* **99,** 414–426.

______ , 1978: Numerical simulations of tropical cyclone development with latent heat by the resolvable scale I: Model description and and preliminary results. *J. Atmos. Sci.,* **35,** 258–271.

Sanders, F., and R.W. Burpee, 1968: Experiments in barotropic hurricane track forecasting. *J. Appl. Meteor.,* **7,** 313–323.

______ , A.C. Pike, and J.C. Gaertner, 1975: A barotropic model for operational tracks of tropical storms. *J. Appl. Meteor.,* **14,** 265–280.

______ , A.L. Adams, N.J.B. Gordon, and W.D. Jensen, 1980: Further development of a barotropic operational model for predicting paths of tropical storms. *Mon. Wea. Rev.,* **108,** 642–654.

Scofield, E., 1938: On the origin of tropical cyclones. *Bull. Amer. Meteor. Soc.,* **19,** 244–256.

Schuman, F.G., 1989: History of numerical weather prediction at the National Meteorological Center. *Wea. Forecasting,* **4,** 286–296.

Shapiro, L.J., and M.T. Montgomery, 1993: A three-dimensional balance theory for rapidly rotating vortices. *J. Atmos. Sci.,* **50,** 3322–3335.

Sheets, R.C., 1986: Hurricane tracking using an envelope approach—impacts upon forecasts. NOAA Tech. Memo. NWS NHC-33, 42 pp.

______ , 1990: The National Hurricane Center—Past, present and future. *Wea. Forecasting,* **5,** 185–232.

Simpson, R.H., 1966: The tracking and observation of hurricanes. *Hurricane Symp.,* Houston, TX, American Society for Oceanography, 34–70.

______ , 1971: The decision process in hurricane forecasting. NOAA Tech. Memo. NWS SR-53, 35 pp.

______ , 1980: Implementation phase of the National Hurricane Research Project 1955–1956. Selected Papers, *13th Tech. Conf. on Hurricanes and Tropical Meteorology,* Miami Beach, FL, Amer. Meteor. Soc., 1–5.

Tannehill, I.R., 1938a: *Hurricanes: Their Nature and History.* Princeton University Press, 257 pp.

———, 1938b: Hurricane of September 16 to 22, 1938. *Mon. Wea. Rev.,* **66,** 286–288.

Tanner, A., C.T. Swift, and P.G. Black, 1987: Operational airbourne remote sensing of windspeeds in hurricanes. Preprints, *17th Conf. on Hurricanes and Tropical Meteorology,* Miami, FL, Amer. Meteor. Soc., 385–387.

Thompson, P.D., 1961: *Numerical Weather Analysis and Prediction.* Macmillan Company, 170 pp.

Toth, Z., and E. Kalnay, 1994: Ensemble forecasting at NMC: The use of the breeding method for generating perturbations. Preprints, *10th Conf. on Numerical Weather Prediction,* Portland, OR, Amer. Meteor. Soc., 202–205.

Tracy, J.D., 1966: Accuracy of Atlantic tropical cyclone forecasts. *Mon. Wea. Rev.,* **94,** 407–418.

U.S. Department of Commerce, 1993a: *Hurricane Andrew: South Florida and Louisiana, August 23–26, 1992.* U.S. Dept. of Commerce, 131 pp.

———, 1993b: *"Hurricane!" A Familiarization Booklet.* U.S. Dept. of Commerce, 36 pp.

Vanderman, L.W., 1961: Verification of JNWP-Unit hurricane and typhoon forecasts for 1959. *Bull. Amer. Meteor. Soc.,* **42,** 239–248.

———, 1962: An improved NWP model for forecasting the paths of tropical cyclones. *Mon. Wea. Rev.,* **90,** 19–22.

Veigas, K.W., 1966: The development of a statistical-physical hurricane model. Final Report, U.S. Weather Bureau Contract Cwb-10966, Travelers Research Center, Inc., Hartford, CT, 19 pp.

Wells, F.E., 1961: Development and status of the Weather Bureau's hurricane tracking radar network. *Proc. Second Tech. Conf. on Hurricanes,* Miami Beach, FL, U.S. Dept. of Commerce, 129–139.

Wexler, H., 1945: The structure of the September 1944 hurricane when off Cape Henry, Virginia. *Bull. Amer. Meteor. Soc.,* **26,** 156–159.

———, 1947: Structure of hurricanes as determined by radar. *Ann. N.Y. Acad. Sci.,* **48,** 821–844.

White, G.F., 1994: A perspective on reducing losses from natural hazards. *Bull. Amer. Meteor. Soc.,* **75,** 1237–1240.

Willoughby, H.E., J.A. Clos, and M.G. Shoreibah, 1982: Concentric eye walls, secondary wind maxima, and the evolution of the hurricane vortex. *J. Atmos. Sci.,* **39,** 395–411.

______ , D.P. Jorgensen, R.A. Black, and S.L. Rosenthal, 1985: Project STORMFURY: A Scientific chronicle 1962–1983. *Bull. Amer. Meteor. Soc.,* **66,** 505–514.

______ , J.M. Masters, and C.W. Landsea, 1989a: A record minimum sea level pressure observed in Hurricane Gilbert. *Mon. Wea. Rev.,* **117,** 2824–2828.

______ , W.P. Barry, and M.E. Rahn, 1989b: Real-time monitoring of hurricane Gilbert. Extended Abstracts, *18th Conf. on Hurricanes and Tropical Meteorology,* San Diego, CA, Amer. Meteor. Soc., 220–221.

Yamasaki, M., 1968: Numerical simulation of tropical cyclone development with the use of primitive equations. *J. Meteor. Soc. Japan,* **55,** 11–31.

Yeh, T.C., 1950: The motion of tropical storms under the influence of a superimposed southerly current. *J. Meteor.,* **7,** 108–113.

CHAPTER 10

Severe Convective Storms: A Brief History of Science and Practice

KENNETH C. CRAWFORD AND EDWIN KESSLER

Introduction

It was the spring of 1953. Three violent tornadoes would soon leave deadly marks on cities well removed from "tornado alley" (an area of the central and southern Great Plains in the United States that has the world's highest frequency of tornado occurrence): 114 dead in Waco, Texas, on 11 May; 116 dead in Flint, Michigan, on 8 June; and 90 dead in Worcester, Massachusetts, on 9 June. Although the respective forecasts noted a prospect for severe thunderstorms, only the citizens of Flint received an advanced "alert" for these devastating tornadoes (Whitnah 1961)—primarily because numerical weather prediction (NWP) models were still in their infancy, and a network of weather surveillance radars was still under development. Consequently, no tornado warnings were issued.

Nearly 40 years later—on 26 April 1991—another violent tornado would leave 17 dead, this time as it slammed into the southern part of Andover, Kansas. Unfortunately, most of these fatalities (13) occurred in a mobile home park, where 223 of its 241 units were destroyed (NOAA 1991). However, the death toll was much lower than it might have been, primarily because a series of weather messages was issued with an increasing tone of urgency. These messages (accurate forecasts 24–36 hours prior to the destructive event, special statements 14 hours in advance advising of a major tornado outbreak, an enhanced tornado watch 6 hours in advance, and a series of warnings that followed the supercell thunderstorm into Andover) were clearly the result of vastly improved numerical forecasts (Shuman 1989), thorough training for meteorologists, and a responsive forecast and warning system composed of the broadcast media and local emergency managers. The introduction and use of new radar technology known as the WSR-88D (NEXRAD) (Crum and Alberty 1993; Klazura and Imy 1993), coupled with the use of trained spotters, enabled the National Weather Service (NWS) to provide site-specific warning information about a mesocyclone moving toward Andover.

As a result, 200 residents of the mobile home park were prepared to respond to the warning, and they moved to a nearby tornado shelter. The 13 fatalities in the mobile home park were the result of individuals choosing not to evacuate in spite of warning sirens and ominous environmental clues. The Andover tragedy dramatically revealed the unfortunate fact that a lack of public education remains as a major obstacle to providing a truly effective forecast and warning service for convective events.

"The last tornado to kill 100 people or more in the United States was in 1953 at Flint," the year that the National Severe Storms Forecast Center (NSSFC) began limited operations in Kansas City, Missouri. Since that time, the combination of improved NSSFC watches, more accurate warnings issued by local offices of the NWS, and the dissemination of information by the media to the public and into schools represents a remarkable accomplishment, effectively "eliminating the massive single-tornado death tolls that once plagued this country." While the number of tornado-related deaths exceeded 300 fatalities in a single year on 16 occasions during the period of 1880–1989, this has occurred only once since 1954 (Grazulis 1991). That year was 1974, when a superoutbreak of tornadoes on 3–4 April produced 148 tornadoes in an 18-h period, resulting in 307 fatalities.

Severe storms—The early days

The documentation of thunderstorm occurrences and the prediction of accompanying forms of severe weather had a slow start in the nineteenth century. The U.S. Army established a weather service as a part of its Signal Service in 1870. Prior to 1873, networks for collecting meteorological observations, while limited by today's standards, were quite ambitious for their time (Fleming 1990). The Signal Service observation network was used to prepare charts for the "indications officer" (forecaster). The primary emphasis was on daily reporting, but some effort was made to compile a data bank for research purposes. For example, Elias Loomis, Cleveland Abbe, and William Ferrel all did meteorological research with Signal Service data. In 1882, Sgt. John P. Finley was transferred to the office of the chief signal officer and placed in charge of a project called "tornado studies". Finley instituted a serious national study of tornadoes, and the first tornado predictions were issued in March 1884 (Abercromby 1888). These early predictions made extensive use of a severe storm and tornado chart developed by Finley (Galway 1985, 1992; Grazulis 1991).

While Finley developed techniques for the prediction of tornadoes, a ban on the use of the word "tornado" in forecasts released by the Signal Service was issued by the chief signal officer in 1883 (U.S. Army 1885). The ban was lifted briefly in 1886 but was reinstated early in 1887; it was revalidated by the U.S. Weather Bureau in 1934 and continued for several more years until it was finally lifted in 1938—the year F.W. Reichelderfer became chief of the U.S. Weather Bureau. Unfortunately, this ban on the use of the word "tornado" proved to be "a deterrent to research on severe local storms in the first four decades of the twentieth century" (Galway 1992).

Controversy in the mid-1880s between groups favoring civilian control of the weather service and those supporting military control became a political issue that ultimately led to a hiatus of research into severe local storms. Even as the Weather Bureau was established in the U.S. Department of Agriculture in 1891, continued political infighting "resulted in an era of almost complete absence of research by the U.S. Weather Bureau on severe local storms. From 1897 until 1916, there was no central collection point for severe local storm reports within the U.S. Weather Bureau" (Galway 1992).

During the first half of the twentieth century, several large tornado outbreaks contributed to a record number of tornado fatalities. The granddaddy of all tornado outbreaks was the infamous tristate tornado of 18 March 1925. On this day, 747 fatalities occurred along a 220-mile path of destruction through Missouri, southern Illinois, and Indiana. Nevertheless, the ban on the word "tornado" continued. As a result of a large number of tornadoes in the central United States in 1942, the public and the press demanded warnings for these events. Yet, the Weather Bureau still stated in a circular letter in June 1943 that these alerts (forecasts) were more trouble than they were worth—even though use of the word "tornado" was sanctioned in 1938. "It was felt that such releases would cause public alarm and panic" (Galway 1992).

On 20 March 1948, a tornado destroyed 32 military aircraft at Tinker Air Force Base near Oklahoma City and caused severe damage to a number of structures on the base. Five days later, another tornado hit Tinker, destroying some of what the first had missed (Bates and Fuller 1986). However, because of their leading-edge work, damage was minimized by a warning of this second tornado issued by Major Ernest J. Fawbush and Captain Robert C. Miller of the domestic wing of the Air Weather Service based at Tinker. Once the existence and success of this warning became known to the public and the local media, the efforts of the air force were lauded, while the Weather Bureau was criticized for its inaction. "The consensus of most bureau officials and senior forecasters was that there was not enough merit in the forecast procedures to justify any change in Weather Bureau policy; that is, the nonrelease of tornado forecasts to the public" (Beebe 1984). The prime objection of bureau officials to the Fawbush–Miller technique was that it failed to pinpoint tornado occurrence, a situation "that is still intrinsic to forecast techniques today" (Galway 1992). However, initiatives like those of Fawbush–Miller were still worthy; no doubt the very low level of bureau funding hindered new techniques from surfacing any sooner than public and media pressure demanded.

Continued criticisms by the media forced Reichelderfer into action, and the Weather Bureau began releasing routine severe storms forecasts from its Washington, D.C., analysis center in May 1952. Additional pressure from the press in the Midwest and plains states, the areas most prone to tornadic activity, forced Reichelderfer to move the new severe storms operation during the summer of 1954 from the East Coast, where tornadoes were less common, to Kansas City, Missouri. Unfortunately, a series of personnel changes and scientific differences of opinion led to a dramatic separation of the research personnel from the forecast group. These lamentable skirmishes led to two unfortunate circumstances: no formal research branch was attached to the NSSFC for almost two decades, and no new forecast methods were developed, although volumes of new data were obtained (Galway 1992).

Major milestones along the road toward a more accurate severe storm morphology and an improved understanding of storm dynamics began in the nineteenth century with the first accurate perception of the function of latent heat in the shower process (Espy 1841). Numerous subsequent contributions to the description, understanding, and prediction of convective storms included Finley's studies of tornadoes and their prediction (1881), Hinrich's emphasis (1888) on derechos (later known as plow-winds and downbursts), the Thunderstorm Project (Byers and Braham 1949), Brooks's identification of the mesocyclone (1949), Browning and Ludlam's identification and study of "supercells" (1962), Fujita's important detailing of microburst and tornado morphology (1963, 1981), and experimental and observational contributions by Ward (1972), Bluestein et al. (1993), and others.

Severe storms and convection—As the twentieth century ends

Forecasts and warnings are issued for severe local storms to protect life and property. These phenomena include severe thunderstorms, tornadoes, flash floods, and paralyzing winter storms, along with high wind and dust storms. To achieve this goal, forecasts and warnings must be accurate and timely; they also must be communicated to the population at risk. To be effective, warnings must contain focused, site-specific information and be received by individuals affected by a specific weather threat. Finally, recipients must understand the warnings and respond accordingly.

The scope of this responsibility is significant when one considers the fact that the United States experiences more severe local storms and flooding than any other nation in the world, including India and China. Also, 85% of all presidentially declared disasters result from severe weather events. On average, each year in the United States, some 10 000 violent thunderstorms, 5000 floods, 1000 tornadoes, and several hurricanes cause considerable loss of life and property damages estimated in the billions of dollars. Major winter storms, such as the Great Blizzard of 1993, can strand thousands, seriously interrupt travel, and curtail daily commerce over wide areas.

Historically, the capability of operational agencies to make skillful short-term forecast and warning decisions has been defined by the state of the science and the availability and effective application of technologies and storm-warning methodologies. The communication of short-fuse, life-saving information has proven time and again to be the weak link in a complex system that begins with scientific decisions made in highly stressful work environments and ends with the population at risk taking proper actions to protect their lives and property.

The advent of weather radar following World War II began a new era in severe storm detection and warning. Surplus radars such as the APQ-13 represented the first in a series of new operational tools. This initial system was quickly followed by the CPS-9, a 3-cm weather radar built by Raytheon; it achieved maximum usage during the 1950s. However, by the mid-1950s, it became necessary to replace these first-generation weather radars (known as the WSR-2 and the WSR-3). At the time, the Weather Bureau had a choice between the WSR-57 radar, which had a detection range of 250 miles, or a Doppler radar, which then was in an experimental state. The WSR-57 was chosen over the Doppler, since it was considered to be more of an all-purpose radar that would meet the bureau's various needs, especially in the warnings area (Galway 1992). As a result of its widespread use, the WSR-57 network became the backbone of the convective forecast and warning program of the NWS from the early 1960s to the 1990s. [See chapter 4 by Rogers and Smith, this volume.]

In a series of developments that extended from the early 1970s to the early 1980s, the NWS provided some offices with a computer-based system for processing radar-generated digital reflectivity data into products that forecasters could use directly (Wilson 1970; McGrew 1972; Devore 1983). These enhancements to the WSR-57 were based on a Digital Video Integrator Processor (DVIP) that was developed and tested at the National Severe Storms Laboratory in the late 1960s. Initial results were promising, leading the NWS to install this new radar data processor at 12 locations in its Digital Radar Experiment (D/RADEX; McGrew 1972). As a result, traditional reflectivity data could be digitized into 5 km by 8 km (3 by 5 nautical miles) grid boxes, manipulated for display purposes, and used as input for various algorithms. Unfortunately, a conflict between routine operational requirements and policy issues and an inadequate training program led to D/RADEX yielding few operational dividends—with one notable exception: the development of Vertically Integrated Liquid (VIL; Greene and Clark 1972), which has since been widely used in the NEXRAD era.

By the mid-1970s, developing computer technology led to improved processing of the voluminous data stream produced by Doppler weather radar and held great promise for improved storm detection (Brown et al. 1971, 1973). The superoutbreak of tornadoes in 1974 only intensified the demand for improved storm detection. As a result, the U.S. Departments of Commerce and Defense conducted a joint test in central Oklahoma to evaluate the operational readiness of Doppler technology. The results became the "official" foundation for the decision to develop and implement the WSR-88D or NEXRAD era (Burgess et al. 1979).

Meanwhile, hardware and software upgrades to D/RADEX in the early 1980s coincided with an increasing number of activities in support of the WSR-88D program. Even though the newer technology was still attached to a conventional WSR-57 radar, the opportunity presented itself to provide an early test of pre-NEXRAD algorithms using the renamed RADAP-II system (Radar Data Processor; Winston and Ruthi 1986). However, the full capabilities of the WSR-57 network were not exploited, even as the network was replaced by the WSR-88D (Kessler 1986a).

In the 1990s, operational agencies developed a strong technological foundation upon which more skillful forecast and warning decisions could be made. Noteworthy improvements in the accuracy and timeliness of forecasts and warnings seemed to rest upon several technological innovations: numerical weather prediction models with their improved spatial and temporal resolution, new forms of local observations combined with more powerful computers to produce rapid update cycles to the basic NWP guidance, and the nationwide implementation of the WSR-88D network.

As late as the early 1980s, average nationwide skill levels [as measured by a critical success index (CSI); Donaldson et al. 1975] for weather warnings remained well below 0.2 for tornadic and severe thunderstorm events. (Note that the CSI ranges from 0 to 1 where 1.0 represents perfection; the CSI penalizes forecasters for false alarms and missed events.) By the mid-1990s, CSI levels for storm warnings improved, exceeding 0.6 at many locations (Polger et al. 1994). Yet as impressive as these improved skill levels may be, they cannot be equated with a notion that the "warning problem" has been solved.

At the same time, *forecasting the occurrence* of severe thunderstorms and tornadoes has continued to improve gradually over the past 30 years. During the 1970s, the so-called Miller techniques (Miller 1972) were in widespread use, and watch areas were valid for 6 hours over areas in excess of 26 000 square miles. Recent trends in watch sizes show that average areas have slowly decreased (to about 23 000 square miles in size in 1990) for both tornado and severe thunderstorms watches (Anthony and Leftwich 1992). In addition, approximately 85% of all watches now record some form of severe weather during their valid period (compared to just under 50% in the late 1960s). More importantly, evidence suggests that skill levels at the NSSFC in forecasting outbreak episodes of numerous tornadoes "is

improving at a faster rate than forecasts for non-outbreak tornadoes"
(Anthony and Leftwich 1992).

Recent improvements in forecast skills for convective events can be
closely tied to advances in numerical modeling of synoptic and mesoscale phenomena and, to some extent, to the deployment of observing
systems that provide greater temporal and spatial resolution of storm-
scale features. Observing systems such as the WSR-88D, the demonstration wind profiler radar network, and local mesonetworks have
allowed the forecaster to monitor the evolution of a wind field through
deep layers of the atmosphere with greater accuracy than ever before.
This knowledge, along with an improved understanding of storm morphology, has led to the forecasting of severe thunderstorms and tornadoes with greater specificity than ever before (e.g., by detecting wind
profiles favorable for tornadoes or by defining the nature of a severe
weather threat—hail versus tornadoes versus high winds—and doing
so before the event).

While real progress had been achieved in improving warnings for
severe thunderstorms and tornadoes, additional problems remain. Today there is almost too much data from which relevant information
must be extracted quickly and efficiently. Data provided by the demonstration wind profiler radar network, local mesonetworks, *GOES-8*
and *GOES-9,* a plethora of numerical guidance information produced
from more frequent update cycles of mesoscale NWP models, and an
apparent "firehose" of observed and derived information from a network of WSR-88Ds are combining to *swamp the process of ingesting
storm-scale data, extracting relevant information, and producing more
skillful warning decisions.* Moreover, there is a serious paucity of
timely knowledge about what is actually happening (or is likely to
happen) at the earth's surface, where people live. For example, mesocyclones indicated by radar do not equate with tornadoes on the
ground. Even when the detected mesocyclones are real, less than one-
half are associated with tornadic thunderstorms. The dilemma posed
in such a case is whether a warning for a severe thunderstorm should
be upgraded to a tornado warning. Thus, some warning decisions still
rely on the use of trained spotters, while 1–2-h forecasts rely on statistical extrapolation procedures.

There is ample evidence to suggest that populations at risk remain
at risk for the following reasons. 1) Improved detection technology has
created a two- to fourfold increase in the number of warnings issued

compared to the number issued 10–15 years ago. 2) Broadcast media outlets have not substantially improved their capacity to deliver more warnings to smaller populations at risk and, at times, appear overwhelmed by the increased number of weather warnings needed. 3) Warnings continue to be issued and received in their historical, textual formats (as opposed to graphical). 4) Warnings remain too generic in nature and are seldom directed to specific user groups. 5) Too often, timely warnings do not reach their intended audience.

Moreover, tools needed to integrate new technology with new science [such as an Advanced Weather Information Processor (AWIPS)] are often missing. Numerical modeling at the storm scale represents one attempt to define better the areas threatened by potentially severe storms and to integrate the information content in data derived from a variety of sensing technologies. Even so, the availability of storm-scale data across the United States for initializing and verifying storm-scale models remains inadequate. A capability to ingest and process storm-scale data for storm-scale models and their output is almost universally absent. Finally, many physical processes in the real world still are parametrized into modern forecast models. While the promise of numerical prediction has been late in coming to fruition with respect to the task of thunderstorm prediction, remarkable increases in computer capacity along with advances in numerical modeling are being used for the first time at the Center for Analysis and Prediction of Storms (at the University of Oklahoma) to forecast some details of thunderstorm development. Such progress has roots in the first successful numerical simulation of the supercell, an achievement of Klemp and Wilhelmson (1978).

Other challenges on the horizon are implicit in the planned shift of responsibility for forecasting severe phenomena to the local office level, presenting a substantial challenge in several areas: 1) coordinating forecasts across numerous political boundaries, 2) decentralizing a critical decision process that will require additional scientific skills at the local level, and 3) assimilating pertinent but increasingly voluminous data from a growing number of local sources and storm-scale models. As Johns and Doswell (1992) noted: "The keys are to focus on those data that are relevant to the situation, and to display them in a useful manner. Development of procedures to do this is an ongoing process that continues as new interpretation techniques and data sources come into operational practice."

Today, there are a number of important books in which these and other events are detailed and that present the developing state of science and practice to virtually the present time. Among these are *Clouds and Storms* (F. Ludlam 1980); the three-volume work *Thunderstorms, A Social, Scientific, and Technological Documentary* (E. Kessler, Ed., 1983, 1986b, 1988); *Mesoscale Meteorology and Forecasting* (P.S. Ray, Ed., 1986); *Radar in Meteorology,* (D. Atlas, Ed., 1990); and *The Tornado: Its Structure, Dynamics, Prediction and Hazards* (C. Church et al., Eds., 1994).

Clearly, there has been growth since the early days of Finley in our ability to understand, detect, and advise the general public of impending severe convective storms. This growth of a scientifically based capability accelerated in the 1990s with the arrival of bigger, faster computers and new detection technology. While new problems loom on the horizon (including soaring budget deficits and a retrenchment of federal support for relevant research and on-going operations), a golden era of mesoscale meteorology appears ready to surface in the opening decade of the twenty-first century. Only time will tell.

REFERENCES

Abercromby, R., 1888: *Seas and Skies in Many Latitudes.* Edward Stanford, 447 pp.

Anthony, R.W., and P.W. Leftwich Jr., 1992: Trends in severe local storm watch verification at the National Severe Storms Forecast Center. *Wea. Forecasting,* **7,** 613–622.

Atlas, D., Ed., 1990: *Radar in Meteorology.* Amer. Meteor. Soc., 806 pp.

Bates, C.C., and J.F. Fuller, 1986: *America's Weather Warriors: 1814–1984.* Texas A&M University Press, 438 pp.

Bluestein, H.B., W.P. Unruh, J. Ladue, H. Stein, and D. Speheger, 1993: Doppler radar wind spectra of supercell tornadoes. *Mon. Wea. Rev.,* **121,** 2200–2221.

Brooks, E.M., 1949: The tornado cyclone. *Weatherwise,* **12,** 32–33.

Brown, R.A., K.C. Crawford, W.G. Bumgarner, and D. Sirmans, 1971: Preliminary Doppler velocity measurements in a developing radar hook echo. *Bull. Amer. Meteor. Soc.,* **52,** 1186–1188.

_____ , D.W. Burgess, and K.C. Crawford, 1973: Twin tornado cyclones within a severe thunderstorm: Single Doppler radar observations. *Weatherwise,* **26,** 63–71.

Browning, K., and F.H. Ludlam, 1962: Airflow in convective storms. *Quart. J. Royal Meteor. Soc.,* **88,** 117–135.

Burgess, D.W., and Coauthors, 1979: Final report on the Joint Doppler Operational Project (JDOP) 1976–78: Part I—Meteorological applications; Part II—Doppler radar engineering. NOAA Tech. Memo ERL-NSSL-86, 84 pp.

Byers, H., and R.R. Braham, 1949: *The Thunderstorm: Report of the Thunderstorm Project.* U.S. Department of Commerce, U.S. Government Printing Office, 287 pp.

Church, C., D. Burgess, C. Doswell, and R. Davies-Jones, Eds., 1994: *The Tornado: Its Structure, Dynamics, Prediction and Hazards.* Amer. Geophys. Union, 637 pp.

Crum, T.D., and R.L. Alberty, 1993: The WSR-88D and the WSR-88D Operational Support Facility. *Bull. Amer. Meteor. Soc.,* **74,** 1669–1687.

Devore, D.R., 1983: Operational use of digital radar data. Preprints, *13th Conf. on Severe Local Storms,* Tulsa, OK, Amer. Meteor. Soc., 21–24.

Donaldson, R.J., R.M. Dyer, and M.J. Kraus, 1975: An objective evaluator of techniques for predicting severe weather events. Preprints, *Ninth Conf. on Severe Local Storms,* Norman, OK, Amer. Meteor. Soc., 321–326.

Espy, J.P., 1841: *The Philosophy of Storms.* Little Brown, 552 pp.

Finley, J.P., 1881: The tornadoes of May 29 and 30, 1879, in Kansas, Nebraska, Missouri, and Iowa. Prof. Paper No. 4, U.S. Signal Services, 116 pp.

Fleming, J.R., 1990: *Meteorology in America, 1800–1870.* The Johns Hopkins University Press, 264 pp.

Fujita, T.T., 1963: Analytical mesometeorology: A review. *Severe Local Storms, Meteor. Monogr.* No. 5, Amer. Meteor. Soc., 77–125.

_____ , 1981: Tornadoes and downbursts in the context of generalized planetary scales. *J. Atmos. Sci.,* **38** 1511–1534.

Galway, J.G., 1985: J.P. Finley: The first severe storms forecaster. *Bull. Amer. Meteor. Soc.,* **66,** 1389–1395.

_____ , 1992: Early severe thunderstorm forecasting and research by the United States Weather Bureau. *Wea. Forecasting,* **7,** 564–587.

Grazulis, T.P., 1991: *Significant Tornadoes, 1880–1989.* Environmental Films, 526 pp.

Greene, D.R., and R.A. Clark, 1972: Vertically integrated liquid water: A new analysis tool. *Mon. Wea. Rev.,* **100,** 548–552.

Hinrich, G., 1888: Tornadoes and derechos. *Amer. Meteor. J.,* **5,** 316–317, 341–349, and 385–393.

Johns, R.H., and C.A. Doswell III, 1992: Severe local storms forecasting. *Wea. Forecasting,* **7,** 588–612.

Kessler, E., Ed., 1983: *The Thunderstorm in Human Affairs.* 2d ed. University of Oklahoma Press, 185 pp.

_____ , 1986a: NEXRAD comments. *Weatherwise,* **39,** 245.

_____ , Ed., 1986b: *Thunderstorm Morphology and Dynamics.* 2d ed. University of Oklahoma Press, 411 pp.

_____ , Ed., 1988: *Instruments and Techniques for Thunderstorm Observation and Analysis.* University of Oklahoma Press, 288 pp.

Klazura, G.E., and D.A. Imy, 1993: A description of the initial set of analysis products available from the NEXRAD WSR-88D system. *Bull. Amer. Meteor. Soc.,* **74,** 1293–1311.

Klemp, J.B., and R.B. Wilhelmson, 1978: The simulation of three-dimensional convective storm dynamics. *J. Atmos. Sci.,* **35,** 1070–1096.

Ludlam, F.H., 1980: *Clouds and Storms: The Behavior and Effect of Water in the Atmosphere.* The Pennsylvania State University Press, 405 pp.

McGrew, R.G., 1972: Project D/RADEX (digitized radar experiments). Preprints, *15th Conf. on Radar Meteorology,* Champaign–Urbana, IL, Amer. Meteor. Soc., 101–106.

Miller, R.C., 1972: Notes on analysis and severe storms forecasting procedures of the Air Force Global Weather Central. Air Weather Service Tech. Rep. 200 (rev.), 181 pp.

National Oceanic and Atmospheric Administration, 1991: *STORM DATA.* Vol. 33, No. 4, National Environmental Satellite, Data, and Information Service, Asheville, NC.

Polger, P.D., B.S. Goldsmith, R.C. Przywarty, and J.R. Bocchieri, 1994: National Weather Service warning performance based on the WSR-88D. *Bull. Amer. Meteor. Soc.,* **75,** 203–214.

Ray, P.S., Ed., 1986: *Mesoscale Meteorology and Forecasting.* Amer. Meteor. Soc., 793 pp.

Shuman, F.G., 1989: History of numerical weather prediction at the National Meteorological Center. *Wea. Forecasting,* **4,** 286–296.

U.S. Army, 1885: Report of the chief signal officer of the Army. 52 pp.

Ward, N.B., 1972: The exploration of certain features of tornado dynamics using a laboratory model. *J. Atmos. Sci.,* **29,** 1194–1204.

Whitnah, D.R., 1961: *A History of the United States Weather Bureau.* University of Illinois Press, 267 pp.

Wilson, J.W., 1970: Operational measurement of rainfall with the WSR-57: Review and recommendations. Preprints, *14th Conf. on Radar Meteorology,* Tucson, AZ, Amer. Meteor. Soc., 257–263.

Winston, H.A., and L.R. Ruthi, 1986: Evaluation of RADAP-II severe storm detection algorithms. *Bull. Amer. Meteor. Soc.,* **67,** 145–150.

CHAPTER 11

75 Years of Research on the Physics of a Lightning Discharge

E. PHILIP KRIDER

Introduction

Lightning is a unique and unusually deleterious weather phenomenon. In the early 1750s, Benjamin Franklin showed that lightning is an electrical discharge and that most flashes originate in clouds that are "most commonly in a negative state of electricity, but sometimes in a positive state" (Franklin 1774). Franklin was able to simulate many types of lightning damage, and he also showed that grounded metallic rods would protect most structures from such damage (Cohen 1990, chapters 6–8). For the next 170 years, lightning research was dominated by qualitative descriptions of different forms of the discharge, examples of lightning damage, and various schemes for lightning protection (see, e.g., de Fonvielle 1867; Flammarion 1905). Summaries of the general state of knowledge about the physical properties of lightning and lightning protection as of about 1920 can be found in Peters (1915), Mathias (1924), Kähler (1924), and Gockel (1925).

In the years following World War I, improvements in photographic and electronic recording technology led to several areas of experimentation that subsequently provided the foundations for modern lightning research. This work was motivated both by a fundamental curiosity about the physics of the process and by the need to ameliorate its effects on electric power and communication systems and aviation. In this review, we will trace research on the luminous development of a cloud-to-ground discharge and the currents that are produced by this process. We will also review lightning spectroscopy and the electric and magnetic fields that are radiated by lightning. We will conclude with a brief discussion of triggered lightning and the methods that are used to detect and locate lightning sources.

The literature on lightning is vast, and interesting papers have appeared in all regions of the world in many languages. Here we will limit our discussion to a very small fraction of the literature on the physical properties of cloud-to-ground flashes. Unfortunately, we will not be able to discuss all of the important work in this area, other types of lightning, lightning protection, cloud electrification, or how lightning affects the larger geophysical environment. We will give quantitative results from some of the original work and also from some recent studies. For more complete bibliographies and discussions of the literature, the reader can consult the sources that are listed in the bibliography; many of these sources will also be cited here.

Luminous development

In 1926, C.V. Boys published a description of a streak camera that he developed in 1900 that was capable of resolving both the spatial and temporal development of a lightning flash (Boys 1926). Prior to this, workers in Britain, Germany, and the United States had determined that lightning is an intermittent discharge with many different components and that discharges to ground are initiated by a downward-propagating leader. In 1933–1938, a particularly important series of papers was published by B.F.J. Schonland, D.J. Malan, and their associates in South Africa (Schonland and Collens 1933, 1934; Malan and Collens 1937; Schonland et al. 1935, 1938a,b). This work utilized Boys (and similar) cameras and led to the discovery of most of what is known today about the luminous development of lightning. Also, most of the terms for the various components of a discharge were assigned in this work.

Schonland (1956) has summarized the main results of the South African work. A typical cloud-to-ground (CG) discharge is called a *flash*. The characteristic "flicker" in the luminosity during a flash is caused by intermittent, partial discharges or *strokes* that generally travel along the same path. A stroke is distinguished from a simple fluctuation in the channel luminosity by the fact that it begins with a downward-moving *leader,* which, on reaching ground, initiates a bright *return stroke.* The branches are almost always directed away from the cloud in the direction of leader propagation. The intervals between strokes range from tens to hundreds of milliseconds, typically 40 ms. The multiplicity of strokes averages about 4, but values as high as 42 have been reported. The duration of the luminosity from a return stroke is typically 1–2 ms, but many strokes are followed by a low-level, continuing luminosity that persists for tens to hundreds of milliseconds. The total duration of the luminosity from a flash is most frequently 200–300 ms and occasionally is more than 1 s.

The leader that precedes the first return stroke is usually negatively charged and it propagates downward in a series of rapid, discontinuous steps with lengths of 10–100 m. The interval between steps ranges from 40 to 100 µs and gets shorter as the leader approaches the ground. The average downward velocity of this *stepped leader* is typically 1.5×10^5 m s^{-1}. Return strokes subsequent to the first are preceded by a negative *dart leader,* a bright pulse of luminosity about 40 m in length, which propagates continuously down the channel that was formed by the first return stroke. The velocities of dart leaders range from about 1.0×10^6 to 2.1×10^7 m s^{-1} and tend to be high when the time interval following the previous stroke is short, and vice versa. The most probable dart velocity is 2.0×10^6 m s^{-1}. When the time interval between strokes is long, a dart leader may propagate continuously down just a portion of the previous return stroke channel and then begin stepping. This process is called a *dart-stepped* leader, and the stepped portion of this process is characterized by short step lengths, short intervals between steps, and a high downward velocity. Sometimes a dart-stepped leader reverts to a normal stepped leader, and in these cases it usually forges a new path to ground. This phenomenon causes the flash to strike ground in two (or more) places, and the channel has a characteristic forked appearance that can be seen in many photographs (see Fig. 11-1).

Schonland also found that when the tip of any branch of the stepped leader gets to within a few tens of meters of the ground, the electric fields become so large that one or more upward discharges are initiated, usually from the tallest object(s) in the local vicinity of that branch. The first return stroke is initiated when one or more of these upward, *connecting discharges* contact the leader.

The return stroke is a very intense positive wave of ionization that propagates up the partially ionized path that was established by the leader at a speed close to the speed of light. Figure 11-2 illustrates the time development of the luminosity in the lower portion of a typical first stroke, as determined from a Boys camera photograph. In this case, the upward, two-dimensional velocity along the main trunk was 1.6×10^8 m s^{-1} from the ground to point A, 2.1×10^8 m s^{-1} from A to B, 9.7×10^7 m s^{-1} from B to C, and 5.5×10^7 m s^{-1} from C to D (Schonland 1956).

FIG. 11-1. Natural cloud-to-ground lightning near Tucson, Arizona. Note that each discharge strikes the ground in more than one place (courtesy Michael S. Leuthold, The University of Arizona).

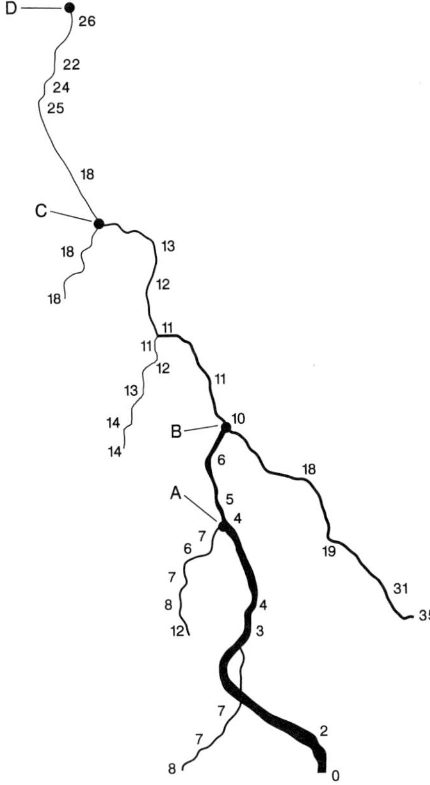

FIG. 11-2. The luminous development of a typical first return stroke. The numbers show the geometrical growth of the luminosity in both the main channel and branches in microseconds (adapted from Schonland 1964).

When the luminosity reached the large branch at B, it appeared to pause for about 5 µs and then it proceeded both down the branch and upward with less intensity. When a return stroke reaches a branch, there is usually a brightening of the channel below that point, and this is termed a *branch component.* The existence of branch components has led to the view that the return stroke effectively removes charge that has been lowered by the leader and that is stored on or near the main channel and its branches. Once the return stroke has entered the cloud, there can be further pulsations in the primary or continuing luminosity that are termed *M components,* and these are thought to be due to branching at high altitudes or to discharge processes within the cloud. The dart leaders that precede subsequent return strokes typically do not have branches; hence, subsequent strokes rarely exhibit branch components. The upward velocities of subsequent return strokes tend to be more uniform than first strokes with values that range from 2.4×10^7 m s^{-1} to 1.1×10^8 m s^{-1} (Schonland 1956).

In the years immediately following the South African work, the General Electric Company conducted an extensive series of experiments on the luminous and electrical properties of lightning. This work focused on both strikes to normal terrain near Pittsfield, Massachusetts (Hagenguth 1947), and also on strikes to the Empire State Building in New York City (McEachron 1939, 1940). The top of the Empire State Building is about 410 m above ground, and between 1935 and 1949 lightning struck the building an average of 22.6 times per year when the experiments were operating. The overall goal of the Empire State Building study was to obtain simultaneous records of the luminous development of the flashes and the associated electric currents at the top of the building. The results show that in about 80% of the cases the tall building actually initiated upward-moving, positive stepped leaders and that in about 50% of these cases there was no return stroke; that is, the leader current just merged smoothly into a low-level continuous current (and luminosity) that was typically about 250 Å (McEachron 1939, 1941; Hagenguth and Anderson 1952). The step lengths in the upward leaders were smaller than those reported in the South African work and ranged from 6 to 23 m, with an average of 8.2 m. The step intervals and velocities were similar to those found in South Africa. In about 50% of the flashes that were initiated by upward leaders, the continuous current fell to small values or zero after 10–20 ms, and then there were one or more downward-moving dart leaders followed by return strokes. The luminous characteristics of the dart leaders and the return strokes that followed were generally similar to the subsequent strokes that were photographed in downward strikes to normal terrain. A photograph of an upward discharge in an urban environment is shown in Fig. 11-3.

In the years following World War II, K. Berger and his associates began a comprehensive study of lightning strikes to instrumented towers on two mountains near Lugano, Switzerland (Berger and Vogelsanger 1965, 1966; Berger 1967). The optical properties of these discharges were measured together with currents in the towers and the associated electric field perturbations. Much information was obtained about the properties of upward and downward leaders, including the discovery that leaders of either polarity can propagate in either direction, as shown in Fig. 11-4. Today, flashes between cloud and ground are usually classified according to the direction of propagation and polarity of the initial leader (Berger 1978). For example, panel (1) in Fig. 11-4 shows the most frequent type of cloud-to-ground lightning, a negative flash initiated by a downward-propagating leader. Here, the total discharge effectively transports negative charge from the cloud to ground; this is the type of lightning that was studied in South Africa. Panel (2) in Fig. 11-4 shows a discharge that is being initiated by a downward-propagating, positive leader. In summer storms, positive discharges are just a few percent of the total, but in winter storms they can be 10%–30% of the total number. Panel (3) in Fig. 11-4 shows an upward-propagating

FIG. 11-3. An upward discharge in Milwaukee, Wisconsin (1977, *Milwaukee Journal,* Robert J. Miller photographer).

ground-to-cloud discharge that begins with a positive leader. This type of lightning is usually initiated by a mountain peak or a tall structure, such as the Empire State Building, and is relatively rare in strikes to natural terrain. Panel (4) shows the rarest form of lightning between cloud and ground, a negative flash initiated by an upward-propagating leader. Berger found that negative stepped leaders exhibit regular, distinct steps throughout their development, whether upward or downward, and in several cases produced a faint corona envelope at the tip of the step. Positive steps are considerably fainter than the negative and frequently can be seen over just a portion of the channel. Rakov et al. (1994) have summarized some recent observations of lightning phenomenology in both Florida and New Mexico.

Evans and Walker (1963) used a high-speed framing camera to examine the small-scale tortuosity of individual return strokes. They

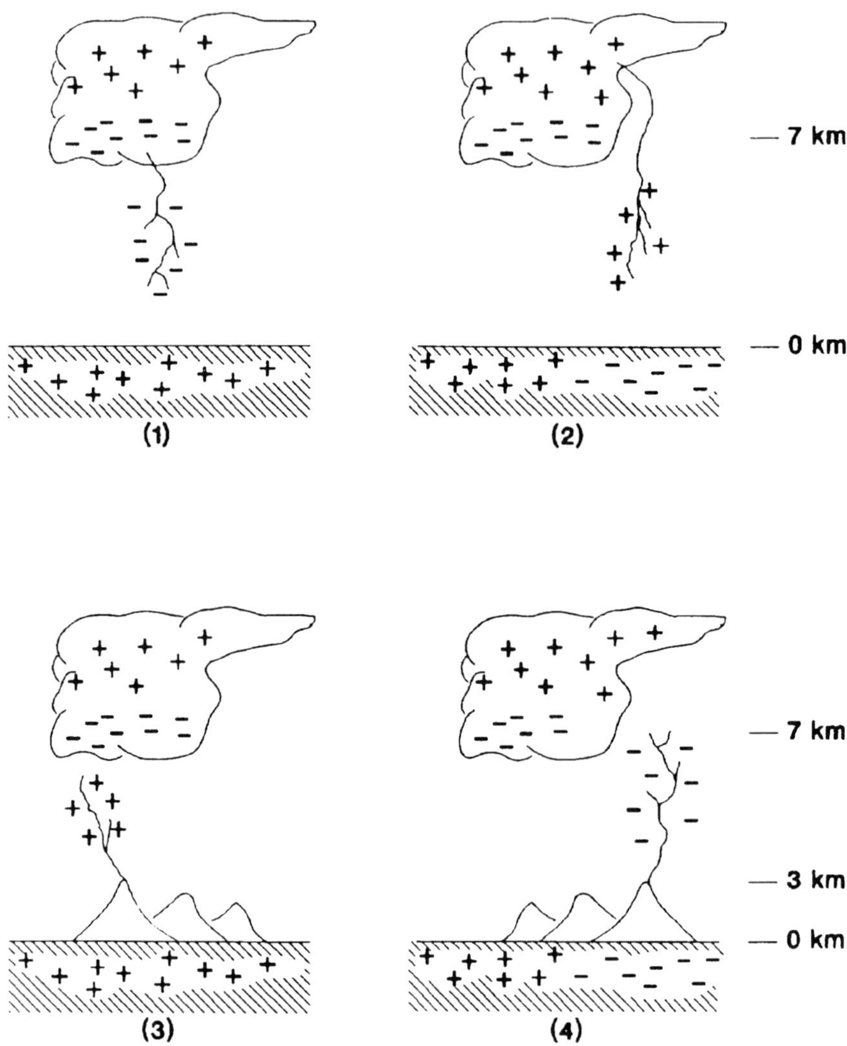

FIG. 11-4. Sketches of the four types of lightning that can occur between cloud and ground (adapted from Berger 1978).

found that the geometry of the channel changed with time within a given stroke and also from stroke to stroke within a given flash. Some parts of the channel appeared to be thicker than others, and there appeared to be a great deal of twisting and bending on a spatial scale of centimeters. Idone and Orville (1988) have quantified the tortuosity in both natural and rocket-triggered flashes and have shown that there can be an abrupt reduction in tortuosity and an increase in velocity during the upward development of a positive leader. Idone (1995) has also shown that different strokes in the same flash can both enhance and reduce the tortuosity of the previous stroke.

Guo and Krider (1982, 1983) used photoelectric sensors to measure the light output from return strokes over a spectral window from 0.4 to 1.1 μm. They obtained an average peak value of 2.3×10^9 W for first strokes and 4.8×10^8 W and 5.4×10^8 W for subsequent strokes preceded by dart and dart-stepped leaders, respectively. The average radiant output from the base of the channel is estimated to peak at about 10^6 W m^{-1}, a small fraction of the total electrical power that is being dissipated in the channel at this time (Krider et al. 1968). Turman (1977, 1978) measured the peak optical power from lightning at satellite altitudes and found a median of 10^9 W with 2% of the pulses exceeding 10^{11} W. The term "superbolt" was used to describe events that had a peak optical power in the range of 10^{11}–10^{13} W. Turman found five superbolts above a threshold of 3×10^{12} W in a sample of 10^7 flashes. The source of superbolt signals is not known, but Turman (1977) and Hill (1978) have suggested that they may be positive return strokes in flashes to ground because this process occasionally produces very large peak currents and charge transfers (see below). Currently, Christian et al. (1989a) are developing a satellite optical sensor that can detect and map lightning flashes in the daytime or at night from synchronous altitudes.

Current measurements

The peak current in a return stroke is an important parameter because, to a first approximation, the voltage that is produced on a resistive load, such as a lightning rod or a power line, is equal to the current multiplied by the impedance of the object that is struck (Ohm's law). The first quantitative estimates of the peak current in lightning were made by Pockels (1900), who placed pieces of basalt a few centimeters from a lightning rod and then measured the residual magnetism that was induced by the discharge. Later, Foust and Kuchni (1932) developed a "surge crest ammeter" or "magnetic link" that was based on the same principle and that was widely applied in the electric power industry (Lewis and Foust 1945). Other devices were also tried [see Uman (1969, chapter 4) for a review of the various current measuring techniques], but in most cases the only current parameter that could be obtained was the peak value, and this was usually from an unknown stroke or flash of unknown polarity.

Direct measurements of lightning current wave shapes were first obtained in the Empire State Building study (Hagenguth 1940; Hagenguth and Anderson 1952). The current at the top of the building was passed through a low-inductance, resistive shunt, and the voltage across the shunt was recorded on one or more cathode ray oscilloscopes. Berger also used the shunt and oscilloscope method to record currents during direct strikes to measuring towers in Switzerland (Berger 1955, 1967; Berger and Vogelsanger 1965). Like the Empire State Building study, about 75% of the strikes in Switzerland were initiated by upward-moving leaders, and during this phase the currents rose to several tens or hundreds of amperes in hundredths to tenths of seconds. When the leader current went to a low value or zero, dart leaders often formed and propagated downward and then initiated a return stroke that propagated upward.

Uman (1969, chapter 4) has summarized and compared currents recorded in the Empire State Building study with those reported by Berger and others. Table 1 summarizes the salient characteristics of currents in strokes that are initiated by downward-propagating, negative leaders. Here the impulse charge is the integral of the stroke current over time, and the action integral is the integral of the current squared. Both of these parameters provide a measure of the heating and explosive damage that can be produced at the point of attachment (Golde 1973; Uman 1988). The currents in negative first strokes typically peak at 30 kA, and subsequent strokes peak at 12 kA. Five percent of negative first strokes peak above 80 kA, but 5% of the positive strokes initiated by downward leaders exceed 250 kA (Berger et al. 1975). The zero-to-peak rise times in first strokes are a few microseconds, and subsequent strokes are somewhat faster. The maximum rate of rise of current $(dI/dt)_{\mathrm{max}}$ during the initial onset of both first and subsequent return strokes is of the order of 100 kA μs^{-1}, and the duration of this maximum is of the order of 100 ns (Leteinturier et al. 1991; Krider et al. 1995). Following the first peak, some strokes exhibit a low-level, continuous current of hundreds of amperes for a few to hundreds of milliseconds. Positive flashes usually produce just a single return stroke and then a period of continuous current. The long continuing currents produce large charge transfers and are frequently the cause of fires. Berger found that 5% of the negative flashes produced

TABLE 11-1. Properties of return strokes initiated by downward-propagating, negative leaders.[a]

Properties	Unit	Percentage of cases exceeding tabulated value		
		95%	50%	5%
Peak current				
(minimum 2 kA):				
First stroke	kA	14	30	80
Subsequent stroke	kA	4.6	12	30
Total charge:				
First stroke	C	1.1	5.2	24
Subsequent stroke	C	0.2	1.4	11
Entire flash	C	1.3	7.5	40
Impulse charge:				
First stroke	C	1.1	4.5	20
Subsequent stroke	C	0.22	0.95	4.0
Stroke duration:				
(2kA to half-value)				
First stroke	μs	30	75	200
Subsequent stroke	μs	6.5	32	140
Action integral:				
First strokes	A^2s	6.0×10^3	5.5×10^4	5.5×10^5
Subsequent strokes	A^2s	5.5×10^2	6.0×10^3	5.2×10^4

[a]Adapted from Berger et al. (1975).

charge transfers that exceeded 40 C, and 5% of the positive flashes exceeded 350 C (Berger et al. 1975). More recent measurements of lightning currents have been summarized by Uman (1987, chapter 7).

Spectroscopy

Slipher (1917) obtained the first photographic record of the spectrum from lightning using an astronomical slit spectrograph and noted that there were both line and continuum emissions. Wallace (1964) obtained similar spectra by time integrating many flashes and reflections from clouds and showed that the spectrum was dominated by the lines of singly ionized and neutral N and O atoms and the molecular bands of N_2^+.

Salanave (1961) obtained the first time-resolved spectrum of an individual return stroke using a slitless spectroscopic technique in conjunction with a rotating film drum (see also Salanave 1980). The wavelength resolution was about 2 Å, and the time resolution was about 20 ms. Salanave et al. (1962) determined that the dominant features in return-stroke spectra are singly ionized and neutral lines of N and O and that doubly ionized emissions and molecular radiations are negligible. Streaks of continuum appear from points where the discharge propagates toward or away from the observer or where regions of the channel have larger radii and remain luminous for a longer time. Prueitt (1963) made the first quantitative temperature estimates for individual strokes by measuring the ratios of the intensities of lines in spectra obtained by Salanave. The values ranged from 24 200 to 28 400 K. Orville (1968a–c) extended this work by measuring the temperature variations *within* return strokes with a resolution of about 5 μs. He found that the temperature rose very rapidly to a peak near 30 000 K and then slowly decayed. Uman (1969, chapter 5) has reviewed these analyses, and many other aspects of lightning spectroscopy, and concludes that the rapid current in a return stroke heats the channel to a peak temperature near 30 000 K and raises the pressure to several tens of atmospheres. The channel then expands behind a shock wave that eventually becomes the acoustic signals that we hear as thunder (Few 1975, 1995).

Electromagnetic fields

Lightning produces large changes in the electric and magnetic field environment, and there is a great deal of literature on this topic. In two classic papers, Wilson (1916, 1920) analyzed changes in the electrostatic field and found that flashes to ground effectively transport 10–50 C of negative charge to ground. Later work by Krehbiel et al. (1979) showed that the individual strokes within a flash neutralize discrete clumps of charge that are located in a narrow range of altitudes where the ambient air temperatures are $-10°$ to $-20°$C. The horizontal extent of these charges is 5–10 km (Krehbiel 1986).

At frequencies of a few to a few tens of kilohertz, Austin (1926), Appleton et al. (1926), Watson Watt (1929), Norinder (1936), Chapman (1939), and others showed that lightning is the primary source of atmospheric radio noise (termed atmospherics, spherics, or sferics), that there is an ionosphere that reflects these waves, and that the waveform of an atmospheric pulse is affected by propagation in the earth–ionosphere cavity. Norinder and Dahle (1945) showed that return strokes produce the largest impulses in the 3–30-kHz frequency range and made an attempt to relate the amplitude and shape of these sig-

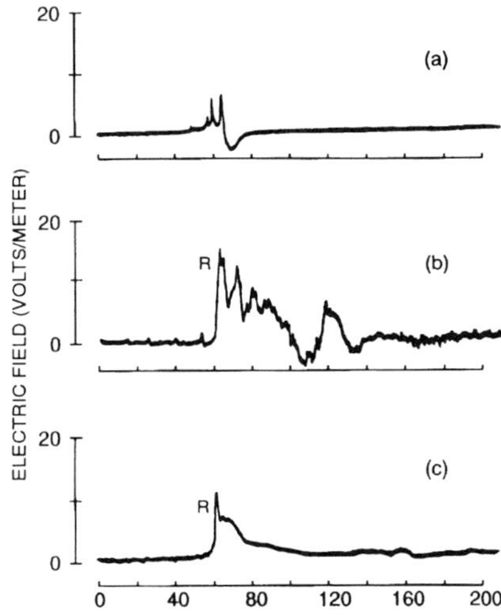

FIG. 11-5. Three of the many electric field impulses radiated by a typical cloud-to-ground discharge at a range of about 60 km.

nals to the current in the channel. Brook and Kitagawa (1964) studied lightning radiations at 420 and 850 MHz and noted that at these higher frequencies the dominant sources of the radiation are not return strokes but what appear to be breakdown processes within the cloud and the stepped leader. [See Horner (1964) and Pierce (1977a) for the sources and characteristics of atmospheric radio noise over the entire spectrum.]

In recent years, it has become clear that the electric and magnetic fields that are radiated by different lightning processes have different but characteristic signatures that are reproduced from flash to flash (Weidman and Krider 1978; Uman and Krider 1982). For example, Fig. 11-5 shows three of the many impulses that were radiated by a typical cloud-to-ground flash at a distance of about 60 km. These particular signatures were measured using a broadband antenna system that covered all frequencies from about 1 kHz to 1 MHz and a transient waveform digitizer with a "pretrigger" recording capability (Krider et al. 1977). Trace (a) shows a pulse that was radiated during the preliminary breakdown process that occurs within the cloud prior to the stepped leader (Beasley et al. 1982), trace (b) shows the waveform that was radiated by the first return stroke, and trace (c) shows the signature from a subsequent return stroke. The small pulses that precede the first return stroke in (b) were produced by the final steps of the leader, just before the connecting discharge and the onset of the return stroke (Krider et al. 1977).

Uman and coworkers have attempted to develop theoretical models that describe the overall shapes of the electric and magnetic fields that are produced by return strokes at different distances (Uman et al. 1975; Uman 1985; Diendorfer and Uman 1990; Nucci et al. 1990; Thottappillil et al. 1991; Thottappillil and Uman 1993, 1994). The goal of this work is to develop a model that can be used to compute the coupling of lightning transients to complex structures such as an electric power system. One important result is that early in the stroke, that is, up to the time of the field peak, the waveform of the distant (radiation) field on the ground can be approximated by the simple transmission line model (Uman et al. 1975)

$$E_{RAD}\,(t) = -\,\frac{\mu_0 v I(t - D/c)}{2\pi D}\,, \tag{1}$$

where E is the vertical electric field at time t, μ_0 the permeability of free space, v the upward velocity of the stroke near the ground, I the current at the base of the channel, c the speed of light, and D the horizontal distance to the flash. A typical first stroke produces a peak field of 5–10 V m^{-1} at a distance of 100 km (Uman 1985) and has a velocity on the order of 1.5×10^8 m s^{-1} near the ground (Mach and Rust 1993). With a peak of 8 V m^{-1}, the above equation predicts a peak current of about 27 kA, a value that is in good agreement with Berger's tower measurements (see Table 1). A peak field of 8 V m^{-1} at 100 km also means that the peak electromagnetic power that is radiated by the stroke is greater than 10^{10} W (Krider 1992). If such a pulse propagates into the middle atmosphere, it probably will be large enough to heat electrons and cause regions of enhanced ionization in the lower ionosphere (Inan et al. 1991; Rodriguez et al. 1992; Taranenko et al. 1993). Evidence for such phenomena has recently been documented by researchers in the United States (Inan et al. 1988) and in New Zealand (Dowden and Adams 1993).

Artificially initiated or triggered lightning

In our discussions of the Empire State Building and Swiss studies, we have noted that tall buildings and mountains can sometimes initiate upward, ground-to-cloud lightning. Aircraft experiments by Fitzgerald (1967) and Pitts et al. (1988) have shown that the majority of lightning strikes to airborne vehicles actually begin at, and hence are triggered by, the airframe (Mazur et al. 1984; Mazur and Ruhnke 1993). The launches of the *Apollo 12* (NASA 1970; Krider et al. 1974) and later the *Atlas-Centaur 67* (Christian et al. 1989b) space vehicles triggered lightning in clouds that otherwise were producing very little or no natural lightning. The fact that such phenomena can occur in weather environments that otherwise appear to be benign makes triggered lightning a particularly difficult hazard to detect or forecast.

Newman et al. (1967) first demonstrated that when a thunderstorm is overhead and the electric fields are high, a flash can be triggered by using a small rocket to carry a grounded wire aloft. This effort followed an experiment by Brook et al. (1961) that showed the idea had promise. Subsequently, the method was improved by researchers in France (Fieux et al. 1975; Fieux and Hubert 1976; SPARG 1982), and there are now several rocket-triggered programs operating in France, Japan, and the United States (Uman 1987, chapter 12; Horii and Nakano 1995). Typically, two out of three launches will trigger a discharge, and most triggers occur when the rocket is at an altitude of just 100–300 m. The initial leader in a triggered flash propagates upward into the cloud and is not like natural lightning. The majority of the subsequent strokes to ground, however, follow the path of the initial leader and are almost identical to their natural counterparts (Le Vine et al. 1989). Figure 11-6 shows the lower portion of a rocket-triggered discharge that has been blown sideways by the wind.

Rocket-triggered lightning is important because it provides the capability of studying both the physics of the discharge and its interactions with structures in a partially controlled environment. It has been used to investigate the luminous development of lightning channels, the waveforms of lightning currents, the velocities of return strokes, the relation between currents and fields, the mechanisms of lightning damage, the performance of lightning-protection systems, and many other questions (Uman 1987, chapter 12; Fisher et al. 1993; Horii and Nakano 1995). Among the more important results to date have been a direct experimental verification that return strokes produce large field and current variations on a submicrosecond timescale and the general validity of Eq. (1) (Willett et al. 1988; Willett et al. 1989). Submicrosecond current variations produce large voltages when lightning strikes an inductive load and can damage any system that is sensitive to high-frequency disturbances. Submicrosecond currents place more stringent requirements on the performance of surge arresters and other devices that are used in lightning protection and also on the engineering tests and test standards that are used to verify that performance (Uman 1988).

Lightning locating systems

Before the development of weather radars, lightning-locating systems were the primary means of identifying and mapping thunder-

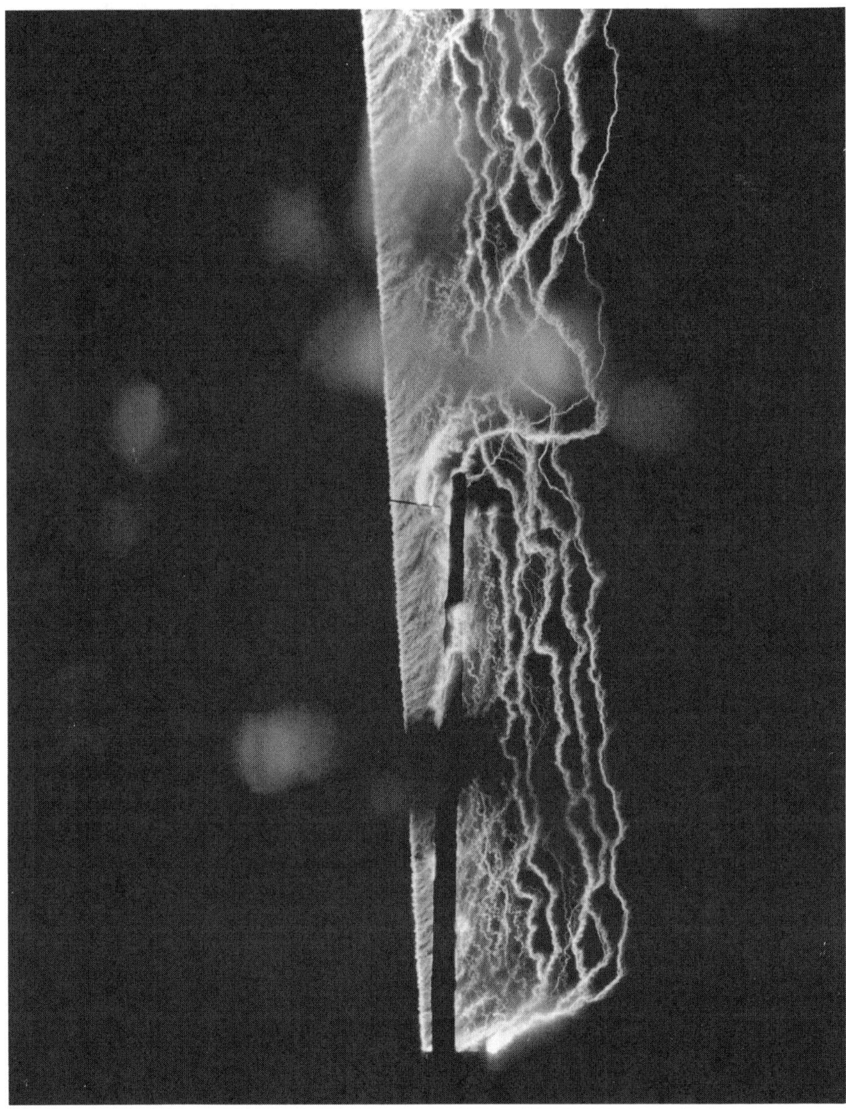

FIG. 11-6. The lower portion of a rocket-triggered discharge at the NASA Kennedy Space Center. The pole that is visible in the center is about 20 m tall (courtesy of W. Jafferis and the NASA Kennedy Space Center).

storms at medium and long ranges. In the 1920s, Watson Watt and Herd (1926) developed a cathode-ray direction finder (CRDF) that used a pair of orthogonal loop antennas tuned to a frequency near 10 kHz, where propagation in the earth–ionosphere waveguide is relatively efficient, to detect the horizontal magnetic field produced by lightning.

The azimuth angle to the discharge was obtained by displaying the north–south and east–west antenna outputs simultaneously on an x–y oscilloscope, so that the resulting vector pointed in the direction of the discharge (Watson Watt et al. 1933). Two or more CRDFs at known positions were sufficient to determine the location of a discharge from the intersection of simultaneous direction vectors. Similar systems were used up to and during World War II in many regions of the world. Magnetic direction finders (DFs) and other methods for locating lightning have been reviewed by Watson Watt (1929), Keen (1938), Adcock and Clarke (1947), Norinder (1953), and Pierce (1977b).

In 1976, an improved version of the magnetic DF system was developed for locating cloud-to-ground lightning within a range of about 500 km (Krider et al. 1976, 1980). This system operates in the time domain (i.e., from about 1 to 500 kHz) and is designed to respond to just those field shapes that are characteristic of return strokes (see Figs. 11-5b and 11-5c). When such a field is detected, the magnetic direction is sampled just at the time of the initial field peak so that the bearing vector points as closely as possible to the strike point. The electric field is also sampled at this time to determine the stroke polarity and to estimate the peak current using Eq. (1). Networks of these gated, wideband DFs are currently used in many locations throughout the world to locate lightning hazards and for research (Holle and López 1993). In the United States, such sensors have been integrated into large networks that provide lightning data to forestry agencies and electric power utilities in real time (Krider et al 1980; Orville and Songster 1987; Orville et al. 1990; Cummins et al. 1992).

Lewis et al. (1960) have described a method for locating lightning using differences in the time of arrival of a radio pulse at several stations. Since radio signals propagate at the speed of light, a constant difference in the arrival time at two stations defines a hyperbola, and multiple stations provide multiple hyperbolas whose intersections define a source location. Time-of-arrival (TOA) methods can provide accurate locations at long ranges (Lee 1989), and if the antennas are properly sited, the systematic errors are minimal. Casper and Bent (1992) have developed a wideband TOA receiver that is suitable for locating lightning sources at medium and long ranges using the hyperbolic method (Bent and Lyons 1984).

The radio-frequency noise that is radiated by lightning in the HF and VHF bands appears in bursts, and within each of these bursts there are hundreds to thousands of separate impulses. Proctor (1971) showed that when the difference in the time of arrival of each pulse is measured at four stations that are precisely synchronized, the location of the source can be computed in three dimensions, and the geometrical development of a burst can be mapped as a function of time. Unfortunately, the physical processes that produce HF and VHF emissions are still not well understood. Proctor (1981) reported that most bursts are characterized by a regular progression of source points and, therefore, he suggested that bursts were produced by processes that create new ionization and extensions of old channels. When the location of each RF pulse within a burst was plotted, the width of the associated "radio image" ranged from about 100 m to 1 km (Proctor 1981, 1983). Figure 11-7 shows the paths of the central cores of six successive lightning discharges that were reconstructed by Proctor (1983). The geometrical forms of intracloud discharges range from concentrated "knots" or "stars" a few kilometers in diameter to extensive branched patterns up to 90 km in length. Proctor et al. (1988) have summarized 16 years of observations of the spatial and temporal development of VHF sources in flashes to ground in South Africa.

In recent years, the National Aeronautics and Space Administration (NASA) Kennedy Space Center has developed a Lightning Detection and Ranging System that is capable of providing 3D locations of several hundred pulses within each lightning flash (Lennon and Maier 1991; Maier et al. 1995). This system is similar to that of Proctor, but the data acquisition is automatic, and the data displays are generated in real time (see also Thomson et al. 1994). Today it is clear that high-frequency TOA methods offer great promise, both for early warnings and for research, particularly for those phases of lightning that occur within the cloud.

Hayenga and Warwick (1981) showed that a radio interferometer could be used to measure the azimuth and elevation angles of lightning radiations at VHF frequencies. Rhodes et al. (1994) and Shao et al. (1995) have developed this technique further and have used interferometers to improve our understanding of cloud-to-ground lightning in both New Mexico and Florida. Richard et al. (1988, 1989) have used a multiple-station network of interferometers to locate and map the sources of VHF radiation in three dimensions with a time resolution approaching 10 μs. A simple version of this system is available commercially and can detect and locate both intracloud and cloud-to-

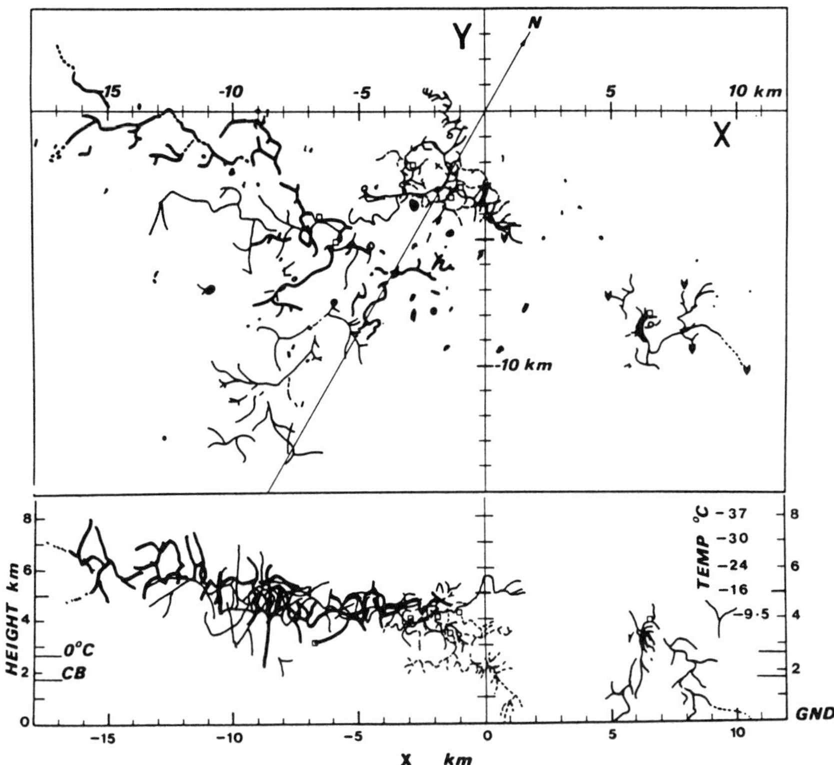

FIG. 11-7. The geometry of six successive lightning discharges in South Africa as measured by a VHF time-of-arrival system. The top panel shows a plan view of the channels as seen from above, and the bottom panel shows an elevation view that is a projection of the same channels on a vertical plane parallel to the x axis (adapted from Proctor 1983).

ground flashes. A recent review of commercial lightning-locating systems has been assembled by Holle and López (1993); this report discusses the overall design of these systems, their performance, and their applications with numerous references.

Summary

Over the past 75 years, applications of high-speed, streak photography have shown that a cloud-to-ground lightning flash contains an ensemble of discharge processes that are reproduced from flash to flash. The most deleterious of these processes is the return stroke, an energetic surge of current that propagates from ground to cloud. Current waveforms have been measured directly during strikes to instrumented towers and inferred indirectly from measurements of the electromagnetic fields. Spectroscopic studies have provided estimates of the stroke temperature and other properties of the lightning plasma. Methods of triggering lightning have been developed that can create a discharge under a thunderstorm at a known place and time. With the advances in analog and digital circuit technology and communication systems have come improved techniques for locating lightning sources. All evidence to date suggests that the characteristics of an individual flash may depend on the meteorological environment in which the storm occurs and when the discharge occurs during the evolution of the storm. Clearly, the next 75 years of lightning research promise to be just as exciting as the first.

Acknowledgments. This paper is dedicated to the memory of Leon Salanave and Karl Berger, who introduced the author to lightning research and who inspired him and many others by their enthusiasm for the subject. The author is grateful to Margaret Sanderson Rae for editorial assistance during the preparation of this manuscript.

BIBLIOGRAPHY

The following books by pioneers in lightning research provide a general introduction to the subject.

Gary, C. *La Foudre: Des mythologies antiques à la recherche moderne.* Paris: Masson, 1994.

Malan, D.J. *Physics of Lightning.* London: English Universities Press, 1963.

McEachron, K.B. *Playing with Lightning.* New York: Random House, 1940.

Salanave, L.E. *Lightning and Its Spectrum: An Atlas of Photographs.* Tucson: The University of Arizona Press, 1980.

Schonland, B.F.J. *The Flight of Thunderbolts.* 2d ed. London: Oxford University Press, 1964.

Uman, M.A. *Understanding Lightning.* Carnegie: Bek Technical Publications, 1971 (reprinted with the title *All About Lightning,* New York: Dover Press, 1986).

Viemeister, P.E. *The Lightning Book.* New York: Doubleday and Co., 1961 (reprint, Cambridge: MIT Press, 1972).

The following works will introduce the reader to the lightning literature on a more technical level.

Baker, R., and E. Kiss. Annotated bibliography on the physics of the lightning flash. *Meteorological and Geoastrophysical Abstracts,* No. 14, (1963):2923–2988.

Golde, R.H. *Lightning.* Vols. 1 and 2. London and New York: Academic Press, Berlin: Springer-Verlag, 1977.

Schonland, B.F.J. The Lightning Discharge. *Encyclopedia of Physics,* Vol. XXII. S. Flugge, Ed. 576–628, 1956.

Uman, M.A. *Lightning.* New York: McGraw-Hill, 1969 (reprint with a supplement, New York: Dover Press, 1984).

_____ . *The Lightning Discharge.* Orlando: Academic Press, 1987.

_____ , and E.P. Krider. A review of natural lightning: Experimental data and modeling. *IEEE Transactions EMC,* **24** (1982):79–112.

_____ , and _____ . Natural and artificially initiated lightning. *Science,* **246** (1989):457–464.

Volland, H. *Handbook of Atmospherics.* Vols. 1 and 2. Boca Raton: CRC Press, 1982.

_____ . *Handbook of Atmospheric Electrodynamics.* Vols. 1 and 2. Boca Raton: CRC Press, 1995.

REFERENCES

Adcock, F., and C. Clarke, 1947: The location of thunderstorms by radio direction-finding. *J. Inst. Elec. Engrs.,* **94B,** 118–125.

Appleton, E.V., R.A. Watson Watt, and J.F. Herd, 1926: On the nature of atmospherics, II. *Proc. Roy. Soc. London, Ser. A,* **111,** 615–677.

Austin, L.W., 1926: The present status of atmospheric disturbances. *Proc. IRE,* **14,** 133–138.

Beasley, W.H., M.A. Uman, and P.L. Rustan, 1982: Electric fields preceding cloud-to-ground lightning flashes. *J. Geophys. Res.,* **87,** 4883–4902.

Bent, R.S., and W.A. Lyons, 1984: Theoretical evaluations and initial operational experiences of LPATS (Lightning Position and Tracking System) to monitor lightning ground strikes using a time-of-arrival (TOA) technique. Preprints, *Seventh Int. Conf. on Atmospheric Electricity,* Albany, NY, Amer. Meteor. Soc., 317–324.

Berger, K., 1955: Die Messeinrichtungen für die Blitzforschung auf dem Monte San Salvatore. *Bull. SEV,* **46,** 193–201.

———, 1967: Novel observations on lightning discharges: Results of research on Mount San Salvatore. *J. Franklin Inst.*, **283**, 478–525.

———, 1978: Blitzstrom-Parameter von Aufwärtsblitzen. *Bull. Schweiz. Elektrotech. Ver.*, **69**, 353–360.

———, and E. Vogelsanger, 1965: Messungen und Resultate der Blitzforschung der Jahre 1955–1963 auf dem Monte San Salvatore. *Bull. SEV*, **56**, 2–22.

———, and ———, 1966: Photographische Blitzuntersuchungen der Jahre 1955–1965 auf dem Monte San Salvatore. *Bull. SEV*, **57**, 1–22.

———, R.B. Anderson, and H. Kroninger, 1975: Parameters of lightning flashes. *Electra*, **80**, 23–37.

Boys, C.V., 1926: Progressive lightning. *Nature*, **118**, 749–750.

Brook, M., and N. Kitagawa, 1964: Radiation from lightning discharges in the frequency range 400 to 1000 Mc/s. *J. Geophys. Res.*, **69**, 2431–2434.

———, G. Armstrong, R.P.H. Winder, B. Vonnegut, and C.B. Moore, 1961: Artificial initiation of lightning discharges. *J. Geophys. Res.*, **66**, 3967–3969.

Casper, P.W., and R.B. Bent, 1992: Results from the LPATS USA national lightning detection and tracking system for the 1991 lightning season. *Proc. 21st Int. Conf. on Lightning Protection*, Berlin, Germany, Association of German Electrical Engineers, 339–342.

Chapman, F.W., 1939: Atmospheric disturbances due to thundercloud discharges. *Proc. Phys. Soc. London*, **51**, 876–894.

Christian, H.J., R.J. Blakeslee, and S.J. Goodman, 1989a: The detection of lightning from geostationary orbit. *J. Geophys. Res.*, **94**, 13 329–13 337.

———, V. Mazur, B.D. Fisher, L.H. Ruhnke, K. Crouch, and R.P. Perala, 1989b: The Atlas/Centaur lightning strike incident. *J. Geophys. Res.*, **94**, 13 169–13 177.

Cohen, I.B., 1990: *Benjamin Franklin's Science*. Harvard University Press, 273 pp.

Cummins, K.L., W.L. Hiscox, R.W. Henderson, R.B. Pyle, and P.K. Blankenberger, 1992: U.S. National Lightning Detection Network. *Proc. 21st Int. Conf. on Lightning Protection*, Berlin, Germany, Association of German Electrical Engineers.

de Fonvielle, W., 1867: *Éclairs et Tonnerre*. Librairie de L. Hachette et Cie, 365 pp.

Diendorfer, G., and M.A. Uman, 1990: An improved return stroke model with specified channel-base current. *J. Geophys. Res.*, **95**, 13 621–13 644.

Dowden, R.L., and C.D.D. Adams, 1993: Size and location of lightning-induced ionization enhancements from measurement of VLF phase and amplitude perturbations on multiple antennas. *J. Atmos. Terr. Phys.*, **55**, 1335–1369.

Evans, W.H., and R.L. Walker, 1963: High speed photographs of lightning at close range. *J. Geophys. Res.*, **68**, 4455–4461.

Few, A.A., 1975: Thunder. *Sci. Amer.*, **233**, 80–90.

_____ , 1995: Acoustic radiations from lightning. *Handbook of Atmospheric Electrodynamics*, Vol. 2, H. Volland, Ed., CRC Press, 1–31.

Fleux, R., and P. Hubert, 1976: Triggered lightning hazards. *Nature*, **260**, 188.

_____ , C. Gary, and P. Hubert, 1975: Artificially triggered lightning above land. *Nature*, **257**, 212–214.

Fisher, R.J., G.H. Schnetzer, R. Thottappillil, V.A. Rakov, M.A. Uman, and J.D. Goldberg, 1993: Parameters of triggered lightning flashes in Florida and Alabama. *J. Geophys. Res.*, **98**, 22 887–22 902.

Fitzgerald, D.R., 1967: Probable aircraft triggering of lightning in certain thunderstorms. *Mon. Wea. Rev.*, **95**, 835–842.

Flammarion, C., 1905: *Thunder and Lightning*. Translated by Walter Mostyn. Chatto and Windus, 281 pp.

Foust, C.M., and H.P. Kuchni, 1932: The surge-crest ammeter. *Gen. Elec. Rev.*, **35**, 644–648.

Franklin, B., 1774: *Experiments and Observations on Electricity, made at Philadelphia in America*. 5th ed. F. Newberry, 530 pp. (Reprint Cohen, I. Bernard, 1941: *Benjamin Franklin's Experiments, A New Edition of Franklin's Experiments and Observations on Electricity*. Harvard University Press, 453 pp.)

Gockel, A., 1925: *Das Gewitter*. Ferd. Dümmlers Verlagsbuchhandlung, 316 pp.

Golde, R.H., 1973: *Lightning Protection*. Edward Arnold, 220 pp.

Guo, C., and E.P. Krider, 1982: The optical and radiation field signatures produced by lightning return strokes. *J. Geophys. Res.*, **87**, 8913–8922.

______ , and ______ , 1983: The optical power radiated by lightning return strokes. *J. Geophys. Res.,* **88,** 8621–8622.

Hagenguth, J.H., 1940: Lightning recording instruments. *Gen. Elec. Rev.,* **43,** 105–201, 248–255.

______ , 1947: Photographic studies of lightning. *Trans. AIEE,* **66,** 577–585.

______ , and J.G. Anderson, 1952: Lightning to the Empire State Building, Pt. 3. *Trans. AIEE,* **71** (Pt. 3), 641–649.

Hayenga, C.O., and J.W. Warwick, 1981: Two-dimensional interferometric positions of VHF lightning sources. *J. Geophys. Res.,* **86,** 7451–7462.

Hill, R.D., 1978: Comment on "Detection of Lightning Superbolts" by B.N. Turman. *J. Geophys. Res.,* **83,** 1381–1382.

Holle, R.L., and R.E. López, 1993: Overview of real-time lightning detection systems and their meteorological uses. NOAA Tech. Memo. ERL NSSL-102, National Severe Storm Laboratory, Norman, OK, 68 pp.

Horii, K., and M. Nakano, 1995: Artificially triggered lightning. *Handbook of Atmospheric Electrodynamics,* Vol. 1, H. Volland, Ed., CRC Press, 151–166.

Horner, F., 1964: Radio noise from thunderstorms. *Advances in Radio Research,* Vol. 2, J.A. Saxton, Ed., Academic Press, 121–204.

Idone, V.P., 1995: Microscale tortuosity and its variation as observed in triggered lightning channels. *J. Geophys. Res.,* **100,** 22 943–22 956.

______ , and R.E. Orville, 1988: Channel tortuosity variation in Florida triggered lightning. *Geophys. Res. Lett.,* **15,** 645–648.

Inan, U.S., D.C. Shafer, W.Y. Yip, and R.E. Orville, 1988: Subionospheric VLF signatures of nighttime D region perturbations in the vicinity of lightning discharges. *J. Geophys. Res.,* **93,** 11 455–11 472.

______ , T.F. Bell, and J.V. Rodriguez, 1991: Heating and ionization of the lower ionosphere by lightning. *Geophys. Res. Lett.,* **18,** 705–708.

Kähler, K., 1924: *Die Elektrizität der Gewitter.* Gebrüder Borntraeger, 148 pp.

Keen, R., 1938: *Wireless Direction Finding.* 3d ed. Iliffe and Sons, 803 pp.

Krehbiel, P.R., 1986: The electrical structure of thunderstorms. *The Earth's Electrical Environment,* National Academy Press, 90–113.

_____ , M. Brook, and R.A. McCrory, 1979: An analysis of the charge structure of lightning discharges to ground. *J. Geophys. Res.,* **84,** 2432–2456.

Krider, E.P., 1992: On the electromagnetic fields, Poynting vector and peak power radiated by lightning return strokes. *J. Geophys. Res.,* **97,** 15 913–15 917.

_____ , G.A. Dawson, and M.A. Uman, 1968: Peak power and energy dissipation in a single-stroke lightning flash. *J. Geophys. Res.,* **73,** 3335–3339.

_____ , R.C. Noggle, M.A. Uman, and R.E. Orville, 1974: Lightning and the Apollo 17/Saturn V exhaust plume. *J. Spacecr. Rockets,* **11,** 72–75.

_____ , _____ , and _____ , 1976: A gated wideband magnetic direction-finder for lightning return strokes. *J. Appl. Meteor.,* **15,** 301–306.

_____ , C.D. Weidman, and R.C. Noggle, 1977: The electric fields produced by lightning stepped-leaders. *J. Geophys. Res.,* **82,** 951–960.

_____ , R.C. Noggle, A.E. Pifer, and D.L. Vance, 1980: Lightning direction finding systems for forest fire detection. *Bull. Amer. Meteor. Soc.,* **61,** 980–986.

_____ , C. Leteinturier, and J.C. Willett, 1995: Submicrosecond fields radiated during the onset of first return strokes in cloud-to-ground lightning. *J. Geophys. Res.,* in press.

Lee, A.C.L., 1989: Ground truth confirmation and theoretical limits of an experimental VLF arrival time difference lightning flash locating system. *Quart. J. Roy. Meteor. Soc.,* **115,** 1147–1166.

Lennon, C., and L. Maier, 1991: Lightning mapping system. *Proc. Int. Aerospace and Ground Conf. on Lightning and Static Electricity,* NASA Conf. Publ. 3106, Vol. II, 89-1–89-10.

Leteinturier, C., J.H. Hamelin, and A. Eybert-Berard, 1991: Submicrosecond characteristics of lightning return-stroke currents. *IEEE Trans. EMC,* **33,** 351–357.

Le Vine, D.M., J.C. Willett, and J.C. Bailey, 1989: Comparison of fast electric field changes from subsequent return strokes of natural and triggered lightning. *J. Geophys. Res.,* **94,** 13 259–13 265.

Lewis, E.A., R.B. Harvey, and J.E. Rasmussen, 1960: Hyperbolic direction finding with sferics of transatlantic origin. *J. Geophys. Res.,* **65,** 1879–1905.

Lewis, W.W., and C.M. Foust, 1945: Lightning investigation on transmission lines, Pt. 7. *Trans. AIEE,* **64,** 107–115.

Mach, D.M., and W.D. Rust, 1993: Two-dimensional velocity, optical risetime, and peak current estimates for natural positive lightning return strokes. *J. Geophys. Res.,* **98,** 2635–2638.

Maier, L., C. Lennon, T. Britt, and S. Schaefer, 1995: Lightning Detection and Ranging (LDAR) system performance analysis. *Proc. Sixth Conf. on Aviation Weather Systems,* Dallas, TX, Amer. Meteor. Soc., 305–309.

Malan, D.J., and H. Collens, 1937: Progressive lightning, Pt. 3: The fine structure of lightning return strokes. *Proc. Roy. Soc. London, Ser. A,* **162,** 175–203.

Mathias, E., 1924: *Traité d'électricité atmosphérique et tellurique.* Les Presses Universitaires de France, 580 pp.

Mazur, V., and L.H. Ruhnke, 1993: Common physical processes in natural and artificially triggered lightning. *J. Geophys. Res.,* **98,** 12 913–12 930.

———, B.D. Fisher, and J.C. Gerlach, 1984: Lightning strikes to an airplane in a thunderstorm. *J. Aircr.,* **21,** 607–611.

McEachron, K.B., 1939: Lightning to the Empire State Building. *J. Franklin Inst.,* **227,** 149–217.

———, 1940: *Playing with Lightning.* Random House, 227 pp.

———, 1941: Lightning to the Empire State Building. *Trans. AIEE,* **60,** 885–889.

NASA, 1970: Analysis of the Apollo 12 lightning incident. National Aeronautics and Space Administration, MSC-01540, 82 pp.

Newman, M.M., J.R. Stahmann, J.D. Robb, E.A. Lewis, S.G. Martin, and S.V. Zinn, 1967: Triggered lightning strokes at very close range. *J. Geophys. Res.,* **72,** 4761–4764.

Norinder, H., 1936: Cathode-ray oscillographic investigations on atmospherics. *Proc. IRE,* **24,** 287–304.

———, 1953: Long-distance location of thunderstorms. *Thunderstorm Electricity,* H.R. Byers, Ed., University of Chicago Press, 276–327.

———, and O. Dahle, 1945: Measurements by frame aerials of current variations in lightning discharges. *Arkiv Mat. Astron. Fysik,* **32A,** 1–70.

Nucci, C.A., G. Diendorfer, M.A. Uman, F. Rachidi, M. Ianoz, and C. Mazzetti, 1990: Lightning return stroke current models with specified channel-base current: A review and comparison. *J. Geophys. Res.,* **95,** 20 395–20 408.

Orville, R.E., 1968a: A high-speed time-resolved spectroscopic study of the lightning return stroke: Part I. Qualitative analysis. *J. Atmos. Sci.*, **25**, 827–839.

______ , 1968b: A high-speed time-resolved spectroscopic study of the lightning return stroke: Part II. A quantitative analysis. *J. Atmos. Sci.*, **25**, 839–851.

______ , 1968c: A high-speed time-resolved spectroscopic study of the lightning return stroke: Part III. A time-dependent model. *J. Atmos. Sci.*, **25**, 852–856.

______ , and H. Songster, 1987: The East Coast lightning detection network. IEEE *Trans. Power Delivery,* **PWDR-2,** 899–904.

______ , R.W. Henderson, and R.B. Pyle, 1990: The National Lightning Detection Network—Severe storm observations. *Proc. 16th Conf. on Severe Local Storms,* Kananaskis Park, AB, Canada, Amer. Meteor. Soc., J27–J30.

Peters, D.S., 1915: Protection of Life and Property against Lightning. Technologic Papers of the Bureau of Standards, No. 56, Washington, DC, 127 pp.

Pierce, E.T., 1977a: Atmospherics and radio noise. *Lightning,* Vol. 1, R.H. Golde, Ed., Academic Press, 351–384.

______ , 1977b: Lightning warning and avoidance. *Lightning,* Vol. 2, R.H. Golde, Ed., Academic Press, 497–519.

Pitts, F.L., B.D. Fisher, V. Mazur, and R.P. Perala, 1988: Aircraft jolts from lightning bolts. *IEEE Spectrum,* **25**, 34–38.

Pockels, F., 1900: Über die Blitzentladungen erreichte Stromstärke. *Phys. Z., 2,* 306–307.

Proctor, D.E., 1971: A hyperbolic system for obtaining VHF radio pictures of lightning. *J. Geophys. Res.,* **76**, 1478–1489.

______ , 1981: VHF radio pictures of cloud flashes. *J. Geophys. Res.,* **86**, 4041–4071.

______ , 1983: Lightning and precipitation in a small multicellular thunderstorm. *J. Geophys. Res.,* **88**, 5421–5440.

______ , R. Uytenbogaardt, and B.M. Meredith, 1988: VHF radio pictures of lightning flashes to ground. *J. Geophys. Res.,* **93**, 12 683–12 727.

Prueitt, M.L., 1963: The excitation temperature of lightning. *J. Geophys. Res.,* **68,** 803–811.

Rakov, V.A., M.A. Uman, and R. Thottappillil, 1994: Review of lightning properties from electric field and TV observations. *J. Geophys. Res.,* **99,** 10 745–10 750.

Rhodes, C.T., X.-M. Shao, P.R. Krehbiel, R.J. Thomas, and C.O. Hayenga, 1994: Observations of lightning phenomena using radio interferometry. *J. Geophys. Res.,* **99,** 13 059–13 082.

Richard, P., A. Soulage, P. Laroche, and J. Appel, 1988: The SAFIR lightning monitoring and warning system: Application to aerospace activities. *Proc. Int. Aerospace and Ground Conf. on Lightning and Static Electricity,* Oklahoma City, OK, National Interagency Coordination Group, 383–390.

——, A. Soulage, and F. Broutet, 1989: The SAFIR lightning warning system. *Proc. 1989 Int. Conf. on Lightning and Static Electricity,* Bath, United Kingdom, Ministry of Defence Procurement Executive, 2 B.1.1–2 B.1.5.

Rodriguez, J.V., U.S. Inan, and T.F. Bell, 1992: D-region disturbances caused by electromagnetic pulses from lightning. *Geophys. Res. Lett.,* **19,** 2067–2070.

Salanave, L.E., 1961: The optical spectrum of lightning. *Science,* **134,** 1395–1399.

——, 1980: *Lightning and Its Spectrum: An Atlas of Photographs.* The University of Arizona Press, 136 pp.

——, R.E. Orville, and C.N. Richards, 1962: Slitless spectra of lightning in the region from 3850 to 6900 Ångstroms. *J. Geophys. Res.,* **67,** 1877–1884.

Schonland, B.F.J., 1956: The lightning discharge. *Handbuch der Physik,* Vol. 22, S. Flugge, Ed., Springer-Verlag, 576–628.

——, and H. Collens, 1933: Development of the lightning discharge. *Nature,* **132,** 407–408.

——, and ——, 1934: Progressive lightning. *Proc. Roy. Soc. London, Ser. A,* **143,** 654–674.

——, D.J. Malan, and H. Collens, 1935: Progressive lightning, Part 2. *Proc. Roy. Soc. London, Ser. A,* **152,** 595–625.

——, D.B. Hodges, and H. Collens, 1938a: Progressive lightning, Part 5: A comparison of photographic and electrical studies of the discharge process. *Proc. Roy. Soc. London, Ser. A,* **166,** 56–75.

——, D.J. Malan, and H. Collens, 1938b: Progressive lightning, Part 6. *Proc. Roy. Soc. London, Ser. A,* **168,** 455–469.

Shao, X.M., P.R. Krehbiel, R.J. Thomas, and W. Rison, 1995: Radio interferometric observations of cloud-to-ground lightning phenomena in Florida. *J. Geophys. Res.,* **100,** 2749–2783.

Slipher, J.M., 1917: The spectrum of lightning. *Lowell Obs. Bull.*, **79,** 55–58.

SPARG 1982: Eight years of lightning experiments at Saint Privat d'Allier. *Rev. Gén. Électr.,* **9,** 561–582.

Taranenko, Y.N., U.S. Inan, and T.F. Bell, 1993: Interaction with the lower ionosphere of electromagnetic pulses from lightning: Heating, attachment, and ionization. *Geophys. Res. Lett.,* **20,** 1539–1542.

Thomson, E.M., P.J. Medelius, and S. Davis, 1994: System for locating the sources of wideband dE/dt from lightning. *J. Geophys. Res.,* **99,** 22 793–22 802.

Thottappillil, R., and M.A. Uman, 1993: Comparison of lightning return-stroke models. *J. Geophys. Res.,* **98,** 22 903–22 914.

_____ , and _____ , 1994: Lightning return stroke model with height-variable discharge time-constant. *J. Geophys. Res.,* **99,** 22 773–22 780.

_____ , D.K. McLain, M.A. Uman, and G. Diendorfer, 1991: Extension of the Diendorfer–Uman lightning return stroke model to the case of a variable upward return stroke speed and a variable downward discharge current speed. *J. Geophys. Res.,* **96,** 17 143–17 150.

Turman, B.N., 1977: Detection of lightning superbolts. *J. Geophys. Res.,* **83,** 2566–2568.

_____ , 1978: Analysis of lightning data from DMSP satellite. *J. Geophys. Res.,* **83,** 5019–5024.

Uman, M.A., 1969: *Lightning.* McGraw-Hill, 264 pp (reprint with supplement, Dover Press, 1984).

_____ , 1985: Lightning return stroke electric and magnetic fields. *J. Geophys. Res.,* **90,** 6121–6130.

_____ , 1987: *The Lightning Discharge.* Academic Press, 377 pp.

_____ , 1988: Natural and artificially initiated lightning and lightning test standards. *Proc. IEEE,* **76,** 1548–1565.

_____ , and E.P. Krider, 1982: A review of natural lightning: Experimental data and modeling. *IEEE Trans. EMC,* **24,** 79–112.

_____ , D.K. McLain, and E.P. Krider, 1975: The electromagnetic radiation from a finite antenna. *Amer. J. Phys.,* **43,** 33–38.

Wallace, L., 1964: The spectrum of lightning. *Astrophys. J.,* **139,** 944–998.

Watson Watt, R.A., 1929: Weather and wireless. *Quart. J. Roy. Meteor. Soc.,* **55,** 273–301.

_____ , and J.F. Herd, 1926: An instantaneous direct-reading radio goniometer. *J. Inst. Elec. Engrs.*, **64,** 611–622.

_____ , _____ , and L.H. Bainbridge-Bell, 1933: *Applications of the Cathode Ray Oscillograph in Radio Research.* His Majesty's Stationary Office, 290 pp.

Weidman, C.D., and E.P. Krider, 1978: The fine structure of lightning return stroke waveforms. *J. Geophys. Res.*, **83,** 6239–6247; Correction, *J. Geophys. Res.*, **87,** 7351.

Willett, J.C., V.P. Idone, R.E. Orville, C. Leteinturier, A. Eybert-Berard, L. Barret, and E.P. Krider, 1988: An experimental test of the "Transmission-Line Model" of electromagnetic radiation from triggered lightning return strokes. *J. Geophys. Res.*, **93,** 3867–3878.

_____ , J.C. Bailey, V.P. Idone, A. Eybert-Berard, and L. Barret, 1989: Submicrosecond intercomparison of radiation fields and currents in triggered lightning return strokes based on the "transmission-line" model. *J. Geophys. Res.*, **94,** 13 275–13 286.

Wilson, C.T.R., 1916: On some determinations of the sign and magnitude of electric discharges in lightning flashes. *Proc. Roy. Soc. London, Ser. A,* **92,** 555–574.

_____ , 1920: Investigations on lightning discharges and on the electric field of thunderstorms. *Phil. Trans. Roy. Soc. London, Ser. A,* **221,** 73–115.

Climatology and Hydrology

Steps in the Evolution of Climatology: From Descriptive to Analytic

JOHN E. KUTZBACH

Introduction

Important developments in our understanding of the earth's climate can be traced in pathlike fashion stretching back over the centuries. Indeed, there are many such paths, each representing a particular area of climatology. This chapter identifies some of the important contributions along two particular paths. The first path traces developments in ideas about the earth's energy budget and about the balance between incoming solar radiation and outgoing terrestrial radiation. Ideas about the earth's energy budget provided the first and most direct explanation of the earth's thermal climate and led to the development of energy budget climate models. The second path traces developments in ideas about the dynamics of climate, including the general circulation of the earth's atmosphere and ocean and the associated hydrologic cycle. These ideas led to the construction of dynamical climate models. Both the energy budget climate models and the dynamical climate models are now being used to address questions about the earth's climate—past, present, and future.

In this brief overview, I highlight some important contributions to these two areas of climatology that I have come across in my own readings. A more thorough or systematic search for historical developments is needed in both of these areas, particularly along the pathway of ideas on the earth's energy budget. Along the second pathway, several important historical surveys exist that trace developments of ideas about the general circulation and dynamical atmospheric models (Lorenz 1967; Smagorinsky 1972, 1983). Many of the scientists mentioned in this overview made contributions to early studies of both weather and climate; aspects of developments in studies of cyclones during the nineteenth century are in Kutzbach (1979, 1987).

This review will reach back to periods prior to the founding of the

American Meteorological Society (AMS) in 1919. However, to set the stage, I mention several of the well-known textbooks on weather and climate that provide a general view of our knowledge of climate in the years before or shortly after the founding of the AMS: J. Hann, *Lehrbuch der Meteorologie* (1915); W.J. Humphreys, *Physics of the Air* (1920); W.G. Kendrew, *The Climate of the Continents* (1922); L.F. Richardson, *Weather Prediction by Numerical Processes* (1922); Sir N. Shaw, *Manual of Meteorology,* four volumes (1926–1931); V. Bjerknes et al., *Physikalische Hydrodynamik* (1933); and W. Köppen and R. Geiger, *Handbuch der Klimatologie* (1936).

Toward development of energy budget climate models

Our word "climate" derives from the Greek "klima," meaning slope or inclination. Thus, the word climate reminds us that the ancient Greeks recognized the fundamental importance for climate of the earth's axial tilt or obliquity. This tilt accounts for the increasing inclination of the sun's rays away from the zenith as one moves from the Tropics toward the poles and explains qualitatively the observed decrease of temperature with increasing latitude.

Edmund Halley (1693) may have been the first to begin to move this qualitative explanation of the relationship between axial tilt and temperature, known for millennia, toward a more quantitative model. Halley used trigonometric relationships to calculate the relative distribution of solar radiation as a function of latitude, from the equator to the pole, for conditions at the solstices and the equinox (Fig. 12-1). He commented that these calculations provided a general idea of the action of the sun throughout the year so that "that part of heat that arises simply from the preference of the sun [can] be brought to a geometrical certainty." Halley, of course, had previously studied how heating produced atmospheric motions; in 1686 he had written an important paper on the cause of the trade winds. Halley is one of many scientists who made early contributions to studies of both weather and climate.

Although Halley's paper points toward a quantitative understanding of the external forcing of climate by solar radiation, an equally important task was the collection and the analysis of basic observations of temperature and precipitation, and wind and pressure. Alex-

Lat.	Sun in $\gamma \triangleq$	Sun in $\mathfrak{S}$	Sun in ϑ
0	20000	18341	18341
10	19696	20290	15834
20	18794	21737	13166
30	17321	22651	10124
40	15321	23048	6944
50	12855	22991	3798
60	10000	22773	1075
70	6840	23543	000
80	3473	24673	000
90	0000	25055	000

FIG. 12-1. Halley's calculations of the relative distribution of incoming solar radiation as a function of latitude at the equinox (left), and summer (middle) and winter solstices (right) (Halley 1693).

ander von Humboldt (1817) was the first to prepare a hemispheric chart of the geographic distribution of annual-average surface temperature (Fig. 12-2). Von Humboldt's analysis was based on observations from only a handful of stations, yet it illustrated clearly the decrease of temperature toward the North Pole and the differences of temperature in the east–west direction produced by differential heating of continents and oceans. Von Humboldt was also among the first scientists to make measurements of solar radiation. Maury (1855) pioneered in early studies of the oceans, publishing charts of the Gulf Stream, Atlantic surface temperature, and global distribution of oceanic surface winds.

In 1884 William Ferrel published a paper on the spatial and temporal distribution of solar radiation, terrestrial radiation, and both surface and free-atmosphere temperature. Ferrel presented his results in mathematical form, using empirical cooling rates to estimate terrestrial radiation. (Stefan and Boltzman were just developing their radiation equations about this time.) Ferrel reported on the earth's thermal budget, on the latitudinal and seasonal response of tempera-

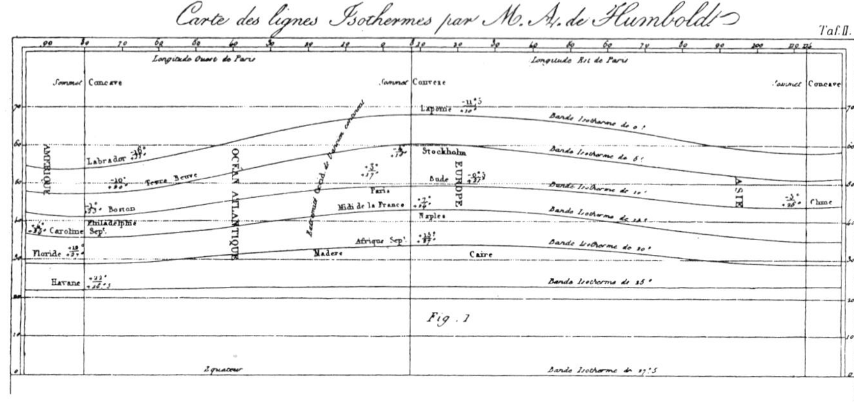

FIG. 12-2. Von Humboldt's chart of annual isotherms for the Northern Hemisphere showing north–south and land–ocean temperature differences; outlines of continents are omitted. Isotherms (°C) are drawn for every 5°. (von Humboldt 1817).

ture to solar radiation, and on the vertical distribution of temperature in the atmosphere. Ferrel had earlier (1859–1860) developed the appropriate equations of motion for application to rotating systems (such as the earth) and applied these equations in a mathematical theory of the general circulation of the atmosphere and ocean.

Ideas about the global energy budget were beginning to assume forms that we recognize today. Wilhelm von Bezold (1892) wrote on the earth's energy budget; he described the domains of heat gain in the Tropics and heat loss at the poles and the requirement for a poleward heat transport to maintain a steady energy balance. He estimated that the boundary between heat gain and heat loss was between 35° and 40° latitude—very close to modern-day estimates of about 30° in the Northern Hemisphere and 35° in the Southern Hemisphere (Peixoto and Oort 1992).

In 1896, Svante Arrhenius published an exceedingly important paper in which he developed what we now describe as the equations for an energy budget climate model. He was apparently the first to do so. His model took into account solar and terrestrial radiation, including the fourth-power relationship between temperature and radiation, and contained estimates of the absorption of terrestrial radiation by water vapor and carbonic acid (carbon dioxide) based upon the work of Tyndall, Langley, and others. He then used his energy budget model to estimate the sensitivity of climate to changes in carbonic acid. He ignored changes in horizontal heat transport and cloud cover but included certain aspects of temperature–albedo feedback related to snow cover. Arrhenius reported that a doubling of CO_2 would raise temperatures by about 3°–3.5°C, while a reduction of CO_2 by one-third would lower temperatures by roughly the same amount (Fig. 12-3). These values happen to be very close to modern-day estimates (IPCC 1990), even though Arrhenius ignored the possible effects of changes of horizontal advection and cloud cover and used a radiative transfer model that was much less detailed than present-day models. Arrhenius also considered the role that these changes might play in helping to explain warm and cold epochs in geologic history. His work certainly helped set the stage for the rapid evolution of climate theory in the twentieth century. Other studies soon followed, with Abbot and Fowle (1908) exploring the sensitivity of climate to changes in solar radiation and Humphreys (1920) reporting on the possible effects on climate of volcanic eruptions.

While these theoretical studies were progressing, there were equally important developments in summarizing the rapidly accumulating climatic observations. By the turn of the century, one can find essentially modern treatments of the seasonal and spatial distribu-

TABLE VII.—*Variation of Temperature caused by a given Variation of Carbonic Acid.*

Latitude	Carbonic Acid = 0·67					Carbonic Acid = 1·5					Carbonic Acid = 2·0					Carbonic Acid = 2·5					Carbonic Acid = 3·0				
	Dec.–Feb.	March–May	June–Aug.	Sept.–Nov.	Mean of the year	Dec.–Feb.	March–May	June–Aug.	Sept.–Nov.	Mean of the year	Dec.–Feb.	March–May	June–Aug.	Sept.–Nov.	Mean of the year	Dec.–Feb.	March–May	June–Aug.	Sept.–Nov.	Mean of the year	Dec.–Feb.	March–May	June–Aug.	Sept.–Nov.	Mean of the year
70																									
60	−2·9	−3·0	−3·4	−3·1	−3·1	3·3	3·4	3·8	3·6	3·52	6·0	6·1	6·0	6·1	6·05	7·9	8·0	7·9	8·0	7·95	9·1	9·3	9·4	9·4	9·3
50	−3·0	−3·2	−3·4	−3·3	−3·22	3·4	3·7	3·6	3·8	3·62	6·1	6·1	5·8	6·1	6·02	8·0	8·0	7·6	7·9	7·87	9·3	9·5	8·9	9·5	9·3
40	−3·2	−3·3	−3·3	−3·4	−3·3	3·7	3·8	3·4	3·7	3·65	6·1	6·1	5·5	6·0	5·92	8·0	7·9	7·0	7·9	7·7	9·5	9·4	8·6	9·2	9·17
30	−3·4	−3·4	−3·2	−3·3	−3·32	3·7	3·6	3·3	3·5	3·52	6·0	5·8	5·4	5·6	5·7	7·9	7·6	6·9	7·3	7·42	9·3	9·0	8·2	8·8	8·82
20	−3·3	−3·2	−3·1	−3·1	−3·17	3·5	3·3	3·2	3·5	3·47	5·6	5·4	5·0	5·2	5·3	7·2	7·0	6·6	6·7	6·87	8·7	8·3	7·5	7·9	8·1
10	−3·1	−3·1	−3·0	−3·1	−3·07	3·5	3·2	3·1	3·2	3·25	5·2	5·0	4·9	5·0	5·02	6·7	6·6	6·3	6·6	6·52	7·9	7·5	7·2	7·5	7·52
0	−3·1	−3·0	−3·0	−3·0	−3·02	3·2	3·2	3·1	3·1	3·15	5·0	5·0	4·9	4·9	4·95	6·6	6·4	6·3	6·4	6·42	7·4	7·3	7·2	7·3	7·3
−10	−3·0	−3·0	−3·1	−3·0	−3·02	3·1	3·1	3·2	3·2	3·15	4·9	4·9	5·0	5·0	4·95	6·4	6·4	6·6	6·6	6·5	7·3	7·3	7·4	7·4	7·35
−20	−3·1	−3·1	−3·2	−3·1	−3·12	3·2	3·2	3·2	3·2	3·2	5·0	5·0	5·2	5·1	5·07	6·6	6·6	6·7	6·7	6·65	7·4	7·5	8·0	7·6	7·62
−30	−3·1	−3·2	−3·3	−3·2	−3·2	3·2	3·2	3·4	3·3	3·27	5·2	5·3	5·5	5·4	5·35	6·7	6·8	7·0	7·0	6·87	7·9	8·1	8·6	8·3	8·22
−40	−3·3	−3·3	−3·4	−3·4	−3·35	3·4	3·5	3·7	3·5	3·52	5·5	5·6	5·8	5·6	5·62	7·0	7·2	7·7	7·4	7·32	8·6	8·7	9·1	8·8	8·8
−50	−3·4	−3·4	−3·3	−3·4	−3·37	3·6	3·7	3·8	3·7	3·7	5·8	6·0	6·0	6·0	5·95	7·7	7·9	7·9	7·9	7·8	9·1	9·2	9·4	9·3	9·25
−60	−3·2	−3·3	—	—	—	3·8	3·7	—	—	—	6·0	6·1	—	—	—	7·9	8·0	—	—	—	9·4	9·5	—	—	—

FIG. 12-3. Arrhenius's calculations of the variation of temperature (°C) caused by a given variation in carbonic acid (carbon dioxide); calculations are presented for various latitudes and for the four seasons and annual average. Carbon dioxide concentrations are varied from 0.67 of the base value (left) to three times the base value (right) (Arrhenius 1896).

tions of temperature, precipitation, and selected components of the earth's energy budget, and descriptions of the mean fields of surface air currents (see, e.g., illustrations from von Hann's *Lehrbuch der Meteorologie;* Fig. 12-4).

Another major contribution to the advancement of energy budget climate modeling was made by Milutin Milankovitch in work begun around 1911 and published in 1920. Milankovitch was the first to use an energy budget climate model to simulate the latitudinal distribution of

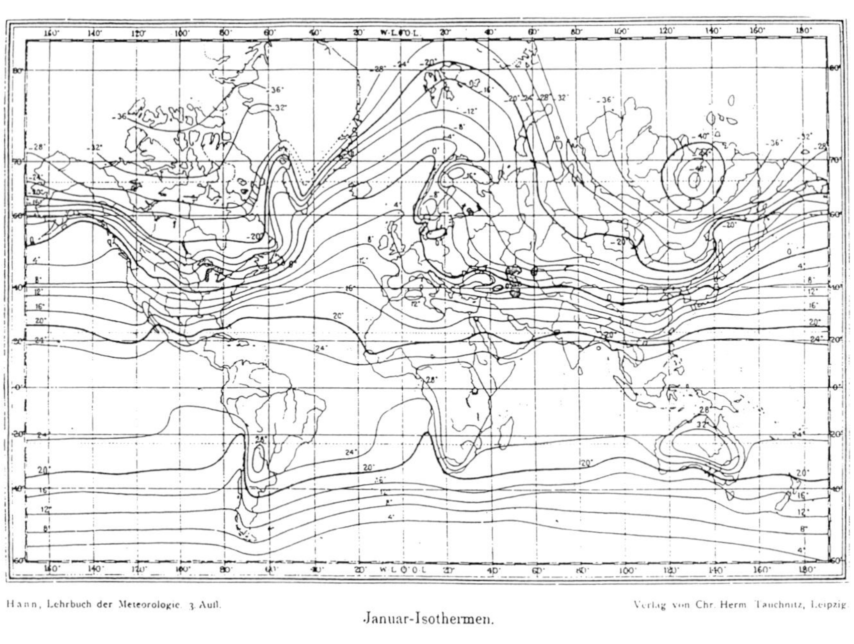

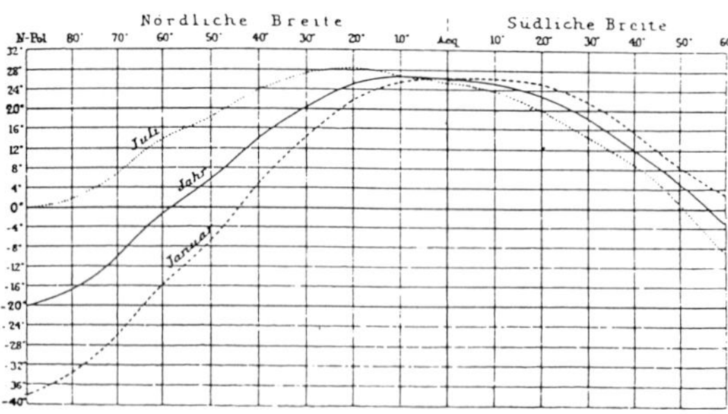

FIG. 12-4. (Top) Von Hann's near-global chart of January temperature (°C). (Bottom) Von Hann's graph of zonal-average temperature (°C) for January, July, and annual average (von Hann 1915).

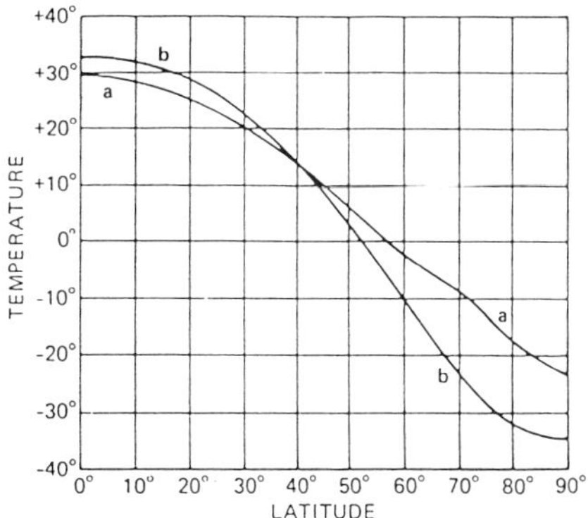

FIG. 12-5. Milankovitch's graph of zonal and annual average temperature (°C) simulated with a zonal-average energy budget climate model for a static atmosphere and for annual-average insolation (line b–b). For comparison, he graphed the observed temperature (line a-a), adjusted to estimate the temperature that would correspond to a 28% land, 72% ocean distribution at each latitude—as assumed in the model calculations (Milankovitch 1920).

temperature in response to annual-average insolation (again ignoring north–south heat exchanges, as had Arrhenius). Milankovitch's model, like that of Arrhenius, was therefore very similar in form and function to the energy budget models developed in the 1960s by Budyko and Sellers, but with the restrictive assumption of local radiative equilibrium. Because of this restriction, Milankovitch's climate model predicted annual-average equatorial temperatures that were too high and polar temperatures that were too low compared to observations (Fig. 12-5). Nevertheless, with this powerful new tool Milankovitch began quantitatively exploring the climatic implications of shifting continents and changing earth orbits. He used his climate model to describe how changes in the earth's orbit could produce glacial–interglacial cycles, an explanation that has now been generally accepted some 50–75 years later (Imbrie et al. 1992, 1993).

At roughly the same time as Milankovitch, Rudolf Spitaler was developing innovative ideas on the thermal response of the atmosphere and ocean to the seasonal cycle of solar radiation (Spitaler 1921). Also in the 1920s, early studies of the general circulation by Defant (1921) introduced ideas about large eddy mixing associated with the movements of cold air toward the equator and warm air toward the pole. These ideas were to provide the rationale for including a simple statistical representation of horizontal heat exchange in Budyko–Sellers-type climate models in the 1960s.

The late 1950s and early 1960s marked the beginning of a strong concentration of effort on studies of the earth's energy budget that has continued to the present. Houghton (1954), London (1957), and Budyko (1958) documented the earth's energy balance based upon observations, and Adem (1963) began to apply energy budget concepts to the modeling of seasonal changes.

Then, building upon the much earlier pioneering work of Arrhenius and Milankovitch, Budyko (1969), and Sellers (1969) independently published descriptions of their energy budget climate models. These models have become the foundation for a wide range of applications to the study of climate and climate sensitivity. Both Budyko and Sellers employed their models to estimate the sensitivity of climate to changes in solar radiation for various assumed snow–ice albedo feedback relationships (Fig. 12-6). Within a few years, this area of research expanded to include studies of the sensitivity of climate to changes in CO_2, studies of cloud and water vapor feedbacks, studies of coupled atmosphere–ocean energy budget models (Schneider and Dickinson

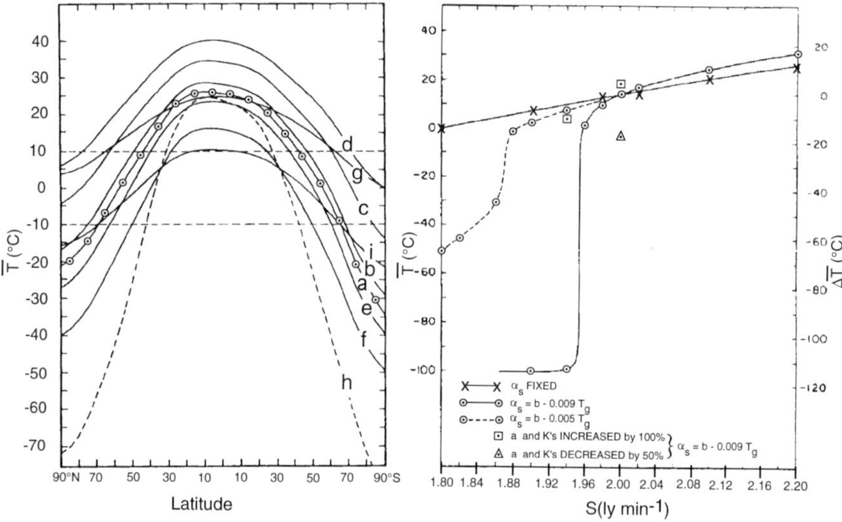

FIG. 12-6. Results from Seller's energy budget climate model. (Left) The predicted latitudinal distribution of the mean annual sea level temperature when the solar constant is increased by (a) 0%, (present conditions), (b) 1%, (c) 5%, and (d) 10%, or reduced by (e) 1% or (f) 2%. Curve (i) shows the distribution resulting from a 3% decrease of the solar constant and a 100% increase in the exchange coefficient and eddy diffusivities. In curves (g) and (h) the solar constant is kept fixed at its present value of 2.00 ly min^{-1}, and the exchange coefficients and eddy diffusivities are increased by 100% and decreased by 50%, respectively. (Right) The mean global sea level temperature $\overline{T}$ as a function of the solar constant S for different representations of the planetary albedo α_s and different values of the meridional exchange coefficient, a, and the eddy diffusivities, K_h, K_w, and K_0 (Sellers 1969).

1974; Schneider 1992), studies of stability and instability of climate states and multiple climate equilibria (North et al. 1981), and studies of biosphere feedbacks, such as illustrated by "Daisyworld" (Lovelock 1988).

Advances in the development and application of energy budget climate models were also a stimulus for the development of a wide range of simplified coupled system models that included selected processes of the atmosphere, biosphere, cryosphere, hydrosphere, and lithosphere. These coupled system models have been applied to modeling of earth system processes on timescales ranging from years to millions of years (Saltzman 1978, 1985).

Toward development of dynamical climate models

This section of my brief historical overview will be covered in less detail than the section on energy budget climate models because there are excellent reviews of early observational and theoretical studies (Lorenz 1967) and of early modeling studies (Smagorinsky 1972, 1983). The three-dimensional structure of the atmosphere's general circulation was studied first with surface observations, including surface-based observations of cloud motions, and then with kites and balloons (Kutzbach 1979). These early observing systems were followed by the introduction of radiosondes (after 1937) and, most recently, satellites (after 1957). Studies of the general circulation of the ocean are much less developed than those of the atmosphere, and programs are being mounted now to obtain critically needed ocean observations (Niiler 1992). [See chapter 3 by Serafin, this volume.]

Quantitative dynamical frameworks for studies of the general circulation of the atmosphere were established by Ferrel (1859–1860, 1884), who formulated the equations of motion in rotating coordinates, developed a theory for the zonally symmetric general circulation of the atmosphere, and derived a form of the thermal wind equation to explain that the midlatitude westerlies increase in strength with height because the temperature of the troposphere decreases toward the poles. Coming much closer to the present, the work of Jeffreys (1926) and the book by Bjerknes et al. (1933), *Physikalische Hydrodynamik,* form good starting points for describing some of the theoretical perspectives on climate that existed at about the time of the founding of the AMS. Jeffreys developed analytical expressions for the zonally symmetric circulation but also argued for the importance of asymmetric eddies for maintaining a steady circulation. The book by Bjerknes et al. combines dynamical and thermodynamical approaches and describes the role of solenoidal fields in north–south vertical planes in generating zonal westerly flows and of solenoids in east–west vertical planes, associated with land–ocean temperature contrasts, in producing east–west perturbations in the westerlies and continent-scale monsoon circulations.

Rossby and collaborators (1939) built upon these circulation concepts of Bjerknes et al. by using a linearized form of the vorticity equation to describe the characteristic stationary wavelengths of "free" perturbations in the zonal westerly circulation. They described how stationary features of the general circulation of the atmosphere, with wavelengths from 3000 to 7000 km in midlatitudes, might occur downstream of "forced" perturbations in the zonal flow. These forced perturbations were assumed to be associated with thermal or topographic effects. This paper by Rossby and earlier calculations by Ferrel, Jeffreys, Bjerknes, and others were important steps in showing how mathematical models could help elucidate observed features of the general circulation.

Charney and Eliassen (1949) and Bolin (1950) extended Rossby's use of vorticity concepts. They used a linearized form of the barotropic vorticity equation to explain how topography (e.g., the Rocky Mountains) could produce standing perturbations in the westerlies (Fig. 12-7). Smagorinsky (1953) combined the linearized vorticity equation and thermodynamic energy equation to form a baroclinic model that he then used to illustrate the role of large-scale heat sources and sinks (e.g., the wintertime heat sources in the western North Atlantic and western North Pacific) in forcing East Coast troughs (Fig. 12-8). These papers by Charney and Eliassen, Bolin, and Smagorinsky clearly demonstrated the power of mathematical models in providing physically consistent and quantitatively accurate descriptions of important climatological features of the atmosphere's general circulation. During roughly the same period, Sverdrup and Munk (1947), Stommel (1948), and Munk (1950) began to use simplified mathematical models to explain features of the wind-driven ocean circulation. Also, in the 1950s, Fultz (1951) and Hide (1953) showed that experiments with rotating fluids could provide important insights into the general circulation.

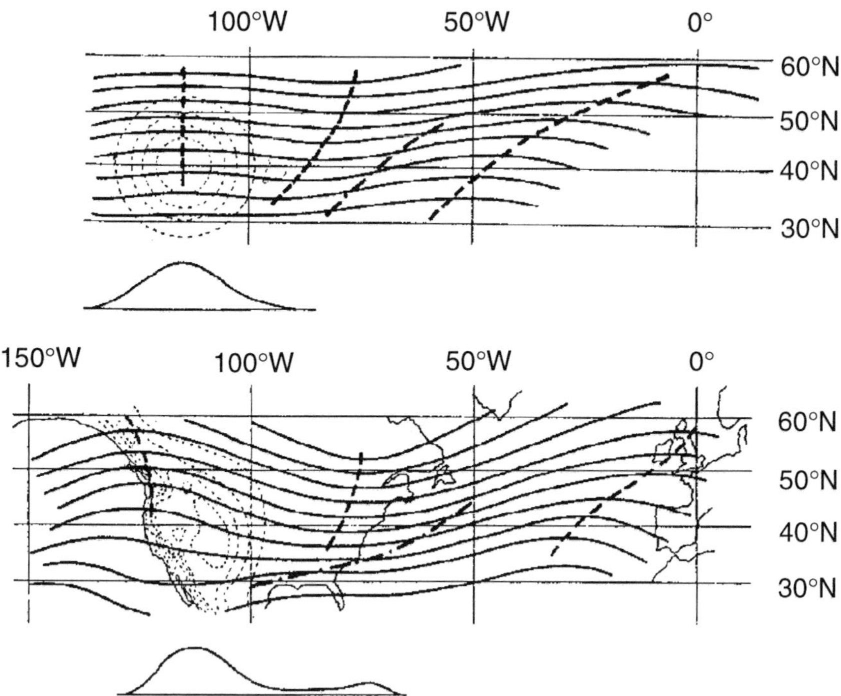

FIG. 12-7. (Top) Results from Bolin's barotropic model illustrating a theoretical flow pattern generated by a circular mountain in a homogeneous basic current. Dashed lines are troughs and ridges, and the dash–dotted line indicates maximum zonal velocity. (Bottom) The observed flow pattern in winter at the 20 000-ft level (Bolin 1950).

During World War II, needs of the military led to improved networks of meteorological observations and to the development of electronic computers. In the late 1940s and during the 1950s, Starr and collaborators used the first few years of observations from the Northern Hemisphere radiosonde network to document, for the first time, the climatology of the three-dimensional atmospheric circulation, including the zonal flow and the meridional fluxes of heat, moisture, and momentum (Starr and White 1951, 1955; Fig. 12-9). This information would prove to be crucial for documenting the importance of both mean and eddy motions in the general circulation and for verifying the accuracy of numerical simulations of atmospheric motions. During these same years, the first numerical integrations of the barotropic vorticity

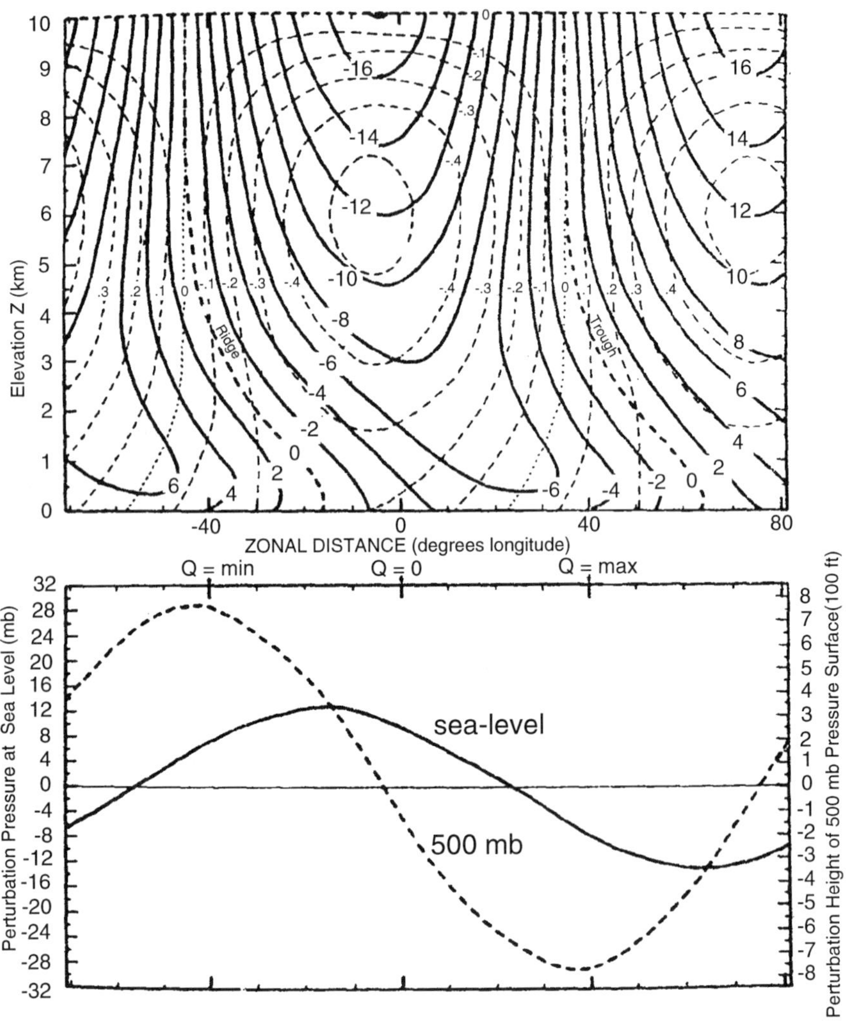

FIG. 12-8. Results from Smagorinsky's general baroclinic model with surface friction. In this example, the heat source–sink function has a wavelength of 160° longitude with the source located at 0° and the sink at ±80°. (Top) Zonal section of perturbation meridional velocity v (solid lines) in meters per second and perturbation vertical velocity w (dashed lines) in centimeters per second. (Bottom) Zonal profile of perturbation pressure at sea level (solid lines) and perturbation height of 500-mb pressure surface (dashed lines) (Smagorinsky 1953).

equation were made using the ENIAC computer at Princeton; the computation time required for the first 24-h forecast was about 24 hours (Charney et al. 1950). [See chapter 2 by Cressman, this volume.]

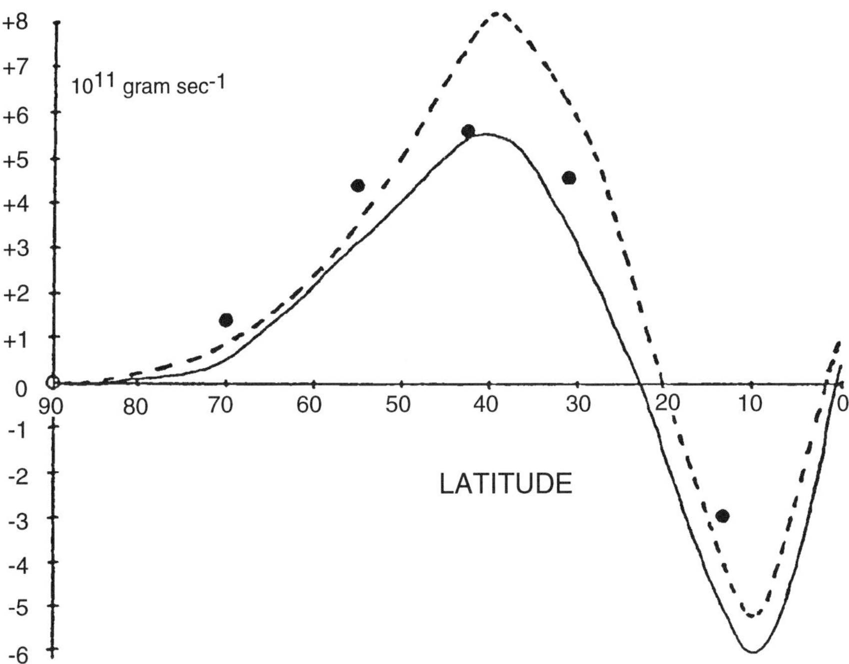

Fig. 12-9. Starr and White's estimate of the meridional distribution of the poleward water vapor flux in the atmosphere. The dashed curve and the solid curve represent estimates of this flux deduced from surface observations. The dots represent the flux computed from atmospheric data for the year 1950. The units are 10^{11}gm s^{-1} (Starr and White 1955).

Smagorinsky (1983), writing about the beginning of numerical weather prediction and general circulation modeling, noted that a new era opened with the work of Phillips, who completed the first general circulation experiments in 1955. Phillips (1956) used a two-level quasi-geostrophic model with very simple specifications of heating and friction to calculate the atmosphere's general circulation for an idealized domain and for a 31-day period. The integration produced realistic features of the general circulation, including a midlatitude westerly wind maximum and baroclinic eddys, with realistic transports and energetics (Fig. 12-10). This successful experiment launched the exciting period of numerical simulation of climate that continues to this day. Smagorinsky recalls that von Neumann recognized the great significance of Phillips's paper and called for a conference on the application of numerical integration technologies to the problem of the general circulation. Von Neumann remarked at the 1955 conference (Pfeffer 1960):

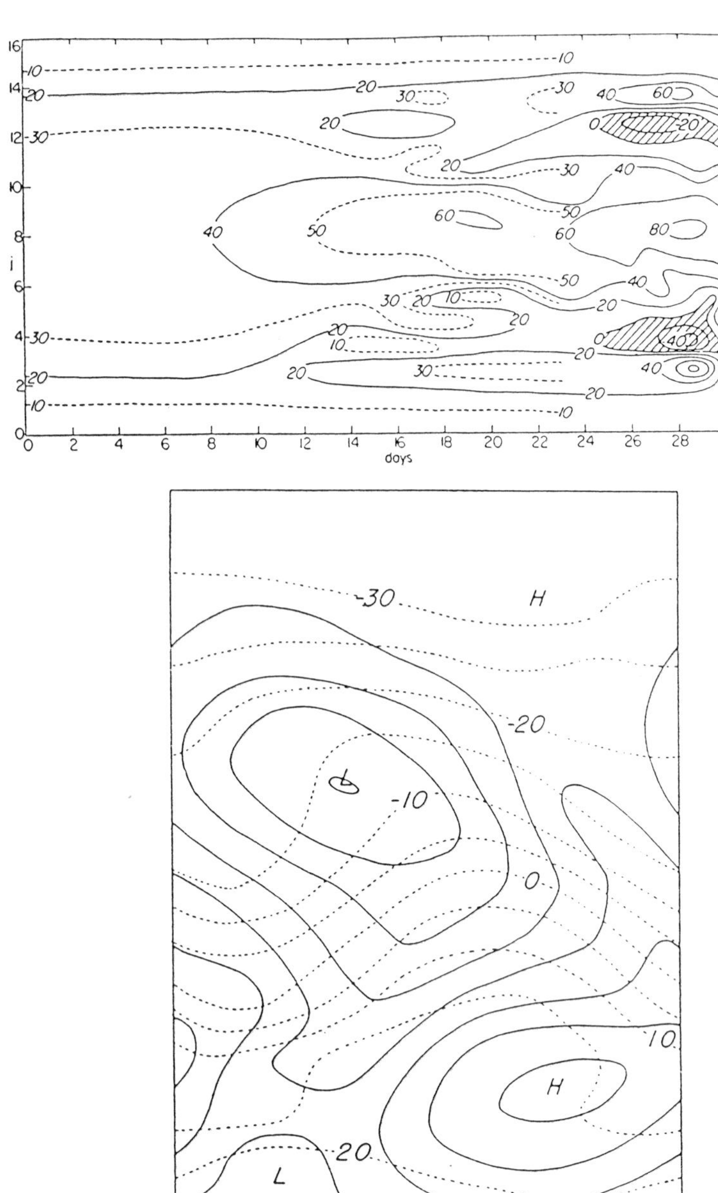

FIG. 12-10. Results from Phillips's numerical general circulation model experiment. (Top) Variation of zonal-average wind at 250 mb with latitude (j) and time (days). Units are meters per second. Regions of easterly winds are shaded. (Bottom) Distribution of 1000-mb contour height at 200-ft intervals (solid lines) and 500-mb temperature at 5°C intervals (dashed lines) at 20 days. The small rectangle in the lower-right corner shows the size of the finite-difference grid intervals Δx and Δy (Phillips 1956).

> . . . it seems quite clear that the problem of forecasting long-term fluctuations of the general circulation will involve a great deal of mathematical analysis and a great deal of calculation. It will not be possible to arrive at the answers by dialectic methods.

Talks by participants at the conference (see the list of speakers, Fig. 12-11) captured the spirit of excitement surrounding the important advances in our understanding of climate that were occurring in the 1950s, including the first numerical simulations of the general circulation by Phillips.

Observations of important features of the earth's climate continued to improve. With the International Geophysical Year, in 1957, came Sputnick and the hope for a space-based observing system. In the same year, Keeling initiated his precise measurements of the atmo-

DYNAMICS OF CLIMATE

*The Proceedings of a Conference on the Application
of Numerical Integration Techniques to the
Problem of the General Circulation
held October 26-28, 1955*

Edited by

RICHARD L. PFEFFER, Ph.D.

Geophysics Research Directorate

I. INTRODUCTORY TALKS

Summary of the welcoming
addresses of

J. ROBERT OPPENHEIMER
and
MILTON GREENBERG

ROBERT M. WHITE: Purpose of the conference

II. NUMERICAL METHODS AND RELATED TOPICS

JOHN VON NEUMANN: Some remarks on the problem of forecasting climatic fluctuations

JULE G. CHARNEY: Numerical prediction and the general circulation

NORMAN A. PHILLIPS: The general circulation of the atmosphere: a numerical experiment

RICHARD L. PFEFFER: On the design of a numerical experiment for the study of the general circulation of the atmosphere

YALE MINTZ: On the incorporation of non-adiabatic effects in numerical integration models for the study of the general circulation

ARNT ELIASSEN: Remarks on the problem of long-range weather prediction

JOSEPH SMAGORINSKY: A synopsis of research on quasi-stationary perturbations of the mean zonal circulation caused by topography and heating

III. OTHER STUDIES OF THE GENERAL CIRCULATION AND CLIMATIC CHANGE

VICTOR P. STARR: Some comments on the general circulation and plans of the M.I.T. project for studies of the general circulation by numerical methods

GEORGE S. BENTON: A review of research conducted by the general circulation project at the Johns Hopkins University

JACOB BJERKNES: Ageostrophic corrections to the computed poleward flux of angular momentum

DAVE FULTZ: Experimental models of rotating fluids and possible avenues for future research

HSIAO-LAN KUO: Theoretical findings concerning the effects of heating and rotation on the mechanism of energy release in rotating fluid systems

EDWARD N. LORENZ: Generation of available potential energy and the intensity of the general circulation

HARRY WEXLER: Possible causes of climatic fluctuations

SIGMUND FRITZ: The heating distribution in the atmosphere and climatic change

IV. RADIATION STUDIES

LEWIS D. KAPLAN: Problems involved in introducing long-wave radiative effects into mathematical models of the atmospheric circulation

JEAN I. KING: A potential theory formulation of radiant-heat transfer

FIG. 12-11. Table of contents from book based upon a conference held in 1955 (Pfeffer 1960).

spheric concentration of CO_2 (Keeling 1960); now, after almost 40 years, these measurements have provided an extremely important and quantitative index of "global change." As satellite observing systems developed, we obtained the first satellite-based estimates of the earth's radiation budget (Vonder Haar and Suomi 1969), estimates that are still being refined and extended (IPCC 1990, 1996).

Following the first general circulation experiments by Phillips, the next 10 years saw rapid development and application of increasingly sophisticated numerical models of the general circulation. Smagorinsky (1963) reported the first experiments with primitive equation models applied at two vertical levels for an idealized geography. Two years later, Smagorinsky et al. (1965) developed a nine-level primitive equation model that incorporated improved representation of physical processes such as radiative transfer and friction; they reported a successful integration extending to 300 days. Mintz (1965, 1968) described a "long term" (285 day) integration of a two-level primitive equation model applied over a spatial grid of realistic geography and topography (Fig. 12-12).

The development of general circulation and climate models offered the promise of quantitative and comprehensive treatment of climate, both present and past. That a new era had begun was underscored a few years later by J. Murray Mitchell Jr., U. S. Weather Bureau climatologist, who prepared a comprehensive statement of theoretical paleoclimatology for the International Union of Quaternary Research meeting in Boulder, Colorado (Mitchell 1965). He stated that:

> . . . the atmospheric scientist is finding it possible to construct remarkably realistic mathematical models of the atmosphere (e.g., Smagorinsky 1963), by which he can explore the nature and extent of various subtle forms of dynamical unrest that underlie the perpetual evolution of climate and circulation from one pattern to another. Indeed, he is now standing at the threshold of a new era in which, for the first time, he will be able to derive *quantitative* evaluations of a number of hypotheses of climatic change by means of suitable controlled experiments with these mathematical stand-ins for the real atmosphere.

This prediction by Mitchell of the rapid development and application of climate models has proved to be accurate.

Beginning in the late 1950s, observational studies by J. Namias and J. Bjerknes had begun to highlight the important role of coupling

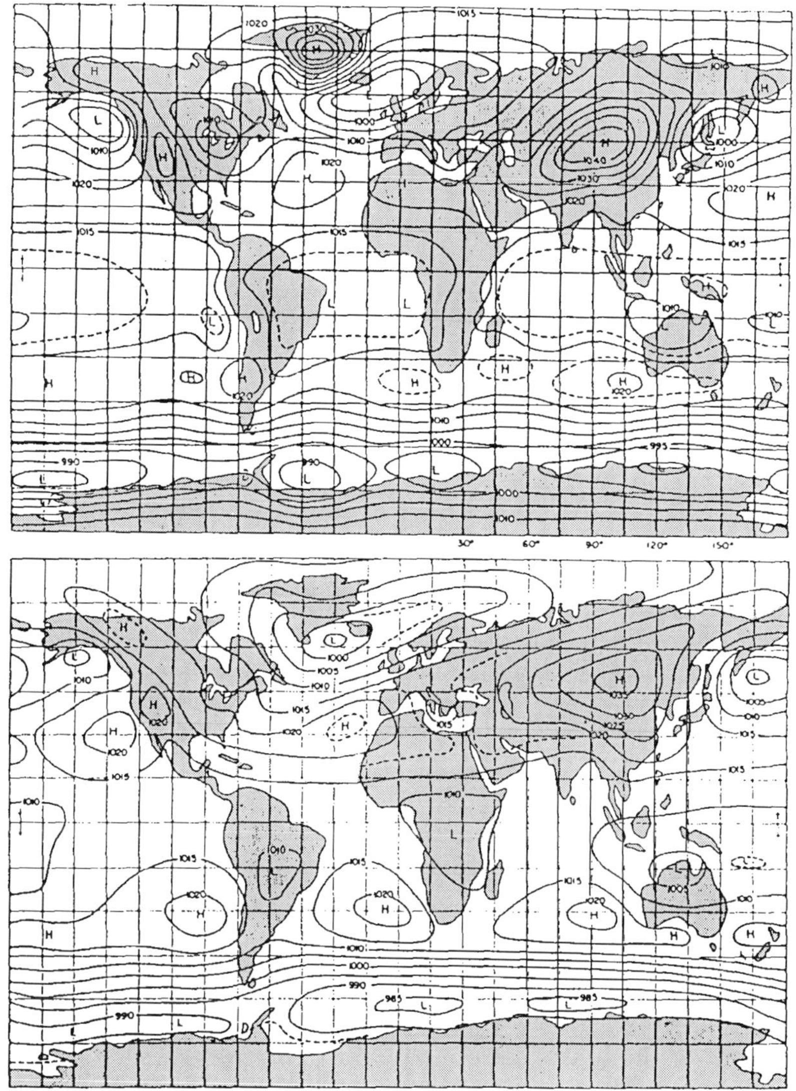

FIG. 12-12. Results from Mintz's numerical general circulation model experiment illustrating surface pressure reduced to sea level at 5-mb intervals for Northern Hemisphere winter and Southern Hemisphere summer. (The broken lines are intermediate 2.5-mb isobars.) The upper figure is the 30-day mean (from days 256 to 285) computed in the numerical experiment with mountains. The lower figure shows the normal January sea level pressure (Mintz 1965).

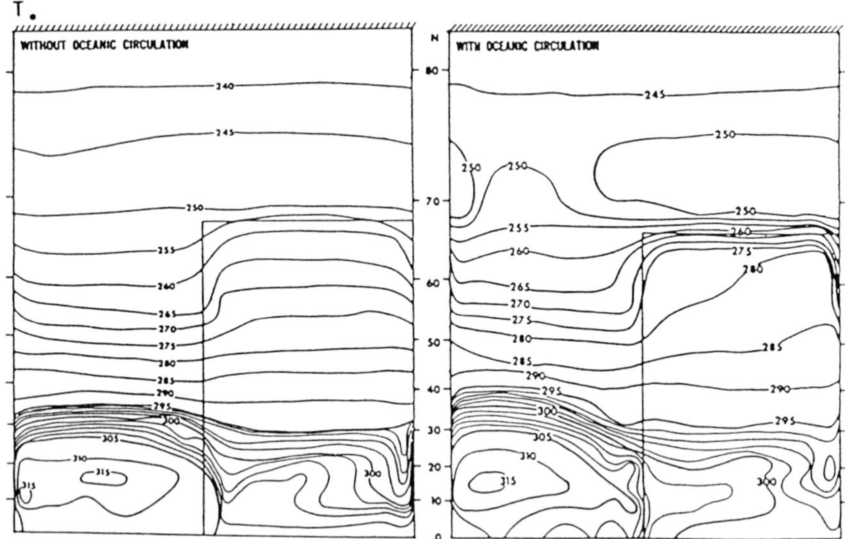

FIG. 12-13. Results from the combined numerical ocean–atmosphere model of Manabe and Bryan. This model used simplified ocean-continent geometry with the small rectangle representing ocean. In the panel with oceanic circulation (right), upwelling produces a weak temperature minimum near the equator; in the northern part of the ocean, the ocean currents produce a southwest–northeast trend of isotherms and generally warmer conditions (Manabe and Bryan 1969).

between the atmosphere and the ocean in producing seasonal to decadal-scale variability (see, e.g., Bjerknes 1959; Namias 1959). Subsequent work has documented some of these atmosphere–ocean interactions in considerable detail (Rasmusson and Carpenter 1982). The work of Namias and Bjerknes helped to make it abundantly clear that numerical models of climate must include treatments of the coupled behavior of the atmosphere and ocean. The first successful climate simulations with a combined ocean–atmosphere model (Manabe and Bryan 1969) illustrated the role of the oceanic upwelling and ocean currents in producing important features of the observed climate (Fig. 12-13).

These advances in modeling of the atmospheric and oceanic circulation helped to spark new efforts to improve observations starting with the Global Weather Experiment (1978–1979) and continuing to include such detailed studies as the Earth's Radiation Budget Experiment and the World Ocean Circulation Experiment. There have also been detailed observational studies of past climates, such as cold glacials (CLIMAP 1976, 1981) and warm interglacials (COHMAP 1988) and modeling studies of glacial climates (Gates 1976), interglacial climates (Kutzbach and Otto-Bliesner 1982; COHMAP 1988), and the climates of much earlier periods of earth history (Barron 1985).

The development of quantitative climate system models is continuing (Washington and Parkinson 1986; Trenberth 1992). Coupled ocean–atmosphere models have simulated dramatically different modes of the oceanic thermohaline circulation (Manabe and Stouffer 1988; Delworth et al. 1993) and certain features of El Niño–Southern Oscillation (Meehl 1990; Philander et al. 1992). Simulated changes in the terrestrial biosphere have suggested important biosphere–climate feedbacks (Dickinson 1992; Bonan et al. 1992; Foley et al. 1994). Simulations of past climates and comparisons of these simulations with observations are providing estimates of the accuracy of climate models (Barron 1985; Kutzbach 1985, 1992). And, almost exactly 100 years after the pioneering work of Arrhenius, climate models are still being called upon to try to provide accurate estimates of the sensitivity of climate to increasing levels of atmospheric CO_2 (IPCC 1990, 1996). However, it is now recognized that the problem of estimating future climate scenarios must be considered in a much broader context involving not only the question of increasing CO_2 but also the climatic effects of other greenhouse gases, of sulfate aerosols, of water vapor and clouds, and of ocean–atmosphere–biosphere interactions.

Looking toward the future

The advances in our understanding of climate in the past 75 years have been extraordinary. W.J. Humphreys, professor of meteorological physics, U.S. Weather Bureau, and vice president of the AMS during its founding year, 1920, published in that same year his text *Physics of the Air*. Examining his text, including the descriptions of his own work on volcanoes and climate, and comparing what was known about climate 75 years ago to what is known now, we can only marvel at the remarkable progress that has been made and look forward with anticipation to further advances.

Acknowledgments. I acknowledge the support of the National Science Foundation, Climate Dynamics Division, during my own years of climate studies.

REFERENCES

Abbott, C.G., and F.E. Fowle Jr., 1908: Variations of solar radiation and their effects on the temperature of the earth. *Annals of the Astrophysical Observatory of the Smithsonian Institution,* Vol. 2, Smithsonian Institution, Washington, DC.

Adem, J., 1963: Preliminary computations on the maintenance and prediction of seasonal temperatures in the troposphere. *Mon. Wea. Rev.,* **91,** 375–386.

Arrhenius, S., 1896: On the influence of carbonic acid in the air upon the temperature of the ground. *The London, Edinburgh, and Dublin Philos. Mag. and J. Sci.,* Fifth Series, **41,** 237–276.

Barron, E.J., 1985: Climate models: Applications for the pre-pleistocene. *Paleoclimate Analysis and Modeling,* A.D. Hecht, Ed., John Wiley & Sons, 341–396.

Bjerknes, J., 1959: The sea in motion. *The Atmosphere and the Sea in Motion: Scientific Contributions to the Rossby Memorial Volume,* B. Bolin, Ed., Rockefeller Institute Press, 65–73.

Bjerknes, V., J. Bjerknes, H. Solberg, and T. Bergeron, 1933: *Physikalische Hydrodynamik, Mit Anwendung auf die Dynamische Meteorologie.* Springer, 797 pp.

Bolin, B., 1950: On the influence of the earth's orography on the general character of the westerlies. *Tellus,* **2,** 184–195.

Bonan, G.B., D. Pollard, and S.L. Thompson, 1992: Effects of boreal forest vegetation on global climate. *Nature,* **359,** 716–718.

Budyko, M.I., 1958: *The Heat Balance of the Earth's Surface.* U.S. Dept. of Commerce, Weather Bureau, 255 pp. (Translated by N.A. Stepanova from *Teplovoi Balans Zemnoi Poverkhnosti,* Gidrometeorologicheskoe izdatel'stvo, Leningrad, 1956.)

______ , 1969: The effect of solar radiation variations on the climate of the earth. *Tellus,* **21,** 611–619.

Charney, J.G., and A. Eliassen, 1949: A numerical method for predicting the perturbations of the middle latitude westerlies. *Tellus,* **1,** 38–54.

______ , R. Fjørtoft, and J. von Neumann, 1950: Numerical integration of the barotropic vorticity equation. *Tellus,* **2,** 237–254.

CLIMAP, 1976: The surface of the ice age earth. *Science,* **191,** 1131–1136.

______ , 1981: Seasonal reconstructions of the earth's surface at the last glacial maximum. Geological Society of America Map Chart Series, MC-36, 18 pp.

COHMAP, 1988: Climatic changes of the last 18 000 years: Observations and model simulations. *Science,* **241,** 1043–1052.

Defant, A., 1921: Die Zirkulation der Atmosphäre in den gemässigten Breiten der Erde. *Geograf. Ann.,* **3,** 209–266.

Delworth, T., S., Manabe, and R.J. Stouffer, 1993: Interdecadal variations of the thermohaline circulation in a coupled ocean–atmosphere model. *J. Climate,* **6,** 1993–2011.

Dickinson, R.E., 1992: Land surface processes. *Climate System Modeling,* K. Trenberth, Ed., Cambridge University Press, 149–171.

Ferrel, W., 1859–1860: The motions of fluids and solids relative to the earth's surface. *Math. Monthly,* **1,** 140–147, 210–216, 300–307, 366–372, 397–406, **2,** 89–97, 339–343, 374–390.

______ , 1884: Temperature of the atmosphere and earth's surface. Signal Service Professional Paper No. XIII, U. S. War Department, Government Printing Office, 69 pp.

Foley, J., J.E. Kutzbach, M.T. Coe, and S. Levis, 1994: Feedbacks between climate and boreal forests during the Holocene epoch. *Nature,* **371,** 52–54.

Fultz, D., 1951: Experimental analogies to atmospheric motions. *Compendium of Meteorology,* T.F. Malone, Ed., Amer. Meteor. Soc., 1235–1248.

Gates, W.L., 1976: Modeling the ice-age climate. *Science,* **191,** 1138–1144.

Halley, E., 1693: A discourse concerning the proportional heat of the sun in all latitudes, with the method of collecting the same. *Philos. Trans. Roy. Soc. London,* **7,** 878–885.

Hann, J., 1915: *Lehrbuch der Meteorologie.* C.H. Tauchnitz, 847 pp.

Hide, R., 1953: Some experiments on thermal convection in a rotating liquid. *Quart. J. Roy. Meteor. Soc.,* **79,** 161.

Houghton, H.G., 1954: On the annual heat balance of the Northern Hemisphere. *J. Meteor.,* **11,** 1–9.

Humphreys, W.J., 1920: *Physics of the Air.* The Franklin Institute, 663 pp.

Imbrie, J., and Coauthors 1992: On the structure and origin of major glaciation cycles: 1. Linear responses to Milankovitch forcing. *Paleoceanography,* **7,** 701–738.

______ , and Coauthors, 1993: On the structure and origin of major glaciation cycles: 2. The 100 000-year cycle. *Paleoceanography,* **8,** 699–735.

IPCC, 1990: *Climate Change: The IPCC [Intergovernmental Panel on Climate Change] Scientific Assessment.* J.T. Houghton, G.J. Jenkins, and J.J. Ephraums, Eds., Cambridge University Press, 365 pp.

———, 1996: *Climate Change: The IPCC [Intergovernmental Panel on Climate Change] Second Scientific Assessment.* J.T. Houghton and L.G.M. Filho, Eds., Cambridge University Press, in press.

Jeffreys, H., 1926: On the dynamics of geostrophic winds. *Quart. J. Roy. Meteor. Soc.,* **52,** 85–104.

Keeling, C.D., 1960: The concentration and isotopic abundances of carbon dioxide in the atmosphere. *Tellus,* **12,** 200–203.

Kendrew, W.G., 1992: *The Climates of the Continents.* The Clarendon Press, 387 pp.

Köppen, W., and R. Geiger, Eds., 1936: *Handbuch der Klimatologie.* Band 1: Allgemeine Klimalehre, Verlag von Gebrüder Borntraeger, 1049 pp.

Kutzbach, G., 1979: *The Thermal Theory of Cyclones: A History of Meteorological Thought in the Nineteenth Century.* Amer. Meteor. Soc., 255 pp.

———, 1987: Concepts of monsoon physics in historical perspective: The Indian monsoon (seventeenth to early twentieth century). *Monsoons,* J.S. Fein and P.L. Stephens, Eds., John Wiley & Sons, 159–209.

Kutzbach, J.E., 1985: Modeling of Paleoclimates. *Advances in Geophysics,* S. Manabe, Ed., Academic Press, 159–196.

———, 1992: Modeling large climatic changes of the past. *Climate System Modeling,* K. Trenberth, Ed., Cambridge University Press, 669–688.

———, and B.L. Otto-Bliesner, 1982: The sensitivity of the African–Asian monsoonal climate to orbital parameter changes for 9000-yr BP in a low-resolution general circulation model. *J. Atmos. Sci.,* **39,** 1177–1188.

London, J., 1957: A study of the atmospheric heat balance. Dept. of Meteorology and Oceanography, New York University, Contract AF19(122)-165 (AFCRC-TR-57-287), 99 pp.

Lorenz, E.N., 1967: *The General Circulation of the Atmosphere.* World Meteorological Organization, 218.TP.115, 161 pp.

Lovelock, J., 1988: *The Ages of Gaia: A Biography of our Living Earth.* Norton, 252 pp.

Manabe, S., and K. Bryan, 1969: Climate calculations with a combined ocean–atmosphere model. *J. Atmos. Sci.,* **26,** 786–789.

———, and R.J. Stouffer, 1988: Two stable equilibria of a coupled ocean–atmosphere model. *J. Climate,* **1,** 841–866.

Maury, M.F., 1855: *The Physical Geography of the Sea.* Harper, 287 pp.

Meehl, G.A., 1990: Seasonal cycle forcing of El Niño Southern Oscillation in a global, coupled ocean–atmosphere GCM. *J. Climate,* **3,** 72–98.

Milankovitch, M., 1920: *Théorie mathématique des phénomenes thermiques produits par la radiation solaire.* Gauthier-Villars, 338 pp.

Mintz, Y., 1965: Very long-term global integration of the primitive equations of atmospheric motion. WMO Tech. Note 66, 141–167.

———, 1968: Very long-term global integration of the primitive equations of atmospheric motion: An experiment in climate simulation. *Causes of Climatic Change, Meteor. Monogr.,* No. 30, 20–36.

Mitchell, J.M., Jr., 1965: Theoretical paleoclimatology. *The Quaternary of the United States,* H.E. Wright Jr. and D.G. Frey, Ed., Princeton University Press, 881–901.

Munk, W.H., 1950: On the wind-driven ocean circulation. *J. Meteor.,* **7,** 79–93.

Namias, J., 1959: Recent seasonal interactions between North Pacific waters and the overlying atmospheric circulation. *J. Geophys. Res.,* **64,** 631–646.

Niiler, P.P., 1992: The ocean circulation. *Climate System Modeling,* K. Trenberth, Ed., Cambridge University Press, 117–148.

North, G.R., R.F. Cahalan, and J.A. Coakley, 1981: Energy balance climate models. *Rev. Geophys. Space Phys.,* **19,** 91–121.

Peixoto, J.P., and A.H. Oort, 1992: *Physics of Climate.* American Institute of Physics, 520 pp.

Pfeffer, R.L., 1960: *Dynamics of Climate.* Pergamon Press, 137 pp.

Philander, S.G.H., R. Pacanowski, N.-C. Lau, and M.J. Nath, 1992: A simulation of the Southern Oscillation with a global atmospheric GCM coupled to a high-resolution, tropical Pacific Ocean GCM. *J. Climate,* **5,** 308–329.

Phillips, N.A., 1956: The general circulation of the atmosphere: A numerical experiment. *Quart. J. Roy. Meteor. Soc.,* **82,** 123–164.

Rasmusson, E.M., and T.H. Carpenter, 1982: Variations in tropical sea surface temperature and surface wind fields associated with the Southern Oscillation/El Niño. *Mon. Wea. Rev.,* **110,** 354–384.

Richardson, L.F., 1992: *Weather Prediction by Numerical Process.* Cambridge University Press, 236 pp.

Rossby, C.-G., and Collaborators, 1939: Relation between variations in the intensity of the zonal circulation of the atmosphere and the displacement of the semi-permanent centers of action. *J. Mar. Res.,* **2,** 38–55.

Saltzman, B., 1978: A survey of statistical-dynamic models of the terrestrial climate. *Advances in Geophysics,* Vol. 20, Academic Press, 184–304.

_____ , 1985: Paleoclimate modeling. *Paleoclimate Analysis and Modeling,* A.D. Hecht, Ed., John Wiley & Sons, 341–396.

Schneider, S.H., 1992: Introduction to climate modeling. *Climate System Modeling,* K. Trenberth, Ed., Cambridge University Press, 3–26.

_____ , and R.E. Dickinson, 1974: Climate modeling. *Rev. Geophys. Space Phys.,* **12,** 447–493.

Sellers, W.D., 1969: A global climatic model based on the energy balance of the earth–atmosphere system. *J. Appl. Meteor.,* **8,** 392–400.

Shaw, N., Sir, 1926–31: *Manual of Meteorology.* Vols. I–IV. Cambridge University Press, 1592 pp.

Smagorinsky, J., 1953: The dynamical influence of large-scale heat sources and sinks on the quasi-stationary mean motions of the atmosphere. *Quart. J. Roy. Meteor. Soc.,* **79,** 342–366.

_____ , 1963: General circulation experiments with the primitive equations: I. The basic experiment. *Mon. Wea. Rev.,* **91,** 99–164.

_____ , 1972: The general circulation of the atmosphere. *Meteorological Challenges: A History.* D.P. McIntyre, Ed., Information Canada, 3–41.

_____ , 1983: The beginnings of numerical weather prediction and general circulation modeling: Early recollections. *Advances in Geophysics,* Vol. 25, Academic Press, 3–37.

_____ , S. Manabe, and J.L. Holloway Jr., 1965: Numerical results from a nine-level general circulation model of the atmosphere. *Mon. Wea. Rev.,* **93,** 727–768.

Spitaler, R., 1921: *Das Klima des Eiszeitalters.* R. Spitaler, 138 pp.

Starr, V.P., and R.M. White, 1951: A hemispherical study of the atmospheric angular-momentum balance. *Quart. J. Roy. Meteor. Soc.,* **77,** 215–225.

______ , and ______ , 1955: Direct measurement of the hemispheric poleward flux of water vapor. *J. Mar. Res.,* **14,** 217–225.

Stommel, H., 1948: The westward intensification of wind-driven currents. *Trans. Amer. Geophys. Union,* **29,** 202–206.

Sverdrup, H.U., and W.H. Munk, 1947: Wind, sea and swell: Theory of relations for forecasting. U.S. Hydrologic Office Publ. 601, 44 pp.

Trenberth, K.E., Ed., 1992: *Climate System Modeling.* Cambridge University Press, 788 pp.

von Bezold, W., 1892: Der Wärmeaustausch an der Erdoberfläche und in der Atmosphäre. *Sitzungsber, K. Preuss. Akad. Wiss.,* 1139–1178.

Vonder Haar, T.H., and V.E. Suomi, 1969: Satellite observations of the earth's radiation budget. *Science,* **163,** 667–668.

von Humboldt, A., 1817: Résearches sur les causes des inflexions de lignes isotherms. *Memoirs de Physique d.l. Soc. d'Arcueil (Paris),* **3,** 462–602.

von Neumann, J., 1960: Some remarks on the problem of forecasting climatic fluctuations. *Dynamics of Climate,* R.L. Pfeffer, Ed., Pergamon, 9–11.

Washington, W.M., and C.L. Parkinson, 1986: *An Introduction to Three-Dimensional Climate Modeling.* University Science Books and Oxford University Press, 422 pp.

CHAPTER 13

Applied Climatology: A Glorious Past, an Uncertain Future

STANLEY A. CHANGNON

The elusive definition of applied climatology

Uncertainty over the role of applied climatologists and the hard-to-track intellectual extent of the field have clouded understanding of the importance of applied climatology. These factors also make defining the field's history complex and challenging. Assessment of the past and the potential of applied climatology requires unraveling the definition of applied climatology. In fact, this investigation of the history of applied climatology has been a study of the evolving definition of an ever-changing field.

One could postulate that the topics covered in conferences on applied climatology define the dimensions of the field. After assessing the materials presented at these conferences, which began in 1980, I find the field to be even larger than what "applied climatologists" outline as their domain. These climatologists are but one part, albeit a significant one, of the large interdisciplinary realm of applied climatology.

The definition of applied climatology remains elusive for many reasons. Activities housed under the broad umbrella of "applied climatology" are spread among many disciplines and also have interfaces with many parts of the atmospheric sciences, forming a smorgasbord of interactive functions.

Over 40 years ago, Landsberg and Jacobs (1951) defined applied climatology as "the scientific analysis of climatic data in the light of a useful application for an operational purpose." Operational was interpreted broadly "as any useful endeavor, such as industrial, manufacturing, agricultural, or technological pursuits." Thus, they viewed applied climatology as a general term, which is now widely used and often misunderstood. According to the *Glossary of Meteorology* (Huschke 1959), applied climatology "includes industrial climatology, agricultural climatology, aeronautical climatology, bioclimatology, and *probably* others" [emphasis added]. Indeed, applied climatology has broadened its domain since 1959 (when the glossary was prepared) to embrace significant parts of urban climatology, hydroclimatology, statistical climatology, and instrumentation.

My interpretation is that the field of applied climatology "describes, defines, interprets, and/or explains" the relationships between climate conditions and thousands of weather-sensitive activities. If viewed in a business sense, applied climatology would involve the marketing, sales, customer services, research support, and delivery arm of the atmospheric sciences and, specifically, climatology. Applied climatology does not include fundamental studies of the climate system but it does embrace causation as a necessary part of explaining climatic relationships to other phenomena. Thus, the field embraces "climatography" (descriptions of the climate for particular applications) and "climatology," the explanation for the climate. For example, an applied climatology study to describe how daily rainfall amounts affect a corn crop may include an explanation for those atmospheric factors that cause the variability of daily rainfalls during the growing season.

Conceptually, applied climatologists function as middlemen within a triangle that encompasses fundamental climatic research, climate data, and the users of climate information. Users include scientists in many disciplines and decision makers in government or business. Important feedback loops exist in the middleman role. For example, the relationship of climate and other phenomena such as water resources and agriculture creates a two-way interaction: climate affects the phenomenon and the phenomenon also affects the climate. Another avenue for feedback is between climate data users and applied climatologists: each may communicate their needs for better data or different information.

It is hoped that this description reveals that applied climatology delves deeply into the atmospheric sciences and particularly into the areas of dynamic climatology, instrumentation, and climate data. In turn, these areas also extend into all weather-sensitive sectors, including human health, ecosystems, water resources, food production, transportation, construction, business, recreation, and industry. Applied climatological research is at the heart of this "complex network." Research develops the informational products for the user community, studies the climate–sector relationships, develops statistical techniques to express climate information (including climatic indexes) more effectively, contributes to weather data collection efforts, and develops databases.

A review of the history of applied climatology illuminates the players in the field. Since about 1940, the field of applied climatology has included scientists focusing on

- assessing climate data (as needed for applications), including its type, quality, collection, archival, representativeness, and access;
- developing techniques, particularly of a statistical nature, to analyze and interpret climate data for applications;
- developing estimates and indexes of climate conditions for specific applications;
- analyzing relationships between climate conditions and other societal and environmental conditions; and
- interpreting user needs for climate information.

Many disciplines are part of applied climatology and include hydrology, agriculture, geography, engineering, architecture, ecology, medicine, commerce and business, recreation, and energy. All of these fields, which embrace all human activities in the United States today, require climatological information to function. How well they function often depends on just how good the information is and how well it is used. The group of professionals who are considered to be applied climatologists is a large one composed of many different kinds of people with widely varying duties. For example, the hydrologist who studies the relationship between climatic conditions and surface water flow is an applied climatologist by function but likely does not identify, careerwise, with applied climatology. Similarly, the load forecaster at a power company, who constantly uses climatic information to develop monthly, seasonal, and annual forecasts, practices applied climatology but is considered a power engineer.

My vision of the field is that since about 1940 it has consisted of interactive functions in three concentric rings. First is the inner core of applied climatology, which focuses on instruments (placement, type, and accuracy) and data (collection, transmission, quality assessment, archival, representativeness in space and time, and access). All of these functions are performed by meteorologists and applied climatologists.

The functions in the second ring relate to the interpretation and generation of climate information. Activities include statistical and physical analyses, the performance of specialized interpretive studies, and the generation of publications and computer-based information. These activities are performed by state climatologists and staffs at the regional climate centers, the National Climatic Data Center, the Climate Prediction Center, and many private firms.

The third ring consists of users of applied climatology and is the largest group: scientists, engineers, and businessmen in nonatmospheric disciplines who use climate data and information for a multitude of purposes. One factor making the field hard to identify has been these practitioners who generally lack extensive education and training about atmospheric sciences and do not consider themselves to be applied climatologists. Applications of climatic information include 1) the design of structures and activities, 2) the assessment of current and past conditions including extreme events, 3) the study of relationships between weather–climate conditions and those in other parts of the physical and social worlds, and 4) the operation of weather-sensitive systems that use climatic information in making management decisions.

A glorious past

Applied climatology has served as the foundation on which this nation's infrastructure has developed. Weather forecasting, relative to applied climatology, is a more recent development in the day-to-day operations of weather-sensitive systems. However, the basis for development of our nation has been in understanding how a crop develops, how to plan a drainage system, or how to build a bridge. This development illustrates the many achievements rooted in applications of climate data and information.

Early period (before 1940)

The beginning of weather observations in the United States early in the nineteenth century marked the initial efforts in applied climatology. The rationale behind these measurements was the desire to define the climate conditions of the nation. Climate conditions and human health were seen as closely linked. Thomas Jefferson was one of the nation's earliest people to apply climatic data. He installed equipment and used his measurements in farming decisions and for building construction (Martin 1952).

In the early years of our nation, as today, there was recognition that climate and weather had an enormous effect on human beings and their activities. It was further understood 200 years ago that data and information about climate conditions were an essential ingredient for successful survival and for human endeavors at any given location.

The expansion and organization of the nation's weather data collection system and science in general in the United States occurred during the nineteenth century. Early leaders in meteorology, such as James Espy and Elias Loomis, recognized regional differences in climate conditions across the nation and, further, that these differences had vastly different impacts on human endeavors. Another key scientist of that era, Joseph Henry, made fundamental studies in applied climatology: for example, in 1855–1856 he analyzed the impacts of climate on U.S. agriculture (Fleming 1990). During this century of enlightenment, James Henry Coffin charted the winds of the globe and William Ferrel explained the general circulation of the atmosphere, while Köppen (1885) and other European scientists worked to describe climates of different regions.

The evolution of climatology from the nineteenth century well into the twentieth century was largely the domain of geographers. Their focus was on describing climate conditions based on visual observations and on the study of early weather data. Geographers pioneered the development of climate classifications to explain global patterns. Some climates were tied to observable relationships between climate and plants, an endeavor that merged the findings of naturalists with emerging quantitative efforts of geographers. As late as 1940, most "climatologists" in the United States were physical geographers within the geography departments of colleges and universities. [See chapter 18 by Koelsch, this volume.]

Further understanding of the climate system and its interactions with the myriad parts of the physical world had to await the collection of sufficient data regarding climate and other segments of the physical world so that a reasonably definitive understanding of climatic effects and relationships could evolve. Certain scientists are noteworthy for their accomplishments in defining these relationships, and most of their works appear in the early decades of the twentieth century. For example, famed hydrologist Robert Horton made basic discoveries about the hydrologic cycle during the 1905–1930 period. He unraveled the complexities of the hydrology of the Great Lakes Basin in the

1920s, but only after there had been sufficient streamflow measurements to assess runoff (Horton 1927).

The drive toward the collection of climatic data to plan and to understand agricultural outcomes was one of the singular national goals behind the establishment of the U.S. Weather Bureau (USWB) as part of the U.S. Department of Agriculture (USDA) (Whitnah 1961). The establishment of agricultural experiment farms at state universities in the late nineteenth century began the data collection effort that was essential to define, in later years, the complex relationships between weather, crops, and animals. Interestingly, one of the pioneers in the sophisticated modeling of climate and crop yield relationships was Henry Wallace (1920), a man who would later become vice president of the United States. Thiessen (1912) was among the first civil engineers to define the emerging uses of climate data in engineering design, and we still use Thiessen polygons to define areas to which point climate data pertain.

Early efforts to develop climate classifications such as those of Köppen (1931) were based on identifying relationships between weather conditions and vegetation. Many geographers wrote definitive climatologies of various parts of the world, and several, including Ward and Brooks (1936), described in detail the climate of North America. These too were efforts in applied climatology. C.F. Brooks (1915) and C.E.P. Brooks (1925) contributed to definitions of climates and various atmospheric conditions. Visher (1943) and Trewartha (1960) were geographers who made contributions to understanding climate.

Others were studying climate relationships and defined the components of the climate. A pioneer in these endeavors was C.W. Thornthwaite, who modeled the hydrologic system to estimate evapotranspiration and soil moisture conditions to provide a better understanding of climatic conditions and their effects on both plants and agriculture (Thornthwaite 1937). Early key studies of climate influences on human behavior and health included those of C.F. Brooks (1925) and Dorno (1933). As an empirical index for monitoring fuel needs, the heating industry developed heating degree-days in the 1930s. The 1920s and 1930s also saw the emergence of mathematicians and statisticians who focused on the use of statistics in climate data analysis (Conrad 1936). The USDA published *Climate and Man* in 1941, which included many good summaries reflecting applied climatological research.

In the pre-1940 period, many of the fundamental relationships between climate conditions and other physical systems had been sufficiently defined to allow the design of safe structures, the selection of appropriate crops and crop varieties, and economically sound management of water systems in the different climatic regimes of the United States. The end of the "early period" in the 1940s was marked by the emergence of scientists labeled as applied climatologists, largely in response to the scientific demands of World War II and the technological advances in data handling and computational capabilities at that time.

Later period (after 1940)

The global-scale logistical demands for climate data and information generated by World War II were an immense challenge. The responses and generation of climate information used by the military brought new emphasis to applied climatology (Jacobs 1947). One example was the thunderstorm and severe storm studies in the 1940s, creating our first good climatology of thunderstorms and hail (USWB 1947). Scientists such as S.D. Flora (1953) also contributed significantly to the climatology of severe storms. Gilbert White and his geographer colleagues at the University of Chicago led the way in studies of weather hazards (White and Haas 1975).

In their monumental treatise on applied climatology, H.E. Landsberg and W.C. Jacobs (1951) referenced 87 papers and books about applied climatology. Interestingly, 65 of these were published in the 1940s, 17 in the 1930s, and 5 in the 1920s, revealing the sudden burst of activity in the field in the 1940s. Most of these 87 publications dealt with agriculture, highways, heating and cooling, human health, and the environment.

This postwar emergence of applied climatology, as a definable field of endeavor, led to more sophisticated climatological products and information. Helmut Landsberg, a pioneer in clarifying the many aspects of climate, was also a pioneer in promoting applied climatology among scientific elites. A revolutionary breakthrough came with the advent of computers during World War II and the use of punch cards as a means of digitizing weather data. Digitization of weather data began in earnest during the 1940s and resulted in the National Weather Records Center in Asheville, North Carolina (now the National Climatic Data Center). In turn, the National Weather Records Center staff began generating a large volume of useful documents, providing a myriad of climate information never before available to the user community.

These new computer facilities and digitized databases allowed major improvements in climate-effect modeling, most notably the development of hydrologic models (Linsley et al. 1958). Other key hydroclimatologists of this era were Walter Langbein (1955), Abel Wolman (1962), and Ven Te Chow (1952). In the study of flood and heavy rain events, major contributors emerged, including Gilbert White (1964), David Hershfield (1961), and Floyd Huff (1986). Wayne Palmer (1965) made a singular contribution in developing a national drought index. His works had been preceded by the fundamental contributions of H.L. Penman (1940), C.H.M. van Bavel (1960), C.W. Thornthwaite (1948), and T.R. Tannehill (1947). L.M. Thompson (1964) reawakened interest in the use of sophisticated climate–crop models to understand yield and production variability. Others made significant contributions to the emerging concept of "impacts of climate" (Maunder 1970; McQuigg 1974; Kates 1985). Major postwar improvements came as a result of the development of an array of statistical techniques and their use in applied climate studies by Conrad and Pollack (1944), Court (1949), and H.C.S. Thom in the 1950s and 1960s.

Research and operational models were generated in various disciplines (primarily in hydrology, agriculture, and ecology) that had weather-sensitive elements. Both data and information became much easier to obtain, and the "golden age" of applied climatology was upon us by the 1950s. Applied climatology had moved to a new plane of recognition and value, particularly in the user community. Atmospheric scientists within the field of applied climatology turned their attention to improved weather-sensing instruments, to better means of assessing data quality and data archival, and to the generation of ever more sophisticated, user-friendly climate informational products. There were notable analyses of instruments and their effects on climate data by Mitchell (1953). A new focus on studies of urban climatology began with contributions from Landsberg (1956), Changnon et al. (1981), and Oke (1984).

Recent developments

The 1980s and 1990s have seen applied climatologists address a new issue: global climate change due to enhancement of the greenhouse effect by ever-increasing trace gases in the atmosphere. Numerous studies have addressed the effects of various changes in climate conditions on agriculture, water resources, ecosystems, and most other facets of human activity.

Institutional growth also occurred. The Weather Bureau's Climate Office under Helmut Landsberg's astute leadership developed a state climatologist program in the 1950s to facilitate information transfer and to enhance understanding of climate among user communities, particularly those in academia. When this federally supported program was terminated in 1972, most of the states undertook the responsibility of maintaining a state climatologist to serve state needs for applied climatological expertise.

The global population grew, and sensitivity increased over impacts of severe climate events, which led to the National Climate Program Act in 1978. Major sections of the act called for efforts in applied climatology and enhanced climate services. It also recognized a new problem: climate data and information being generated and made available often were not used effectively. Education about climate in the other disciplines had apparently fallen behind the advances in climatological knowledge, including methods of data analysis.

The dilemma over the lack of climate services led to the establishment of regional climate centers during the 1980s (Changnon et al. 1990). The aim of these centers was to produce new information and enhance its use by the user communities, particularly through near-real-time availability, access, and applied research.

The last 15 years have seen new institutions, advances in computers and data handling, improved statistical techniques to interpret data, and inexpensive and easy access to near-real-time climate products. Yet usage of this information in key sectors such as agriculture and energy production lags (Changnon et al. 1988, 1995).

The epilogue: An uncertain future

There have been many notable achievements in the field of applied climatology, as reflected in the successful development of our nation. Central national issues can and should embrace applied climatology 1) to improve our environment, 2) to manage our resources more effectively, and 3) to enhance the nation's economy through applications of climate data and information. The bad news is that applied climatological information is not being used effectively to address these issues and our database is in danger. The good news is that advanced climatological products (data and information) are now easily accessible and research has us on the verge of improved long-range climate forecasts.

One of the major concerns reflected in many assessments of the use of climate information in agriculture, engineering, and energy generation is that the improved informational products available are not being used effectively. Examination of research relating to climate in other fields suggests that scientists, engineers, and economists are not being properly trained in the use of climate data and information. Further, few in posteducational training programs are being informed of new and vital climate products. One result is that we often lack the complex relationship models needed to assess risk (Riebsame et al. 1991).

Among the major concerns of applied climatologists is the evolution of data collection. For example, there are great concerns over the erosion of the 100-year-old cooperative network and its data, the backbone of many applied climatological studies. Of great concern is the advent of new weather-measuring systems that do not have adequate calibration against older instruments or the accuracy levels required by many users. Evidence also exists that many researchers and users of climate data have insufficient knowledge of data quality problems and also lack awareness of new analytical techniques for climate applications.

Another concern relates to the fact that assessments of climate forecast users reveal few are employing climate forecasts, largely because they do not understand the forecasts' dimensions or how to integrate the forecasts into decision making (Changnon 1992; Changnon et al. 1995). These studies further reveal that those who develop climate data and information for the user community often do not understand the users and their needs. The net result is a decided gap between information and its application. In summary, these various concerns include questions about data quality, education, and training in the use of climate information for research and decision making; the service–extension arena; and the transmission of useful products.

The future holds great promise for applied climatology: applications of historical climate data will continue to be the heart of the field's utility. Furthermore, two recent changes will alter the way climate applications are achieved. The first change relates to improved communications and computer resources, which now allows the monitoring of climate anomalies in near real time (Rodenhuis 1994). The second change is the emergence of improved understanding of the climate system leading to improved climate forecasts. This is partly the result of research driven by concerns over global climate change and is providing new insights into how climate functions. The enormous value of information about conditions in the months, seasons, and years ahead means that even meager improvements in longer-term forecasts can vastly improve management decisions.

The further development of real-time climate information systems being pioneered at the regional climate centers will allow users to access a wide variety of information inexpensively (Kunkel et al. 1990). Recent increases in inexpensive high-speed data transmission allow near-real-time assessments of today's conditions in light of historical conditions, either in space or time. Operational models in user-accessible systems now allow real-time assessment of potential outcomes in crop yields and streamflows. Such advances will lead to major breakthroughs in decision making by managers of natural resources systems or economic operations.

Ironically, the field has developed an identity crisis because of the diversity of its inherent multidisciplinary nature and the persons involved. A key issue in recognizing the importance of applied climatology is the understanding of the breadth of activities it encompasses, from the data core to the final applications: Where, who, and which components belong to applied climatology within a "discipline oriented" world? The broad user community is an integral part of the field of applied climatology.

Within university curricula and in the funding arena, the field of applied climatology has received too little attention. This is reflected in the lack of adequate climate training of many scientists in other disciplines, with the consequence being poor use of climate products throughout most weather-sensitive sectors. Nationally, we are failing to increase understanding among users. Without greater education, outreach, and delivery of climate information, the finest discoveries about climate will go unused or misunderstood.

Atmospheric scientists can perform three fundamentally different activities in the service of society. Underlying all of these is the analysis of atmospheric conditions to derive fundamental understanding, which can be used to 1) predict the weather, 2) modify the weather and possibly the climate for humanity's benefit, and 3) measure, define, and describe the climate. The latter activity is the purview of applied climatology, which along with weather forecasting and severe weather warnings, represents the major contributions of the atmospheric sciences to society. Applied climatology is the oldest activity of the atmospheric sciences in service to society. Use of climatological data and information has been central to the successful development of our nation and will remain critical in our nation's future.

Acknowledgments. I thank the Applied Climate Committee for encouraging me to perform this historical analysis. My 45-year career in applied climatology has benefited enormously from the influence of Helmut E. Landsberg, J. Murray Mitchell, William C. Ackermann, E. Ray Fosse, Floyd A. Huff, and many others. They taught invaluable lessons about the field, how to analyze and interpret data, and how to interact with users. I appreciate the help of Eva Kingston who edited the manuscript and Jean Dennison who typed the manuscript.

REFERENCES

Brooks, C.E.P., 1925: The distribution of thunderstorms over the globe. *Geophys. Mem.,* **3,** 36–47.

Brooks, C.F., 1915: The snowfall of the eastern United States. *Mon. Wea. Rev.,* **43,** 9–15.

———, 1925: The cooling of man under various weather conditions. *Mon. Wea. Rev.,* **53,** 28–33.

Changnon, S.A., 1992: Contents of climate predictions desired by agricultural decision makers. *J. Appl. Meteor.,* **31,** 1488–1491.

———, R.G. Semonin, A.H. Auer, R.R. Braham, and J. Hales, 1981: *METROMEX: A Review and Summary. Meteor. Monogr.,* No. 40, Amer. Meteor. Soc., 291 pp.

———, S. Sonka, and S. Hofing, 1988: Assessing climate information use in agribusiness. Part 1: Actual and potential use and impediments to use. *J. Climate,* **1,** 757–765.

———, P.J. Lamb, and K. Hubbard, 1990: Regional climate centers: New institutions for climate services and climate impact records. *Bull. Amer. Meteor. Soc.,* **71,** 527–537.

———, J. Changnon, and D. Changnon, 1995: Assessment of uses of climate forecasts in the utility industry in the central United States. *Bull. Amer. Meteor. Soc.,* **76,** 711–720.

Chow, V.T., 1953: Frequency analysis of hydrologic data with special application to rainfall intensities. Bulletin 444, University of Illinois Engineering Experiment Station, Urbana, IL 83 pp.

Conrad, V., 1936: Die klimatologischen Elemente und ihre Obhangigket von terrestreschen Einflussen. *Handbuchder Klimatologie,* Vol. 1, W. Köppen and R. Geiger, Eds., Berlin University, 1–48.

_______ , and L.W. Pollack, 1944: *Methods in Climatology.* Harvard University Press, 459 pp.

Court, A., 1949: Separating frequency distributions into two normal components. *Science,* **110,** 500–501.

Dorno, C., 1933: Physiological meteorology. *Bull. Amer. Meteor. Soc.,* **14,** 69–70.

Fleming, J.R., 1990: *Meteorology in America, 1800–1870.* The Johns Hopkins University Press, 264 pp.

Flora, S.D., 1953: *Tornadoes of the United States.* University of Oklahoma, 201 pp.

Hershfield, D.M., 1961: *Rainfall Frequency Atlas of the United States.* U.S. Weather Bureau, 61 pp.

Horton, R.E., 1927: *On the Hydrology of the Great Lakes.* Engineering Board of Review, Sanitary District of Chicago, 341 pp.

Huff, F.A., 1986: Urban hydrometeorology review. *Bull. Amer. Meteor. Soc.,* **67,** 703–711.

Huschke, R.E., Ed., 1959: *Glossary of Meteorology.* Amer. Meteor. Soc., 638 pp.

Jacobs, W.C., 1947: *Wartime Developments in Applied Climatology. Meteor. Monogr.,* No. 1, Amer. Meteor. Soc., 52 pp.

Kates, R.W., 1985: *Climate Impacts Assessment: Studies of the Interaction of Climate and Society.* John Wiley & Sons, 403 pp.

Köppen, W., 1885: Zur Charakteristik der Regen in NW Europa und Nordamevika. *Meteor. Z.,* 10–24.

_______ , 1931: *Grundriss der Klimakunde.* Walter de Gruyter & Co.

Kunkel, K.E., S.A. Changnon, C.G. Lonnquist, and J.R. Angel, 1990: A real-time climate information system for the midwestern United States. *Bull. Amer. Meteor. Soc.,* **71,** 601–609.

Landsberg, H.E., 1956: The climate of towns. *Man's Role in Changing the Face of the Earth,* University of Chicago Press, 584–606.

_______ , and W.C. Jacobs, 1951: Applied climatology. *Compendium of Meteorology,* T.F. Malone, Ed., Amer. Meteor. Soc., 976–992.

Langbein, W.B., 1955: *Floods.* Princeton University Press, 469 pp.

Linsley, R.K., M.A. Kohler, and J. Paulhus, 1958: *Hydrology for Engineers.* McGraw-Hill, 340 pp.

Martin, E.T., 1952: *Thomas Jefferson: Scientist.* H. Schuman, 89 pp.

Maunder, W.J. 1970: *The Value of Weather.* Methuen & Co., 388 pp.

McQuigg, J.D., 1974: The use of meteorological information in economic development. Applications of meteorology to economic and social development. World Meteorological Organization Tech. Note 132, 7–59.

Mitchell, J.M., 1953: On the causes of instrumentally observed secular temperature trends. *J. Meteor.,* **10,** 244–261.

Oke, T.R., 1984: Methods in urban climatology. *Appl. Climatol.,* **14,** 19–29.

Palmer, W.C., 1965: Meteorological drought. U.S. Weather Bureau Research Paper 45, 61 pp.

Penman, H.L., 1940: Meteorological and soil factors affecting evaporation from soil. *Quart. J. Roy. Meteor. Soc.,* **66,** 401–410.

Riebsame, W.E., S.A. Changnon, and T.R. Karl, 1991: *Drought and Natural Resources Management in the United States.* Westview Press, 174 pp.

Rodenhuis, D.R., 1994: *Climate Services of the Future.* Climate Analysis Center, 8 pp.

Tannehill, T.R., 1947: *Droughts, Its Causes and Effects.* Princeton University Press, 412 pp.

Thiessen, A.H., 1912: Using weather data in engineering problems. *Mon. Wea. Rev.,* **40,** 1565–1567.

Thompson, L.M., 1964: Our recent high yields—How much due to weather? *Research on Water,* Soil Science Society of America, 74–84.

Thornthwaite, C.W., 1937: The hydrologic cycle re-examined. *Soil Conserv.,* **3,** 85–91.

______ , 1948: An approach toward a national classification of climate. *Geophys. Rev.,* **38,** 55–94.

Trewartha, G.T., 1960: *The Earth's Problem Climates.* University of Wisconsin Press, 334 pp.

U.S. Department of Agriculture, 1941: *Climate and Man.* 1789 pp.

USWB: Thunderstorm rainfall. Hydromet. Rep. 5, U.S. Weather Bureau, 183 pp.

van Bavel, C.H.M., 1960: Use of climate data in guiding water management on farms. *Water Agric.,* **62,** 89–100.

Visher, S.S., 1943: The season's arrivals and lengths. *Ann. Amer. Assoc. Geograph.,* **33,** 129–134.

Wallace, H.A., 1920: Mathematical inquiry into the effect of weather on corn yield in the eight corn belt states. *Mon. Wea. Rev.,* **48,** 439–446.

Ward, R., C.F. Brooks, and A. J. Conner, 1936: The climates of North America. *Handbuch der Klimatologie,* Vol. 2, W. Köppen and R. Geiger, Eds., Berlin University, 264–289.

White, G.F., 1964: *Choice of Adjustment to Floods.* University of Chicago Press, 108 pp.

______ , and J.E. Haas, 1975: *Assessment of Research on Natural Hazards.* MIT Press, 487 pp.

Whitnah, D., 1961: *A History of the United States Weather Bureau.* University of Illinois Press, 288 pp.

Wolman, A., 1962: Water resources. National Academy of Sciences, Publ. 1000-B, 69 pp.

CHAPTER 14

Hydrology in the Twentieth Century

EDWIN T. ENGMAN

Introduction

Hydrology's history is rooted in early civilizations when humanity began to manage water in their transformation from a nomadic existence to an agrarian society. The development of civilization as we know it has been directly related to humanity's success in managing the rivers and developing ground water. However, hydrology did not develop as a scientific discipline until measurements, theories, and design equations were developed.

This chapter traces the development of hydrology from ancient attempts to harness nature to today's current research trends. It discusses its transformation from an engineering discipline to a newly defined geoscience. Last, this chapter addresses some of the scientific issues facing hydrologists; it discusses some of our shortcomings, including the inability to model processes and the need for more and better data.

History of hydrology

There are several histories of hydrology (Chow 1964; Biswas 1972; Nace 1974; NRC 1991; Maidment 1993) that have traced its development and the principal contributors to its development. Of these, the work by Biswas (1972) is by far the most thorough from a purely historical perspective. Other accounts have presented the history as a background to current hydrologic thinking and issues. One thing about the history of hydrology that is very clear is its parallel development with the history of civilization itself.

The transformation from a nomadic, hunter–gatherer society to a village, agrarian society was possible only when humanity began to master water supply problems during the new stone age. Great societies rose and fell with the successes of their ability to control water resources. The ancient city of Leptis Magna, east of modern Tripoli, which was once an important supply port for ancient Rome, is now barren ruins brought about by the decline and failure of its water supply systems (Leopold and Davis 1964). Mesa Verde in the United States and more recently Fatehpur Sikri in India are among other deserted cities that are thought to have declined at least in part due to a lack of or change in available water resources.

Not all early civilizations perished after a short period of prosperity. Some of the sites are still flourishing and are more or less successfully battling the water resource issues that their ancestors faced. Four geographically widely separated societies developed in approximately the same era and are seemingly independent of each other. These societies that developed along the Nile, the Tigris–Euphrates, the Indus, and the Hwang-Ho Rivers (Leopold and Davis 1964) all began to flourish at roughly the same time—2500 to 3000 years ago.

These civilizations did not simply exist on the banks of these great rivers; it was at these locations that humans first began to control the floods and drought flows and develop transportation systems and urban water supplies. The Egyptians were masters of canal building for irrigation and swamp drainage, and the Nile's annual flood cycle was so consistent that the 365-day calendar the Egyptians developed reflects its regularity. The Mesopotamians had to contend with the much more unpredictable and variable flooding of the Tigrus–Euphrates. They succeeded not only in controlling the floods but also in developing an urban water supply system and a canal-based transportation system. Although less archeological evidence exists from early life along the Indus River, there are indications that there was extensive irrigation and trade, as well as frequent flooding that apparently was never fully controlled. Even less is known about early life in the Hwang-Ho River Basin. People began to settle there as early as 5000 B.C., but their river was even more unpredictable than the other three, and they also had to cope with a much more severe climate. One artifact that has survived the centuries is much of a 1000-mile north–south canal that crossed the major east–west rivers and linked the political centers in the north of China to the agricultural regions of the south (Leopold and Davis 1964).

Most of these ancient engineering works were developed by trial and error. Ancient knowledge of the river systems consisted of a great deal of nonquantitative observations, folk tales, and an innate understanding of nature. Hydrology as a discipline cannot be thought to have

TABLE 14-1. Periodization of the history of hydrology, antiquity to 1964 (adapted from Chow 1964).

Period	Dates	Representative figures
Speculation	antiquity–1400	Plato Aristotle Homer
Observation	1400–1600	Bernard Palissy Leonardo da Vinci
Measurement	1600–1700	Pierre Perrault Edme Mariotte Edmond Halley
Experimentation	1700–1800	Henri Pitot Jean d'Alembert Antoine de Chezy Daniel Bernoulli
Modernization	1800–1900	Henry Darcy Jules Dupuit A. Thiem
Empiricism	1900–1930	U.S. Bureau of Reclamation Miami Ohio Conservancy District
Rationalization	1930–1950	L.K. Sherman E.J. Gumble C.V. Theis R.E. Horton A. Hazen
Theorization	1950–1964	J.R. Philip G.P. Kalinin P.I. Milyukov J. Martinec

begun until humans first started to develop theories and make measurements. Chow (1964) chose to periodize the history of hydrology by eight epochs, starting in antiquity and ending in his own time (Table 14-1).

This is in concert with most other descriptions of the increase in the quantity of knowledge—typically an exponential curve. In the period of speculation (antiquity–1400), Plato, Homer, and Aristotle recognized some form of hydrologic cycle. Although these early philosophers and scientists did not have a quantitative understanding of hydrology, a great number of practical hydraulic structures, such as aqueducts and irrigation systems, illustrated humankind's desire and need to control water resources as a prerequisite for civilization as we know it. During the Renaissance, in the period of observation (1400–

1600), Bernard Palissy and Leonardo da Vinci described a hydrologic cycle in which water moved from the oceans to rain on the land and then returned to the oceans. Quantitative hydrology probably began during the period of measurement (1600–1700), in which scientists such as Pierre Perrault, Edme Mariotte, and Edmond Halley made measurements of different hydrologic components. It is interesting to note that Perrault and Mariotte were physicists and Halley was an astronomer. The period of experimentation (1700–1800) gave us a long list of familiar names that includes Henri Daniel Pitot, Bernoulli, Jean d'Alembert, and Antoine de Chezy, along with discoveries that bear their names.

The nineteenth century, which Chow called the period of modernization (1800–1900), saw the establishment of the science of hydrology as we now know it. Many significant advances in hydrology were recorded in the nineteenth century, especially in the areas of ground water and surface water. The foray into quantitative hydrology was extended in the period of empiricism (1900–1930). A lack of good scientific understanding of hydrology led to a large number of empirical formulas developed to solve site-specific problems. It is interesting that most references associated with this epoch are institutions (Bureau of Reclamation, Miami Conservancy District, etc.) rather than people. This is not the case for the period of rationalization (1930–1950), during which we find the true gurus of modern hydrology. Advances like L.K. Sherman's unit hydrograph (1932), R.E. Horton's work on infiltration (1933), groundwater studies by C.V. Theis (1935), and flood frequency statistics by E.J. Gumble (1941) and A. Hazen (1930) are examples of contributions during this period. These pioneers published their research and developed procedures that are still very much in use today (Chow 1964).

Chow's last epoch is the period of theorization (1950–1964), during which hydrologists attempted to use theoretical approaches to solve hydrologic problems. Examples of advances included in the period are J.R. Philip's (1957) infiltration equation, G.P. Kalinin and P.I. Milyukov's (1957) flood routing, and J. Martinec's degree-day snowmelt work (1960). These studies and others during the period of theorization provided a great deal of insight into the complexities of hydrology but in some cases only limited advances for the practicing hydrologist. Were this list to be updated to the present, several additional developments could be added, including the computer, multidisciplinary research, systems analysis, environmental quality, modeling, stochastic hydrology, etc. I think we could also add remote sensing applications to this list. I will discuss the reasons why later in this chapter.

In the updated *Handbook of Hydrology*, Maidment (1993) extended Chow's historical eras to 1980. Although not identifying specific periods, Maidment selected representative books and papers or studies that have advanced hydrology from its engineering heritage to its more scientific posture. Tables 14-2 and 14-3 illustrate developments in understanding the hydrologic cycle and in advancing hydrologic technology since about 1960.

TABLE 14-2. Historical steps advancing the understanding of the hydrologic cycle 1960–1980 (adapted from Maidment 1993).

Budyko's heat balance of the earth	1963
Brooks–Corey model for soil water properties	1964
Shreve's randomized channel networks	1966
Hewlett and Hibbert's forest hydrology study	1967
Muskingum–Cunge routing	1969
Priestley–Taylor evaporation equation	1972
Mein and Larson infiltration under rainfall	1973
Dunne's concept of saturated overland flow	1975
Doorenbos and Pruitt evaporation report	1977
Soviet Union study of world water balance	1978
Freeze and Cherry's "Ground water"	1979
Geomorphic unit hydrograph	1979

The interesting thing to notice about post-1960 developments is that they generally widen the breadth of hydrology as a science. Since about 1970, both the breadth and depth of hydrology have increased rather dramatically. This growth has been the result of two general external forces: technology and the needs defined by other disciplines. Examples of new technology include, perhaps most importantly, the availability of ever larger and ever cheaper computers and their peripherals. Other innovations include new and better instrumentation; methods for data handling, recording, and transmission; and remote sensing. The needs defined by other disciplines, specifically those related to environmental quality research, started with water quality needs, in particular the needs defined by nonpoint source pollution, which required a good definition of storm runoff and subsurface flow paths. Today, many of the new directions in hydrologic research are driven by the atmospheric sciences and their need to understand and

TABLE 14-3. Historical steps advancing hydrologic technology 1960–1980 (adapted from Maidment 1993).

ASCE–WPCF design manual for storm sewers	1960
Preissman's scheme for numerical routing	1961
Hershfield's study of extreme rainfall (TP-40)	1962
Road Research Laboratory runoff method	1962
Palmer drought index	1965
HEC-1 developed	1966
Stanford watershed model IV	1966
Howe–Lineweaver water demand equations	1967
Instream flow requirements	1970
Matheron's regionalized variables	1971
Storm water management model (SWMM)	1971
Prickett–Lonnquist groundwater flow model	1972
Fread's dynamic routing model	1973
Groundwater quality modeling	1974
Three-dimensional groundwater flow model (Trescott)	1975
Box–Jenkins time series analysis (first edition)	1976
HYDRO-35 for rainfall estimation	1977
Journel and Huigbregt's mining geostatistics	1978
HMR-51 for probable maximum precipitation	1978
Landsat—snowmelt runoff (Martinec, Rango)	1979

parameterize the transport of energy and water across the land–atmosphere boundary at scales ranging from the mesoscale to the global circulation.

It is from this historical context that hydrology has changed from a purely problem-solving engineering discipline to a distinct geoscience that deals with the occurrence, distribution, circulation, and properties of water on the earth (NRC 1991).

Scientific hydrology

It is clear that the professional discipline of hydrology has deep historical roots brought about by humanity's need to control and manage water resources. The development of hydrology as a recognized discipline, however, has really taken place only in the twentieth century. It was first recognized as a distinct discipline by the International Union of Geodesy and Geophysics when it established the International Association of Hydrological Sciences (IAHS) in 1922. Eight years later the American Geophysical Union created the Section of Hydrology (Hornberger 1992). Previous to these actions, hydrology was, for the most part, simply a part of civil and agricultural engineering.

The recognition that hydrology and hydrologic issues encompassed more than just the engineering aspects of water management has been evolving during the last half of the twentieth century. In 1982, the National Academy of Sciences, responding to concerns expressed by hydrologists, created the Water Science and Technology Board (WTSB) as a unit of the newly reorganized National Research Council (NRC). This step was taken specifically to recognize the importance of water resources, to emphasize the complexity of water science and technology issues, and to provide a scientific foundation to guide hydrology into the future. In 1982 the WTSB convened the Committee on Opportunities in the Hydrological Sciences (COHS) composed of prominent hydrologists. This committee produced the *Opportunities in the Hydrological Sciences* (NRC 1991), which has become the definitive text that defines hydrological science as a distinct geoscience. In this book the COHS dealt with the need to define hydrological science as a distinct discipline:

> Hydrological science deals with the occurrence, distribution, circulation, and properties of water on the earth. It is clearly a multidisciplinary science, as water is important to and affected by physical, chemical, and biological processes within all the compartments of the earth system: the atmosphere, glaciers and ice sheets, solid earth, rivers, lakes and oceans. Because of this geophysical ubiquity, concern for issues of hydrological science has been distributed among the traditional geoscience disciplines. As a result, an infrastructure of hydrologic science (i.e., a distinct discipline with a clear identity and supporting educational programs, research grant programs, and research institutions) has not been developed, and a coherent understanding of water's role in planetary-scale behavior of the earth system is missing (NRC 1991).

The NRC book provides a comprehensive background in sections titled "Water and Life" and "The Earth's Hydrologic Cycle." It also offers a number of recommendations to establish hydrological sciences as a distinct and identifiable geoscience. Perhaps the most significant manner in which hydrology has been developing toward a true geoscience is in addressing ever-increasing time and space scales. In the last half of the twentieth century we have seen hydrology move from river basin management to continental water processes and global water balance issues. Some of the impetus for this has come from the needs of climate modelers. According to Chahine (1992), the hydrologic cycle within the framework of climate change encompasses much more than the classic surface hydrologic framework (precipitation, evaporation, runoff, etc.). In addition to these land surface processes, scientific hydrology must focus on the interactive processes of clouds and radiation, precipitation, oceans, and atmospheric moisture. This conception of the hydrologic cycle addresses not only the transport and storage of water in the global system but also the energy needed and released through the phase changes.

The NRC book also examines "Unsolved Problems" and discusses "Data Issues" and "Educational Issues." "Unsolved Problems" provides a comprehensive list of research issues that cover the full spectrum of the new hydrological sciences. Examples of unsolved problems include questions such as the following.

- How do we aggregate the dynamic behavior of hydrologic processes at various time and space scales in the presence of great natural heterogeneity?
- What are the feedback sensitivities of atmospheric dynamics and climate to changes in land surface hydrology, and how do these vary with season and geography?
- What can we learn about the equilibrium and stability of moisture states and vegetation patterns? Is "chaotic" behavior a possibility?
- What are the states and space–time variabilities of the global water reservoirs and their associated water fluxes?

If the list of unsolved problems defining scientific hydrology could be distilled into one statement it would be that "the major problem facing scientific hydrology is the tremendous spatial and temporal variability of hydrologic processes across the globe" (NRC 1991). Thus, it is primarily a scale effect that separates engineering hydrology from today's needs in scientific hydrology. However, it is more than just that. We are now asking more detailed questions about the physical processes of the hydrologic cycle in contrast to the engineering questions that have been frequently answered empirically. Thus, in scientific hydrology we literally are asking where each drop of water resides and for how long and how this drop moves through the earth's system. In addition, we are attempting to quantify the feedback mechanisms resulting from the land–surface coupling to the atmosphere and the role of hydrology in ecosystem dynamics and biogeochemical cycles.

Effectiveness of new methods

One interesting aspect of recent hydrologic research is that although we feel we know more about the physical process, use sophisticated analysis techniques, and produce very elaborate output, we have not been able to demonstrate consistently improved accuracy or reproducibility. Is it possible that the recent development of complex and sophisticated process-based models work no better than the older, simple models? A case can be made to support this argument.

The volume of published techniques describing commonly used procedures (e.g., the rational formula and Soil Conservation Service curve number technique) is not in proportion to their use (McCuen and Rawls 1979). In addition, newer or complex (sophisticated?) methods have not been adopted widely by practicing engineers. Presumably this is because the newer techniques have not been shown to give demonstrably better results, and in general, their use requires more data and usually a large computer.

An interagency working group of the Hydrology Committee of the Water Resources Council was assigned the task of developing consistent national guidelines for defining peak-flow frequencies at ungauged stream locations (Water Resources Council 1981). The many differing procedures used in practice and the lack of agreement about their use precluded selecting procedures to include in a national guide without developing objective information about procedure performance.

Naef (1981) addressed how well various models succeeded in reproducing measured discharge. His conclusions are based on two projects: the World Meteorological Organization's intercomparison of conceptual models used in operational hydrological forecasting (1975) and a study of rainfall–run off models using data from small basins in Switzerland. The results show that simple models can give satisfactory results; however, neither the simple nor the more complex models tested were free from failure in certain cases, because none of them adequately describe the rainfall–runoff process. In addition, it could not be proved that complex models give better results than simpler ones.

A third study by Loague and Freeze (1985) presented model performance calculations for three event-based rainfall–runoff models on three datasets involving 269 events from small upland catchments. The models include a regression model, a unit-hydrograph model, and a quasi–physically based model. The results of the study show surprisingly poor model efficiencies for all models on all datasets on an event by event basis. The poor performance of the quasi–physically based model could probably be ascribed to a combination of model error and input error. The authors speculated that the primary barrier to the successful application of physically based models in the field may lie in the scale problems that are associated with the unmeasurable spatial variability of rainfall and soil hydraulic properties. The fact that simpler, less data intensive models provided as good or better predictions than physically based models is food for thought.

In a recent series of papers, Grayson et al. (1992a,b) questioned the value of distributed parameter models to represent hydrologic processes accurately if the fundamental algorithms and assumptions cannot be validated. In their conclusions they state "the misperception that model complexity is positively correlated with confidence in the results is exacerbated by the lack of full and frank discussion of a model's capability and limitations and the reticence to publish poor results." They go on to conclude that "the seductive attraction of the more complex models is their ability to provide information about points within the catchment, but it is concluded that the representations used in current process-based models are often too crude to enable accurate, a priori application to predictive problems."

In a more recent paper, Jakeman and Hornberger (1993) question "what limits the observed data place on the allowable complexity of rainfall–runoff models." They further state, "conceptual and physically based models developed and used for describing rainfall–runoff processes tend to be over parameterized. They are no more useful for prediction than are simpler models whose parameters are identifiable from available data."

The point is that the development of more complex and sophisticated models and innovative analysis techniques has not resulted in an overwhelmingly improved ability to predict hydrology performance or to represent hydrologic processes. If one accepts these studies as indicators of the effectiveness of recent hydrologic research results, one should ask why. Why is it that more complex and more physically based models do not give us better results?

There is perhaps no clear answer, but I would speculate that lack of the proper amounts and types of data may be a large part of the answer. I will try to show how remote sensing may provide some new types of data that will help make the complex models easier to use as well as improve their performance. For example, remote sensing may be the only viable approach to handling spatial variability of water-shed properties because the basic data are spatial in nature.

Data issues in hydrology

Given that 1) hydrology has developed into a state-of-the-art computer-dependent discipline, 2) the current stage of model development has not really provided tools that give us better answers in spite of their complexity, and 3) the new hydrological sciences are demanding even more answers from a very complex system, the following questions should be asked: What is needed most in our discipline, and where should our energies and resources be directed as we are poised on the threshold of the twenty-first century? My simple one-word answer is *data*. I will step farther out on the limb and state that neither new analysis techniques nor faster computers will lead the way to answering the twenty-first century questions. We have seen a number of new techniques applied to hydrology: linear systems, stochastic hydrology, finite elements, krigging, 4DDA, GIS, and remote sensing, to name a few. Have these advanced the discipline and science of hydrology? Yes, but not in quantum leaps, only in baby steps, and perhaps not even going in the same direction or perhaps retracing other's steps. If we look at the development of hydrology historically, I think we are marking time and not marching ahead. The one thing preventing significant progress is the lack of data. Not just numbers of measurements but the correct measurements and all the variables. How can we truly believe a so-called physically based, distributed model that supposedly accounts for all the processes but is driven only by rainfall and validated only by a single measured hydrograph?

These issues have been addressed in *Opportunities in the Hydrological Sciences* (NRC 1991), which bases its discussions on the premise that "hydrologic science is currently data limited." The report concedes that traditionally hydrologic data have been collected to address water resource problems and not the emerging hydrologic sciences. Dozier (1992) correctly points out that our historical data have not been able to represent the wide variety of spatial and temporal scales encountered in hydrology and that the resulting models reflect a simple, homogeneous view of the natural world. He concludes that "this forced oversimplification impedes scientific understanding and management of water resources."

If the emerging hydrological sciences are to break away from traditional engineering hydrology, a number a general and specific data needs are going to have to be addressed and solved. According to *Opportunities in the Hydrological Sciences* (NRC 1991), they are the following.

- Hydrologic data are needed to measure fluxes and reservoirs in the hydrologic cycle and to monitor hydrologic change over a variety of temporal and spatial scales.
- Detection of hydrologic change requires a committed, international, long-term effort and requires also that the data meet rigorous standards for accuracy.
- Synergism between models and data is necessary to design effective data-collection efforts to answer scientific questions.
- A fundamental block to progress in using most hydrologic data is our poor knowledge of how to interpolate between measurement points.

The NRC book convincingly argues that the collection of hydrologic data should not be an afterthought. In addition, the issue of data accessibility and management is addressed as part of the general data issue. This is a particularly important issue because all the data in the world are useless unless one can access them, and the technology for storing, transmitting, displaying, and analyzing data is changing very rapidly. Other than to encourage hydrologists to keep abreast of these rapidly changing developments, further discussion of this topic is beyond the scope of this chapter.

The question must now be raised: If one accepts the premise that "hydrologic science is data limited," what do we do about it? Do we lobby for more stream gauging stations and rain gauges? Perhaps a few, particularly if we locate them in data-sparse areas. However, I think that it is obvious that the answer to the hydrologic data issue is not more of the same. Two issues beyond the scope of contemporary instrumentation are the spatial variability of hydrologic processes and the wide disparity of time and space scales that scientific hydrology must address. Clearly, some innovative solutions are needed to solve these problems.

Remote sensing may meet some of these needs. Remote sensing can address the spatial heterogeneity and the scale disparity problems. For example, remote sensing may be the only viable approach to handle spatial variability of drainage basin properties and hydrologic process because the basic data are inherently spatial in nature. Space-borne instruments also have the ability to make measurements that are truly global in nature. In my opinion, advances in remote sensing may result in a very large and significant step in the knowledge curve. This is based on the uniqueness of remote sensing to obtain spatially distributed information as well as some entirely new forms of measurement.

Two recent research results will illustrate how remote sensing may provide the types of nontraditional data needed to address the questions of the hydrological sciences. During the recent NASA Space Shuttle Imaging Radar mission, Synthetic Aperture Radar (SAR) images of snow at Mammoth Mountain in California were obtained. However, it is not the image that is important, but the information contained in the data that makes up the image. In this case, different polarizations of the C-band data were used to estimate the snow wetness. Free water in the snowpack affects the dielectric constant of the snow and thus the reflectivity of the radar signal (Fig. 14-1). These are new data and they are spatial data of very high resolution, approximately 20 m. Snow hydrologists may react with some interest but will really want snow water equivalent, not snow wetness. Why? Because those are the data they have traditionally tried to use in their models. Those data are collected from snow tubes and snow courses, and their models have been developed to use those data because that is all that is available. Snow course data are generally good data but they are essentially point data in a 4D world.

A second example is illustrated in Fig. 14-2, which is a time series of passive microwave measurements taken over the Little Washita Watershed in Chickasha, Oklahoma. In reality these day-by-day images are presenting a hydrologic signature of the watershed. Starting from a uniformly wet condition, we can observe how the land surface is responding to a uniform drying from an atmospheric perspective. These images are integrating drainage, lateral redistribution of soil

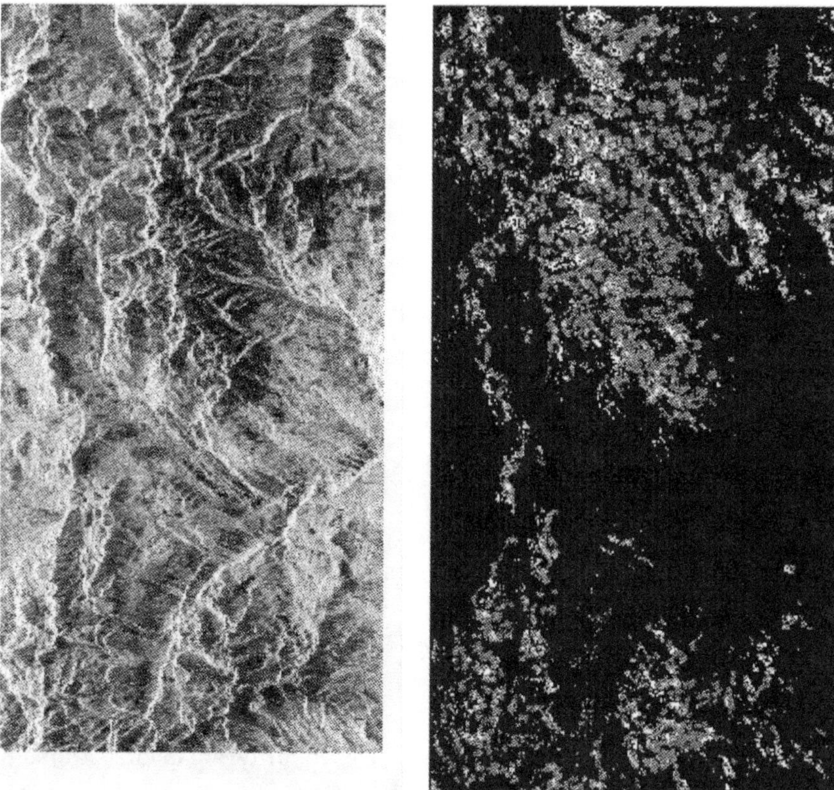

FIG. 14-1. Space shuttle imaging radar image of snow on Mammoth Mountain. (a) Terrain and (b) wetness. The image illustrates a new data form (snow wetness) and its spatial distribution.

moisture, and evapotranspiration. They reflect vegetation, topography, geology, and soil properties. The key is to figure out a new way to use these new data, not to try to derive a familiar parameter that is currently called for in our model. For example, these images should not be used to try to abstract a value of saturated hydraulic conductivity (Ks). First of all, Ks is a laboratory-derived point measurement, and there have been very few successful uses of such data in hydrologic models. The reason it has been used at all is that is has been available, even if only on a limited basis. Figure 14-3 shows the soil texture map of the Little Washita Watershed. The patterns are reproduced with remarkable accuracy. The challenge here is how to use these new spatial data.

I agree with the argument that, "modeling and data collection are not independent processes. Ideally, each drives and directs the other.

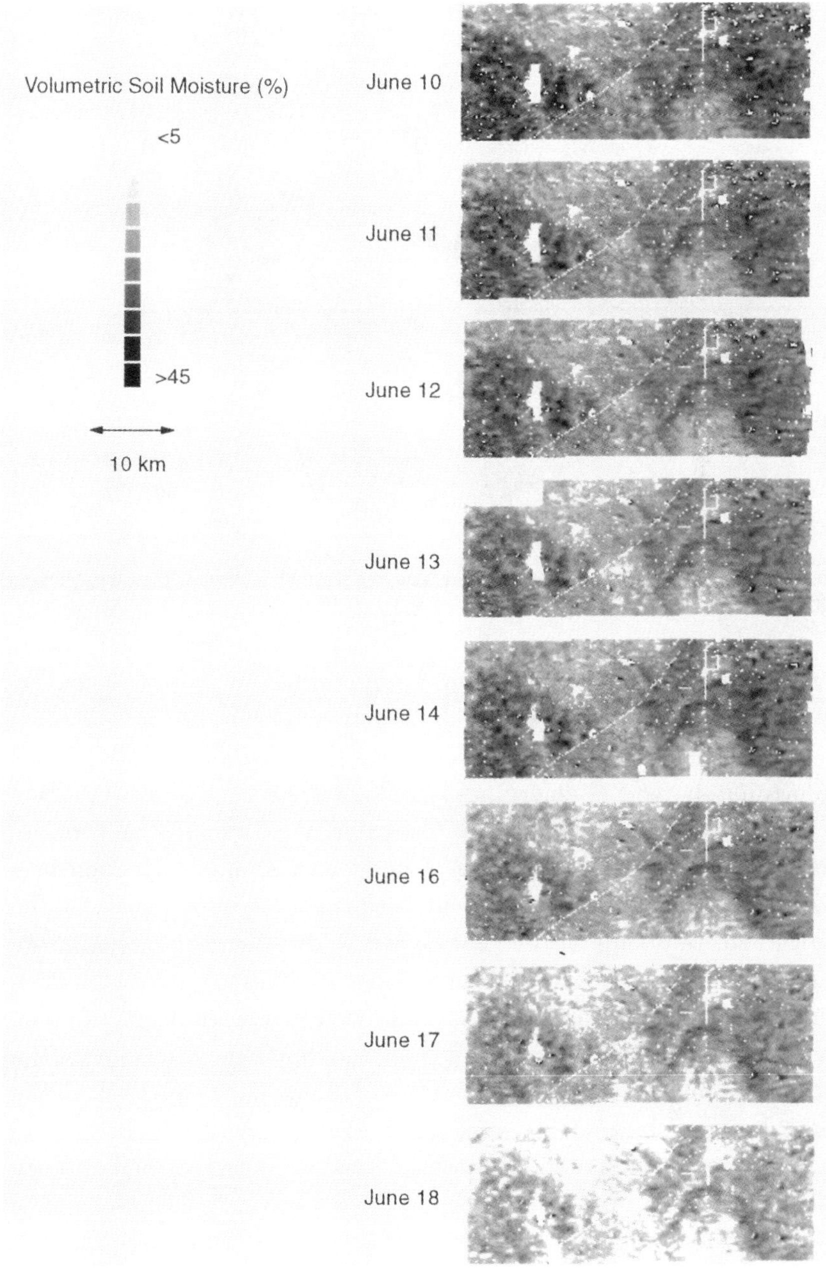

FIG. 14-2. Time sequence of passive microwave brightness temperatures from the Little Washita Watershed near Chickasha. The developing light gray (warm) from the black (cold) illustrates sequential and nonuniform drying of soil moisture. This is a new spatial data form that provides a hydrologic signature of the Little Washita Watershed.

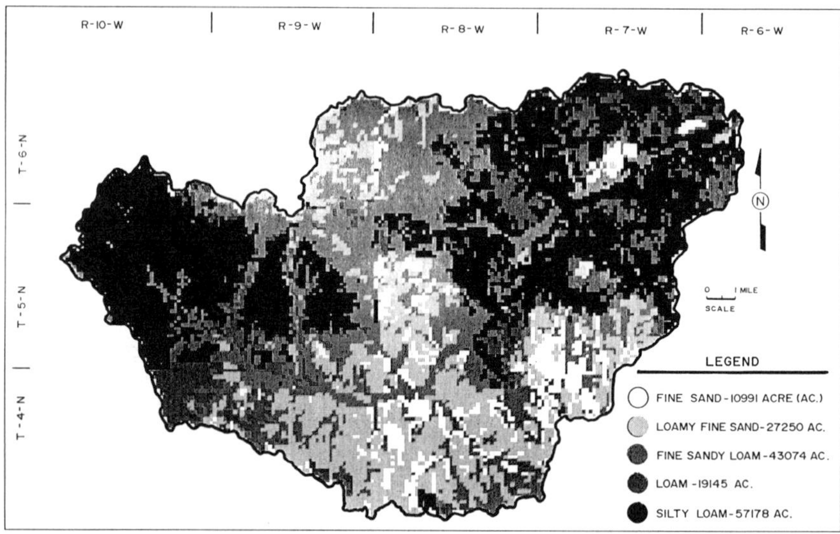

FIG. 14-3. Soil texture map of the Little Washita Watershed. Patterns closely resemble those that developed in Fig. 14-2.

Better models illuminate the type and quantity of data that are required to test hypotheses. Better data, in turn, permit the development of better and more complete models and new hypotheses" (NRC 1991). Unfortunately for the emerging hydrological sciences, data collection and archiving assume relatively low priority, politically, institutionally, and educationally. In time of budget cuts, data collection programs are often the first to be cut because their importance for the health and well-being of the general public is difficult to demonstrate in the beginning years of their operation. In addition, data collection activities are not looked upon as high-level intellectual activities by even our own research community—even though they depend on the data. The universities and research institutions have signed on to the "publish or perish" model for proving one's scientific mettle. In such an academic climate, data collection activities are not rewarded and are even looked upon negatively in some cases. One very positive sign to break this stereotype has been the initiation of the "Data and Analysis Note" in *Water Resources Research* (Hornberger 1994).

Summary

Hydrology has developed historically from an innate and generally successful approach to water resource management to an effective engineering discipline that has, through water management, raised our living standards to an unprecedented level. Now it is an emerging geoscience attempting to address questions heretofore unthought of in scale and complexity. I have attempted to make the point, through a rather simple tracing of history, that progress in scientific hydrology will be impeded by a lack of proper and adequate data. Specific points that I have tried to emphasize are as follows.

- Following a period of seemingly exponential growth, hydrologic research advances have almost leveled off.
- Performance of recent hydrologic models and analysis techniques are not demonstrably better than older, simpler techniques.
- The new hydrological sciences are asking questions that our traditional approaches may be unable to address. The types and amounts of data to address these questions certainly do not exist.
- The major problem facing scientific hydrology is the tremendous spatial and temporal variability of hydrologic processes across the globe.
- Future advances in hydrology will be inhibited by a lack of data.

Remote sensing is but one approach to solving this problem, and it is not the only approach. What is needed is some innovative thinking on how to make some of the measurements needed to reflect the spatial and temporal heterogeneity that exists in hydrology. The hydrologic community needs to support the institutions that collect data. The community also needs to acknowledge data collection as a professional necessity.

REFERENCES

Biswas, A.K., 1972: *A History of Hydrology.* 2d ed. North-Holland Publishing, 336 pp.

Chahine, M.T., 1992: The hydrologic cycle and its influence on climate. *Nature,* **359,** 373–380.

Chow, V.T., Ed., 1964: *Handbook of Applied Hydrology.* McGraw-Hill.

Dozier, J., 1992: Opportunities to improve hydrologic data. *Rev. Geophys.*, **30** (4), 315–331.

Federal Council for Science and Technology, 1962: *Scientific Hydrology.* U.S. Government Printing Office, 37 pp.

Grayson, R.B., I.D. Moore, and T.A. McMahon, 1992a: Physically based hydrologic modeling. 1. A terrain-based model for investigative purposes. *Water Resour. Res.* **28,** 2639–2658.

———, ———, and ———, 1992b: Physically based hydrologic modeling. 2. Is the concept realistic? *Water Resour. Res.* **28,** 2659–2666.

Gumbel, E.J., 1941: The return period of flood flows. *Ann. Math. Stat.,* **12,** 163–190.

Hazen, A., 1930: *Flood Flows, a Study of Frequencies and Magnitudes.* John Wiley & Sons, 199 pp.

Hornberger, G.M., 1992: Hydrologic science: Keeping pace with changing values and perceptions. *Proc. 10th Symp. on Sustaining Our Water Resources,* National Academy Press, 43–58.

———, 1994: Data and analysis note: A new type of article for water resources research. *Water Resour. Res.,* **30,** 3241–3242.

Horton, R.E., 1933: The role of infiltration in the hydrologic cycle. *Trans. Amer. Geophys. Union,* **14,** 446–460.

Jackson, T.J., D.M. LeVine, C.T. Swift, T.J. Schmugge, and F.R. Schiebe, 1995: Large area mapping of soil moisture using the ESTAR passive microwave radiometer in Washita '92. *Remote Sens. Environ.,* **53,** 27–37.

Jakeman, A.J., and G.M. Hornberger, 1993: How much complexity is warranted in a rainfall–runoff model? *Water Resour. Res.,* **29,** 2637–2649.

Kalinin, G.P., and P.I. Miljukov, 1957: On the computation of unsteady flow in open channels. *Meterol. Gidrol. Z.,* **10,** 10–80.

Leopold, L.B., and K.S. Davis, 1966: *Water.* Time Incorporated, 200 pp.

Loague, K.M., and R.A. Freeze, 1985: A comparison of rainfall–runoff-modeling techniques on small upland catchments. *Water Resour. Res.,* **21,** 229–248.

Maidment, D.R., Ed., 1993: Hydrology. *Handbook of Hydrology,* McGraw-Hill, 1.1–1.15.

Martinec, J., 1960: The degree day factor for snowmelt runoff forecasting. *Surface Waters,* International Union Geodesy Geophys. Gen. Assembly of Helsinki, International Assoc. Hydrol. Sci. Comm., IAHS Publ. 51, 468–477.

McCuen, R.H., and W.J. Rawls, 1979: Classification of evaluation of flood flow frequency estimation techniques. *Water Res. Bull.*, **15**, 88–93.

Nace, R.L., 1974: General evolution of the concept of the hydrologic cycle. *Three Centuries of Scientific Hydrology*, UNESCO, WMO, IAHS, 40–44.

Naef, F., 1981: Can we model the rainfall–runoff process today? *Hydrologic Sci. Bull.*, **26**, 281–289.

NRC, 1991: *Opportunities in the Hydrologic Sciences.* National Academy Press, 348 pp.

Philip, J.R., 1957: The theory of infiltration: 1. The infiltration equation and its solution. *Soil Sci.*, **83**, 345–357.

Sherman, L.K., 1932: Stream flow from rainfall by the unit-graph method. *Eng. News. Rec.*, **108**, 501–505.

Theis, C.V., 1935: The relation between lowering of the piezometric surface and the rate and duration of discharge of a well using groundwater storage. *Trans. Amer. Geophys. Union, 16th Annual Meeting*, 519–524.

Water Resources Council, 1981: *Estimating Peak Flow Frequencies for Natural Ungaged Watersheds—A Proposed Nationwide Test.* U.S. Water Resources Council Hydrology Committee, 346 pp.

The Private Sector

CHAPTER 15

A History of Private Sector Meteorology

DAVID B. SPIEGLER

Introduction

Shortly after World War II, several former U.S. Army Air Force and Navy meteorologists/forecasters independently, but at nearly the same time, decided to start businesses that would offer customized forecasts to weather-sensitive industries. Their idea for doing this stemmed from their role during the war in which they were responsible for providing weather forecasts that had to be very specific for particular times and locations, in support of weather-sensitive military operations and missions. Mission planners depended on these specific forecasts, which were more valuable than "traditional" forecasts for general areas and time periods. The early pioneers of the private sector believed that there was a business in supplying forecasts that were more detailed in time and space than the traditional forecasts available from the U.S. Weather Bureau to industries affected by the weather. As a result, the origin of private sector meteorology, excluding weather instrumentation companies, is usually traced to the period just after the war when some of the currently existing private weather forecasting companies began operations and the private weather services sector became an identifiable part of the meteorological profession.

Private sector meteorology historically has been synonymous with weather forecasting services but has many other segments today, including instrumentation companies, data and graphics providers, staff meteorologists in industrial companies, general meteorological consulting services, forensic meteorologists, specialty companies, environmental consultants, weather system developers, and media meteorology.

The growth of the private meteorology sector has been remarkable, particularly since the early years after the second world war. Total volume of business for the few companies in existence at that time is estimated to have been in the $200,000–$400,000 range. (In 1995 dollars, it would be approximately \$2–4 million.) *Today, the total business volume for the hundreds of companies in the private sector is estimated to be in the vicinity of \$1 billion.* Amazingly, several of the early pioneer weather forecasting companies have stood the test of time and have thriving businesses today. In the following sections, the continuing history and the growth of the private sector are linked to people and companies, significant events, and important factors that led to the growth.

The overall approach to develop the information on the history and growth was the following.

1) Identify companies and individuals involved in private sector meteorology through listings in the *Bulletin of the American Meteorological Society (Bulletin)* and the *Certified Consulting Meteorologist (CCM) Directory,* and advertisements in the *Bulletin* and *Weatherwise;* corporation members are listed in the *Bulletin.*

2) Prepare a questionnaire and send it to a representative sampling of companies in each segment of the private sector; time and resource constraints precluded it being sent to all companies.

3) Interview selected private sector leaders and representatives.

4) Analyze and interpret information obtained.

Important dates, factors, people, and companies

Analysis and assimilation of the data and information obtained has resulted in identifying the key dates/times, factors, people, and companies that were significant in the development and growth of the private meteorology sector over the years. This part of the analysis focuses on companies other than the instrumentation companies.

The defining dates and events that triggered expansion of the private sector are as follows.

1934: The first private sector meteorologist is hired by a utility company.

1946: The first group of private weather service forecast companies begin operations.

1955: The start of operational numerical weather prediction and wider use of meteorological professionals in the private sector media.

1960: The launch of the first weather satellite—TIROS (Television and Infrared Observation Satellite), and the initial thrust for major meteorological consulting/research companies.

1968–1970: The emergence of environmental companies in response to an increasing concern about environmental pollution.

1975: The development of interactive, automated, weather information systems (AWISs) by private companies under government contract is proven as viable for operations.

1976: The launching of the first operational geostationary satellites that opened up new areas for research in all sectors (government, private, and university), product development, and weather forecasting.

Early 1980s: The acceleration of technology for weather data/graphics begins, and weather system developers/system integrators produce AWISs.

Mid-1980s: The development by private sector companies under government contract of automated, remote, weather-observing systems begins an era of a reduced number of human weather observers in government.

Early 1990s: The WSR-88D (Weather Surveillance Radar—1988 Doppler) weather radar [formerly NEXRAD, (Next Generation Weather Radar)], real-time operational lightning detection, and relatively low-cost computer access to weather data and graphics creates new opportunities and challenges in the private sector.

The major external forcing, or driving factors, in the growth of the private meteorology sector are *economic*—cost versus benefits; *technological*—advances in computer technology, increased capabilities of weather and communications satellites, and the growth of the aviation industry; *political*—the Clean Air Act and amendments; and *military*—the need for ever more rapid access to up-to-date weather data and graphics. Over the years 10 different business segments for the private sector have developed, exclusive of the local TV meteorologists. They are the following:

1) weather instrumentation/remote sensing;
2) weather forecasting services;
3) weather data/graphics providers;

4) staff meteorologists in industrial companies (public utility, oil and gas exploration, airlines, etc.);

5) general meteorological consulting services (research and development, studies, modeling, market analyses and surveys, strategies, and plans, excluding environmental-only businesses);

6) forensic meteorology;

7) specialty companies (weather modification, lightning detection, weather radars, wind profilers, wind tunnels, hot air ballooning, etc.);

8) environmental consulting only (air-quality studies, modeling, analyses, monitoring, permitting, environmental impact statements);

9) weather system developers, providers, system integrators (weather workstations, software, system integration); and

10) media meteorology (excluding local TV meteorologists only).

Early events

In the course of researching the history of the private sector and communicating with others who know the early history, I learned that the era of private sector forecasting can be traced back to the mid-1930s and even as early as 1918 if one considers the Bergen Establishment of Vilhelm Bjerknes in which Jacob Bjerknes worked. The very first event that could be considered private sector meteorology in the United States was when the Cincinnati Chamber of Commerce hired Cleveland Abbe in 1868 (Fleming 1990).

According to some, the famous (and to others, the infamous) Dr. Irving P. Krick illustrates the relatively obscure beginnings of private sector meteorology in the 1930s in the United States (J.C. Freeman and Loren W. Crow 1995, personal communication). Krick provided private forecasting services to Hollywood filmmakers in 1935 while he was at the California Institute of Technology. He is credited with picking the night when weather conditions would be "right" for the burning of Atlanta in *Gone With The Wind,* and his reputation in the industry was made because there could be no retakes (Boesen 1978). Special effects technology was not what it is today.

Pacific Gas & Electric (PG&E) also employed a meteorologist, Charles Pennypacker Smith, as early as 1934 (R.N. Swanson 1995, personal communication). Now many utilities have their own meteorologists on staff, as do commercial airlines. Thus, the evidence indicates that the actual origins of private sector meteorology date back more than 60 years.

An early pioneer in what may be considered the earliest segment of the private sector, weather instrumentation companies, is Belfort Instrument of Baltimore, Maryland. Julian Friez founded the Belfort (weather) Observatory in 1876—only six years after the establishment of a national weather service in the Army Signal Office. Friez was a precision instrument maker, and the Belfort Observatory manufactured nearly all the weather instrumentation devices used by the government in the 1920s and the weather balloons that began to be sent up in the 1930s. In 1938, shortly after the famous Orson Welles *War of the Worlds* radio hoax broadcast, the observatory received back a weather balloon transmitter full of bullet holes. The person who found it tangled in a tree thought it was a Martian bomb and had the state police shoot it down! More about the weather instrumentation companies will be presented later in this review.

The 1940s through the 1960s

The year 1946 marked the real significant beginnings and growth of private sector meteorology companies with the emergence of two segments of the private sector at nearly the same time—the weather forecasting services and the weather data/graphics providers. The *economic factor* was prominent here. Forecasts were provided by the companies that were more detailed in time and space than those available from government forecasts. In addition, the forecasts were tailored to specific customer operations. The benefit–cost ratio for the companies subscribing to the forecast service was high. To do business, the private forecasting companies needed the weather data and weather maps that they received via weather teletype and facsimile machines from the U.S. Weather Bureau.

After the second world war, Krick expanded his private forecasting service to include utilities, oil companies, construction firms, etc. But Krick began to have problems with the Weather Bureau, and later with the AMS, when he claimed the ability to make skilled forecasts for a year in advance; he also got involved in weather-modification (rainmaking) activities, claiming credit for rainfall, which others believed was questionable.

The other early pioneers that were prominent in founding weather forecasting service companies were Alfred H. Glenn, who founded A. H. Glenn and Associates; John E. Wallace, who started Northeast Weather Services—since renamed Weather Services Corporation (WSC); John R. Murray and Dennis W. Trettel, principals of Murray and Trettel; Bill Hartnett, who formed Weather Corporation of America; and Robert D. Elliott, founder of North American Weather Consultants. All of these companies have remained in business continuously for nearly 50 years and have established excellent reputations in the field. These five companies collectively received the AMS award for *Outstanding Services to Meteorology by a Corporation* in 1970. Each was cited for "pioneering the practice of private meteorology in the United States, thereby helping to develop an industry which has made significant contributions to the American economy." In addition, four of the founders of these companies have individually received the AMS award for *Outstanding Contribution to the Advance of Applied Meteorology*—Elliott (1961), Glenn (1962), Wallace (1977), and Trettel (1986).

An interview with John Wallace, who retired from WSC in 1993, revealed that the early clients of what was then Northeast Weather Services were public works and highway departments and municipalities. Wallace's first account was a candy company in Boston and resulted in a profit of $30 in the first month! In 1946, Wallace and one employee worked 18-hour days. Additional clients that came on board in the late 1940s were electric and gas utilities and commodity trading companies. Typical rates charged clients in those days were from $250 to $500 *per year!* (about $2,500 to $5,000 in 1995 dollars). It was much more labor intensive to run a private weather service in those days. Maps were plotted and analyzed manually from teletype observations that were printed at a speed of 60 words per minute. Sometimes the teletypewriter carriage would fall off the machine to the floor and data were missed.

The two companies that supplied weather data and Weather Bureau forecasts via teletype were Western Union and Weather Fotocast, a subsidiary of United Press International. Western Union has a long history in the weather business, having carried the weather reports from the Signal Corps in the 1870s. The Signal Corps was a predecessor to the Weather Bureau and the National Weather Service (NWS).

The company that pioneered the offering of weather analyses and forecasts produced at the central office of the Weather Bureau in Washington, D.C., was Alden Electronics, Inc., of Westborough, Massachusetts. Milton Alden developed the chemically treated facsimile paper and methodology for routinely transmitting the maps over telephone lines to facsimile recorders at the offices of the private weather forecasting services and to government forecast offices.[1] His son, John Alden, built the company into an internationally known and respected company. J. Alden ran the company for decades until he died in 1993. Alden Electronics, Inc., was recognized by the AMS in 1992 when it received the *Award for Outstanding Services to Meteorology by a Corporation* "for more than forty years of leadership in the development of weather map recording, transmission, and display." A second company transmitting weather maps via facsimile was Times Facsimile. Times used a carbon process that "burned" black lines on electrostatic paper.

A fourth segment of the private sector—staff meteorologists in industrial companies—had meteorologists in the utility as well as in the aviation industry. PG&E was the first utility to hire a meteorologist and has kept that tradition to this day. C.P. Smith of PG&E received the AMS award for *Outstanding Contribution to the Advance of Applied Meteorology* in 1967 for "his pioneering contributions to the ethical development of the field of industrial meteorology, and his continued support of the field of consulting meteorology." PG&E received the AMS award for *Outstanding Services to Meteorology by a Corporation* in 1961, which stated, in part, that PG&E has maintained "an active support of private weather services since 1937."

Several airlines had their own staff of meteorologists from the early days of scheduled airline flights in the 1930s. [See chapter 16 by Cartwright and Sprinkle, this volume.] Among those airlines with meteorologists were American Airlines, Trans World Airlines, United Airlines, Northwest Airlines, Eastern Airlines, and Pan American World Airways. These latter two airlines are no longer in business, but the practice of airlines having staff meteorologists continues to this day, with other major airlines such as Delta, American, and Northwest Airlines maintaining their own staffs. Other airlines use commercial weather services.

---

[1]The telegraph "fax" concept was actually demonstrated at the centennial exposition in Philadelphia back in 1876 but was not used operationally by the Weather Bureau until the mid-1940s.

Joseph J. George, a former general in the army air force and head of Eastern Airlines weather forecasting group in the late 1940s and 1950s developed many operational forecasting techniques that proved their worth over the years, not only in the aviation industry but throughout the overall weather industry. For his numerous contributions, Joe George received the first AMS award for *Outstanding Contribution to the Advance of Applied Meteorology* in 1956.

A fifth segment of the private sector, general meteorological consulting services, also had its origins about this time. A.H. Glenn's company has always done engineering research and development studies in meteorology and oceanography for planning, design, and operation of coastal and offshore facilities for offshore drilling companies, tanker terminals, and nuclear power stations, among others. The *economic factor* was the driver.

And Loren Crow established his own consulting business in the 1950s. He had previously been a consultant with Irving P. Krick Associates. Crow continued to be very active in the consulting service business for more than 40 years. He received the AMS award for *Outstanding Contribution to the Advance of Applied Meteorology* in 1965 for being "one of the first meteorologists to establish a successful business based solely on the private practice of consulting" and for having "greatly improved the stature of the Consulting Meteorologist."

In 1955, numerical weather prediction (NWP) became operational after extensive testing of atmospheric barotropic and baroclinic models. The barotropic model for predictions of heights and temperatures at the 500-mb level was the first operational NWP model. It began a revolution in weather forecasting capabilities in both the private and the government sectors. It also provided modelers in all sectors the opportunity to research and develop ever more sophisticated models of the atmosphere. [See chapter 2 by Cressman, this volume.]

In 1960, two events occurred that were to have significant impact on the meteorological profession, including the private sector. One was the launch of the TIROS weather satellite, which began an era of unprecedented *technological* advances for observing the atmosphere from space globally. Satellites provided opportunities for research that lead to greatly enhanced understanding of atmospheric behavior and processes as well as to improvements in diagnosis and forecasting. [See chapter 5 by Purdom and Menzel, this volume.]

The second important event in 1960 was the formation of what is believed to be the first major meteorological consulting and research and development company—the Travelers Research Center (TRC). Dr. Thomas F. Malone, the first chairman of the board, and Dr. Robert M. White, the first president of TRC, are two people who are recognized internationally in the meteorological profession as "giants." Both men are Honorary Members of the AMS, the highest distinction of membership; both have served as president of the AMS; and both have received the AMS *Cleveland Abbe Award for Distinguished Services to Atmospheric Sciences by an Individual* and the *Charles Franklin Brooks Award for Outstanding Services to the Society.*

TRC, as a wholly owned subsidiary of the Travelers Corporation, had an interesting origin. In 1954 and 1955, there were a total of five hurricanes that caused massive destruction along the eastern seaboard from the Carolinas to New York and New England. The Travelers Insurance Company (TIC) suffered significant losses through claims filed for the damages the hurricanes wrought. The TIC management decided to implement a course of action that might help to reduce future losses from hurricanes and from adverse weather in general. TIC owned a radio and a TV station, WTIC, and wanted to establish a weather service that would provide weather presentations around the clock on radio and TV. Thomas Eaton, the news director for WTIC, went to AMS Headquarters and met with Dr. Kenneth Spengler, executive director, asking who Spengler would recommend to head such an endeavor. Spengler suggested that Eaton speak with Malone—and "the rest is history" as they say.

Malone had more ideas than the Travelers Weather Service in mind for the insurance company. He logically and effectively made the case for research that needed to be done regarding hurricanes that would help TIC in setting appropriate rates for insurance, as well as in providing an understanding that might assist in forecasting the paths and intensities of hurricanes. Malone organized a group of researchers that included Dr. Don G. Friedman, Dr. Joseph G. Bryan, Dr. Robert G. Miller, and Bernard Schorr. The Weather System Division of the TIC was born. Friedman was recognized by the AMS in 1976 when he was given the *Award for Outstanding Contribution to the Advance of Applied Meteorology* "for his astute contributions to the quantitative estimation of the risks of natural geophysical and environmental hazards." And Miller, together with Keith W. Veigas, pioneered in the development of statistical predictions of hurricane motion that formed the basis of hurricane prediction equations later used by the National Hurricane Center. [See chapter 9 by DeMaria, this volume.]

In 1959, there was a cooperative initiative between the U.S. Air Force, the Federal Aviation Administration (FAA), and the Weather Bureau for a program called Weather System 433L to develop new techniques for analysis and prediction of weather that would be operational on the large computers at weather centrals. United Aircraft (UAC, now United Technologies) submitted a bid as prime contractor with the Weather System Division (WSD) of TIC as a subcontractor. UAC/TIC won the contract award. The program was very large in scope and time and required the hiring of many meteorologists and computer programmers by WSD/TIC and UAC.

Fresh out of school after receiving a masters degree from New York University, I was interviewed by Malone and hired into the WSD in the insurance company. Malone brought on White to be director of research for WSD and was instrumental in having the TIC form TRC—with White as president. Meteorologists at TRC developed many new techniques that were implemented operationally on the computers at weather centrals, particularly the Air Force Global Weather Central at Offutt Air Force Base, Nebraska. TRC was the forerunner of a number of meteorological consulting companies that became established in the next three decades. And in 1963, White was selected by President John F. Kennedy to become chief of the Weather Bureau.

In the late 1950s and early 1960s, another part of the private sector began to develop and become an identifiable sixth segment that would continue as a private sector business area—forensic meteorological services. Walter F. Zeltman was one of the early experts in this field that focuses on "reconstructing" the weather at the time and location of an accident or event that is in litigation. The weather reconstruction is performed using a combination of conventional weather observations, observations from cooperative observers, weather analysis maps, and weather radar and satellite data. Zeltman formed National Weather Corporation in 1958 and subsequently changed the name to International Weather Corporation. He says that beginning in the middle to late 1960s, he averaged 300 cases per year and has maintained that level up to the present time. He made a conscious decision not to expand beyond that level.

The forensic segment of the private sector, however, has grown steadily as evidenced by the number of listings for forensic services in the *Bulletin* Professional Directories for Certified Consulting Meteorologists and Professional Members. There are likely many more forensic meteorologists that are not listed. Prominent among those that are listed is the Climatological Consulting Corporation (CCC), headed by William H. Haggard. CCC has about 20 representatives listed in the *Bulletin* who provide services in their locales.

In the 1950s and 1960s there was another segment of the private sector that gained attention both inside and outside the meteorological profession but has since lost much of the attention and the associated business volume. That segment is weather-modification companies. Because it is so much smaller now than it was in its heyday, it is being combined with other companies that specialize in certain areas such as lightning detection, wind tunnels, hot air ballooning, etc. This seventh private sector segment will be referred to as specialty companies.

Weather modification received initial impetus from the pioneering efforts of Vincent J. Schaefer in artificial nucleation beginning in 1947, for which he received the AMS award for *Outstanding Contribution to the Advance of Applied Meteorology* in 1957. There was a need for more rain over farmlands that were experiencing below-average rainfall amounts over an extended period, as well as a need for more snow in the Sierra Nevada in California and in the Rockies to provide water from runoff.

In addition to Krick, who was mentioned earlier as being involved in weather-modification activities, there were many others who attempted to use cloud-seeding techniques to increase precipitation. Among them were North American Weather Consultants, under Bob Elliott, and Dr. Wallace Howell, of Wallace E. Howell Associates, both of whom established a reputation for trying to increase precipitation in a scientifically acceptable manner.

There were many state and federally sponsored programs in weather modification through the 1960s and into the early 1970s, as well as interest in foreign countries, particularly in the Caribbean and Central and South America, Israel, and India. The state of California and the Bureau of Reclamation in Denver, Colorado, were two of the biggest sponsors of weather-modification experiments that used private sector companies to carry out the work. Companies had weather modification contracts that ranged from hundreds of thousands to millions of dollars. Over a long period of time, results were mixed. Perhaps because the consistency of results was lacking and no really new technology came into existence, the weather-modification budgets declined quickly. In addition, there was controversy and occasionally litigious events. These factors combined to cause the role of the private sector in this segment to become much smaller in the 1980s and 1990s.

More and more concern about our environment in the 1960s led to some programs sponsored mostly by state and local governments for measuring the pollutants in the atmosphere and tracing the transport and diffusion of pollutants over time and space. The increasing advancement of computer technology allowed modelers to develop atmospheric diffusion models that would predict the concentration of pollutants as a function of source data and meteorologically observed and predicted data. "Environmental consulting companies" emerged in the late 1960s as the eighth segment of the private sector in meteorology.

The 1970s through the 1990s

A spin-off of TRC that had been an environmental division within TRC was formed in 1969 and eventually was known as TRC of New England. Today it is called TRC Environmental Corporation and is headed by Dr. Vin Rocco. TRC, along with Environmental Research and Technology, Inc., (ERT) and Geophysics Corporation of America, were among the early pioneer companies in the environmental segment.[2]

The Clean Air Act of 1970, which had been anticipated by industry for at least a few years, produced very rapid growth of the environmental segment over a short period of time. The first pure environmental company that became the leading and fastest-growing company in the early 1970s was ERT. Dr. Norman E. Gaut and Dr. James R. Mahoney were confounders of ERT in December 1968 with initial financing provided by Norman's father, Marvin J. Gaut. In the first full year of operation, 1969, ERT business volume was about $190,000. It was about $250,000 in 1970 and then about doubled in size every year for the next several years, to over $20 million by 1977. This writer had the good fortune to be part of ERT during the exciting spinup growth period in the early and mid-1970s. In 1979, ERT won the AMS *Award for Outstanding Services to Meteorology by a Corporation* "for the sig-

---

[2] ERT was renamed ENSR in 1985.

nificant role it has played in facilitating the solution of environmental problems by the meteorological profession."

The demand by the federal and state governments for monitoring and modeling air quality, performing environmental analyses, and preparing impact statements led to a large market for companies that could provide these capabilities. The number of new companies that were founded to meet the large demand was considerable. Some remained small, but others grew quickly into a force in the industry. In addition, several engineering construction firms that build power plants formed divisions within the company to focus on the environmental aspects. Among the first to hire professional meteorologists were Stone and Webster and Dames and Moore in 1969. To this day, both companies maintain environmental divisions and employ meteorologists. After a slowdown in the 1980s, new environmental legislation in the 1990s is producing new growth.

The atmospheric portion of the environmental business segment has a number of significant-sized subsegments. They include air quality studies, atmospheric dispersion modeling, field studies, analyses, air quality monitoring, permitting, and environmental impact statements. Each of these segments is of considerable size. It is beyond the scope of this history of the private sector of meteorology to treat the environmental segments in depth in this chapter.

Another segment benefiting from the growth of environmental companies is the weather instrumentation segment, the first identifiable segment of the private sector. In the 1970s hundreds of programs for monitoring air quality always included meteorological instrumentation, because it is important to know local atmospheric conditions for the analysis of the impact of pollution sources on the environment. In addition to conventional weather instruments, a device that measures vertical temperature profiles and infers atmospheric turbulence—the acoustic sounder—gained acceptance and use in many of the air monitoring programs. Dr. Paul MacCready, chairman of Aerovironment, was instrumental in producing the first commercial acoustic sounder based on research work done at the Wave Propagation Laboratory in Boulder, Colorado. REMTECH of Longmont, Colorado, is also a prominent company in the manufacture of acoustic sounders.

Other methods for determining the vertical structure of the atmosphere that are sometimes used in monitoring programs are radiosondes and tethersondes. The tethersonde has similar measurement equipment to the radiosonde, but the inflated dirigible-like balloon is tethered with wires to be nearly stationary over an area and reports meteorological variables at the height at which it is installed.

Atmospheric Instrumentation Research (AIR) and VIZ Manufacturing produce radiosondes. The former also offers tethersondes. In total there are at least 30 weather instrumentation companies that sell their products nationally, and there are probably many, many more local companies. The best-known national weather instrumentation companies, besides Belfort that was mentioned earlier, are Vaisala, Inc., a subsidiary of Vaisala, OY, founded in 1936 by Dr. Vilho Vaisala; Campbell Scientific; Climatronics; Qualimetrics; Rotronic; Scientific Technology; and Rosemount Aerospace. Growth of this segment of the private sector in the 1970s and 1980s was substantial.

In the 1970s, technology advanced at a rapid pace with respect to both computers and satellites for weather and communications. The technological advances included the development of interactive AWISs in three main thrusts. One was the Man–computer Interactive Data Acquisition System (McIDAS) developed at the University of Wisconsin under contract to the National Aeronautics and Space Administration for about $10 million. McIDAS was originally intended to be a research tool focused on satellite imagery–derived information, but a combination of "lunch bag" computer programming to add the software capabilities to process, plot, analyze, and otherwise manipulate conventional meteorological data and the high interest of forecasters in its potential capabilities caused redirection in the use of McIDAS. It began to be employed by weather forecasters as a total AWIS.

A second thrust was by the navy. The Naval Environmental Display System was the first operational AWIS in 1977. It was at the Fleet Numerical Meteorological and Oceanographic Center in Monterey, California. And the third thrust, the Automated Field Operations and Services under development by the NWS in the 1970s, became operational in 1980.

The dramatic advances in computer technology and the demonstration of operational AWISs directly impacted the private sector in several ways. First, it allowed for a ninth segment of the private sector—weather system provider companies. (Details on this segment are discussed later in this section.) Second, it expanded the capabilities of weather data/graphics provider companies, while encouraging new entries into that segment. And third, it changed the way private weather forecasting service companies do business and greatly increased their capabilities to provide more services, faster and more efficiently. It also resulted in more private weather service companies.

Several private forecasting companies that were either newly formed or experienced new growth related to technology in the 1970s include the following: Oceanroutes, Inc., which began by providing optimum ship-routing services and now provides forecasting services for many other industries; Accu-Weather, Inc., founded in 1971 by Dr. Joel Meyers with one full-time employee—Elliot Abrams—now has a staff of about 270 employees, with more than 70 meteorologists and thousands of clients worldwide; and Surface Systems, Inc., which specializes in forecasting services for ground transportation.

The first operational geostationary weather satellite was launched in 1976. It provided immediate benefits to the private weather forecasting companies by allowing forecasters to see continuous changes in, and the movement of, the atmospheric features of interest for prediction for their clients. It also became a rich source of research for the private research and development companies and it supplied a new stream of products to weather graphics providers for transmission to their subscribers. The main subscribers were the private weather service forecasting companies and the television weather presenters at TV stations.

The history of the 10th segment in the private sector, media meteorology, is being treated in depth by Roy Leep in this volume [chapter 17], with the focus on TV meteorologists. The first major private sector group in this area was the Travelers Weather Service (TWS) that began in 1955 under Malone. TWS did both radio and TV broadcasts in Hartford, Connecticut, and gradually expanded to radio stations outside the Hartford area. There are also several private weather service companies that have the broadcast and print media as a significant component of their business. Some do radio broadcasts only, and others do radio and TV forecasts. For example, WSC and Accu-Weather, Inc., both of whom provide private forecasting services, each have radio stations as clients nationally, and Accu-Weather also provides TV stations with forecasts and graphics. Both companies serve the print media domestically and internationally. WSC began its radio broadcasts in the 1960s, growing modestly at first, but in the last two years it has expanded by about 50% to provide services to over 120 radio stations. WSC also does weather radio broadcasts for a station in Puerto Rico.

Another company that has become a major force in this segment was the brainchild of John Coleman, a TV meteorologist in Chicago in the early 1980s. With the advent of expanded cable TV capabilities at that time, Coleman conceived of a 24-hour round-the-clock TV weather program that featured on-camera meteorologists describing the weather nationally and regionally, with local text forecasts from the NWS on a very frequent basis. The Weather Channel (TWC) was born in 1982. It took some time for the concept and its benefits to take hold, both with the viewing public and with advertisers. TWC lost money for the first five years, but the rapidly increasing number of homes with cable TV and the need for updated weather information on a frequent basis during rapidly changing and adverse weather conditions, combined with the technology to provide it, has resulted in wide acceptance of TWC nationally. It is now viewable in over 53 million homes!

From the early 1980s to the present time, technological advancements have occurred at a dizzying pace. This is particularly true for computer technology. Yesterday's computer technology breakthroughs have become today's "dinosaurs." Yesterday's ceiling prices have become today's floor prices. Processing speed keeps improving by orders of magnitude in relatively short periods of time. RAM (random access memory) and ROM (read only memory) capacities continue to expand rapidly, as do storage capacities on hard disks. These mind-boggling technological advances have had an impact on the private sector greater than that caused by the technological advances of the 1970s.

The weather system provider companies flooded the private sector market with weather workstations and systems that are interactive, have the capabilities to perform many functions, and allow the user to derive products easily that would be very labor intensive or nearly impossible manually. There are large markets for these weather systems. Among them are the Department of Defense; the federal civilian government, including the NWS and the FAA; state and local governments; weather-sensitive industries; the broadcast industry; and international markets, both governmental and commercial. Because the market opportunities were great, some weather data/graphics provider companies decided to develop and market weather systems also. Alden Electronics bought Zephyr Weather Information Service in 1986. Led by Jimmie Smith as president, Zephyr had started operations in 1983 as one of the first providers of weather data and graphics via satellite communications.

For the sophisticated weather systems, meteorologists/forecasters were hired to specify the products, functions, and features needed, and software analysts developed the computer programs to carry out the specs. System engineers are also heavily involved in installation, communications, and integration of hardware components. The private sector companies involved were enjoying a business boon.

Even private weather forecast services participated in the expanding weather systems market. WSC—one of the pioneer weather forecasting companies—formed a subsidiary company in 1983: WSI, which offers weather data and graphics on personal computers. Peter R. Leavitt, an executive vice president of WSC, was the first president of WSI.[3] WSI was subsequently bought by The Analytical Sciences Corporation (TASC), a consulting company in Reading, Massachusetts, supported by government contracts. A weather consulting division was part of TASC, and WSI became a wholly owned subsidiary. TASC also purchased a company that offered weather satellite and radar imagery on computer systems—Environmental Satellite Data (ESD) in 1989. The WSI and ESD systems were merged, giving WSI enhanced capabilities. Accu-Weather also began offering weather data and graphics on personal computers. For the most part, these systems were not interactive.

Another prominent company providing weather graphics is Kavouras, Inc., started by Steve Kavouras in 1973. The company provides weather graphics primarily to TV stations and also has a private weather forecasting services business. Kavouras, WSI, and Accu-Weather have graphic artists that create value-added products specifically designed to be attractive and relatively simple to understand by the average viewer.

As most TV stations use the value-added weather graphics of one of the three companies, the content of the graphics seen on TV weathercasts is similar from one station to another within a viewing area and also from one viewing area to another. Television meteorologists are open to new value-added graphic products such as 3D graphics animations, combined with cloud motions from satellite imagery, and creative software to display a simulation of a "fly through" and/or "fly under" the clouds. Two companies provide these new graphics that add

---

[3]Mr. Leavitt is now the chairman and chief executive officer of WSC Corporation.

some pizzazz to the presentation of weather on TV. One is Weathernews Inc. with their "Skywatch" (Weathernews Inc. is headquartered in Japan), and another is Earthwatch, Inc., founded in 1990 by Paul Douglas, a TV meteorologist in Minneapolis, Minnesota. And I formed Impact Weather, Inc., to develop multimedia products that emphasize, in an innovative way that personalizes it for the audience, how weather, climate, and the environment impact people's lives.

The interactive weather information systems of the 1980s became quite sophisticated in the late 1980s and early 1990s. The most prominent ones in the United States are the Automated Weather Distribution System (AWDS) for the air force, the Tactical Environmental Support System (TESS) for the navy, and the Meteorological Weather Processor (MWP) for the FAA. Planning Research Corp. (PRC) is currently building the Advanced Weather Information and Processing System (AWIPS) under contract with the NWS.

AWDS has been installed at 163 air force bases worldwide during the period from 1990 through 1993. GTE Government Systems was the production contractor for AWDS. The division of GTE that built the AWDS has had a long history in the weather data distribution business. The predecessors of that particular group were Western Union, American Satellite Corporation, Siscorp, Eaton, and CONTEL dating back to 1966 when the group was part of Western Union.

TESS was designed and developed by the Lockheed Missiles and Space Co., Inc. (now Lockheed-Martin). The Harris Corporation produced the MWP. In the late 1980s and very early 1990s the cost for these sophisticated systems ranged upward to $500,000 apiece. But the plummeting costs for computer hardware have driven system costs for even some of the more sophisticated systems to under $100,000. The software that in the early weather workstation system days was proprietary to particular hardware has more and more become independent of the computing platform. This has led to a noticeable trend in the past two years: potential customers are less interested in the hardware but more interested in the software associated with the systems. Many customers either have or prefer to purchase their own computers and related hardware and simply pay for a license to use the software.

The technological revolution related to an expanded private meteorological sector since the 1970s was not confined to computers. Remote sensing capabilities of weather satellites have constantly improved since the launches of the first TIROS in 1960 and the first geostationary satellite in 1976. The enhanced capabilities have led to research by consulting companies in the private sector such as Atmospheric and Environmental Research, Inc. (AER), of Cambridge, Massachusetts, started by Dr. Dak Sze in 1979, and several companies in the Washington, D.C., area including Science and Technology Corporation (STC), Science Applications International Corporation (SAIC), and General Sciences Corporation (GSC). Government agencies and universities, of course, have also had increased research opportunities.

Large general consulting and engineering companies engaged in meteorological research and development have also experienced growth in the 1980s and 1990s related to both performing research and to contributing to technological advances. Among the leading U.S. companies with meteorologists on staff to perform the research and development are TASC, the MITRE Corporation, and Aerospace Corporation. The new technology enhances the abilities to make further progress in the understanding of atmospheric behavior and processes, as well as more improvements in diagnosis and forecasting.

Other technological advances for the atmospheric sciences producing growth of the private sector in the 1980s and 1990s include the WSR-88D Doppler radar, wind profilers, and automated weather observation stations (ASOS). Unisys Corporation built and has been installing the WSR-88D radar network. Unisys sold the part of the company that owns the WSR-88D—the Government Systems Division—to Loral in 1995. Loral is now doing the installation.

Sperry Corporation and Tyco Technology, Inc., a subsidiary of Vaisala, Inc.,[4] manufacture wind profilers. Sperry was selected by NOAA to develop and establish a 30-system network in the central United States to measure atmospheric turbulence by detecting Doppler frequency shifts. The frequency shifts, in conjunction with positions of vertical and oblique beams produced by the profiler, are translated into a vertical profile of wind speed and direction. Eventually NOAA had planned to have a nationwide grid of wind profilers, but budget constraints have put the plans on hold.

The ASOSs are being installed by AAI/Systems Management Inc. across the country. AAI/SMI is an example of a company with capa-

---

[4]Tyco Technology is no longer a separate company. It was absorbed into Vaisalia, Inc.

bilities in several segments of the private sector. AAI/SMI consider themselves to be in the weather specialty segment first, but they also do business as weather system providers, provide weather instrumentation, and conduct environmental monitoring. As technologies have advanced and businesses have expanded, it has become common for most private sector companies to diversify and have capabilities in several of the business segments. Unisys, for example, is expanding into the weather workstation market by retaining a Weather Information Systems group that offers weather data and graphics via personal computers.

One of the more interesting specialty companies that has developed a dramatic and very valuable product for both research and practical applications for a wide variety of activities is Lightning Location and Protection, Inc. (LLP). Dr. E.P. Krider of The University of Arizona (UA) and Dr. M.A. Uman of the University of Florida, and formerly of Westinghouse Research Laboratories, founded LLP in 1975. [See chapter 11 by Krider, this volume.] Both Krider and Uman had done pioneering work in measuring electric fields of lightning strikes. The Bureau of Land Management had funded Krider's work in 1975 to develop an automatic lightning-direction finder at UA because of the bureau's interest in the early detection of lightning-caused fires. Dr. A. Pifer, the third principal of LLP, joined the company in 1976 and was instrumental in developing the LLP "position analyzer" that receives the outputs of two or more direction finders (usually via standard telephone lines), checks for time coincidence, and then computes lightning locations.

Another technology for lightning detection, time-of-arrival technology, was advanced by Dr. Walt Lyons, CCM, in 1982. Three or more receivers listen to the static that arrives at antennas placed several hundred miles apart, and the locations of lightning strikes are computed from the time of arrival and the distances between antennas. Lyons believed there was a real need for a national lightning-detection network. Lightning, in fact, is the leading severe weather phenomenon cause of death in the United States. In 1983, Lyons formed a company, R-SCAN, to implement his concept of a national network for lightning detection operated by the private sector. At that time, some people had difficulty adapting to the idea of a lightning-detection network with data not provided through government agencies. Nevertheless, he persisted and gradually was instrumental in building a national network for lightning detection over the period 1983–1989, which operated for the last two of those years on a nationwide basis. This network represented a prototype of the current national lightning-detection network.

The two technologies for lightning detection were combined, and now networks of direction finders, position analyzers, and display devices cover a good fraction of the total land surface of the earth. Applications of these networks include research, power utility reliability analysis and crew placement, forest fire detection, weather forecasting, and safety and hazard warning, especially for aviation and space vehicle operations.

The federal government and the private sector

One of the important topics to cover in a history of private sector meteorology is the relationship of the private sector with the federal government regarding the division of responsibilities as they apply to weather data and information and forecasting services. Since Krick was one of the first individuals to offer private forecasting services, it is not surprising that the first government–private sector confrontation was between Krick and the Weather Bureau back in 1935. According to Boesen (1978), the Weather Bureau added special services of their own each time Krick offered a new type of service for industry. Krick lost some customers in the process. The *Los Angeles Times* reported that the Weather Bureau became "the first government forecasting bureau in the nation to offer a teletype system whereby information of interest to a particular group is 'piped' directly into the offices of the latter." Boesen states that many who defected to the government service were soon back as customers of Krick, apparently because of the greater detail of his forecasts in time and space. This points to the fundamental difference in the types of services the private and public sectors provide. Private forecasting companies can supply customized forecasts for their many different types of clients, while, by necessity, the government provides the public with forecasts that are more general in nature, most of the time.

A policy statement issued by the NWS in 1991 (Gross 1992) defines the respective roles and functions of the NWS and the private sector and the general criteria upon which the statement is based. The basic criteria are as follows.

- The primary mission of the NWS is the protection of life and property and the enhancement of the national economy. Hence, the basic functions of the NWS are the provision of forecasts and warnings of severe weather, flooding, hurricanes, and tsunami events; the collection and distribution of meteorological, hydrologic, climatic, and oceanographic data and information; and the preparation of hydrometeorological guidance and core forecast information. The NWS is the single "official" voice when issuing warnings for life-threatening situations.
- The NWS will not compete with the private sector when a service is currently provided or can be provided by commercial enterprises.
- The private weather industry is ideally suited to put the basic data from the NWS into a form and detail that can be utilized by specific weather and water resources–sensitive users. The private weather industry provides general and tailored hydrometeorological forecasts and value-added products and services to segments of the population with specialized needs.

The above policy statement represents a public–private partnership, but as noted earlier, it was not always that way. In the Compton Report, prepared for the secretary of agriculture in 1940, the recommended policy was for the government not to provide basic weather data to any person or organization that issues its own independent weather forecasts and to allow access to weather data "only to those having approved access which can be accomplished by control of the teletype privilege." In 1947, the Weather Bureau began to provide specialized service on a fee basis in direct competition with the private sector.

In the 1950s the philosophy changed to allow free access to all who requested it through extension hookups to Weather Bureau teletype and facsimile services. The new policy was a result of an initiative of Secretary of Commerce Sinclair Weeks, who appointed an Advisory Committee on Weather Services chaired by General Joe George to encourage the development of private sector meteorology. Recommendations of the George committee were adopted. The policy in the 1950s was similar to the current policy regarding division of responsibilities, except that the Weather Bureau personnel did weather broadcasting on radio and were even allowed to appear on television for a while. It was illegal and continues to be illegal for government employees to be paid for broadcasting the weather on radio or TV. A complaint filed by the American Federation of Television and Radio Artists about Weather Bureau employees not being members of the union led to government meteorologists being barred from regular appearances on commercial TV. That was not the case for radio for some unexplained reason, and government meteorologists did direct broadcasts on commercial radio until 1990 when the practice was stopped by the NWS. During times of hazardous weather or anticipated hazardous weather, government meteorologists may be interviewed on commercial broadcast media. There is generally no objections from the private sector in these instances.

In 1977, the NWS recognized the importance of the role of private sector meteorologists serving industry and created the position of Special Assistant for Industrial Meteorology to maintain close relationships between the government and the private sector. The special assistant handles the interpretation of policy with regard to the relationship of NWS to the private sector.

The NWS historically disseminated weather information and forecasts via direct commercial radio broadcasts, newspapers, and recorded telephone messages. (At one time it provided over 1500 routine weather broadcasts on commercial stations.) The policy of the NWS with respect to mass dissemination of weather information has gradually been changing as reflected by its decisions to discontinue direct commercial radio broadcasts, support to newspapers, and weather information by telephone. This has opened up additional business opportunities for the private sector.

One important point to mention is that the media must relay NWS warnings as written. Individual TV and radio meteorologists may elaborate, but to avoid confusion on the part of the public, they should not downplay the official warnings.

The private sector today—Estimates of current size

In this historical review of the private sector, I have identified 10 private sector business segments for meteorological products and services. From the sources mentioned at the outset, and from the responses received from a representative sampling of companies, I have attempted to estimate the number of companies and the total volume of business within each segment. The database assembled was very large. Built-in time and resource constraints for assessing the information, both with respect to me and to companies and individuals who were asked to respond, necessarily limited the detail available for estimating the size of the segments. Nevertheless, enough information was available to produce the estimates that follow with an accuracy of about ±20%. Table 15-1 provides the estimates of the number of companies within each segment and their aggregate volume of business per year. There is duplication in the number of companies because many of the companies have business in more than one segment. There is no duplication for the total volume for a segment. For companies with more than one type of market segment, estimates were developed for the business in each segment.

The approximate number of companies with meteorologists exceeds 200, not counting "companies" with one meteorologist. The approximate number of private sector meteorologists is 2200. This history has traced a period of accelerating growth of the private sector for the past 50 plus years. With technology continuing to advance, and

TABLE 15-1. Categories of private sector companies and estimated business volume.

Business segment category	Estimated number of companies	Estimated volume per year in millions of dollars
1) Weather instrumentation/remote sensing	23	150–200
2) Weather forecasting services	29	70
3) Weather data/graphics providers	9	50–60
4) Staff meteorologists in industrial companies	5+	15
5) Consulting (except environmental)	41+	60–70
6) Forensic meteorology	21+	3–6
7) Specialty companies (e.g., lightning detection, wind profilers, weather modification)	9	8–12
8) Environmental consulting (only)	29+	200–250
9) Weather system developers/integrators	23	200–400
10) Media meteorology (nonlocal TV meteorologists)	11	20–30
Total		$780–1,100

serious talk of privatization of at least some of the functions performed by the NWS, there is reason to believe that the outlook for the private sector is for continued growth for the foreseeable future.

APPENDIX: Corporate acronyms

AER—Atmospheric and Environmental Research
AIR—Atmospheric Instrumentation Research
ENSR—Environmental Services
ERT—Environmental Research and Technology, Inc.
GCA—Geophysics Corporation of America
GSC—General Sciences Corporation
LLP—Lightning Location and Protection, Inc.
PG&E—Pacific Gas & Electric
PRC—Planning Research Corporation
SAIC—Scientific Applications International Corporation
STC—Scientific technology Corporation
TASC—The Analytical Sciences Corporation
TIC—Travelers Insurance Company
TRC—Travelers Research Center; The Research Center (of New England)
UAC—United Aircraft Corporation
WCA—Weather Corporation of America
WSC—Weather Services Corporation

REFERENCES

Boesen, V., 1978: *Storm: Irving Krick vs. the U.S. Weather Bureaucracy.* G.P. Putnam & Sons, 159 pp.

Fleming, J.R., 1990: *Meteorology in America, 1800–1870.* The Johns Hopkins University Press, 264 pp.

Gross, E.M., 1992: The National Weather Service and the private weather industry: A public–private partnership. NWS Note, Industrial Meteorology Division, 21 pp.

Spence, C. 1980: *The Rainmakers: American "Pluviculture" to World War II.* University of Nebraska Press, 181 pp.

CHAPTER 16

A History of Aeronautical Meteorology: Personal Perspectives, 1903–1995

GORDON D. CARTWRIGHT AND CHARLES H. SPRINKLE

Introduction

Ever since humans looked up to the sky and wondered where the clouds came from and where they went, there has been a symbiosis between weather and flying.

The progress in aviation over the past nine decades is almost beyond imagining. Likewise, its impact on our commercial, industrial, and social systems could scarcely be guessed at the beginning of the twentieth century. To compare a routine international flight in a Boeing 767 to the first flights of man is to bridge a gap in technology and organization almost equivalent to the transition from the Stone Age to the military might of the Roman Empire at the beginning of the Christian era.

The first controlled powered flight at Kitty Hawk carried only the pilot, lasted one minute, and traveled a distance of 852 feet at an air speed of 35 MPH. Contrast these figures with those of a scheduled nonstop flight from Paris to San Francisco. An Air France Boeing 767 two-engine aircraft flying at an altitude of 37 000 feet and at a ground speed of 550 mph carries 250 passengers a total of 5500 miles in 12 hours without landing. It travels a great circle course well inside the Arctic Circle over essentially trackless terrain and oceans. This remarkable development over nine decades was the product of many minds, many billions of dollars of research and development, and a substantial loss of life.

This technological accomplishment that we now accept almost casually has changed our concept of space and time. It has made possible whole new industries, fostered new methods of doing business, and brought the most remote communities into regular contact. In effect, it is at the heart of the global community, along with its essential partner, high-speed communications. It triggered humanity's most daring adventure—the leap into outer space.

How did meteorology follow the spectacular progress of aviation? They were surprisingly harmonious partners, dependent on one another both for resources and innovations arising out of the largely independent technical and scientific growth of each. And for both, progress was greatly accelerated by wars and military demands. As described more fully in other chapters of this volume, meteorology has had a long but scarcely spectacular history. The atmosphere has been too vast a laboratory in which to manage experiments, its ranges of moods too great, its intricacies of motion and changes too varied to submit readily to detailed scrutiny. But its ever-changing presence, its obvious life-giving properties, and its often spectacular and sometimes frightening moods captured the human imagination from the earliest times.

But how to penetrate this nebulous mysterious blanket that both protected and threatened man's comfort and sometimes his very existence? One of the great aviation pioneers, Dr. Albert Plesman, saw the atmosphere as a medium of transport when he said, "The ocean of air unites all peoples"—words inscribed on his bust at the international airport that bears his name on the Dutch Island of Curacao. It may be more pertinent to those who devote much of their professional skills to unraveling these atmospheric mysteries to note that humans can live a month without food, a week without water, but only a few minutes without air. The remarkable fact is that the atmosphere serves as the common thread of our planet, uniting all living creatures irrevocably connected to it. Perhaps Plesman's words should be modified to serve as the underlying motif behind all atmospheric science: the ocean of air sustains [connects] all the earth's creatures large and small.

Weather and aviation to 1939

The Weather Bureau finds a home

At the end of the nineteenth century, meteorology—or rather "the official weather service"—had only recently found a more or less permanent home. After some 20 years as a branch of the Army Signal Corps, responsibility for providing weather information to the American people was transferred in 1890 to the Department of Agriculture, where it remained until 1940.

In the 1890s the westward movement of the population and particularly the farmers brought new and sometime difficult impacts from weather events over the Great Plains and the Mississippi Valley. The Department of Agriculture had expanded to assist these pioneers to begin a new life in an environment often quite different from the climate and weather patterns of the country east of the Appalachians.

As had been learned from several earlier attempts, both private and governmental, to set up an organized weather service, the basic need was for an institution with a broad national organization able to build and sustain the network of weather and climate observations across a very broad land. Recognition of the need for warnings of severe storms on the extensive coastlines and on the Great Lakes was the Army Signal Corp's initial responsibility. Soon, the weather service added warnings of sudden cold waves and heavy snows in the great heartland of the nation, with special attention to the vulnerability of both cattle and crops.

Professor Harrington, the first chief of the U.S. Weather Bureau, was well aware of these needs and set out to organize his service accordingly. Consequently, the Weather Bureau's forecast structure was organized around three major forecast centers, where the principal public forecasts for all regions of the country were prepared and issued. Initially these centers were in Washington, D.C., Chicago, and San Francisco. Others were added as the nation's needs for more and more specific weather information grew.

Local and regional information was provided by the weather offices in most of the larger cities. These offices were most commonly located in federal buildings, where daily maps were prepared on the basis of reports collected by observers and redistributed by the Western Union Telegraph Company to each Weather Bureau office. Forecasts for "tonight and tomorrow" were issued by the local official in charge, based essentially on the broad guidance forecasts for the area issued by one of the three major forecast centers.

Before radio and extensive telephone services were available, these forecasts were posted in a few public places and appeared in the local newspaper. Warnings of severe weather were displayed by flags on towers in the city or at strategic ports. When one considers the character of this national system, it is evident how limited the amount of weather information available to the general public was, at least until the beginning of regular public radio broadcasts. It is interesting that

KDKA Pittsburgh, starting in the early 1920s, prided itself as the "Pioneer Broadcast Station in America." Obviously, a major change was needed to meet the highly specific needs of a growing national aviation system.

Humanity takes to the air

It was the morning of 17 December 1903 when Orville and Wilbur Wright placed their fragile wood and canvas biplane on the little wheeled cradle that served as the takeoff mechanism. They were facing a 20-MPH wind under the influence of a high pressure system centered offshore to the northeast. The Kitty Hawk Weather Bureau had provided them with climate information that convinced them that Kitty Hawk was a desirable place to begin the history of manned flight. The little Wright Flyer, with Orville lying prone like a modern hang glider pilot and their own Taylor-built engine whirling two, two-bladed propellers, the first *controlled and sustained* flight carrying a man, flew for 12 seconds. The fourth flight that same day lasted 59 seconds and covered a ground distance of 852 feet. A sudden gust of wind overturned the plane, ending further flights that day. Although the press was not on hand, world history was made that day. And there has been no turning back since.

Nearly three years elapsed after the Kitty Hawk flights before the Wright brothers had designed and built a practical airplane. Manned flight was now a proven fact. In 1909 the Army Signal Corps bought its first airplane from the Wright brothers for the grand sum of $25,000. Military use of the airplane had begun. Since that date, the military has been a major contributor to aeronautical research and development, as well as to meteorological science and practice.

World War I and airmail

Two events speeded the transition to a commercial civil aviation transportation system in the United States. The first was the result of U.S. participation in the war in Europe, when a surprisingly large number of airplanes were built to support the American Expeditionary Force in France. A large cadre of new pilots was needed to fly them.

By the end of World War I, the U.S. military had 14 000 airplanes with some 250 000 personnel to support their operation. In many respects it was this large reserve of both airplanes and pilots that enabled civil aviation to become established in America soon after the end of the European war. We now accept without question the intimate support that meteorology provides to all versions of the aviation industry, from military aviation to the casual sports pilot. Such remarkable personalities as pilots Billy Mitchell and Eddie Rickenbacher became well-known public figures as the war ended. Much of the aviation material and planes, as well as the thousands of trained pilots, found their way back to the United States where they became invaluable resources needed to respond to the growing public interest in flying.

But most of these military airplanes were scarcely designed to carry passengers. Fortunately, the second major influence came into play as the post office saw in this resource the possibility of speeding up airmail deliveries between major U.S. cities. In mid-1919 the first airmail was carried by Army Signal Corps planes between Washington, D.C., and New York City. In spite of the lack of sound operational procedures and support, it was immediately recognized as a viable means to speed mail services in a nation as large as the United States. Consequently, airmail service was soon started between New York, Cleveland, and Chicago.

With this growth it became evident that such activity was hardly an appropriate military function. Thus, in 1925 the U.S. Congress authorized the post office to grant contracts to civil pilots to carry U.S. mail. This was a landmark decision that provided a financial base for the fledgling airlines and pilots who got their start from the early airmail contracts. These contracts are still a source of some funds for today's commercial airlines, including most international carriers.

It is not entirely clear how these early airmail pilots got weather information for their flights. In some cases, the Weather Bureau city offices were called upon to give weather information. It is assumed that pilots themselves exchanged notes on weather along their routes for pilots on later flights, or used the telephone to call their destination airfields.

The Air Commerce Act of 1926 and a national air route structure

By 1926, it was evident that formal action by the U.S. Government was needed in order to set up a suitable national air route system, with its supporting services. Thus, passage by Congress of the Air Commerce Act of 1926 provided this essential support funding. Besides providing safety service along the established air routes (there was now a transcontinental route from Newark, New Jersey, to Oakland, California), it authorized the Weather Bureau to provide weather information in support of air commerce in the United States and to carry out necessary research for the improvement of that service.

The established air routes were equipped for night flying by use of rotating light beacons 10 miles apart on routes between major cities. Intermediate airfields were also built to provide for weather observations and emergency landing facilities. There were no regular radio navigation aids. Air-to-ground radio systems were added as the technology became operational. Traffic control at airports was provided by the local airport authority. Air route traffic control was started by a few cooperating airlines for whom congestion of converging routes raised some serious safety problems (e.g., approaches into the New York area). Soon it became a government function.

Aviation weather forecast centers

As the air route system grew, full-time aviation forecast centers were established at the leading airports, and a system of hourly airway weather reports gradually emerged. In support of this, a dedicated system of teletype circuits was organized by the Civil Aeronautics Administration (CAA). The circuits paralleled all the major routes and carried all kinds of information for aviation, including NOTAMS (Notices to Airmen) and airway weather data.

Several aspects of this communications system may be of interest to meteorologists. The "airway weather reports," as they were termed to distinguish them from the twice-daily map or synoptic reports, were designed for ease of reading by aviation personnel and for fast transmission over teletype circuits. They contained mostly visual observations, as the instrumental data from the standard weather instruments of the day were quite limited. The observations normally consisted of the following information: cloud cover, ceiling, visibility, present weather, temperature and dewpoint, wind direction and speed, barometer reading or altimeter setting data, and plain language remarks.

Upper-air data were limited to that from optically tracked pilot balloons, and so-called ceiling balloons, for determining the height of lower clouds. Both of these systems relied on a nominal rate of ascent of the balloon. Errors were often introduced by deviations from these rates in unsettled weather. The lack of information on upper-air pressure levels made it impractical to draw charts of upper winds or their height distribution. The few upper-air observations by airplane were still too scattered to make upper-air charts practicable. Pilot balloon data were shown in tabular form for each major route.

By the late 1920s, aviation forecast centers had been established at major air route terminals such as Newark, Cleveland, Chicago, Kansas City, Denver, and Oakland and were in full operation 24 hours a day, with forecasts issued every 6 or 12 hours. Other centers were set up as the route structure grew.

Even at this early date in the development of an airway system, the need to provide weather information beyond the limits of the forecast office itself was growing. The concept of the dispatcher and operations office was steadily taking hold, even though most weather briefing of pilots took place in the Weather Bureau offices.

Weather maps over teletype

As a consequence of this trend, an important development was undertaken by the Weather Bureau and the CAA: the regular transmission of aviation weather maps over the teletype circuits, an innovation that preceded by several years the use of facsimile for transmissions of weather charts. It was a simple and ingenious arrangement that worked as follows. A base map covering the forecast area was designed to fit into the regular page teletype machine. The duty forecaster at the aviation forecast center (e.g., Newark, Cleveland, etc.) plotted on this special base map the current weather data in airway symbols for all the important airports and then drew in the fronts and isobars. This plotted base map was then inserted by the airways communicator into a page printer. He then proceeded to overprint the data in the usual teletype symbols, using periods and plus signs to delineate the isobars and fronts. This resulted in a punched teletype tape. Before the scheduled transmission time, each station on the circuit inserted a similar base map in their page printer. On a given bell signal, the communications center turned on the tape transmitter . . . and presto! In a few minutes, every station on that teletype circuit had a printed weather map for the area. All they had to do was connect the dots and pluses to get the isobaric and frontal pattern. The base map and the data typed thereon were in ditto ink, so copies could be made for distribution to the airline offices and the airport personnel.

Central analyses and the forecast system

Because the weather patterns from adjoining forecast centers preparing these maps did not always match perfectly, it was decided that a small center would be set up in Washington, D.C., to provide countrywide analysis of fronts and isobars that could be sent in coded form over the teletype circuits and used by each aviation forecast center as guidance. It was hoped this practice would eliminate significant differences in each center's maps. Thus, the simple teletype weather map was, in effect, the instigator of a much more fundamental concept, the central analysis system. By the time World War II started, the Washington Central Analysis Unit was an important adjunct to the whole forecast system in the United States. It has now grown to a global feature of fundamental importance in international meteorology, with three world centers serving the World Meteorological Organization's (WMO) World Weather Watch System.

What has been described so far is largely the structure of the aviation forecast system that evolved during the formative years of civil aviation and the air route system in the United States in the late 1920s. During this period, the daily public forecast system, based on the three major forecast centers and the daily collection of synoptic reports by the Western Union Telegraph Company, continued as a separate program. More will be said herein about how these systems eventually merged as communications steadily improved and the organization of the public and aviation forecast responsibilities were progressively integrated.

What was the scientific basis of these two systems and how did they eventually move to a common ground? During the era of civil aviation's rapid development, virtually all weather service was provided by the government, either by the Weather Bureau or one of the two military services—the U.S. Navy Aerological Service or the Army Air Corps Weather Service. In these three services, during the period of the early growth of aviation, most of the training of weather personnel was by the service itself. There were only one or two universities offering courses in meteorology, mostly in the geography departments, where the emphasis was on climatology. While there was a growing body of scientific theory on the structure of storms, little of this had been formalized to be of use to the daily forecaster.

Knowledge of the revolutionary work of the Bergen School was scarcely available to the regular duty forecaster, even though a number of Bergen School articles had been published in U.S. scientific journals, including the *Monthly Weather Review*. The average forecaster, both in the public forecast field as well as in aviation, worked essentially on the basis of weather analogs. That is, they depended on their study of previous weather systems and local experience as an observer or forecaster in a given locality to guide them in predicting the course of the day's weather systems depicted on their 0800 LT chart. The outstanding practitioner of this system was Charles L. Mitchell, who came into the Weather Bureau as an observer early in the century and progressively worked his way up to become lead forecaster at the Washington, D.C., center.

The forecasting routine at the major Weather Bureau centers worked as follows. Each center had only two regular forecasters on duty. There was a small group of map plotters and assistants, but only the designated forecaster was permitted to issue a forecast for public distribution to the press and other users. Each forecaster was on duty for a full month, making all the official twice-daily forecasts for the entire month. The following month, his counterpart took over, while he was relieved of further official forecasting duties. The practice however, was for the person off duty to spend whatever time he felt necessary studying the daily maps of past years for the upcoming month. In other words, he was renewing his store of weather map "analogues" that would be called forth from memory each time he examined the weather map for the current day. During the years leading up to World War II, the two Washington forecasters were Mitchell and Richard Hanson Weightman, both men of long experience in eastern U.S. weather events. Indeed, they collaborated in writing a text containing the essence of their experience in the study of weather systems and the prediction of the weather for "tonight and tomorrow" (Weightman 1940).

Further developments in commercial aviation

Let us return for the moment to the further development of commercial aviation and its effects on the aviation weather service. In the

1920s, it was still more of an adventure than a pleasure to "go by air." However, the need for speed was a compelling incentive for American business. As the nation recovered from the Great Depression, commercial air travel expanded rapidly. To keep up the safety standards, the airways weather service (as it was usually called) had to grow with it, with official observations, forecasts, and briefings required for all commercial civil flights.

One of the pleasant features of these early days was the absence of the airline operations office, the pilots being pretty much in control. Pilots got their weather briefings by going personally to the Weather Bureau office. There the government teletype system carried the hourly and special airways reports from all stations on the air route, as well as route and terminal forecasts (in an abbreviated language and format). The radiosonde did not come into regular use until the late 1930s, so upper-winds data depended on the optical pilot balloon measurements. Nor had instrumental measurement of cloud ceilings or visibility been developed.

In fact, many airports, including an important terminal such as Cleveland, had no established runways. There were field boundary lights and a few special flood lights for night landings and takeoffs. The modern airport was still a decade in the future. Most airports had small passenger buildings and a couple of hangars, with an office for the airport manager, a weather office, and a simple airport control tower. A fence kept passengers from wandering onto the field. The landing aircraft taxied directly up to a simple gate, blasting dust or rain onto the 5–10 passengers ready to board the flight. Since the planes were small—they carried only 10–15 passengers—and of tailwheel design, there was no need for a special boarding ramp: a simple step was adequate. The crew unloaded the mail sacks and baggage for delivery to the local airmail post office. It is recalled that the elite transport planes of the 1930s were the Boeing 247 and the Douglas DC-1 or -2. Simple instrument flying was possible, but there were only rudimentary radio navigation systems, and air-to-ground radio communication was just coming into regular use. (Lindbergh made his historic nonstop flight from New York to Paris in 1927 without radio contact to save weight during his 33-h flight.)

Changes and improvements came mostly from local pressures by the primitive airlines, by pilots, and by local airport authorities who wanted to boost the reputation of their local airport, which they had largely financed. Experienced weather personnel at major forecast offices also provided incentives for improvement. (C.G. Andrus at Cleveland, a former Signal Corps balloonist, was such an innovator and critic.)

In the early years of aviation, turbulence was an even greater hazard than it is today, as the aircraft were smaller, with lighter wing loading, and were therefore more sensitive to turbulence. Clear-air turbulence was not yet a concern, since cruising altitudes were still well below the levels at which such turbulence is common. On the other hand, aircraft icing was a major hazard in winter and in more northern latitudes during all seasons. Since most aircraft were of fabric construction, an inflatable boot on the leading edge of the wings and de-icing fluids for propellers were common solutions to in-flight icing. Hail and lightning were significant hazards, but most pilots were wise enough to detour around large thunderstorm areas.

Recalling the pace of development at that time, it is surprising that there were virtually no printed formal instructions or procedures for making observations or forecasts, or other technical manuals for guidance at aviation offices. (The standard instructions in the Weather Bureau came out in a simple eight-page printed circular called *Topics and Personnel,* which contained mostly administrative instructions.) Technical procedures were covered mostly in letters from the responsible division in the central office in Washington, D.C. This explains in part the difficulty of tracing the evolution of observing and forecasting practices during these early decades. It was only when involvement in World War II faced the United States that the need for formal manuals for both the civil and the military weather units became pressing. Specific action was taken to provide this unifying material for the rapidly expanding military and civil weather services.

Polar fronts and airmass analysis reach America

However valuable the memories of forecasters like Mitchell and Weightman were in the day-to-day business of weather forecasting, it was difficult to pass on their techniques, and the newer technology was proceeding without them, particularly for the much more specific demands of the aviator. As will be seen from other chapters of this volume, the Norwegian School techniques could not be kept from U.S. forecasters. The Guggenheim Fund tackled the matter in a forthright manner and brought Professor Carl-G. Rossby, one of the outstanding members of the Bergen School, to the United States to help develop improved forecast services for aviation. In 1927, Rossby gave advice on the characteristics of a "Model Airway Weather System." A West Coast north–south route was selected for the experiment, which lasted about 6 months. The results were useful for further development of the national aviation weather service. By the early 1930s, the Bergen School was bringing significant changes in forecasting techniques to some parts of the U.S. weather community, particularly to aviation weather forecasters such as H. T. Harrison, then stationed at the Cleveland Aviation (Airways) Forecast Center.

The military weather services were also alert to improving the forecasts for their aviation and for their general military needs. This process was greatly hastened by the efforts of a young navy aerologist, Lt. F. W. Reichelderfer. Remnants of the weather staff that served with the Army Signal Corps in France continued to have an influence on developments in the United States, especially in aviation weather support. For example, Reichelderfer and W. R. Gregg of the Weather Bureau had provided weather support for the flight of the Navy NC-4s, four of which in 1919 attempted a mass flight from North America to England via Newfoundland, the Azores, and Portugal. This early attempt at a trans-Atlantic crossing supplied further evidence of the urgent need for more weather information, especially in the upper air. But this need had to await a decade or more of technological development to find a satisfactory, practical answer.

Before World War II, the big impact on forecast techniques came when Reichelderfer returned from a year of studying the Norwegian methods at the Bergen School. In order to improve forecasting in the U.S. Navy, Reichelderfer wrote a set of practical guidelines on the Norwegian analyses and prediction that was titled *Reichelderfer's Notes*. These were marked "Restricted," a classification that greatly speeded their informal circulation among Weather Bureau and airline meteorologists. Later, Reichelderfer commented that he had done this not for any security reasons but because he regarded the notes as quite incomplete and hoped they could be properly enlarged before final printing. In any case, they became the aviation forecaster's bible and had a major influence on bringing the Norwegian methods into common use in the United States (Reichelderfer 1932).

By 1938, the threat of war in Europe had grown to ominous proportions and the possibility of U.S. involvement could not be overlooked. All U.S. military services began to expand, including the weather services. The need for additional weather support was obvious, as was the need for a large number of scientifically trained forecasters. The formation of meteorological departments in several of the leading universities and technical schools was followed soon after by the formation of special highly concentrated courses to accelerate this training need. Soon, a number of the leading practitioners of the Norwegian methods came to the United States to teach in the new departments. By the time the United States actually entered the war in 1941, thousands of meteorological cadets had been given advanced training in the use of Norwegian forecast methods.

Another major influence in the rapid introduction of newer forecasting methods into the Weather Bureau came with the appointment in December 1938 of Cmdr. Reichelderfer as the chief of the Weather Bureau to fill the post left vacant by the death of Dr. Wills Ray Gregg. Reichelderfer was chosen by a panel of top scientists, headed by Dr. Andrew Millikan, who had been active during World War I in the formation of a weather service to support the AEF in France. Reichelderfer's close associations with the Bergen School and most of the premier weather scientists in Europe were invaluable assets in the rapid assimilation of new methods into the Weather Bureau system. As a first step, he appointed Rossby as his deputy. This immediately added significant scientific stature to the position of the Weather Bureau and accelerated further developments in the whole field of meteorology in the United States.

Fortunately for Reichelderfer's new scientific programs in the Weather Bureau, the radiosonde was being introduced rapidly into the basic upper-air network in the United States, Europe, and the Soviet Union. It soon replaced the costly and limited airplane soundings. Perhaps more than any other single development, the radiosonde brought a major breakthrough in weather analysis and forecasting. With the establishment of routine upper-air analyses into the stratosphere, weather services around the world were now in a better position to meet the newer challenges of air commerce for more reliable forecasts at higher and higher flight levels. These new data were also laying the foundation for new analysis techniques (e.g., Rossby waves) that would further improve weather predictions.

But to return to the aviation side of this equation, rapid development of newer aircraft that were larger, faster, and designed to fly at higher levels added to the airline needs for an expanded weather service but also brought an additional source of weather information on upper-air conditions. Improvements in landing systems increased the pressure for all weather operations and thus for more refined forecasts of cloud and visibility at major terminals. Longer flight ranges also meant longer forecast validity periods.

OMNI Range navaids tended to reduce the adherence to fixed route structures, and the introduction of air-route traffic control brought new requirements on the Weather Bureau to assist in the reduction of delays around major airports and to reduce the hazards of landings in minimum weather conditions. Observations on multiple runways and under night lighting made it necessary to review the visibility and cloud measurement and reporting systems.

Trans-Pacific flights

In 1936 Pan American Airways (Pan Am) was pioneering air routes across the Pacific from San Francisco to Honolulu, Guam, the Philippines, and the southeast coast of China. Soon thereafter, Pan Am began exploratory flights to New Zealand via Honolulu, Kingman's Reef, Fiji, and Aukland. Because of the lack of organized weather services in many of these areas, Pan Am hired its own meteorologists and set up its own observing network to augment the official observations, where these might exist. It even designed and had fabricated simple observing equipment that could be quickly set up in remote areas. A pilot balloon system was developed that included a hydrogen generator of French design and a one-man theodolite of German design. The basic Pan Am observing packages were complete and could be readily installed for operation in jungles, on islands, or on major land areas. A system of company logistics ensured that supplies were always available (E. B. Buxton, personal communication). By 1936, Pan Am meteorology divisions had been established in Latin America and the Pacific regions.

It was Pan Am's aggressive approach to international air routes that gave it a head start over other American airlines. It was only after they had served as part of the Military Air Transport System in World War II that other U.S. airlines moved into the international field. These initiatives led Pan Am to seek U.S. Government designation as "the chosen instrument," that is, the U.S. flag carrier on all international flights. Other major U.S. air carriers strongly resisted such a move. As a consequence, there are now half a dozen U.S. companies vying for business on global air routes.

World War II and global aviation

As the pace of war preparations increased, aviation was in the forefront of U.S. plans, should the country be drawn into the war on the side of the Allies. By the time the war had started in Europe in September of 1939, trans-Atlantic flights were beginning, initially with flying boats. LaGuardia, the most modern airport in the United States and the first to be located within one of the New York boroughs, had the foresight to provide a marine terminal for these trans-Atlantic flights. These flights brought a whole new dimension of responsibilities and practices for the Weather Bureau.

At this stage, Canada, the United Kingdom, and Ireland were primarily concerned with supporting flights. Together with the United States, an agreement was worked out to provide the essential support for these new adventures into air transport. The result was the Trans-Atlantic Air Safety Service Organization (TASSO), which laid down international procedures for the support of these unique flights. These procedures became precedents for the work of the International Civil Aviation Organization (ICAO) after the war.

The Atlantic was a harder nut to crack than the Pacific, and it was due only to the enormous financial and logistic support made possible by the preparations for U.S. entry into World War II that U.S. commercial airlines operators were brought into the competition. Demands of the Allied commands forced the establishment of route facilities that would permit the ferrying of all types of military airplanes from North America to Europe. By 1941, hundreds of flights a week of land planes were being made across the North Atlantic. At the same time regular civil flights by seaplanes were being made on routes from New York via Bermuda to Lagens, in the Azores, and then to Lisbon, Portugal (Tagus River), and Foynes, Ireland. It was clearly essential to get more upper-air data from the open ocean itself. Fixed ocean weather stations were put in place by the United States, Canada, and the United Kingdom, reaching a total of some 22 locations before the war ended.

This can be illustrated by considering the efforts made by the U.S.

Navy to support the mass NC-4 flights. To do this, U.S. Navy ships were stationed along the expected route at intervals of about every 50 miles. Four major ships of the line were stationed on either side of the route to come to the assistance of any downed aircraft. In the light of these "requirements," one can understand the wartime solution of some 22 fixed weather stations, spotted strategically over the North Atlantic at the height of World War II, which provided both regular upper-air and surface observations, as well as ground-to-air communications and a search and rescue facility.[1]

It was not only in the North Atlantic that the war brought about enormous expansion of air routes. With both the European and the Pacific theaters expanding virtually in all directions, the war zones took on global dimensions. Consequently, aviation support facilities also became a global need, with new routes being extended to virtually every country in the world. An immense expansion of military and civil transport had to be made in an incredibly short span of time. Military weather services expanded accordingly, with thousands of new staff in the American services alone. Allied services were under similar pressures. In the space of two or three years, the entire globe was the operating area of the most advanced transport aircraft, with improved designs coming out each year from the burgeoning aircraft factories in many countries.

Throughout the war, weather data and forecasts were classified as military security information. A complex set of coding devices was quickly prepared and distributed to all Allied Forces. In a rare gesture of military confidence, most of these codes were printed by the Weather Bureau in its own printing plant at 24th and M Streets NW in Washington, D.C.

Developments in communication

By the early 1940s, the domestic airway weather support structure was well in place. There were 12–15 aviation forecast centers (depend-

---

[1]This system of fixed ocean weather stations was taken over by the ICAO after the war, with an initial network of 13 stations. Later it was progressively reduced as the reliability of large commercial air transport aircraft improved and the radio communications system was able to reach across the entire span of ocean. The only civil aircraft to crash in the mid-Atlantic was a charter flight Boeing 314 seaplane, the *Bermuda Sky Queen,* which ditched safely near one of the U.S. Ocean Weather Ships. All passengers were safely taken aboard the Coast Guard cutter that had steamed to the rescue site. The seaplane remained afloat so long that it had to be sunk by gunfire from the frigate.

ing on the period), with dedicated 24-h teletype circuits operating at 75 words per minute and an experienced central analysis unit in Washington, D.C., sending coded frontal and isobaric analyses to all aviation centers. The teletype symbol–based airways codes were used for the distribution of all U.S. aviation weather information. The exchange of synoptic reports for the preparation of public forecasts was carried on, but the Weather Bureau had abandoned the old word code in favor of the international synoptic code adopted by the International Meteorological Organization (IMO) just before the war.[2] The airways weather teletype codes continued to be used long after the war, as they were easily read by all U.S. aviation users and were based on the characteristics of our teletype keyboards.[3]

Introduction of radio communications, both for air–ground and for radio aides to navigation, obviously had a major impact on the safety and efficiency of U.S. airline operations. But how did this development affect the character of weather support for these operations? Perhaps the principal impact was the great expansion of all-weather flying. This put much more pressure on the weather offices for more details on the vertical distribution of clouds. The tops and bottoms of multiple layers became a more significant operational matter, especially in winter when aircraft icing was more likely. It also made it possible to alert pilots to new developments and changes in the forecasts provided before takeoff. Use of radio navigation aids, especially for landing and takeoff, and improved runway lighting brought about a lowering of the minimum operating limits at well-equipped airports.

These new operating limits made it necessary for the forecasts to provide information on a more and more refined character at terminals. With the increase in size and range and operating altitudes, the forecasts had to predict winds, cloud, and visibility conditions at the new operating altitudes. Air-to-ground radio communications also made possible regular in-flight reporting of weather en route, a major

---

[2]This action was taken at a meeting of the IMO Commission for Synoptic Weather Information (CSWI) in a conference in Berlin attended by U.S. representatives. (I. R. Tannehill, USWB, was the principal delegate.)

[3]The United States has finally agreed to introduce the international METAR figure code format beginning 1 June 1996. The United States will continue to use the present format for terminal forecasts for domestic airports until that same date, when it will begin the use of the international terminal forecast figure code, TAF, for all terminals. It has been using this format for international airports for some time.

improvement in the data sources that even today is a fundamental part of the aviation weather picture. The reliability of the in-flight reports depended to a great degree on the character and experience of the individual pilot and at what stage in the flight route he happened to be.

Discipline enters the air route system

Because of the rapid growth of the U.S. air route structure, Air Route Traffic Control Centers (ARTCCs) came into operation on a national basis at about the same time the United States entered the war in Europe. Twenty centers were set up at key route intersections, and a system for "flight following" was put into effect. That is, all flights operating under Instrument Flight Rules (IFR) were required to file flight plans and to keep ARTCC informed of their progress. As a backup to this information, the center itself followed the progress of every IFR flight by a simple system of "flight progress boards" organized for each air route.

It was obvious that weather conditions had an important effect on the orderly movement of air traffic, particularly the conditions at major air terminals. In 1941, an agreement was reached between the CAA and the Weather Bureau to provide a small forecasting unit in the ARTCCs to be available for interpretation of current weather information in terms of their impact on traffic and to alert the controllers of impending weather events that could cause major disruption of traffic flow. This arrangement lasted until 1947, when expanded Flight Advisory Weather Service (FAWS) units were set up in selected Weather Bureau facilities to issue such information as flash advisories (later termed AIRMETs and SIGMETs). They also prepared area and terminal forecasts and hourly pilot weather report summaries, as well as provided pilot weather briefings. This arrangement lasted until 1982, when selected forecast services for aviation were centralized in Kansas City.

Documentation for ocean flights

Commencement of long over-ocean flights as part of the U.S. war effort and the adoption of the TASSO procedures for civil flights required a new set of analyses and forecast procedures. New base weather charts to cover the entire North Atlantic and adjacent continents were needed, as were new forms for flight documentation. In contrast to the practice for domestic U.S. flights, the flight documents had to be issued to each overseas crew. In the initial days of these operations, the documentation for each crew consisted of a copy of the appropriate segment of the surface synoptic chart, a vertical cross section for the particular route that could be divided into segments chosen by the forecaster on the basis of the current synoptic picture, and a set of terminal forecasts for the destination airport and selected alternates covering the total period of the flight plus several hours in case of delays in the flight. Destination airdrome forecasts were usually based on exchanges from the cooperating weather services, that is, Ireland, the United Kingdom, and Canada. Each part of the cross section, which was divided vertically to picture the highest altitudes expected for the flight, included cloud distribution, frontal positions, wind and temperature averages for the zone, icing, turbulence, and plain language remarks. A confidence figure in percentages for each zone could be indicated by the forecaster.

The aircrews came to the weather office about two hours before takeoff for oral briefings by the duty forecaster. These briefings were a valuable means for informal personal assessments of the accuracy of the forecast services and the kinds of problems being encountered. In a few instances, these briefings took on a very personal nature. In one exciting case an aircraft commander physically assaulted the forecaster for an erroneous forecast for a flight to Lisbon that could have resulted in disaster. These exchanges were not easily forgotten and enhanced the forecasters sense of responsibility.

Because the flights relied on celestial navigation, a navigator was an essential part of the crew. The principal tracks used across the North Atlantic were the Great Circle and the Rhumb Line. The former was used mainly on westbound flights since it carried the aircraft on a more northerly track, where the westerly winds were normally lighter. The Rhumb Line resulted in a more southerly track, often preferable for eastbound flights.

Experience during the early days of seaplane flights over the North Atlantic showed the urgent need for more accurate forecasts of sea and swell conditions at the landing sites, especially at Bermuda and the Azores. The constraint of sea surface conditions unsuitable for landing was one of the inhibiting factors in the use of seaplanes for long ocean flights, as well as their less efficient design in comparison to wheeled aircraft with retractable landing gears. Pan Am, faced with long and costly delays at the Azores, especially, sought the help of Dr. Walter Munk, a well-known oceanographer of that period. Munk developed a set of rules for predicting the sea state at Horta that was suitable for use by the weather forecaster on the basis of predicted pressure gradients in the area.

Land operations

At the same time as flying boat operations were being carried out over the central Atlantic, air routes were being developed to permit shorter-range wheeled aircraft such as fighters to get from North America to Europe. These routes took advantage of the shorter gaps between potential landing spots on land areas. They usually started in eastern Canada (Gander, Newfoundland, and Goose Bay, Labrador); proceeded to Greenland and Reykjavik, Iceland; and then on to Ireland (Shannon) or Scotland (Prestwick). From there, they fanned out to the military dispersing points in England. The full cooperation of the Canadian authorities, both military and civil, was vital to the building of this route structure and its supporting weather services. Forecasting responsibilities were shared between the civil and military services of Canada, the United States, the United Kingdom, and Ireland.

The base in Gander was used for longer-range planes—bombers or military transport—or Allied civil transport aircraft under contract with one or the other military services. These routes thus became the normal transport means for land planes, and the majority of air passengers flew these routes throughout the war. Many passengers will remember refueling stops at Gander on all east- and westbound flights. After some 18 hours of uncomfortable flying westbound in a noisy propeller-driven DC-4 or Canadair, the one or two hours waiting at the boring Gander terminal for the onward flight to Montreal or New York was scarcely a memorable experience. On the other hand, the Gander base was a critical link in the Allied transport system from North America to Europe throughout the war. It was said that the Canadian meteorologist in charge of the center, Dr. P.D. McTaggart Cowan, made the trans-Atlantic forecasts for all these flights during a period of some two years! He later became controller of the Canadian Weather Service.

When the Soviet Union came into the war on the Allied side, routes to Alaska and on to Siberia were also established. Here again, the full cooperation of the Canadian authorities greatly aided these air operations. This land route system required the expansion of basic meteorological and forecast services, as well as other defense and development measures. This brought about a rapid buildup of the weather observing networks throughout the whole of these northern latitudes, including Alaska and Greenland. In order to serve Soviet needs for military supplies in their eastern states, routes were also opened from Elizabeth City, North Carolina, across the South Atlantic to Liberia, Fish Lake, and beyond.

The analog forecast system supports the military units in the field

Another major program that illustrates the character of weather support for aviation, and for military actions in particular, was the Historical Northern Hemisphere Map Series. Experience had shown that communications were often disrupted or nonexistent in the battle areas. For weather units supporting strategic actions, this rendered them virtually useless, except for the meager deductions that could be drawn from single-station forecasting. Various war plans were made to minimize these deficiencies in weather support. The Historical Northern Hemisphere Map Series covering a period of 40 years of weather maps, from 1899 to 1939, provided a useful collection of charts that could be used for analogs by the isolated field weather units. Teams of map plotters were set up in Washington, D.C., and all possible sources of weather information were called upon to provide the most complete coverage of data ever put together on a single set of charts. These charts were then farmed out to a small group of highly skilled analysts to complete the more than 14 000 daily charts covering the entire Northern Hemisphere. The final series was then published and distributed widely to all U.S. civil and military units in the field.

The next step was to classify each portion of the chart covering each important war theater as portraying a specific synoptic situation or "weather type" for that area. Then the dates for all similar weather situations (weather types) were placed in a special file for each major war theater. Thus, when the current map was analyzed at one of the major weather centers for that theater, the snyopticians classified the current map by types and then searched the analog files over the past

40 years. The dates of similar weather types were then put into a secret coded message and transmitted to the weather units concerned. The forecaster in the theater could then go through his historical map series to find the analogs that best fit the current date. By comparing these maps with whatever data he was able to gather locally and studying the maps for subsequent days, he could make useful inference on the likely weather situation for the next few days.[4]

The post–World War II era

International institutions

As a consequence of these far-reaching changes in global air operations, even before the war ended, policy makers of the Allied governments foresaw a new era dawning for civil aviation on a truly global scale. The United States, now clearly the major world power both in military and industrial resources, recognized that this new instrument for commerce would have to be put on a sound international footing. The following is a quote from a recent account of the 50th anniversary of the ICAO:

> From the very beginning, the development of commercial aviation had been notable for the remarkable foresight of everyone concerned, from designers to manufacturers to airlines to government regulators. There was perhaps no greater act of foresight than the convening of the International Aviation Conference in Chicago in November 1944, when World War II was still very much underway.
>
> The gathering of 52 nations recognized that there would be an irresistible surge of commercial aviation activity in the aftermath of the war, which would transform the world as people then knew it. In this "new world order," international cooperation would be more crucial than ever. The Chicago Convention promulgated the "freedom of the air," by which multilateral relationships in aviation have been ruled ever since.
>
> Out of the Convention was born the International Civil Aviation Organization (ICAO), which is now celebrating its 50th Anniversary. The Montreal based United Nations Agency is dedicated to maintaining orderly and safe commerce on the world's airways, on which it

---

[4]It is recalled that weather typing of synoptic situations was extensively used in making predictions for five days or a month. Such a system was used by Dr. I. P. Krick as a basis for his forecasts for commercial users. For example, he was hired by the White House to make the forecast several weeks in advance for the weather in Washington, D.C., on Eisenhower's inauguration day.

regulates operating practices and traffic control. Governed by an Assembly and a Council of Nations, ICAO sets standards for the licensing and infrastructure of world aviation, provides technical assistance, keeps track of key statistics and mediates international disputes.

As in pre-war days, there was a need for a parallel organization of commercial carriers to work with that of governments. Thus, IATA [International Air Transport Association], which had become a casualty of the war, was revived in 1945. The new organization has it headquarters in Montreal, in close proximity to its governmental counterpart. It has since grown vastly to embrace 224 member airlines which fly to 133 nations. Today, ICAO and IATA work closely together on many matters of mutual concern.

At virtually the same time, IMO, a body composed of the directors of the world's weather services, also prepared to meet the burgeoning demands of aviation, as well as those arising from the new United Nations organization. With ICAO well on the way to setting up procedures and organizing facilities on a global scale for support to the world's air routes, the IMO convened in Toronto in the autumn of 1947 a conference of all its technical commissions to review the state of their programs and the impact of the global war on earlier procedures and practices. The work of three IMO commissions was of specific importance to aviation: aerology (upper air), synoptic weather information, and aviation.

Once these technical matters had been fully reviewed, the IMO members met in Washington, D.C., in September–October 1947 in a conference of directors to draft a convention for a World Meteorological Organization. This draft convention was subsequently ratified by the required number of states (22) and came into force in 1950. The former IMO Secretariat moved into new quarters in Geneva. Thus, the international institutions for building and coordinating the meteorological systems for support of the new global aviation system were well in place by the mid-1950s.

In the beginning of international operations, forecast offices at departure airports prepared the weather information for the entire route of flight. Then on 12 January 1959 the Weather Bureau implemented a new international forecast scheme. Two offices, the National Weather Analysis Center and the Weather Bureau Forecast Office at Idlewild Airport (now John F. Kennedy International Airport), began preparing the operational area forecasts and distributing them in chart form by analog facsimile to airports with scheduled international departures. Within a few months, offices at Miami, San Juan, San Francisco, Anchorage, and Honolulu were added to the system. Each office prepared all required forecasts for its assigned area of responsibility and entered them onto facsimile for transmission. These operations were later centralized at the National Meteorological Center (NMC) in 1972.

The full transition of these wartime actions into the present global air route system is a remarkable story of the actions of many individuals, agencies (both government and private), and the air carriers themselves. Foresight, good sense, and general concern for safety and efficiency have brought about a logically managed global air transportation industry in peacetime that is unparalleled in any other industry. Many leaders shared the views that such a result was a fundamental requirement for a peaceful world. And this philosophy was incorporated into the preamble of the ICAO Convention.

Needless to say, meteorology and most weather services benefited greatly from these international developments. In fact, it was the enterprising work at ICAO international conferences that made clear to the community of weather services under the IMO that speeded up the transition of that body of weather directors into a full-fledged international agency.

Standards and recommended practices in aviation weather

The structure of the ICAO, with both global and regional bodies, was a logical pattern for meteorology as well. In the ICAO system, all operating standards for the different technical branches of international aviation are, when finally approved by the council and the assembly, incorporated as annexes to the ICAO convention. These annexes are identified as "Standards and Recommended Practices," where the distinction between the two categories is easily identified by the active adverbs *shall* and *should:* the former is a *standard;* the latter a *recommended practice.* In the parlance of the Secretariat, they were known as "SARPS" and were applicable to all aviation activities worldwide. Those relating to meteorology are contained in annex 3 of the convention.

To apply these SARPS to actual operations over the world's air routes, ICAO regional air navigation conferences were held to plan and implement the global air route structure, doing this progressively region by region. As further indication of its importance in global air transportation, the North Atlantic Regional Air Navigation Meeting was the first of such meetings, held in Dublin, Ireland, in the winter of 1946.

What was important for the weather services was that these meetings proceed vigorously to lay out the aviation needs for *basic* weather information as well, since the synoptic networks were in a devastated state in many parts of the world. Aviation could not wait for the weak IMO to get these restored quickly. Furthermore, ICAO had recognized the need to build up the aviation services themselves in many parts of the world and so formed a Joint Support Program, under which groups of concerned states could agree to provide funds and equipment to build the missing parts of the air route structure. Meteorology was one of these. As one of its first actions under this program, ICAO took responsibility for the North Atlantic Ocean Station (NAOS) network, as the NAOS Agreement. Soon after, the military weather and other aviation facilities in Iceland and Greenland were put on a peacetime basis under the DEN-ICE Agreement.

These forthright actions were a major stimulant as well as a challenge to weather services and helped immeasurably in the conversion to civil operation of the enormous global spread of military facilities of all kinds developed during the war: airfields, communications, radio navigation aids, servicing and repair equipment, radars, and other weather facilities. All became part of the global transition from wartime to peacetime operation. ICAO was a major instrument in ensuring an orderly transition for air transport, helping to preserve large amounts of this vast store of military materiel so that it could be used in peacetime operations.

Once this orderly transition was established, the airlines themselves, and the thousands of eager travelers pushed the further development at a rapid pace. IMO became the WMO in 1950 and soon thereafter was brought into the United Nations system of specialized agencies. Both organizations, ICAO and WMO, recognized the importance of close collaboration in the support of global air transportation. An agreement to this end was signed by the two presidents (both at that time were American and long-time friends).

The World Area Forecast System

As would be expected in a rapidly developing field such as aviation, there were a number of differences in the responsibilities of the two organizations. Much discussion between secretariats and between experts and policy leaders in the member nations have resolved these to the benefit of all concerned. Joint meetings of both the ICAO Met Division and the Commission for Aeronautical Meteorology are held every few years where common problems are sorted out in very practical ways. Cooperation in basic services led to the present high level of development, culminating in the jointly supported World Area Forecast System.

This system, which started in joint studies by the two bodies nearly 20 years ago, is now being implemented on a global scale. The United Kingdom will begin regular broadcasts via an ICAO-managed geostationary satellite that will provide a comprehensive array of weather data to all countries in the Europe, Africa, Asia, and the Middle East that install the relatively inexpensive ground equipment needed to receive the continuous 24-hour-a-day broadcasts. The completeness of these data will meet all their needs to provide a thoroughly up-to-date service to all aviation activities.

A similar satellite broadcast will be implemented by the United States for the Central American and the Caribbean areas to provide weather collection system that will overcome many of the deficiencies that have plagued the weather networks in these areas for decades. It is, simply, a return uplink to the satellite over which the weather observations that are needed to make the whole system work can be collected quickly at weather centers for redistribution over the same satellite by broadcast to all the ground-receiving stations. The evolutionary steps that led to this apparently "ideal" solution are too numerous to recite. Also, the United States will provide a one-way satellite distribution system for the rest of the Americas and for the Pacific Ocean region.

Airline meteorologists

An important development for aviation weather support was the establishment by several major U.S. airline companies of their own weather staffs. The staffs were familiar with the operating policies of the airline and knew the constraints arising from the different equipment being used, the route patterns, and the company priorities, as well as the particular experience of each air crew. As the operating routines became better established, equipment more reliable, and the government support services more readily available, these private weather groups were gradually phased out. Today, only four carriers, American, Delta, Northwest, and United, have weather staffs beyond that of a special consultant or two.

U.S. Air Transport Association

Another element in the array of groups interested in weather support for U.S. commercial aviation was the U.S. Air Transport Association (ATA), the national counterpart of IATA, funded by the airlines. ATA has a number of technical groups to study problems affecting safety and efficiency of airline operations. One of these is the ATA Met Committee. This group has taken a very active part in advising both the airlines and the Weather Bureau on company weather needs. In fact, the voice of the ATA was often heard during the briefings in Congress on the Weather Bureau's annual budgets.

Coordination

Prior to World War II, requirements for aviation weather services were determined by joint action of the Weather Bureau and the CAA [now the Federal Aviation Administration (FAA)]. These needs were documented in a yearly exchange of correspondence between the two agencies. After World War II and until the about the mid-1960s the requirements were handled by a complex of committee actions and by personal contacts with the aviation industry. In late 1963, a Federal Coordinator for Meteorological Services and Supporting Research was established and became the mechanism for coordinating aviation weather requirements among the federal agencies. The MET Committee of the ATA took on the role of coordinating air carrier requirements. A General Aviation Association Council was useful for a time but is no longer a major figure in presenting the aviation requirements of general aviation. The Aircraft Owners and Pilots Association and the National Business Aircraft Association, with the Regional Airline Association and several helicopter organizations, are now more active.

Under the FAA Act of 1958, agreement was reached to the effect that pilot weather briefings by FAA staff members would be on the basis of actual weather situations (reports) and on forecasts furnished by the National Weather Service (NWS). This agreement is still in general force.

Earlier we referred to the synergy between aviation and meteorology. This is clearly illustrated by the mergers and transfers of responsibilities within the Weather Bureau. As the airport offices grew in size and functions, they gradually absorbed the work of the city offices, merging most of the two staffs. It is almost a case of the tail wagging the dog when one looks back over the manner in which a particular industry forced major adjustments in the operation of a primary government service. A careful study of the different legislative actions to meet the demands of aviation makes this quite clear. One can look at this synergy from another point of view as well. While aviation demanded more and more information on cloud distribution, vertical wind, and temperature structures, as well as detailed wind effects close to the surface, the operation of aircraft at higher altitudes provided much more specific information on the atmosphere well beyond the reach of the standard weather instruments. Thus, aviation has been a major factor in improving our understanding of the atmosphere.[5]

Organizational structure

In the 1950s, the Aviation Weather Service was developed, implemented, and administered out of the Weather Bureau's National Headquarters at 24th and M Streets NW in Washington, D.C. Regional offices were essentially administrative offices responsible for personnel to implement and run the programs developed in Washington, D.C. During the 1970s, regional offices took on a much larger role. They implemented and managed the national guidelines passed down from

---

[5]Research is now under way to perfect a practical airborne humidity measuring device. If this is achieved, it may be that the frequency of routine air reports and vertical soundings during takeoff and landings will provide much more detailed upper-air data with much higher frequency than is possible with the current models of radiosondes. If this proves to be a practical approach, this would be a significant contribution of the airplane to a better understanding of the atmosphere and a daily contribution to greatly improved forecasts for all purposes.

the red brick tower known as the Gramax Building in Silver Spring, Maryland.

Field structure

In the late 1940s and early 1950s, the bureau's aviation forecast service was located primarily in the Air Route Traffic Control Centers. After World War II and into the last part of the 1950s, the Weather Bureau Airport Station (WBAS) was the bureau's interface with the aviation community. WBAS's programs normally consisted of surface airways observations, pilot weather briefings, and preparation of terminal forecasts, along with advice and counsel to the airport control tower and other FAA entities at the airport. In the early 1960s, a significant restructuring of the field services occurred. Nine Flight Advisory Weather Service units were created to serve the entire country. Each unit had a defined group of states as its major responsibility, providing area and terminal forecasts, preparing hourly pilot report summaries, briefing pilots, and issuing AIRMETS and SIGMETS for its area. A quality control officer visited the airport stations within his area of responsibility. Changes began in the late 1970s. Center weather service units were operational in 13 ARTCCs and later expanded to all of the ARTCCs. They remain an integral part of the national aviation weather services. In the early 1980s, however, the FAWS centers disappeared, and the SIGMET function was transferred to the National Severe Storms Forecast Center (NSSFC) in Kansas City. The continuing need for a general flight advisory service was provided by the National Aviation Weather Advisory Unit within the NFSSC.

Research and development

Research and development for aviation weather services has had a rather unusual history. Much of the early R&D in meteorology was justified essentially on the basis of meeting aviation needs, although it was frequently applicable to basic meteorological services as well. Since World War II, most specific aviation weather R&D (turbulence, icing, visibility, airport instrumentation, and observations) has been accomplished outside the NWS primarily by the FAA or the National Aeronautics and Space Administration (NASA) or the National Oceanic and Atmospheric Administration (NOAA) on a reimbursable basis. Prior to 1969, there was no specific aviation R&D program in the Weather Bureau except for Weather Bureau involvement in an air force weather program prior to 1960. In 1964, the Weather Bureau went forward to Congress with an aviation R&D program of approximately $2.5 million. This program was primarily to take over from the FAA the aviation weather development program in data acquisition, data processing, and engineering test and evaluation. It was disallowed; since 1966, there has been virtually no aviation R&D line item in the Weather Bureau or in the NWS budgets.

The airplane as a weather research tool

There are innumerable examples of the use of aircraft for monitoring the atmosphere and for making new research measurements. One of the earliest and most important uses of aircraft was in tracking tropical storms and investigating their structure and intensity. It may be recalled that the first flight into a tropical storm was by Col. Joseph B. Duckworth during World War II. Flying out of Bryan, Texas, in an AT6 Trainer, Duckworth encountered a hurricane between Galveston and Houston. He returned the aircraft to Bryan safely and without damage and repeated the flight later the same day.

This daring act demonstrated the effectiveness and practicability of regular aircraft reconnaissance of tropical storms and led to the organization of regular aircraft tracking of tropical storms. Such flights were set up in the Atlantic and Caribbean, as well as in several areas of the Pacific. Carrying out these reconnaisance missions on a regular basis eventually involved several squadrons of aircraft and thousands of personnel. Until the availability of weather satellite cloud imagery, these programs by both the military and the Weather Bureau continued to be the most valuable tool in the hurricane forecaster's book. But in the end, high costs and improved use of satellite data resulted in the shrinking of a remarkable program, one that stemmed from the very earliest use of airplanes in World War I—reconnaissance of enemy troop movements and battle impact.

Even today, most of the important field experiments of the atmosphere—for example, the WMO Global Atmospheric Research Program—have made extensive use of research aircraft. One of the more recent of these, the Tropical Ocean and Global Atmosphere program, focused mainly on the central Tropics of the Pacific Ocean and depended heavily on wide-ranging aircraft observations that could not be made in any other way.

Aircraft weather reports as part of the basic network

In-flight reports from commercial aircraft have been a requirement almost since the early airmail flights and continue to be a part of the basic weather-observing networks. Air reports from U.S. commercial flights are organized by Aeronautical Radio, Inc., under a system called the ARINC Communications and Reporting System. The frequency of these reports has grown to large proportions. (More than 10 000 are made available every day throughout the United States with a similar number posted daily in Australia.) The National Centers for Environmental Prediction (NCEP, formerly the National Meteorological Center) has a special program to screen and evaluate these reports for their particular value on a given day.

In contrast to the national program, international air traffic control procedures normally require position reports every hour, to which the weather reports are appended. This has meant that over the North Atlantic air reports were bunched on the hourly position reporting times, normally at major intersections with longitude lines.

To increase the number of air reports on less-traveled air routes, an automatic system was developed by NASA, whereby temperature and related position information would be automatically taken from the flight navigation system and an air report would be formatted and transmitted automatically to one of the polar-orbiting weather satellites and then to the appropriate ground station, without intervention by the air crew. The system is designated Aircraft to Satellite Data Relay (ASDAR). To exploit the potential of this system, various groups were formed under WMO at different times to work out arrangements for the production of the aircraft units and for assigning them to commercial airlines willing to carry them at no or minimum cost. This proved to be a much more complex effort than originally anticipated: first it was difficult to find a manufacturer willing to invest in construction of the units without much guarantee of suitable profit, and second, once a number of the units were finally produced (after trying several different manufacturers), it was more difficult than originally thought to persuade carriers with suitable route structures to have the units installed. Furthermore, certification of the air-worthy acceptability of the installations had to be made for each type of aircraft involved.

It was primarily through the dedicated efforts of a few people (particularly Jim Giraytys of the WMO Secretariat and Derek Painting of the U.K. Meteorological Office) that attention to this need was kept alive. An operating consortium of ASDAR participants, with continuing support from a small number of WMO members, has now purchased 22 ASDAR units, and 17 [as of January 1996] have been installed on commercial carriers with aircraft flying on routes where such data are especially useful. The resulting data have been highly praised, both that derived from the cruise period of the flight and, especially, the vertical sounding data taken during the takeoff and landing parts of the flights. The cost per observation under current arrangements is said to be lower than for other types of upper-air observation now routinely available. The present ASDAR system is expected to be phased out by the year 2001.

Technological innovation

As a result of advice from its own staff and several outside aviation bodies, the Weather Bureau responded to the changing requirements arising from new aircraft types, better navaids, and airport landing aids. Because of constrained budgets, the bureau was sometimes forced to take advantage of surplus World War II equipment, such as airborne radars adapted for ground weather surveillance. The Signal Corps–designed radio-tracking system WRSC (or the "bed spring antenna" as it was called by its later users) was the primary wind-finding equipment used by the bureau for a number of years after the war. (By 1952, the antenna on Wake Island was said to be held together mostly by the 20 or so layers of paint applied to it to keep it from rusting.) It was a day of celebration, therefore, when a no-year budget of $3 million plus was granted to the Weather Bureau to enter into contracts for a new ground weather radar, the WSR-57.

Equally important were the improvements in types of ground-observing equipments and their more useful locations within the expanding runway complex. Powerful fixed-beam ceilometers with highly sensitive detectors and integrating recorders gave much more representative ceiling heights and distribution of cloud layers over the airport. Visibility measurements along the runway and in the approach zones gave more reliable indications for landing aircraft, with the runway visual range as a standard observation. (The concept of "slant range visibility," studied extensively at London's Heathrow Airport after the war, is no longer considered important by the IATA.) Snow and ice on runways has become a greater hazard for the much larger airplanes having higher landing and takeoff speeds. Furthermore, the long lines at busy airports of airplanes waiting to take off has become an important element in today's flight operations equation. The present overall aerodynamic designs, with low-slung engine pods, present special problems under certain weather conditions.

ANALYSIS CENTERS AND HIGH-SPEED COMPUTERS CHANGE THE PACE

From the point of view of broad synoptic analysis and prediction, introduction of numerical weather prediction methods in the early 1950s produced a major improvement in both weather and upper-wind forecasts for aviation. As the techniques and models improved and their applications were extended, it became possible for the operational offices of the airlines to tap directly into the NMC digital data stream and use it as the basis for their own flight planning purposes (see computerized wind and temperature forecasts, below). The constant demand for economy of operation in the air transport business brought into play partial mergers of operating groups of several airlines as well as contract arrangements between airlines for common services, such as wind and temperature forecasts for major international routes. Such arrangements made it more efficient to draw on a common source for the upper-air predictions used as the basis of the combined operations planning.

The launch of the first Television Infrared Observation Satellite (TIROS) weather satellite on 1 April 1960 was a significant milestone in aviation weather support. But as old problems were resolved, new ones arose. They were a function of newer and more sophisticated aircraft, bigger airports, and advanced navaids and communications systems—particularly satellite communications.

DOPPLER RADAR

Improvements in ground-based observations are highlighted by the progressive implementation on a nationwide basis of the NEXRAD

Doppler weather radar system. Coupled with the FAA's runway installation of modified wind shear and downburst detectors, U.S. aviation will have completed one of the most important steps in the protection of aircraft that land at major U.S. airports. The more general application of NEXRAD data for thunderstorm and other violent storm detection and prediction demonstrates once more the important contributions that aviation weather needs have made to meteorological services in general.

On the other hand, it is sometimes argued that technical improvements in other parts of the aviation ground system and in the performance of the aircraft themselves are steadily reducing the need for additional weather information for aviation. In the past, this view has often proved to be wishful thinking. Remarkable technical changes in aircraft design and operation have generally put new and more exacting demands on weather services for aviation. One of the better examples of this is the continuing, and even increased, need for information on the state of runways and on the occurrence of snowfall, even of modest intensity, at time of takeoff. These factors have been shown to be the major elements in a series of disastrous accidents.

Most of these advances in technology were scarcely thought of in the early days of the airmail flights and the DC-1 airliner. Many problems involved with flying were first experienced by the military as high performance fighters and bombers came on-line. One example is the recognition of hydroplaning on rain-flooded runways as a serious impact on the braking distances of landing aircraft. Similarly, ice, snow, and slush on both landing and takeoff surfaces and their effects on payloads became a concern for both safety and efficiency of day-to-day operations. Major accidents also brought attention to new hazards not properly evaluated in earlier operations. One of the best examples is the accumulation of snow and ice on wings of aircraft while the aircraft are in queue for takeoff—the cause of several major accidents. The effect of higher temperatures on takeoff performance of heavily loaded aircraft now must be taken into account. The risks from strong gusts and downbursts were known from the earliest days, but more recent detailed knowledge of their structure and their detection, prediction, and avoidance has been gained primarily as a result of a number of serious plane crashes due to these phenomena. Application of frontal analysis techniques in the early 1930s greatly sped up the understanding of the development of violent phenomena such as squall lines and thunderstorms. However, the effect of such highly localized wind structures was still to be fully analyzed, the study of which has been greatly enlarged since the introduction of Doppler radar.

Clear-air turbulence, indeed, severe turbulence of any origin, continues to pose concern for the largest intercontinental aircraft. Now, with even larger aircraft on the drawing board, this level of concern is likely to rise. Furthermore, the supersonic aircraft is certainly a future possibility, as nonstop route lengths assume global dimensions. The environmental impacts of such flights will no doubt add a further dimension to the their weather forecast needs. But this is a look into the future.

COMPUTERIZED WIND AND TEMPERATURE FORECASTS

One of the most significant advances in aeronautical meteorology was computer-generated winds aloft forecasts. In 1964, the Weather Bureau began receiving inquiries from commerical companies interested in the automated generation of computer-prepared flight plans. They requested that the Weather Bureau consider the possibility of providing computer-to-computer winds and temperatures aloft in digital format. Several airlines had developed computer programs for flight planning. They wanted to hasten the receipt of forecasts and to eliminate the tedious task of manually extracting the winds, temperatures, and tropopause heights from the data.

Optimum route selection is that process whereby a flight path and altitude are selected to meet the operational objectives of an operator in the most cost-effective manner. These objectives include the comsumption of the least amount of fuel, the least amount of flight time, and arriving at the scheduled time, or some combination of these objectives that would result in the most cost-effective operation consistent with the safety of the aircraft and passengers. The air traffic control system must be flexible enough to allow the operator a choice in selection of their route.

It was concluded that since provision of such data would be redundant, as it was already available in many other formats, the cost of transmission and further distribution would be borne by the users. This policy was published as Weather Bureau Circular Letter 1-66, dated January 1966. Thus, in response to user requests, a program was developed by NMC to provide wind, temperture, and tropopause forecasts in digital form and implemented on 1 October 1965. However, it was not until 1978 that the forecasts were truly global. The values of the forecasts can scarely be overstated. According to the ATA, the U.S. airlines alone used an estimated 15.8 billion gallons of fuel in the 12 months ending 30 September 1992. If it were possible to save only one minute of flight time each hour through improved forecasting and use of digital data, a savings of $150 million per year could be realized by U.S. carriers alone.

The "designated track system," adopted for the North Atlantic in the early 1960s is still in use, even though modern satellite positioning can readily provide for accurate navigation on multiple tracks. It is designed to simplify the air route traffic control problems on the heaviest over-ocean air traffic anywhere in the world. There are five designated tracks eastbound and five westbound, with a minimum horizontal separation between each track. The eastbound and westbound tracks are themselves adequately separated as a group from one another, so there is no possible conflict between aircraft going in one direction with those going in the opposite direction. The eastbound tracks are essentially based on the old Rhumb Line routes, while the westbound are essentially Great Circle routes. Altitude assignments are made on the basis of the current traffic flow and on the flight characteristics of the airplanes on each track. The track system is clearly not the most efficient arrangement for all flights and especially those that have departed or are arriving at terminals well outside the track area, as they must blend their flight plans into the fixed North Atlantic patterns.

Some conclusions

This factor must be taken into account when assessing benefits from wind forecast for the track areas alone, as is often pointed out by IATA when faced with optimistic statements on the economic benefits of accurate wind forecast for flight-planning purposes. The growth of international air commerce came as an inevitable consequence of World War II, when the demands of global military operations forced the establishment of air routes to virtually all parts of the globe. Design and construction of large numbers of airplanes suited to long-haul operations was primarily to meet the needs of the different war theaters for rapid transport of both men and materials. The great acceleration that took place in both these matters would likely have taken a decade longer if it not had been for the pressures of military needs. This is, indeed, a further example of the great impact that military requirements have had throughout the entire history of aviation. And, as has been argued in other parts of this history, there were the logical consequences on the growth of weather services and the whole science of meteorology.

It is evident from this brief history of aeronautical meteorology that the services generated in support of the aviation industry as a whole, and particularly for air commerce, have had a profound impact on the whole field of meteorology and on the development of national weather services in all parts of the world. Further it is clear that the military services, and particiulary the military aviation services, have been in the forefront of developments that have greatly aided the progress in meteorological sciences and in the expansion of weather services. Perhaps we might even say that one of the more positive sides of military science has been this outstanding contribution to meteorology and weather forecasting in all its aspects.

It has been true that the weather services in many developing countries have, likewise, been built up primarily to meet aviation requirements. It is only fair to note, however, that environmental and related concerns have begun to find long-delayed support for their improvement. It has been said that in the next decades meteorologists are likely to concentrate heavily on the chemistry and radiative properties of the atmosphere as a balance to the emphasis long given to the atmosphere's physical and dynamical aspects. Looking back over the history of the last nine decades, it would appear that this has been the proper and logical sequence of development. Perhaps the time has come that the nations of the world, having developed what is essentially a global economy, should now look to the health of the atmosphere and to the consequences of environmental disregard on humanity's longer-term future. In this new approach, we can be grateful for the dedicated work done by our predecessors in pursuit of a safer and more efficient global transport system that brings all nations closer together.

REFERENCES

Bates, C.C., and J.F. Fuller, 1986: *America's Weather Warriors*. Texas A&M University Press, 360 pp.

Canby, C., 1963: *A History of Flight.* Hawthorne Books, 113 pp.

Fitzpatrick, T.A., 1992: *Weather and War.* Pentland Press, 146 pp.

Flying the World, 1994: *Royal Bank Letter.* Vol. 75, Montreal, PQ, Canada.

Hosler, C.L., 1994: *Weather for Those Who Fly.* National Weather Service Modernization Committee, National Academy Press, 100 pp.

Josephy, A.M., Jr., Ed., 1962: *History of Flight.* American Heritage Publishing, 416 pp.

Kitty Hawk Weather Bureau Office, 1900: Correspondence to Orville and Wilber Wright, 18 August 1900. Wright Brothers Papers, Library of Congress Manuscript Division, Washington, D.C.

Reichelderfer, F.W., 1932: Norwegian methods of weather analysis. U.S. Weather Bureau (mimeographed copy).

Taba, H., 1988: *The Bulletin Interviews.* World Meteorological Organization, 405 pp.

Weightman, R.H., 1940: Forecasting from synoptic weather charts. USDA Misc. Publ., Washington, DC, 52 pp.

CHAPTER 17

The American Meteorological Society and the Development of Broadcast Meteorology

ROY LEEP

Introduction

Forty years of my professional career have been spent in weathercasting. That in itself is not unique. However, 38 of these years have been spent at the same television station, and that may be unique. I had never been inside a television station prior to starting my current job. (I give this brief history so that you can understand my perspective for the review that follows.) Although I have worked in only one area of the country, I have been closely involved with the AMS Seal of Approval program since its inception and with the broadcasting industry since my entry into it. From my "stable" job position I have been able to view the ever-changing emphasis and direction of two industries, broadcasting and meteorology. Thus despite the fact that my view may be rather myopic, I have attempted to gather and correlate, in some meaningful manner, information on broadcast meteorology spanning 50 years of the 75-year history of the AMS—a specialty that I believe has been the greatest single influence on our Society and its public perception.

The professional weathercaster is often accused of being an overpaid, eccentric showperson, but the true professional can be a bridge by which our complex science can be reached and understood by the public. When I became fascinated with meteorology, there was no television, but today many fine students received their first taste of the science through the weathercast.

I was a product of the U.S. Weather Bureau, having interned there in the 1940s, but today there are many professionals who are shaped largely by television weathercasting. This is a lasting legacy that the professional weathercaster can point to with pride. The AMS has played an important role in the development of this remarkable specialty. The standards of the Society have been accepted by and large by the broadcast industry and hold at bay the attempts of some to return to an unprofessional approach.

What the future will hold can be discussed at another time. Broadcast meteorologists today have an established, important role in radio and television newscasts. It is difficult to foresee a future in which they are absent.

The early years

Genesis

In the beginning there was radio, and no doubt some of the early ad-lib conversations in this new medium in the 1920s made reference to weather. Aside from unique governmental efforts on special frequencies, we may never know who first remarked about the weather over the air. However, we have some evidence suggesting who this person was and when and where forecasting was first put into a daily format. Many contemporary weathercasters owe their own interest in the profession to E. B. Rideout, who was weathercasting as early as 1923. Rideout, a former Weather Bureau forecaster, was hired full time by WEEI in Boston, in 1926. He remained on the air for some 40 years. Don Kent, later of WBZ, Boston fame, was enthralled at the age of five by Rideout's broadcasts as he listened on his crystal set in 1923 (D. Kent 1994, personal communication). Harry Volkman, another resident of Boston and later an AMS Sealholder in Chicago, visited Rideout in his office several times in the early 1940s. Volkman states that Rideout gathered information from the short-wave radio stations WEK, New York City, and WSY, New Orleans, which broadcast in Morse code (H. Volkman 1994, personal communication). Of course, both Kent and Volkman have had illustrious careers to which I will refer later. It seems fitting that Boston was the incubator for future weathercasters and that the seeds of this specialty were planted so near the birthplace of the AMS. There is no way to trace the genealogical tree that spread from these early radio broadcasts. Although it seems safe to say that Rideout was the father of weathercasting, Rideout's name is not found in the AMS membership files.

1930–1950 era

Despite the fact that Rideout was the pioneer of weathercasting, there were only a select few of his followers that matured into full-time weathercasters prior to 1950. This was partly due to the occurrence of World War II, when all weather broadcasts were restricted. I can point, however, to a handful of exceptions and the years of their weathercasting debuts:

- Jim Fidler, WLBC, Muncie, Indiana, radio 1934; Weather Bureau Dumont Network, TV 1947
- Don Kent, WBZ, Boston, radio 1935; WBZ, Boston, TV 1955
- Harold Taft, radio 1946; KXAS, Dallas, Texas, TV 1949
- Nash Roberts, WDSU, New Orleans, TV 1948
- Louis Allen, WNBW, Washington, D.C., TV 1948
- Francis Davis, WFIL, Philadelphia, TV 1948
- David Ludlum, Philadelphia, TV 1948
- Clint Youle, NBC, Chicago, TV 1949
- Wally Kinnan, WKY, Oklahoma City, Oklahoma, TV 1950

Undoubtedly, there were others, but the emphasis here is on AMS professional members who were full-time weathercasters in the private sector. This was the core group of meteorologists who would lay a foundation of standards for others to follow and exceed.

1950s growth

The explosive growth of television stations during the 1950s provided impetus for private sector meteorologists to multiply rapidly. World War II and the Korean War also provided training for a number of meteorologists in the 1950s. As the broadcast industry began to grapple with the issue of formatting of news, weather, and sports, weather proved to be most controversial. The controversy stemmed from well-meaning journalists or program directors who knew little about meteorology or its importance to the public. If a journalist had high standards, he quickly realized the importance of a qualified broadcast meteorologist but was usually unable to find one. Most program directors, on the other hand, knew they had a certain number of minutes to fill and concentrated on glitz and showbiz rather than content to hold the attention of the viewers. Some television stations took the view that any journalist could just read the weather. Others wanted to hype the presentation with attractive but uninformed and unqualified presenters. Still others wanted to treat the weather as either the "throw away line" for the news anchor ("It will be fair and warmer tomorrow—good night.") or present it as entertainment. It was the entertainment aspect that created great concern for early professional weathercasters. As stated earlier, only a few professionals had a toehold on the national scene.

There were a few television companies strongly committed to good journalism and armed with an Edward R. Murrow–type credibility that envisioned only a qualified meteorologist dispensing the weather report. These companies were certainly in the minority in the 1950s. Qualified weathercasters had to prove that they were the ones viewers trusted and watched. Time and time again in a variety of television markets, breakthroughs were made when severe weather became the main story. A credible, reliable report from a familiar face became the public's demand. From reporting on tornadoes in the Midwest, to hurricanes on the coast, to blizzards in the West, to nor'easters in the East, it was the professional who stood out. In every case, the public overwhelmingly demanded reliability, accuracy, and an articulate, educated communicator to provide information and assurance during an emergency. Radio came close to fulfilling this demand but was faceless. Thus only television could do it, and only the meteorologist was fit for the task of providing accurate, dependable, necessary information on a daily basis, thereby developing a rapport with viewers that inspired trust when severe weather threatened. It did not take long for news directors and station managers to grasp this fact. Wherever there was one meteorologist in a market, competing stations were soon looking for their own, and so the multiplication process began in both large and small markets. If one person did the job at night, why not have a second for the mornings and a third for weekends? Network affiliates seemed to have sufficient funds for this type of staffing, and thus by the 1960s, the demand could be as high as nine qualified weathercasters, in every market served by CBS, ABC, and NBC. The frenzy was on through the 1960s and 1970s.

What makes an outstanding weathercaster? Not necessarily a Ph.D. or a Masters degree, not necessarily a glib tongue or pleasing profile. All of these attributes help, but in the final analysis, the viewing audience and professional peers will decide what they prefer and why. I think to begin with, there must be a love of the subject and the method of communication—that indefinable mystic charisma that embodies certain qualities that are natural and cannot be learned: intensity, professionalism, friendliness, and sensitivity. Precision yes, but human vulnerability and genuine sincerity are traits that viewers inherently trust. Perhaps the qualities of a successful salesperson combined with those of a wise scientist could go a long way in defining what makes weathercasters outstanding. If we rush to rank individuals by market, city size, or years or experience, we may very well miss the mark, since I believe outstanding weathercasters can be found in both large and small markets and with 5 or 40 years of experience. There are many who are good, but only a few are outstanding. Do they command above-average salaries? They should and usually do.

Weathercaster versus government

Policy evolution

Jim Fidler (1948) and Francis Davis (1976), among others, have alluded to an adversarial relationship between federal agencies and private sector broadcast meteorologist. I, too, have experienced this at several levels over my years in the profession. The problem is that big government begets bigger government, and the private sector usually suffers as a result. Throughout the tenures of different chiefs of the Weather Bureau or National Weather Service (NWS), there have been various policies regarding relationship of the government to the private broadcaster. The NWS, of course, has a Congressional mandate to serve the public, especially in regard to life-threatening weather. The NWS generates data, archives it, and is the sole source of most basic observations and refined computer forecasts globally. Obviously, the broadcast meteorologist or private sector consultant must rely and build upon many of these data for their work. Jealousies have arisen in the past over crediting the source of forecast information, accuracy claims, equipment capabilities, suspected errors in basic data, and, above all, the overshadowing public attention given to weathercasters. Policy has enlarged, ebbed, and flowed over the last 60 years as the interpretation of the Congressional mandate for service and technologies in the marketplace change. The Freedom of Information Act solved some long-time points of contention involving availability of data in a timely fashion. Initially, broadcast policies were nonexistent, and the old Weather Bureau freely supplied data to meteorologists for television and radio on commercial stations. Policies prohibiting this are now in place. However, the shadows of old policy are still in the background and may arise again in new areas, such as the new National Oceanic and Atmospheric Administration (NOAA) Weather Radio transmission and even "broadcasting" on the Internet. The government sector will probably never be completely pleased by the broadcast community's role in this relationship because of several factors. For example, I believe there is an inherent, and understandable, envy on the part of some government meteorologists toward the highly paid and visible private sector. Nonetheless, this problem seems to have lessened since the 1970s, and the most recent policy (1991) formulated under Director Elbert "Joe" Friday has gone a long way toward revitalizing a spirit of cooperation and understanding between the two sectors. With all of the past difficulties and problems between them, the total system successfully works like no other in the world. No other nation has such a large, diverse, and efficient network of radio and television stations staffed by qualified broadcast meteorologists who are able to interpret and relay life-saving weather information supplied by the government service. This appears to be the bottom line for continuing a fruitful cooperation.

Private sector concerns

Francis Davis (1990) and Don Kent (1994, personal communication) have alluded to battles fought with the old Weather Bureau over allowing private meteorologists access to teletype circuits in the 1940s and 1950s. The AMS played a role in settling this dispute, which cracked the door for today's flood of data and services. The freedom of information that we enjoy today was not always in vogue, and there are still pressures from foreign governments to restrict certain data and/or computer outputs. Many foreign governments severely restrict the operation of private meteorologists for fear of direct competition. Broadcast meteorologists rely upon increasing amounts of global data, as models produce more accurate forecasts.

Most weathercasters on television ad-lib their discussions because of the very timely nature of the data. In weather emergency situations, this can be a formidable task and, during the undertaking, may strain or even break the strict policy of the NWS. There is an increasing belief that the NWS severe weather bulletin, whether right or wrong, must be aired in total. Could it be wrong? Yes. With real-time radar and satellite pictures displayed in front of the public, written warnings may be out of date. There have been several instances of outside tower cameras observing tornadoes and waterspouts for which no official warning has been issued. Surely the professional weathercaster has that duty and must exercise it responsibly. Another area of concern relates to the unique relationship between the NWS and private universities. Training aids provided at a low cost to universities are provided at a much higher cost to private broadcasters. Although the present-day spirit of cooperation between private broadcasters and the NWS is good, there will always be problem areas ranging from the local forecast office to the regional office to headquarters because decisions made at any of these three levels can adversely affect the private broadcaster. Better communications seem to be ameliorating these difficulties.

Encroachments

Anywhere the NWS is broadcasting is an encroachment on the private sector. In the late 1960s when the NOAA Weather Radio (NWR) was established, I was opposed to it as I am today, on the single principle that it was and is clearly outside of the NWS's realm of responsibility. While the NWR has had mixed reviews from its audience over the years and its equipment has become antiquated, I am certain that it has been a source of pride and promotion for the NWS. One can make all kinds of arguments for its continuation. My problem arises from the trend toward the NWS getting out of surface observations and into broadcasting. If subcontracted work is an option, then broadcasting should have been contracted to professional broadcasters using state-of-the-art equipment. Many private broadcast meteorologists could have obtained a job, and overburdening NWS employees (who are not broadcasters) could have been avoided.

The AMS commitment

Beginnings

As mentioned previously, one of the earliest concerns of the few pioneering professional broadcasters was the deplorable spread of "weather shows" that ranged from light comic entertainment to weathercasts performed by announcers passing themselves off as meteorologists. Members of the AMS wanted to have the science portrayed in the proper light and started procedures toward attaining that goal. Early chronology of the AMS providing concern and support for the weathercasting industry is described in its own documents and by two of the original committee members, Francis Davis (1976) and Kenneth Jehn (1956).

Organized committees

In January 1954, the council voted to establish an ad hoc Committee on Radio and Television. Francis Davis was the chairperson, and Howard Taft, Jim Fidler, Eugene Bollay, Louis Allen, Clint Youle, and Richard Reed were members. At the AMS meeting in Washington, D.C., in May 1955, the Council accepted the ad hoc committee report consisting of nine points and voted to establish a permanent Committee on Radio and Television to adopt a Seal of Approval to be given to individual weathercasters meeting minimum requirements (AMS 1955a). In September 1955, the Society put out a request for names of all members interested in radio and television (AMS 1955b).

In his address before the national meeting in September 1956, President Fletcher said, "The best opportunity in our history is now available through the medium of television" (Fletcher 1956). The ad hoc committee was dissolved, and in its place the same members were appointed to a committee that made final recommendations to the Council in April 1957 on establishing the Seal of Approval program. Shortly thereafter, the Board on Radio and Television Weathercasting was established and was composed of Kenneth Jehn (chairman), Davis, Fidler, Wally Kinnan, and Lawrence Mahar. This board continues to operate with regular changes in membership.

Organized conferences

In October 1955 at the national meeting, Jehn presented a paper envisioning a Society program that would encourage and promote the professionalism of weathercasters (AMS 1955b). In March 1956, the first *National Conference and Workshop on Radio and Television Presentation of Weather* was held in Hartford, Connecticut, cosponsored by the AMS and the NWS (AMS 1956). This first meeting evolved into an annual conference. However, the idea of a regularly scheduled national broadcaster's meeting was overlooked until 1967. I was privileged to serve as Chairman of the Board on Radio and Television Weathercasting that year and, with Executive Director Ken Spengler's help, convinced the Council that this growing specialty needed its own regular meeting. Subsequently, the first solely AMS-sponsored regularly scheduled meetings began in Tampa, Florida, on 8 December 1967 (AMS 1967). The most recent was the *24th Conference on Broadcast Meteorology* held at Dallas with the diamond anniversary observance. Only radar meteorology has had more conferences.

Seal of Approval

Aside from Honorary Seals that were issued in 1959 to the original Board members, the first AMS Seals of Approval were issued at the national meeting in January 1960. Qualifications for membership in the AMS have changed since the inception of the seal program. For the most part, these changes have been to require higher levels of formal training (AMS 1994a). The seal program in earlier years allowed periods of "grandfathering" members to enter the program providing other criteria were met, such as a written examination in some cases. Today, the seal applicant must be a full member of the Society. The seal requirements today are as they were in 1960, which is much to the credit of the original committee. The Board on Radio and Television reviews applicant's tapes on the basis of four criteria: technical competence, information value, explanatory value, and communication skills. The Board and the AMS reserve the right to suspend or revoke the seal, so that a continuing standard is expected of those who receive and display it (AMS 1994b).

Probably the most gratifying example of AMS membership recognition and support of weathercasting was when Robert Ryan, weathercaster at WRC-TV, Washington, D.C., was elected president of the AMS in 1994—the first weathercaster to hold that position.

Professional growth

The growth era

I have previously alluded to the growth era of the 1960s, 1970s, and 1980s. Throughout these decades, over 700 seals were distributed from

coast to coast. The rapid increase in the number of of weathercasters
continued at break neck speed. In some regions, the number grew faster
than in others. For example, the eastern states had more weathercasters than the West. However, Florida, New England, Texas, and Oklahoma seemed to have the greatest number of active sealholders.

In Florida, the Tampa Bay area exemplified the nation's rapid
growth. Professional weathercasting started there in 1955, and the
number of weathercasters continued to grow. As of today, there are 13
active sealholders on television in Tampa Bay, including five apiece for
two television stations. Generally speaking I think of sealholders at
three categories of stations:

1) *basic,* where one meteorologist works with a minimum of value-
   added data;
2) *weather service,* where two or more meteorologists work with
   value-added data and additional station-owned equipment; and
3) *departmental,* where three or more meteorologists work in their
   own department separate from the news department and report
   directly to the general manager.

The departmental service would be expected to have a wide range
of equipment, an operating budget, capital expense budget, and perhaps a consulting service. There are few such departmental stations.
WTVT in Tampa became such a station in 1962, and two of its trained
meteorologists have established similar departments in New Orleans
(Bob Breck) and San Francisco (Pete Giddings). The WTVT department today has a staff of nine.

Growing pains

In the 1970s, cable television grew tremendously, and by 1980
CNN began with a limited amount of weather information. Valerie
Voss who started in 1979 and was one of the first female professional
weathercasters, now heads up the CNN weather staff. Cable finally
generated a dedicated weather channel. It was the brain child of John
Coleman, who took his idea to Landmark Communications. The
Weather Channel was launched in May 1982. In 1983, Coleman was
awarded the AMS Award for Outstanding Service by a Broadcast Meteorologist, due in large part to his development of The Weather Chan-

nel. But sadly, later that year, Coleman and Landmark parted ways because of the initial lack of profit, and he never was part of the huge success of the venture he founded. The Weather Channel has had no more of a detrimental effect on local weathercasters than CNN has had on newscasters. In fact, both have probably stimulated viewing for each other during severe weather. It is disappointing, however, that although some of The Weather Channel staff have been awarded the AMS Seal of Approval, it is never displayed on this nationally seen broadcast.

Directions and goals

The major networks, ABC, CBS, and NBC, have only dabbled with using meteorologists, and none have ever consistently done so in their evening newscasts since Clint Youle was featured in the 1949–1959 era. I think it is disappointing that the networks run hot and cold on the importance of weathercasting in their morning programming and are currently using weather reporters instead of meteorologists. Joe Witte has been the most successful sealholder, appearing on the early *NBC News at Sunrise* program since 1983. It appears to me that what the local station affiliates learned well is still unknown at the network level. The fact that all networks have used professional meteorologists during periods of severe weather, such as hurricanes, illustrates the esteem in which they are held by their respective news organizations. Yet often the morning network programs are not under the total jurisdiction of the news department. Furthermore, weather personalities are used for commercials and other duties that are not allowed for the pure news anchor. This turbulent area is still ripe for qualified on-air meteorologists and should be a goal for the near future.

Cable continues to fill an insatiable appetite for weathercasting. Many local stations are supplementing their broadcast agendas with some type of continuing weathercast. This relatively new area will supply new job opportunities and new formats to explore. As more sophistication arrives with greater technology, one can see that some type of interactive weather information will be available by cable. Whether this will be graphics on demand or actual one-on-one briefings will be determined by the marketplace and available technology. But it is certain that during the next 10–20 years these will be the new areas in which broadcast meteorologists can become involved.

Concerns

While the AMS seal program today is largely recognized throughout the broadcast industry, having the seal is no guarantee of employment. On the other hand, not having it could deter employment. Since its inception, the different memberships of various boards have wrestled with a multitude of fears and problems. The seal program itself seems to be alive and well, but the concerns of its membership range from a perceived lack of AMS support, to the ever-present threat from non-professional or unqualified broadcasters, to increasing pressure from television management to be less scientific or to "dumb down" presentations, eliminating educational value. All of these problems are real, and the younger, less established sealholder may be tempted to compromise instead of defending the science. As a result, the AMS needs to continue its strong endorsement of the Board on Broadcast Meteorology in its pursuit of educational content and scientific excellence in weathercasts.

Three problems in weathercasting that are surfacing more these days are as follows.

NATIONAL WEATHER SERVICE. When there is one exclusive national source for global weather information, then it is obvious that such a source could coerce private broadcasters into policies that may not be in their best interest. I have touched on and explained some of the problem areas earlier. While the trend seems to be toward improvement, there continues, on behalf of many broadcasters, to be an incomplete trust of current NWS policy.

NEWS DIRECTORS. Some, but not all, news directors treat their weather staff as second-class citizens. This can be reflected in budgets and on-air time. This is an internal industry-wide problem.

CONSULTANTS. Related to the two problems above is the fact that television stations hire consultants to assist them in programming. Among their many tasks is that of viewing weather program content and making recommendations for change to the news director. Almost invariably, the consultants advise shorter and simpler weathercasts. According to them, maps should not have isobars or perhaps even fronts. There are constant disagreements at many stations between the consultant–news director team and the meteorologist. Since the 1970s, the use of consultants has increased, and many weather programs have been forced to dumb down their presentations.

In his presidential address before the AMS, Fletcher related a story about a forecaster who was forced to state his forecast in one word before the city paper would print it. Fletcher stated, "The imposition of such short-sighted limitation on the forecaster certainly no longer exists—at least for the TV weathercaster" (Fletcher 1956). Consultants today are recommending one word or no words for each day of the five-day forecast!

Concerns and danger signs for the 1990s have been raised by Paul Joseph, a venerable broadcast meteorologist, of WTMJ, Milwaukee, Wisconsin. He cites what he calls the "Willard Scott syndrome," where the weathercaster's primary function is promotion and he or she ends up with little time for the weather program. The second danger sign Joseph mentions is the *"USA Today* approach," in which beautiful graphics are illustrated with little meteorological discussion. The final signs as seen by Joseph are "super-duper advanced graphic systems," which require an inordinate amount of time for preparation, although the images are only on the air for seconds. Joseph's concerns are ones that most broadcast meteorologists grapple with almost daily.

The digital revolution and presentation

Broadcast industry

Perhaps there has been no industry that has evolved in the public's living room as much television has. The evolution has progressed from the days in the 1940s and 1950s when a 4-in. screen produced a fuzzy black and white picture, provided the antenna was pointed properly, to today when a 100-channel cable-ready 5-ft screen provides brilliant color, resolution, and stereo sound. With such remarkable advances in electronics, the weathercaster has seen preparations become more time consuming and complex.

Presentation development

The early presentations by Fidler, Davis, Kent, and others employed blackboard and chalk, poster paint and plastic maps, or grease pencils and paper. Fred Ostby, the current director of the National Severe Storms Forecast Center, started in television with the Travelers Weather Service in Hartford, Connecticut, in 1957. He stated, "Our gimmick at the time was to present the forecast in the making. We sat on a stool at the drawing board and would draw some features (with squeaky marking pens) on the map . . ." (F. Ostby 1994, personal communication).

Maps and forecast presentation made up one aspect, but another that was to prove equally important was the presentation of instrumentation and data. In the early 1950s, Bendix–Friez was one of the companies that produced a complement of dials for use in bank lobbies and other public places. With minor adaptation, these could be used directly on television. The wind dials were the only ones that showed much movement. Many stations had a set of mock dials that were preset prior to the program. In fact, Bendix even manufactured such a set. These were the earliest display devices.

All meteorologists realized the value of the relatively new tool, "radar." Since it generated an image on a screen, it was a natural for television. However, there just were not any radars being produced at an affordable price for television. Who was the first to use radar on a weathercast? Nash Roberts claims that he included radar for the New Orleans area in 1954, and Fidler had it for the Cincinnati area in 1955. Most of these early sets were converted shipboard navigational 3-cm radars with a range of about 25–30 miles. By the late 1950s, however, both Bendix and Collins were making 5-cm weather radars for aircraft with modification for ground use. Range was on the order of 150–200 miles, and most used about a 2–3-ft dish. Another technique of the time was to send the Weather Bureau radar picture to the television studio via microwave transmission.

Weather dials have not changed much over the years, but radar presentations have. By 1970, a few commercial vendors were making radar sets for television. WTVT in Tampa installed one of the new, solid-state radars including a Video Integrator Processor (VIP) encoder. The VIP installation was a copy of the NWS device but preceded their field installations. By the early 1970s, remote dialing into distant NWS radars was a reality, and by 1976, colorization of the radar VIP rainfall intensity levels was available. In 1981, a full Doppler radar was installed at KWTV in Oklahoma City, the first such on-air device. By 1984, pseudo-Doppler or wind shear radars were being produced, and KSTP in Minneapolis claimed to have the first. A customized full Doppler radar with a 200-ft support tower and a 20-ft antenna was built at WTVT in Tampa in 1988. Called SkyTower, it produces 12.5 gigawatts of effective radiated power and has a 450-mile reflectivity range. In addition to local installations, value-added companies have been producing composite radar maps and digital images providing single-site WSR 88D radar pictures that have joined the mix of available radar presentations in the 1990s.

Computerization

The 1980s brought the most tremendous changes to television weathercasting: Computers had arrived. WTVT in Tampa followed intently the development of the Man–computer Interactive Direct Access System (McIDAS) at the University of Wisconsin in 1977. In 1978, they purchased a McIDAS system, including a mainframe computer, communication ports, and Geostationary Operational Environmental Satellite interface. Special software was developed so that all weather circuits could be archived and data recalled. The McIDAS system was customized for direct television output. In 1977, the first partial broadcasts were made, and by 1978 the entire broadcast was digital. WTVT is now using their third-generation McIDAS, which archives and generates all the data locally without value-added assistance. Terry Kelly, a Madison, Wisconsin, weathercaster had also followed McIDAS's development. He and Richard Daley of the University of Wisconsin spun off the first weather commercial graphics system for television, called ColorGraphics. This system was the forerunner of more complex computer devices with increasingly greater adaptability and power. The on-air paper map was replaced forever, and presentations would never be the same. By the mid-1980s, most television stations had converted to some weather computer display device. At the same time, the entire broadcast industry was joining the computer revolution. The competitive drive of television stations led to a fast changeover to new technologies—a drive that continues in many areas of the industry.

Nowcasting

With the advent and use of radar in the mid-1950s, the first nowcasts were presented to the public. Since those early days, the whole purpose of the forecasting profession has been to gather, display, and quickly disseminate highly accurate short-range forecasts. Since radio and television allow immediate access to users, much of this responsibility lies with the weathercaster. Radar has been the steadfast tool for 40 years, accomplishing much of the desired goal. With faster computers analyzing more data in near real time, the radar picture is being supplemented and merged with other pertinent pictorial representations. Satellite pictures, composite radar, graphical analysis and lightning data—all of this and more are available, with a few computer key strokes, to present an accurate and immediate visualization of any weather situation. Much of this type of operational development can be credited to the weathercasting community. Even before the Automation of Field Observations and Service (AFOS) and the Automated Weather Information Processing System (AWIPS), broadcasters had, through either their own development or a value-added company, most of the currently available nowcasting capability plus a built-in communication channel.

The spin-offs

Value-added industries

GROWTH

One company, Alden Electronics, has had a long track record with meteorologists. There are probably few forecasters who have not worked with a map or product produced by their famous 19-in. facsimile machine. Most television weather offices in the 1950s, 1960s, and 1970s had at least one and perhaps two of these recorders. With the launch of the Automatic Picture Taking (APT) TIROS and Nimbus satellites in the 1960s, it was again Alden that offered a complete earth station. As software replaced hardware and more value-added companies arrived on the scene in the 1980s, Alden made the shift to personal computer (PC) display devices. Lawrence Farrington, former Alden president, stated, "The 1960s and 1970s saw a coalescing of three forces—technology, products, and markets. Technology advances made possible new products, there was an industry ready to manufacture those products, and there was a ready made market for the new products—TV weathercasters" (L. Farrington 1994, personal communication).

Since the 1970s, three companies have remained progressive and competitive enough to share the Alden market: Accu-Weather, Kavouras, and WSI. Each of these companies supply value-added graphics and data to the majority of television stations. While each started and grew in different ways, all benefitted from the coalescing of technology, products, and markets.

MARKETS

All of the companies that serve the broadcast industry serve other related markets as well. Some have contracts with NOAA, NWS, FAA, or other government agencies, colleges, and universities. Although sophisticated and expensive terminals are required for broadcast quality, these companies are now offering PC versions to aviation or other weather-sensitive clients. However, these companies are a direct result of the broadcast industry. Most were formed by broadcasters and are run, to a degree, by broadcasters. When walking through the AMS Annual Meeting exhibit display, one is struck by the large percentage of hardware and software on display of which an earlier version was first seen on television weathercasts. It appears that the high quality standards originally established by weathercasts are now reflected in many video applications.

Education

The broadcast meteorologist has always occupied a position of teacher–instructor. Most of the time, forecasts are sugarcoated for the average viewer. However, just watching a professional weathercaster on a routine basis is educational. There have also been special educational projects that have been carried out over the years. One of the earliest projects using television was described by Ira Geer (1975). Sealholders have been providing types of direct education in on-air specials or as guests in schools or science museums for many years. In 1967, there was a Tampa station that provided a weekly 15-minute program called *Behind the Weather* (WTVT Weather Service 1991).

LOCAL AMATEUR NETWORKS

Perhaps the most novel and widely used tools for science education in broadcasting are the amateur volunteer networks that many broadcast meteorologists use to some degree. Whether formal or informal, these networks offer a source of weather data at a county or city scale and stimulate the interests of both adults and students to learn more about weather while providing information. Long before SkyWarn, televisions stations were establishing amateur viewer networks. One of the earliest and best organized is the West Central Florida Severe Weather Network (Leep 1969). Organized in 1959, it still operates today with 200 observers. Participants receive educational materials and a monthly newsletter. Severe storm reports are instantly relayed from the network to the local NWS office.

Viewer–observer networks are in operation around the nation, offering practical education while performing a public service. Many young people have an opportunity to participate in the weathercast by learning to report accurate and timely weather information.

K–12 school stimulus

During the days of teletype reports and paper maps, students would stand in line to receive what the weathercaster would eventually throw out. Having perceived this need for materials, weathercasters started to save what once filled the wastebaskets. This would be sent to or handed out at schools. Today with PC displays, there are not as many materials to answer this need. Television meteorologists, however, are still in great demand in the classroom, and they take whatever materials they can, or they host field trips to their stations for a firsthand look at the weather operations. In most areas, this could be a full-time occupation.

University programs

An increasing number of television stations are making internships available in their weather departments. Programs exist in which college credit is given for a term of interning with a qualified meteorologist. Many meteorology majors already have some on-the-job training obtained at a local television weather service. Some of these students are employed by TV Weathercasting after graduation.

In 1955, the AMS ad hoc Committee on Radio and Television recommended and the Council approved the following resolution: "Voted that the Society call to the attention of the professional schools of meteorology the desirability of their offering instruction in the presentation of weather information to the public via radio and television" (AMS 1955a). Colleges and universities were slow to respond to this appeal, probably because of the economics of the situation. After all, most of the jobs were elsewhere. There were some exceptions, however. For example, The Pennsylvania State University began to give meteorology students the opportunity to work before the camera. Charles Hosler (sealholder 1960) and John Cahir (1969), both professors there, were largely responsible. Some universities looked down at such a "showbiz" occupation as demeaning to the science. Today, this attitude is changing, and many schools have combined studies with on-campus radio and television broadcasting to allow promising students an opportunity to test their capabilities. As in my case, most early weathercasters had no such advantage. I auditioned cold on camera, before videotape was invented, at the first television station I was ever in and the only one in which I have worked. Those days are gone forever. Now, when I look for an entry-level on-air meteorologist, the one who has had formal training and has videotapes is the one who usually has the advantage.

Epilogue: Personal observations

This historical review is not nearly complete due to the lack of space and information. Most papers presented by weathercasters at conferences are never published, and because of the nature of their demanding operational schedules, few have an opportunity to write articles for the Society's publication. Consequently, I have listed in the appendix conferences and their *Bulletin* references for further reading. I conclude with a series of historical photographs.

Broadcast meteorologists are often the recognized local representative of the AMS, and, for that reason, I dedicate this brief historical review to their professionalism and hope that it will provide current generations with a look at our specialty from the inside out. AMS Headquarters, especially Dr. Kenneth Spengler, has always been a strong supporter and ally of the weathercaster. Certainly it has helped provide the proper political atmosphere for the survival and growth or what more than a few think is a maverick specialty. The Conferences on Broadcast Meteorologist and the Seal of Approval program are alive and well. At the 1985 Conference on Broadcast Meteorology, the silver anniversary of the AMS Seal of Approval Program, 11 practicing broadcast meteorologists were recognized for their long continuous service. The following are some of the pioneers:

> Bob Copeland—Boston
> Francis Davis—Philadelphia
> Al Duckworth—New Orleans
> Jim Fidler—Cincinnati
> Don Kent—Boston
> Roy Leep—Tampa
> Jim Smith—Cleveland
> Milton Strauss—Madison
> Harry Volkman—Chicago
> Bob Weltie—Salt Lake City
> George Winterling—Jacksonville.

In 1996, there are three remaining that have been continuously active: Leep, Volkman, and Winterling. Kent continues to broadcast part-time from his home office in New Hampshire.

Acknowledgments. My thanks to the many weathercasters, service companies, and headquarters staff who contributed to this history. My assistant, Anne-Marie Fagler, has tirelessly researched, typed, and edited. For an excellent reference, I recommend Robert Henson's *Television Weathercasting: A History* (McFarland & Co., 1990).

FIG. 17-1. Jim Fidler while broadcasting at WLWT in Cincinnati in 1954. Note the analog dials and hand-drawn maps. The concept of bringing the camera into the weather office can be as fresh and exciting today as it was in 1954.

FIG. 17-2. Don Kent, in his Boston office in 1956, claimed the first teletype license in the late 1940s and the second fax license in the early 1950s. Kent continues to broadcast part-time from his home office in New Hampshire.

FIG. 17-3. Nash Roberts of New Orleans in his studio office in the 1950s. Roberts, like Fidler, used his office as the weather set, and while cluttered, it brought the viewer into where the action was.

FIG. 17-4. Francis Davis on camera at WFIL-TV in Philadelphia around 1950. This was one of the early studio sets. Note the commercial logo on the map. Today's purist would have trouble with that, but in the 1950s, meteorologists were glad to be sponsored!

FIG. 17-5. Wally Kinnan presents one of the first color weathercasts in the nation at WKY-TV in Oklahoma City in 1955. WKY-TV's color cameras were the first employed by a local affiliate. Note the old 60-wpm teletype machine on which many early meteorologists cut their teeth.

FIG. 17-6. Fred Ostby, current director of NSSFC, at the WTIC-TV Travelers Weather Center in Hartford in the 1950s. There were many notable meteorologists with Travelers. There is even an old kinescope recording of Dick Hallgren, current AMS executive director, making the rounds.

FIG. 17-7. The 2-ft antenna for the WP101 Collins weather radar used in 1959 at WTVT in Tampa. Its range was 150 miles. These small dishes, originally intended to go in the nose of an aircraft, were modified for 360° rotation.

Fig. 17-8. The 200-ft concrete tower that supports the 20-ft antenna for the Doppler radar in use at WTVT in Tampa since 1989. Its reflectivity range is 450 miles. A two-man elevator can lift technicians to the air-conditioned receiver– transmitter room under the radar dome.

Fig. 17-9. An Alden exhibit booth at an AMS meeting in the 1970s, displaying two facsimile machines used in that era. During the 1950s and 1960s, Alden was often the only exhibitor at AMS meetings. Their fax machine usually drew a crowd of out of town meteorologists checking the weather back home.

FIG. 17-10. In this 1970-era picture of the WTVT Weather Center, Tampa, there are four Alden systems being used, including two 18 in. machines for NAFAC, DIFAX, and radio fax; an APT system for TIROS and Nimbus satellites; and a dial-up remote radar recorder to receive NWS radar pictures.

FIG. 17-11. The WTVT Weather Department in Tampa as it looked in 1989 shortly after its inauguration. The operational area includes the use of McIDAS and its several workstations modified for television production. The large elevated monitor is an on-air briefing site. An automated camera can view any portion of this area for scheduled or nonscheduled nowcasting.

APPENDIX
Chronology of Weathercasting Conferences

Name	Date	Place	*Bulletin* reference
National Conference and Workshop on Radio and Television Presentation of Weather	3–4 March 1956	Hartford, CT	Vol. **37**, No. 2
2d Conference on Weathercasting	8–9 December 1967	Tampa, FL	Vol. **48**, No. 10
3d Conference on Weathercasting	7–8 November 1969	Chicago, IL	Vol. **50**, No. 8
4th Conference on Weathercasting	9–10 January 1972	New Orleans, LA	Vol. **52**, No. 10
5th Conference on Weathercasting	9–10 June 1973	Harwichport, MA	Vol. **54**, No. 4
6th Conference on Weathercasting	19–20 January 1975	Denver, CO	Vol. **55**, No. 11
7th Conference on Weathercasting	26–27 June 1976	Toronto, ON, Canada	Vol. **57**, No. 3
8th Conference on Weathercasting	29 January–2 February 1978	Savannah, GA	Vol. **58**, No. 10
9th Conference on Weathercasting	16–17 June 1979	San Francisco, CA	Vol. **59**, No. 12
10th Conference on Weathercasting	14–15 June 1980	Denver, CO	Vol. **61**, No. 3
11th Conference on Weathercasting	26–18 June 1981	Boston, MA	Vol. **62**, No. 2
12th Conference on Weathercasting	25–27 June 1982	Seattle, WA	Vol. **63**, No. 3
13th Conference on Broadcast Meteorology	17–19 June 1983	Minneapolis, MN	Vol. **64**, No. 4
14th Conference on Broadcast Meteorology	22–24 June 1984	Clearwater Bch, FL	Vol. **65**, No. 3
15th Conference on Broadcast Meteorology	9–12 April 1985	Honolulu, HI	Vol. **66**, No. 1
16th Conference on Broadcast Meteorology	3–6 April 1986	Atlanta, GA	Vol. **67**, No. 1
17th Conference on Broadcast Meteorology	25–27 June 1987	Reno, NV	Vol. **68**, No. 2
18th Conference on Broadcast Meteorology	17–19 June 1988	Chicago, IL	Vol. **69**, No. 2
19th Conference on Broadcast Meteorology	1–4 June 1989	Bal Harbour, FL	Vol. **70**, No. 3
20th Conference on Broadcast Meteorology	21–24 June 1990	Boulder, CO	Vol. **71**, No. 3
21st Conference on Broadcast Meteorology	18–19 October 1991	Crystal City, VA	Vol. **72**, No. 9
22nd Conference on Broadcast Meteorology	11–13 June 1992	San Diego, CA	Vol. **73**, No. 4
23rd Conference on Broadcast Meteorology	29 June–1 July 1993	Charleston, SC	Vol. **74**, No. 4
24th Conference on Broadcast Meteorology	15–20 January 1995	Dallas, TX	Vol. **75**, No. 10

REFERENCES

AMS, 1955a: Minutes of the Council. Washington meeting, 4–5 May 1955. *Bull. Amer. Meteor. Soc.* **36,** 296–297.

______ , 1955b: News and notes. Radio and television. *Bull. Amer. Meteor. Soc.,* **36,** 317.

______ , 1956: National Conference and Workshop on Radio and Television Presentation of Weather Program. *Bull. Amer. Meteor. Soc.,* **37,** 81–82.

______ , 1967: Program: Second Conference on Weathercasting. *Bull. Amer. Meteor. Soc.,* **48,** 781–784.

______ , 1994a: Membership qualifications. *Bull. Amer. Meteor. Soc.,* **75,** 1485.

______ , 1994b: Seal of Approval Program for Radio and Television. *Bull. Amer. Meteor. Soc.,* **75,** 1484–1488.

Davis, F.K., 1976: Weather and the media. *Bull. Amer. Meteor. Soc.,* **57,** 1331–1333.

______ , 1990: Book review of *Television Weathercasting: A History. Bull. Amer. Meteor. Soc.,* **71,** 1774–1776.

Fidler, J.C., 1948: Weather via television. *Bull. Amer. Meteor. Soc.,* **29,** 329–331.

Fletcher, R.D., 1956: Two outstanding problems of modern meteorology. *Bull. Amer. Meteor. Soc.* **37,** 473–476.

Geer, I., 1975: The establishment, operation, and evaluation of a school weather service. *Weatherwise,* **28,** 114–117.

Jehn, K.H., 1956: The challenge of television weather programs. *Bull. Amer. Meteor. Soc.,* **37,** 351–353.

Leep, R.L., 1969: A broadcaster's severe weather network. *Bull. Amer. Meteor. Soc.,* **50,** 630–632.

WTVT Weather Service, 1991: *Thirty-Five Years of History, 1956–1991.*

Education

From Geo- to Physical Science: Meteorology and the American University, 1919–1945

WILLIAM A. KOELSCH

Introduction

In the year that the American Meteorological Society (AMS) was founded, meteorology in America was sadly lacking in opportunities for advanced training and for the sustained production of professional meteorologists. A U.S. Bureau of Education study undertaken for the U.S. Weather Bureau in 1919 found that of 433 institutions surveyed, only 70 offered instruction in meteorology or climatology. Of these, only eight had more than two courses in this area (Brooks 1919).

This situation was a legacy of the period before World War I. Meteorologists and others had failed to secure the subject's legitimation in the new and emerging research universities of the 1880s and 1890s. That effort foundered in part over the question of placement: Was the appropriate context for the subject geology, geography, physics, or some applied science such as agriculture? This chapter will examine the alternative ways meteorology was institutionalized at an advanced level in the 1920s and 1930s until it emerged in the early 1940s as a recognized university discipline.

Models of scientific legitimacy: Meteorological education before 1919

First, a brief look at earlier education in meteorology. Fleming (1990, xvii) has termed the period prior to 1800 "the era of individual, isolated diarists," who by and large were making private compilations of local weather data (see also, e.g., Henry 1893; Miller 1933; Brown 1940; Berman 1986). Limited meteorological observation and some teaching had also existed early on in American colleges, in connection with the teaching of natural philosophy (e.g., Varney 1908; Morton 1940; Hornberger 1945; Cohen 1950). Beginning in 1838 at Williams

FIG. 18-1. Alexander McAdie as Signal Corps cadet, Ft. Myer, Virginia, 1882 (from McAdie 1949).

College and in 1841 at Harvard University, student meteorological societies emerged, to take observations in connection with natural philosophy courses, as part of the work of newly established college astronomical observatories, or in connection with the Smithsonian network of weather stations (Koelsch 1966, chapter 2; Guralnick 1975, 69–70; Fleming 1990, 10–12, 63–64). Yet even the growing systematization of observations in the colleges and academies remained voluntary and to a degree amateur; Fleming is correct in characterizing the period before 1870 as "'organized' but not yet professional" (Fleming 1990, 170–171). The meteorological activity of students in the college or academy was either a minor adjunct to their academic work or a voluntary extracurricular activity.

Another model of meteorological education, emerging in the early 1880s, was advanced in-service training. The Army's Signal Service, then the nation's weather service, established a "study room" for research purposes under Cleveland Abbe in 1880, and the following year began recruiting promising young science graduates for specialized training in meteorology and signaling at a new Signal Corps school at Ft. Myer, Virginia (Fig. 18-1). In a sense, the Signal Corps school was America's first postgraduate program in meteorology, and the military continued to use this in-service model through the 1919–1945 period (Reichelderfer 1970). Out of the Signal Service venture also came cooperative research and study arrangements with physical scientists at several universities (Whitnah 1961; Koelsch 1986).

The rise of American universities and a parallel race to incorporate specialized academic disciplines in them proved problematic for meteorology, however. Abbe raised the issue most sharply in 1906 in an article in one of the geographical journals (Abbe 1906): Was meteorology a branch of natural history and thus primarily empirical in approach? Or, given the earlier American contribution to the theory of storms, was it a physical science, tied to theory, instrumentation, and the search for general laws? [See also Nebeker (1989).] Government scientists, by and large, took the latter view, as exemplified by the sponsorship of academic research and training in atmospheric electricity in emerging research physics departments. But in the universities, meteorology more often fell into the natural history tradition that, by the late nineteenth century, had broken into two major segments: zoology–botany and geology–physical geography. An academic placement closely tied to the interests of geologists and physical geographers in climatology was generally unsatisfactory to Abbe and other Weather Bureau scientists.

Each of these ad hoc arrangements had worked to a limited degree, however, to nurture a handful of trained meteorologists in the pre-1919 period. Two of the young Signal Service scientists assigned to Boston, Alexander McAdie and Austin McRae, studied atmospheric electricity under John Trowbridge and received graduate degrees at Harvard. Another of the bright young Signal Service recruits interested in atmospheric electricity, Park Morrill, had been detailed to The Johns Hopkins University in 1882 to work with the physicist Henry Rowland. In 1889 McAdie began a program of study for the Ph.D. degree in atmospheric electricity at the newly opened Clark University under the physicist Albert A. Michelson but returned to the Signal Service after a year (Koelsch 1986).

When Columbian University [now George Washington University (GWU)] opened a graduate school in 1893, Abbe was made professor of meteorology there and he and other Weather Bureau officials with similar part-time academic appointments began supervising candidates for the M.S. degree. Courses continued to be offered there by the scientists of the Weather Bureau, primarily for the bureau's own employees. In 1910, GWU awarded a Ph.D. in solar physics to H. H. Kimball, a longtime bureau scientist; the university's first Ph.D. in meteorology proper was awarded to C. LeRoy Meisinger in 1922 (Lewis 1995). The Weather Bureau encouraged other scientists in its employ to offer courses or special lectures in meteorology at nearby universities and even high schools. In the *Annual Report* of the chief of the Weather Bureau for 1907–1908, for example, Chief Willis Moore listed some 15 universities and colleges at which station scientists were giving regular courses in meteorology. None of the Weather Bureau's cooperative activity with colleges and universities, however, yielded academic programs capable of sustaining both pure research and the production of a sustained series of M.A. and Ph.D. degrees in meteorology (see, e.g., Kutzbach 1979).

The principal geography–geology programs were those developed at Harvard and Johns Hopkins. During the 1880s, the Harvard geographer William Morris Davis organized a program of advanced meteorological education, termed by contemporaries the "Harvard School of Meteorology," on an institutional base encompassing several Harvard departments, the Boston weather station, the New England Meteorological Society, and Abbott Lawrence Rotch's new Blue Hill Meteorological Observatory (Koelsch 1981a). As part of another ad hoc program in physical geography pulled together by Johns Hopkins geologist William Clark (who was also head of the Maryland Weather Service), Abbe gave his private library to The Johns Hopkins University and was made lecturer in meteorology in 1898. The following year a young Weather Bureau scientist who had previously studied in Germany, Oliver L. Fassig, was awarded the first American Ph.D. in meteorology and climatology, under the aegis of the geology department. Fassig then began a decade of irregularly offered lecture courses in meteorology and climatology, but after 1909 the subject disappeared from the Hopkins curriculum (Koelsch 1981b).

After the turn of the century, an attenuated form of the Harvard School of Meteorology was offered by Robert DeCourcy Ward, who taught in Harvard's geology and geography department from 1895 until his death in 1931. Made professor of climatology in 1910, Ward turned out the second and third Ph.D.'s in meteorology and climatology: Charles F. Brooks (Ph.D. 1914), later first secretary of the AMS, and Co-Ching Chu (Ph.D. 1917), later first director of the Academia Sinica's Institute of Meteorology. Ward also supervised several Harvard M.A. students who earned their doctorates at other institutions but became well known in meteorology or climatology (Koelsch 1983).

Ward's primary interest and impact was in teaching an introductory meteorology course to undergraduates, for whom it served as a relatively easy way to meet the Harvard science requirement. Ward's advanced courses, largely in regional climatology, drew far fewer enrollees, but the serious student could learn much from Ward's extraordinarily broad command of the American and foreign meteorological literature and from his talent for extracting the best of the relevant material in his lectures. Ward's six courses in the field reported in the 1919 survey were more than anyone else in the country was offering at the time. More importantly, after 1908 Harvard graduate students could earn research credits for observational experience in meteorology, as well as access to the latest research publications, at the Blue Hill Meteorological Observatory under its director, Rotch, or his successor, McAdie. Both Brooks and George Porter Paine (A.B. 1905), later the fourth American Ph.D. in meteorology (University of Wisconsin 1918), were among those serving as research assistants at Blue Hill in the period just before World War I (Rotch 1908; McAdie 1930; Conover 1990).

Programmatic affiliation I: Geography's last hurrah

America's entry into World War I had given meteorology a fresh impetus because of its relationship to flight, navigation, and long-range artillery needs (Gregg 1933). The federal government needed 600 (later 1000) meteorologists well trained in mathematics and physics. Since no such pool existed, the military took emergency measures to increase the number of qualified weather observers. About 200 men were inducted into the Signal Service and given crash courses of 8–10-week duration at Weather Bureau stations throughout the country. In May 1918 the Signal Service established a special school of meteorology in cooperation with the Weather Bureau, first at an army camp near Waco, Texas, and then at the Texas Agricultural and Mechanical College.

About 300 men from over 100 colleges, most with degrees in science or engineering, were given an intensive three-month course in meteorology under Fassig's direction; Brooks was among the instructors (Fassig 1918; Millikan 1919, 1920). One of the trainees was astronomy student C. LeRoy Meisinger, who became interested in meteorology as a result of Brooks's enthusiasm. Another, Clarence Koeppe, was to become one of Brooks's own Ph.D. recipients at Clark University in 1929. A third, Sgt. Perez Etkes, concerned about the lack of meteorological knowledge among military aviators, wrote a series of letters to Brooks early in 1919 urging him to form a new organization to promote both academic and popular interest in the subject. The result was the birth of the American Meteorological Society (Brooks 1930).

At Blue Hill, 54 Naval Reserve officers, graduates of 27 American universities and colleges, were trained in meteorology during the war, among them was Francis L. Reichelderfer, who later became chief of the Weather Bureau (Conover 1990). Wartime Student Army Training Corps programs at various colleges and universities also included systematic instruction in meteorology. At some colleges and universities, courses in this field seem to have continued into the postwar period (Brooks 1919).

In 1919 Charles F. Brooks was editor of the *Monthly Weather Review.* He had become interested in meteorology as a result of his preparation for the Harvard entrance examinations in physical geography, for which William Morris Davis's *Elementary Meteorology* (1894) was a required text (Brooks 1925). Discouraged by postwar budget cuts in the Weather Bureau and the *Monthly Weather Review,* Brooks, who had previously taught at Yale, began looking again for academic employment. In June 1921 he became associate professor of meteorology and climatology in the newly established Graduate School of Geography at Clark University (Koelsch 1980). The new school was the centerpiece of Clark's reorganized graduate programs under a new president, Wallace W. Atwood, who as professor of physiography at Harvard after 1913 had known Brooks and indeed served on his final examination board. The American Meteorological Society, whose office was effectively under Brooks's hat, moved with him to Clark, where he was given two graduate assistants and the part-time services of a stenographer to help produce the *Bulletin of the American Meteorological Society (Bulletin),* which was first published in January 1920. Material sent by exchange was deposited in the Clark library, where it remained until its transfer to Blue Hill in 1951 (Brooks and Richardson 1958; Conover 1990).

Like his Harvard mentors, Brooks maintained that meteorology and climatology should be taught in association with the "outdoor" sciences of geography and geology (Brooks 1918, 555). His Clark appointment gave him the chance to elaborate that model in a full undergraduate and graduate sequence of some eight courses, one of them taught jointly with Clark physicist (and rocket pioneer) Robert H. Goddard. A weather station set up to Weather Bureau standards was established atop the university's physics building. At Clark, Brooks (Fig. 18-2) supervised the research of seven Ph.D. students in meteorology and climatology, beginning with Burton Varney (Ph.D.

1925) and ending after Brook's death, with Phil E. Church (M.A. 1932, Ph.D. 1937), whose dissertation, though signed by Brooks's successor, was clearly under Brooks's direction "from beginning to end," as Church put it in his acknowledgements (Church 1937).

Brooks also supervised several M.A. degree candidates in this area, as well as the climatological work for Ph.D. and M.A. candidates in regional geography. Other graduate students, such as Brooks's Blue Hill assistant Robert F. Stone, studied with him at Clark, as did some undergraduates. When Antioch College student Henry S. Pierce,

FIG. 18-2. Professor Charles F. Brooks, Graduate School of Geography, Clark University, 1920s (courtesy Clark University Archives).

later a longtime Weather Bureau forecaster, was looking for a school to transfer to in order to prepare himself for Weather Bureau employment, a bureau official told him that Clark was "the best college in the East" for that purpose (Pierce 1930).

Brooks's research projects gave graduate students employment and research training, often providing them opportunities for publication. The projects included studies of New England rainfall, the marine meteorology of the western Atlantic and the Caribbean, and research for the climatology of North America in the famous Köppen–Geiger series. Brooks also found time to popularize meteorology and climatology, most successfully through his "Why the Weather?" columns, syndicated through the Science Service to a large number of American and Canadian newspapers. Beginning in 1921 Brooks offered summer school courses in meteorology and climatology, largely for teachers, some of whom became his future graduate students. He also devised four correspondence courses in meteorology and climatology offered through the Home Study Department of Clark University. One of Brooks's students, Jerome Namias, took his first meteorology course by that method and later went on to study meteorology at the Massachusetts Institute of Technology (MIT) (Basu 1984; Conover 1990, 384–386; Namias 1986).

In 1931, anticipating Alexander McAdie's retirement, President Lowell asked Brooks to return to Harvard as professor of meteorology and director of the Blue Hill Meterological Observatory, initially on a half-time basis. Robert DeCourcy Ward died unexpectedly in November and, in addition to administering the observatory, Brooks finished out both Ward's attenuate Harvard courses and his own courses at Clark. Brooks had been able to breathe new life into the geographically based program model at Clark at a time when meteorology and climatology were gradually pulling apart. Under his successor, Samuel van Valkenburg, however, the emphasis shifted markedly from meteorology to regional climatology. The total number of courses offered in meteorology and climatology in the School of Geography at Clark was cut to four, and three of the courses (including the two graduate courses) were in world regional climatology. van Valkenburg continued to supervise climatological master's theses and occasional dissertations in climatology. Brooks assisted a number of Clark University M.A. and Ph.D. candidates (van Valkenburg 1959), notably Jen-hu Chang (Ph.D. 1954), whom Brooks later employed at Blue Hill. But after 1932, Clark University no longer offered a model program for training meteorologists.

At Harvard, Brooks was drawn ever more heavily into administration and fundraising for Blue Hill, the new high-altitude observatory on Mt. Washington, and various research projects (Conover 1990). Brooks had hoped that Harvard would renew its early lead in meteorology and climatology through cooperation with MIT, with Harvard offering only one course in meteorology proper (taught for some years by Hurd Willett of MIT) but a number of advanced courses and research opportunities in climatology. Owing to Harvard's fiscal problems during the 1930s and to President Conant's skepticism concerning both geography and atmospheric science, Brooks's hopes that Harvard would produce a stream of Ph.D.'s in meteorology and climatology were not to be realized (Stone 1958). The courses developed by Ward continued to be listed for some years in the Harvard catalog, though never offered, since his vacant professorship in climatology was never filled. During the 1940s Victor Conrad, formerly professor of meteorology and climatology at the University of Vienna, served as research associate in climatology, working out of the Institute of Geographic Exploration and doing some of his research at Blue Hill (Conover 1990). Although Brooks, Conrad, and others offered research supervision at the observatory to both Harvard and MIT students, by 1945 Brooks had supervised only one Harvard dissertation in climatology, that of Edmund Schulman (Ph.D. 1944).

During the 1930s the interests and approaches of geography and meteorology diverged more sharply than before. Some meteorologists denied that climatology was worthy of the meteorologist's attention. "Climatology has its own place in science," opined H. H. Kimball in 1933, "but it is not meteorology" (Kimball 1933). Others pointed to the lack of adequate background among geography students; Willett noted a decline in the number of geography students taking his Harvard general meteorology course as he stiffened the mathematics and science prerequisites to meet the needs of the physics and engineering students in the course (Committee on Meteorological Education 1946).

Even in the overlapping area of climatology, geographers and meteorologists often spoke in different tongues. When Stephen Jones, a Harvard-trained geographer (Ph.D. 1934), tried his hand at a climatology article, he received a sharp letter from Blue Hill's Robert Stone "in which he laid both [Wladimir] Köppen and [Warren] Thornthwaite by the ears." Somewhat on the defensive, Jones replied, and received a bibliography of references to "dynamical climatology" in return. "There is no doubt," wrote the chagrined Jones, "that the climatology of geographers is a long way from the climatology of meteorologists" (Jones 1937).

American geography departments together produced fewer than a dozen doctorates in climatology between 1932 and 1945. Except for an introductory weather and climate course, meteorology proper had been largely dropped from the programs. Even climatology, at the graduate level, had become a service course for regional and economic geographers. Mikesell is undoubtedly right in arguing that by 1945 the average Ph.D. in geography could be expected to know the Köppen system of world climatic regions (or some variant) and little else of the broad meteorological field (Mikesell 1984, 187).

Programmatic affiliation II: Meteorology and aeronautics

Within a few years after 1919, widely publicized attempts at trans-Atlantic flight, popular interest in "stunt" flying, the development of airmail service, increased recognition of meteorology's contributions to military aviation, some unhappy experiments with dirigibles, and a growing awareness that many Americans might become patrons of reliable commercial airline services all led to public demands for safer flying (Gregg 1933; Whitnah 1966; Bilstein 1983). A by-product of this demand was increased support for advanced meteorological research and training. The Air Commerce Act of 1926 required the Weather Bureau to aid flight safety by establishing additional weather stations and conducting research in atmospheric phenomena. This and subsequent legislation enabled the bureau to develop a new job market for trained forecasters in airport stations. Private airlines, too, began to hire weather forecasters (Whitnah 1966). [See chapter 16 by Cartwright and Sprinkle, this volume.]

The Weather Bureau, however, largely continued along traditional lines in the 1920s and 1930s. It had only three Ph.D.'s on its staff, two of them in their seventies by 1932. The bureau meteorologists for the most part ignored upper-air phenomena, and their standard cartographic and analytic methods in preparing daily forecasts were not much of an advance over practices of the 1890s. Most importantly, the bureau's senior scientists turned their backs on the new methods, models, and theories coming out of the Norwegian Weather Service and the Geophysics Institute at Bergen under Vilhelm Bjerknes and his legendary team of young Norwegian and Swedish scientists (Friedman 1987, 1989; Jewell 1981, 1984). The aging William J. Humphreys, the bureau's "meteorological physicist," was opposed to much of polar front theory, and his widely used text, *The Physics of the Air* (1920), devoted less than one page to airmass and frontal analysis. Although the library at the Weather Bureau received the publications of the Bergen School, they remained largely unread, and the occasional younger bureau meteorologist who expressed an interest in newer methods was actively discouraged from pursuing them (Namias 1983; Lewis 1995).

Elsewhere in the federal government there was more receptivity to new ideas. Reichelderfer, who knew the research of Bjerknes and his assistants, incorporated the new Bergen methods into the U.S. Navy's aerological training programs, which included an advanced training course for enlisted men at the Naval Air Station at Anacostia and a year of mathematics and physics for officers at the Naval Postgraduate School at Annapolis, Maryland. Reichelderfer evidently approached Brooks in the mid-1920s with the proposal of making Clark University's School of Geography the site for a naval officers' advanced program in meteorology. Brooks, a Quaker, declined the affiliation. Reichelderfer then talked with Harvard officials and sent his officers to study meteorology with Ward and McAdie during 1927–1928. Beginning the following year, naval officers in the program were sent to MIT for their year of meteorological study. In 1931 Reichelderfer himself went to Bergen for six months to study the new methods. He and his naval trainees were thus the first Americans to use the new Norwegian forecasting system and were among its earliest American advocates (Reichelderfer 1928, 1970; O'Brien 1935; Hughes 1981; Bates 1989; Namias 1991).

Funds for the institutionalization of Scandinavian meteorological concepts and practices in American universities came out of the aviation interests of philanthropist Daniel Guggenheim and his son Harry, a World War I naval aviator. In 1925 the senior Guggenheim endowed the Daniel F. Guggenheim School of Aeronautics at New York University (NYU), the first such institution in America. The following year he created the Daniel Guggenheim Fund for the Promotion of Aeronautics, which sponsored a wide range of activities directed toward flight safety, public education, and aeronautical research. Owing to Harry Guggenheim's experience as a licensed pilot, the fund (which Harry headed during its four-year life) placed special emphasis on understanding the environmental conditions affecting flight (Hallion 1977; Bilstein 1983).

The Guggenheim Fund supported project-oriented research into problems of fog flying, inaugurated a model weather service along the Western Air Express San Francisco to Los Angeles air travel route, and funded a meteorologist for Richard E. Byrd's first Antarctic expedition. The fund also financed a University of Michigan expedition to Greenland (whose team included a Clark-trained meteorologist) to study atmospheric conditions and the feasibility of establishing a system of weather stations there in anticipation of trans-Atlantic commercial flights. Beyond that academic project, the Guggenheim Fund announced its support of a program of advanced research and training in meteorology that would test "a theory of weather forecasting which maintains that weather is manufactured near the Poles"—a clear endorsement of the early model of weather genesis expounded by the Bergen group (Daniel Guggenheim Fund 1925–1929).

This latter commitment was initiated in August 1927 by the establishment of the Daniel Guggenheim Committee on Aeronautical Meteorology. The committee chair was 28-year-old Carl-Gustaf Rossby, who after working briefly with Bjerknes as an assistant at the Bergen Geophysical Institute, had come to the United States in 1926 to do research at the Weather Bureau (Bergeron 1959). When the Weather Bureau, unresponsive to his new ideas, actually declared Rossby persona non grata in a letter sent to all weather stations, the Guggenheim Fund snapped up his services and put Rossby in charge both of the committee and the model aeronautical weather service in California (Byers 1959, 1960).

In June 1928 the Guggenheim Fund granted MIT a special appropriation of $34,000 to support the first three years of a program of meteorological research and advanced training. The U.S. Navy guaranteed an initial cadre of students; six officers came the first year, following their year at the Naval Postgraduate School. Rossby was released by the Guggenheim Fund to organize the new program, initially administered through MIT's Department of Aeronautical Engineering. An undergraduate course was also offered under the aegis of MIT's Department of Physics. Rossby remained in the United States for the next 20 years, becoming the primary organizer of university teaching and research in his chosen subject in America (Byers 1959; Reichelderfer 1960; Bates 1989).

By contrast with the descriptive climatological tradition practiced by the geographical and agricultural climatologists, the MIT program emphasized the further development and application of the Bergen-initiated polar front theories. In addition to empirical work on the movement of air masses, instruction at MIT was to explain "by means of established physical laws, those of the observed weather phenomena which are at present susceptible to theoretical treatment." Rossby and his successor Sverre Petterssen gathered together an exceptionally able group of investigators (Haurwitz 1986), including the young Weather Bureau meteorologist Hurd C. Willett (Nielson 1993), and a number of superior graduate students (e.g., Horace Byers, Jerome Namias, Chaim Pekeris, Harry Wexler), through whom the techniques of airmass analysis were disseminated to the American meteorological community. Building on the work of Rossby and the Bjerknes group, Willett worked out a scheme for the analysis of North American air masses that became the foundation of U.S. aviation weather forecasting (Hallion 1977, 219–221).

FIG. 18-3. Professors C.-G. Rossby, K.O. Lange, and D.C. Sayre (research pilot) examining meteorograph used on MIT's "flying laboratory," early 1930s (courtesy the MIT Museum).

Rockefeller grants supported the cost of an airplane for three years between 1931 and 1934, so that the MIT scientists could work out practicable methods for in-flight observations (Fig. 18-3). A satellite research station was established at South Dartmouth, Massachusetts, for fog studies. The South Dartmouth facility, Rossby hoped, would be the first step in establishing a detailed weather reporting system for New England under MIT control. MIT, not the Weather Bureau, produced the only published daily analysis of North American air masses during this period. A joint seminar with Blue Hill meteorologists provided a forum at which local professionals and visiting foreign scholars, often Scandinavian, could present and hear the latest research concepts; Jacob Bjerknes himself gave several lectures in the fall of 1933. In December 1933 the AMS met in Cambridge, Massachusetts, and heard Rossby describe the work of his department. In February 1934 a major symposium on the new meteorology was held at MIT, followed in June by a similar one at the California Institute of Technology (Cal Tech).

Two new publications series were begun, the *Professional Notes* of the MIT meteorology course in 1929, and MIT *Meteorological Papers* in

1930. The series were established because, according to Rossby, "at the present time there is no proper forum in the United States for the publication of more abstract investigations in the meteorological field." All of this, Rossby argued, presupposed a concept of meteorological research and education "shared by most authorities in the subject" but only now, at MIT, being implemented by "the first systematic effort towards its realization in the United States" (Rossby 1929).

In his first annual report, Rossby had thrown down the gauntlet on behalf of new concepts of meteorological education. Descriptive and even statistical climatology of the type common in geography departments was to be shunted aside in the new two-year graduate program, which would instead study "the fundamental principles underlying the behavior of the atmosphere." The program initially consisted of four courses. First came Willett's introductory course, based on Humphreys' text *The Physics of the Air* and modern synoptic charts. This was followed by a nonmathematical synoptic course "based on the Bjerknes' polar front theory," also taught by Willett. These were succeeded by a course in "dynamic meteorology," emphasizing mathematical methods, and a seminar reviewing and discussing current literature, both offered by Rossby (Rossby 1929).

Although the first M.S. degrees awarded meteorological students were given without field specification, by 1931 MIT had approved the award of graduate degrees specifically in meteorology. The first MIT doctorate in the field, an Sc.D., was awarded in 1933 to Pekeris, who later became a well-known geophysical theorist. Namias was principal author of a series of articles for the AMS *Bulletin* during 1934 that explained the new approaches (Namias 1982). Published separately in an inexpensive paper format, the book version, *An Introduction to the Study of Air Mass Analysis,* went to five editions and eventually sold about 50 000 copies (Namias et al. 1935; Namias 1980–1981, 1982, 1983, 1986; Basu 1984).

MIT's President Karl Compton and the academic administration supported Rossby's venture with enthusiasm, as fitting their concept of transforming the institute from a traditional engineering school, which emphasized service to industry, into a major research-based technological university (Lecuyer 1992). By 1945, MIT had granted 11 doctorates in meteorology, and the meteorology faculty had emancipated themselves from aeronautical engineering to become a full-fledged department. Beginning as an adjunct of the demands of aviation technology, the MIT program in meteorology had quickly become, as one historian has put it, "the most important academic center of weather studies in the United States" (Hallion 1977, 221).

A small but steady stream of bright, well-trained young civilian and military meteorologists went on from MIT to fill positions in the army, the navy, the civilian aviation sector, and even the hitherto recalcitrant Weather Bureau (O'Brien 1935; Whitnah 1961). Chief Willis Gregg, appointed in 1934, hired a dozen new MIT-trained men and invited Sverre Petterssen, then director of the forecast center at Bergen, to give lectures on the new meteorology to Weather Bureau forecasters in Washington and elsewhere (Bundgaard 1979). Reichelderfer, who succeeded Gregg in late 1938, took advantage of a clause in the Civil Aeronautics Act of 1938 to subsidize Weather Bureau personnel studying "advanced methods of meteorological science" at MIT, Cal Tech, and NYU. In 1939 Rossby took a three-year leave of absence from MIT to join the bureau as assistant chief for research and education, where he supervised a greatly expanded program of research and training. Enrollments in the MIT program tripled the next year "as a result of government requests to train meterologists for the Army Air Corps and the Weather Bureau" (Whitnah 1961; Bates 1989).

Institutional expansion in the 1930s

At the conclusion of its 1933 Cambridge meeting, the AMS passed a number of resolutions, among them an expression of its "great interest in and appreciation of" MIT's work in airmass and polar front analysis, and called for promotion by the Society of "systematic meteorological instruction and research" in other American universities. Four years later, the Society again urged major universities to establish graduate programs in meteorology, funded through gifts from corporations and individuals in amounts sufficient not only to offer instruction but also to carry on research. In the last years of the Great Depression, four other American institutions were able to develop as centers of graduate training and research in this science.

The Guggenheims had given NYU a $500,000 grant in 1925 to help launch a School of Aeronautics. In 1928 the fund approved a grant to NYU to begin a course in aeronautical meteorology, initially housed in the geology department. In 1935, the university created an Institute of Aeronautical Meteorology to organize research in the upper air and forecasting for the New York City region, as well as to administer a meteorological observatory and provide support for the professionally oriented master of science program. Two years later Athelstan Spilhaus, who had studied with Rossby at MIT and had been Rossby's research assistant at the Woods Hole Oceanographic Institution, became assistant professor of meteorology; Spilhaus had previously organized the first upper-air observations in South Africa. In 1938 a department of meteorology was created at NYU under Spilhaus's direction, and in 1946, in cooperation with the physics department, the program began offering the doctorate in meteorology. (*Weatherwise* 1953; New York University 1956; Bilstein 1983).

Graduate work in meteorology at Cal Tech, also a professionally oriented program, began in the 1934–1935 academic year. It featured a one-year course that, with an additional thesis, might lead to the M.S. degree in meteorology (Goodstein 1991; Lewis 1994). The Cal Tech program was directed by Irving Krick, who had worked on problems of air turbulence with the director of Cal Tech's Guggenheim Aeronautical Laboratory, the Hungarian-born Theodore von Kármán. Though not a meteorologist by training, von Kármán had taught a meteorology course at Cal Tech in 1933–1934. He valued its potential contribution to commercial aviation and military technology, though admitting that "very few academicians accepted meteorology because it was regarded as a guessing science." Von Kármán moved to add meteorology to Cal Tech's permanent curriculum, and at his suggestion Cal Tech sent Krick to Europe to study the newest meteorology research methods (von Kármán 1967; Lewis 1994).

On his return to Cal Tech, Krick was made chair of an independent department of meteorology. Many of the department's first students were Army Air Corps officers, thanks to the interest of Col. H. H. Arnold. In 1935 Cal Tech got its own Scandinavian meteorologist when Sverre Petterssen lectured there for three months as a visiting professor. The renowned Cal Tech geophysicist Beno Gutenberg offered a course on the structure of the atmosphere (Haurwitz 1985). But on the whole, the program was oriented toward training meteorologists for the burgeoning California airline industry and other applied activities; many of the graduates found careers as consultants, and Krick himself ran a lucrative consulting business on the side. The absence of a strong theoretical component, and the controversial nature of some of Krick's methods and claims, weakened the standing of the program with the

Weather Bureau and, ultimately, with the Cal Tech administration. After the head of Cal Tech, Robert Millikan, a supporter of meteorology from his days in the military during World War I, was replaced in 1945 by Lee A. DuBridge, the meteorology program was cancelled in 1948 (Boesen 1978; Lewis 1994).

Full scientific acceptance of meteorology as an advanced university discipline in America came about in 1940 when both the University of California, Los Angeles (UCLA) and the University of Chicago established major programs, each under the leadership of a distinguished Scandinavian meteorologist. At UCLA, meteorology courses had previously been offered under the aegis of the department of geography, which had also maintained a weather observatory (Dunbar 1981). For a decade, beginning in 1919, Ford A. Carpenter, designated lecturer in meteorology, had offered courses on "Weather Science in Relation to Aeronautics and Industry" and "Weather Science in Relation to Agriculture and Engineering." After 1929, however, UCLA courses in weather and climate were oriented toward training geographers.

In 1938 Joseph Kaplan, a physicist who had worked on atmospheric physics and taught courses in astrophysics, became chair of the UCLA Physics Department and began to press for expansion into the field of meteorology. He was joined by Harald U. Sverdrup, Vilhelm Bjerknes's successor at Bergen, who since 1936 had been director of the Scripps Institution of Oceanography in La Jolla, California (Jacobs 1936). The following year Jacob Bjerknes was engaged by the Weather Bureau, now in the hands of Rossby and Reichelderfer, to lecture for three months at Weather Bureau training centers. J. Bjerknes preceded this assignment with month-long visits or groups of lectures at several university centers, including the University of California, Berkeley (Berkeley) and Cal Tech. At Sverdrup's initiative, Bjerknes was invited to give three lectures at UCLA in December 1939 (University of California 1977).

From his Weather Bureau perch Rossby had tried to engineer a post for J. Bjerknes in the geography department at Berkeley as a threshold for expansion into the premier West Coast research university. A simultaneous one-year exchange was to be arranged between Willett and Berkeley's climatologist, John Leighly, but these plans, and possibly the whole idea, were vetoed by the geography department's autocratic chair, Carl Sauer (Rossby 1939–1940). Early in 1940 interested UCLA faculty and Sverdrup joined the dean of UCLA's graduate division, Vern Knudsen, in pressing the University of California system president, Robert G. Sproul, to approve a new graduate program in meteorology at UCLA. Their case rested, they argued, on the fact that MIT at the time offered the only truly outstanding graduate department of meteorology in the United States and therefore had become the chief center for the advanced training of meteorological personnel for the Weather Bureau, civil aviation, and the military. Implicit in their argument (along with an implied critique of the nearby Cal Tech graduate program) was the inference that if UCLA could secure J. Bjerknes, it could both match MIT in prestige and gain a guaranteed, tuition-paying clientele.

All concerned were agreed (in the words of Burton Varney, the geography department's weather and climate specialist) that "to attempt to prepare young men for research careers in meteorology in an atmosphere not reeking with physics and mathematics would . . . be a mistake of the first magnitude" (Varney 1940). Since no one could suggest that UCLA's geography department so reeked, it was proposed that the program initially be housed in the department of physics. While these deliberations were in progress, the Germans invaded Norway, fortuitously cutting off Bjerknes's ability to return to Bergen. The UCLA plan received hearty endorsements from Reichelderfer and Rossby, along with a tacit understanding that the Weather Bureau would contract with UCLA for advanced training programs. President Sproul approved the expansion after receiving Dean Knudsen's assurance that it could initially be funded with monies previously earmarked for professorships in chemistry and French.

In June, J. Bjerknes was offered and accepted the post, no doubt to the chagrin of the chemistry and French departments, to begin 1 July 1940. An Advisory Committee on Meteorology was established, and both the Weather Bureau and the army air force began to supply students for the new program. J. Bjerknes's principal assistant, Jörgen Holmboe, was a Norwegian who had previously taught at MIT, where other early UCLA faculty meteorologists had studied. UCLA conferred its first bachelor's and master's degrees in meteorology in 1941, although the regular program was soon to be overwhelmed by wartime training needs. In 1944 meteorology attained independent departmental status, and two years later its first doctoral candidate, Jule Charney, received the Ph.D. degree (Sekera 1967; Mintz 1975).

Almost as soon as he had taken up his Weather Bureau post,

Rossby began to lobby for the establishment of a major graduate program in meteorology in the Midwest, preferably at the University of Chicago. Not only was Chicago's location critical to the study of upper-air phenomena affecting the weather of the East Coast, but the university had outstanding departments of physics and mathematics, as well as a general reputation for a combination of interdisciplinary innovation and sound scholarship. Early in 1939 Rossby solicited the endorsement of President Compton of MIT, who was also a member of the advisory committee to the Weather Bureau, for such a department (Rossby 1939). Compton forwarded the proposal to his brother Arthur, then professor of physics at Chicago.

Although the idea was initially rejected on financial grounds, after Arthur Compton had become dean of the physical sciences division a private donor (Sewell Avery) was located. An institute of meteorology was then inaugurated within the physics department, beginning instruction in the fall quarter 1940 (University of Chicago 1940; Compton 1940; Hutchins 1940–1943). Horace Byers, an early MIT graduate (Ph.D. 1935) then with the Weather Bureau in Chicago, became secretary of the institute. The following year Rossby himself came to Chicago, first as a visiting professor and subsequently (beginning in 1942) as head of an independent department of meteorology (Hutchins 1940–1943; Byers 1976). The department's first Ph.D., Morris Neiberger, received his degree in 1946.

At Chicago, Rossby had an unprecedented free hand. In 1895 Albert A. Michelson, Chicago's first professor of physics, had added "terrestrial physics" to his department's offerings, an action hailed by the *American Meteorological Journal* as inaugurating "a new era in the development of the study of meteorology in the United States." Geologist T.C. Chamberlin focused his methods course on the geological agency of the atmosphere in 1896. But these initiatives, like others of the 1890s, were stillborn.

Meteorology and climatology courses were offered after 1903 in the newly created department of geography by J. Paul Goode, an enthusiastic teacher of the subject but primarily a cartographer (Martin 1984). No meteorology courses had been regularly offered at Chicago since the departure in 1935 of the Australian geographer Griffith Taylor for Toronto, though Brooks had taught a graduate course in climatology in the summer of 1939 (Conover 1990). Thus, not only were there no existing departmental interests to be circumvented, but the program at Chicago was free of any pressures to meet the service demands of aeronautical engineers, as it had been, initially at least, at MIT, Cal Tech, and NYU.

The result was a department bearing Rossby's distinctive stamp, the ability to visualize the dynamics of the atmosphere "from the top down," which became the hallmark of the young scientists he brought together to form the so-called Chicago School of meteorology (Lewis 1992). As at MIT, Rossby quickly established a monograph series, the department's *Miscellaneous Reports,* in 1942. In 1943 he established the Institute of Tropical Meteorology at the University of Puerto Rico, which, especially under Herbert Riehl, was to do for the understanding of tropical weather systems what the V. Bjerknes group had done for polar meteorology in Bergen during World War I. At both Puerto Rico and Chicago the weather and climates of the war zones were extensively studied, and from the beginning, military and naval meteorologists were among the students (Bergeron 1969).

Although Rossby's primary interest in going to Chicago was to set up a program of research and education around a few, carefully selected, theoretically and mathematically oriented doctoral candidates, wartime needs soon shaped Rossby's efforts and those of his department. During the war Rossby used his formidable powers of persuasion to convince the military that its advanced instructional needs for meteorology were best met not primarily through military training programs, as in 1918, but through training programs sited in university departments (Byers 1959, 1970). Then Rossby organized the University Meteorological Committee, which designed an accelerated curriculum for the "Big Five" graduate programs. The committee instituted feeder programs at other institutions that included basic physical science and mathematics, as well as physical geography and meteorology. [Despite the differences from his own approach, Rossby thought every student in the advanced meteorology program should have had previous instruction in physical geography (Dicken 1988).]

Because of wartime demands, the number of meteorology students at Chicago doubled every quarter, from 20 in 1940 to 1000 by 1943 (Fig. 18-4). About 1700 students were trained at Chicago itself, and others studied low-latitude problems at the Institute for Tropical Meteorology at the University of Puerto Rico. Although estimates differ, perhaps 7000–10 000 men (and not a few women) received instruction in meteorology at Chicago and elsewhere in the course of the war. The work

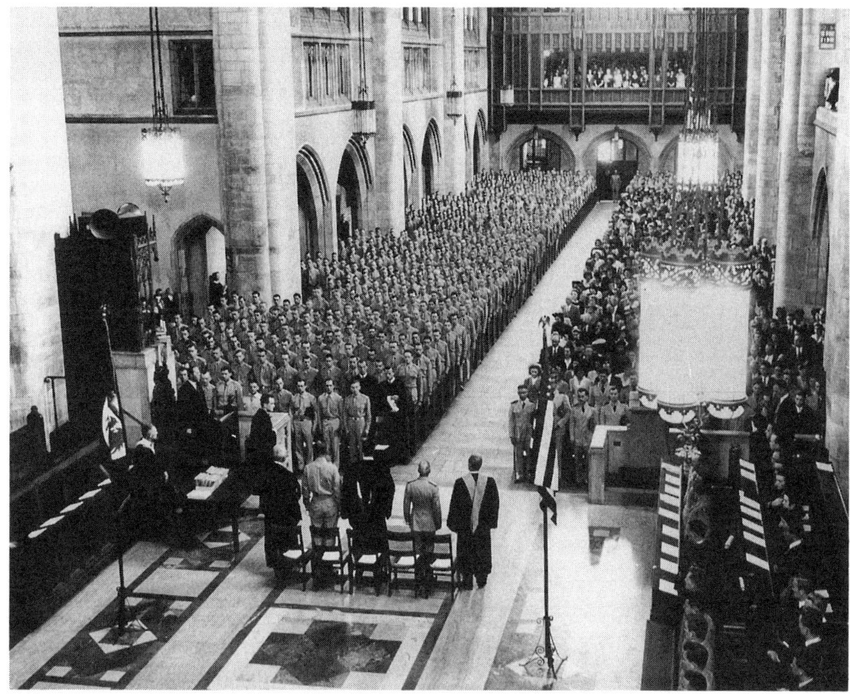

FIG. 18-4. Graduation ceremony for meteorological students, Rockefeller Chapel, the University of Chicago, 1943 (courtesy University of Chicago Department of Special Collections).

of the University Meteorological Committee was professional meteorological education on the grandest scale ever. By all accounts, Rossby found it to be among the most satisfying enterprises of his varied American career (Byers 1959).

By the close of World War II, American meteorological education had entered the modern age. In a small group of institutions meteorology had become legitimated as a graduate and research field on a par with other physical sciences, grounded in theoretical and mathematical methods, and having achieved well-worked-out programs of research and teaching. A pool of trained young meteorologists had been created that may have been 20 times the number professionally qualified either through advanced university work or in-service instruction prior to 1940. Many, probably most, of these returned to their prewar occupations after 1945. Others, however, were available for the many new positions in meteorology that opened up in government, private industry, and academia.

Looked at in longer view, the establishment of programs and departments of meteorology in the 1930s and 1940s represents a "second wave" of attempts to find an institutional home for the subject by piggybacking it onto other disciplines. In the late nineteenth century American university, meteorology had fallen by the wayside as several other disciplines (geography, anthropology, zoology, physics, sociology) had emerged from older groupings of academic subjects. It acquired among academics the "ignoble reputation" of being, according to von Kármán, the "guessing science," lacking theory, precision, and status.

The initial step toward achieving full academic recognition for meteorology was to ride the 1920s wave of interest in aeronautics in the established engineering schools, under the patronage of civil and military aviation decision makers who would supply the funding and the initial cadres of students. But to invade the more scientifically exclusive (and thus higher status) universities required the prior capture of the Weather Bureau by adherents of the new teaching. From that potential (and indeed, after the Civil Aviation Act of 1938, actual) patronage bastion, the final assault on the American university could be staged. Thanks to the fact that two leading practitioners of the new meteorology were available to head new graduate programs at UCLA and Chicago from 1940 onward, and to Rossby's successful insistence that wartime meteorological training be carried on to the largest degree possible within university precincts and in close association with front-line investigators, significant further steps in meteorology's legitimation as a university science had been taken by 1945.

In the year World War II ended, there were five major producers of professional meteorologists in America. Despite the closing of the program at one of the Big Five (Cal Tech) in 1948, by 1983 graduate programs in meteorology existed in some 55 American universities. In the 20 years before 1919, only four doctorates in meteorology had been awarded in America; by 1983 production of meteorology doctorates had reached an astounding 100 per year (Cressman 1983). Such numbers alone suggested that meteorology, in one configuration or another, would never again be excluded from the banqueting halls of American academic science.

Notes and Acknowledgements. Because this paper covers a long time span and several institutions, short items from many sources are rarely separately referenced. These include information drawn from institutional bulletins and catalogs, manuscript and published dissertation lists, standard biographical dictionaries, and short, though invaluable, notes and obituary notices from such journals of record as the *American Meteorological Journal,* the *Bulletin of the American Meteorological Society,* the *Monthly Weather Review,* and *Science,* as well as annual summaries entitled "Meteorology and Climatology" in the *American Year Book* and annual reports and special bulletins of the Army Signal Service and Weather Bureau for the period.

Similarly, I have built upon but rarely separately referenced many individual letters and other manuscript materials examined in the following repositories: Clark University Archives (Wallace W. Atwood Papers, Records of the Graduate School of Geography, Records of the Office of the Registrar, individual theses and dissertations); Harvard University Archives (Records of the Blue Hill Meteorological Observatory, Records of the Division of Geology, Robert DeCourcy Ward Papers, Derwent Whittlesey Papers); Houghton Library, Harvard University (William Morris Davis Papers); The Johns Hopkins University (Cleveland Abbe Papers, Daniel Coit Gilman Papers, Records of the Department of Geology); Institute Archives and Special Collections, MIT (Records of the Office of the President: Karl T. Compton, and the Hurd C. Willett papers); U.S. National Archives and Records Service (Records of the Weather Bureau); the UCLA Archives (Records of the Office of the Chancellor); and the University of Chicago Archives (Records of the Office of the President, 1940–1946: Robert M. Hutchins).

I am indebted for access to and assistance with these materials to Philip C. Bantin, formerly university archivist, UCLA; Clark A. Elliott, associate archivist, Harvard University Archives; Elizabeth Falsey and the staff of Houghton Library, Harvard University; Daniel Meyer, university archivist, and his staff at the University of Chicago; Julia Morgan, archivist of The Johns Hopkins University, and staff members of the Department of Manuscripts; Dorothy Mosakowski of the Goddard Library, Clark University; Helen Samuels, institute archivist, and her staff at the Institute Archives and Special Collections, MIT; and a number of staff members at the National Archives, where as an archivist as well as a researcher I was granted the rare privilege of direct access to manuscript material in the stacks. In addition, I wish to thank Deborah Cozart Day, formerly at MIT and now archivist of the Scripps Institution of Oceanography, La Jolla, California, for searching and sending copies of relevant materials from her collections, and Michael Yeates of the MIT Museum for the photo used as Fig. 18-3.

I also thank the editor, and Professors Mildred Berman (Emerita, Salem State College), Gary S. Dunbar (Emeritus, UCLA), and Paul Todhunter (University of North Dakota) for critical readings. As always, I am indebted to my secretary, Madeleine Grinkis, for preparing the manuscript for transmission on the information highway.

REFERENCES

Abbe, C., 1906: The present condition in our schools and colleges of the study of climatology as a branch of geography and of meteorology as a branch of physics. *Bull. Amer. Geogr. Soc.,* **38,** 121–123.

Basu, J.E., 1984: Jerome Namias: Pioneering the science of forecasting. *Weatherwise,* **37,** 190–201.

Bates, C.C., 1989: The formative Rossby–Reichelderfer period in American meteorology, 1926–1940. *Wea. Forecasting,* **4,** 593–603.

Bergeron, T., 1959: Methods in scientific weather analysis and forecasting: An outline in the history of ideas and hints at a program. *The Atmosphere and the Sea in Motion,* B. Bolin, Ed., Rockefeller Institute Press, 440–474.

Berman, M., 1986: Salem's physician–meteorologist: Dr. Edward A. Holyoke. *Essex Inst. Hist. Colls.,* **122,** 237–245.

Bilstein, R.E., 1983: *Flight Patterns: Trends of Aeronautical Development in the United States, 1918–1929.* University of Georgia Press, 236 pp.

Boesen, V., 1978: *STORM: Irving Krick vs. the U.S. Weather Bureaucracy.* G. P. Putnam's Sons, 159 pp.

Brooks, C.F., 1918: Collegiate instruction in meteorology. *Mon. Wea. Rev.,* **46,** 555–560.

———, 1919: General extent of collegiate instruction in meteorology and climatology in the United States. *Mon. Wea. Rev.,* **47,** 169–170.

———, 1925: C.F. Brooks to W.M. Davis, 6 February 1925. William Morris Davis Papers, Houghton Library, Harvard University.

———, 1930: Our Society's first decade. *Bull. Amer. Meteor. Soc.,* **11,** 8–12.

———, and S.J. Richardson, 1958: The Blue Hill Meteorological Observatory library. *Harv. Lib. Bull.,* **12,** 271–281.

Brown, R.H., 1940: The first century of meteorological data in America. *Mon. Wea. Rev.,* **68,** 130–133.

Bundgaard, R.C., 1979: Sverre Petterssen, weather forecaster. *Bull. Amer. Meteor. Soc.,* **60,** 182–195.

Byers, H.R., 1959: Carl-Gustaf Rossby, the organizer. *The Atmosphere and the Sea in Motion,* B. Bolin, Ed., Rockefeller Institute Press, 51–59.

______ , 1960: Carl-Gustaf Arvid Rossby. *Biog. Mem. Natl. Acad. Sci.,* **34,** 249–270.

______ , 1970: Recollections of the war years. *Bull. Amer. Meteor. Soc.,* **51,** 214–217.

______ , 1976: The founding of the Institute of Meteorology at the University of Chicago. *Bull. Amer. Meteor. Soc.,* **57,** 1343–1345.

Church, P.E., 1937: Surface temperatures of the Western Atlantic. Ph.D. dissertation, Clark University, 362 pp.

Cohen, I.B., 1950: *Some Early Tools of American Science.* Harvard University Press, 201 pp.

Committee on Meteorological Education, 1946: The teaching of meteorology in colleges and universities. *Bull. Amer. Meteor. Soc.,* **27,** 95–98.

Compton, A.H., 1940: Arthur H. Compton to Emory T. Filbey, 9 July 1940. Records of the Office of the President, 1940–1946, University of Chicago Archives, Chicago, IL.

Conover, J.H., 1990: *The Blue Hill Meteorological Observatory: The First Hundred Years, 1885–1985.* Amer. Meteor. Soc., 514 pp.

Cressman, G.P., 1983: Francis W. Reichelderfer, 1895–1983. *Bull. Amer. Meteor. Soc.,* **64,** 398–400.

Daniel Guggenheim Fund 1925–1929: *Reports.* Daniel Guggenheim Fund for the Promotion of Aeronautics.

Dicken, S.N., 1988: *The Education of a Hillbilly.* Lane County Geographical Society, 93 pp.

Dunbar, G.S., 1981: *Geography in the University of California (Berkeley and Los Angeles), 1868–1941.* De Vorss and Company, 18 pp.

Fassig, O.L., 1918: A Signal Corps school of meteorology. *Mon. Wea. Rev.,* **46,** 560–562.

Fleming, J.R., 1990: *Meteorology in America, 1800–1870.* The Johns Hopkins University Press, 264 pp.

Friedman, R.M., 1987: Constituting the polar front, 1918–1919. *Isis,* **73,** 343–362.

______ , 1989: *Appropriating the Weather: Vilhelm Bjerknes and the Construction of a Modern Meteorology.* Cornell University Press, 280 pp.

Goodstein, J.R., 1991: *Millikan's School: A History of the California Institute of Technology.* W. W. Norton and Co., 317 pp.

Gregg, W.R., 1933: History of the application of meteorology to aeronautics with special reference to the United States. *Mon. Wea. Rev.,* **61,** 165–169.

Guralnick, S.M., 1975: Science and the antebellum college. *Amer. Philos. Soc. Mem.,* **109,** 227 pp.

Hallion, R.P., 1977: *Legacy of Flight: The Guggenheim Contribution to American Aviation.* University of Washington Press, 292 pp.

Haurwitz, B., 1986: Meteorology in the 20th century: A participant's view. *Bull. Amer. Meteor. Soc.,* **66,** 282–291, 424–431, 498–504, 628–633.

Henry, A.J., 1893: Early individual observers in the United States. Report of the International Meteorological Congress held at Chicago, August 21–24, 1893, *U.S. Weather Bureau Bulletin,* No. 11 Pt. 2, 291–302.

Hornberger, T., 1945: *Scientific Thought in the American Colleges, 1638–1800.* University of Texas Press, 108 pp.

Hughes, P., 1981: Francis W. Reichelderfer, Part I: Aerologists and airdevils. *Weatherwise,* **34,** 52–59.

Hutchins, R.M., 1940–1943: The state of the university. Annual Reports, 1940–1943, University of Chicago, Chicago, IL.

Jacobs, W.C., 1936: Harald Ulrik Sverdrup—New director of the Scripps Institution of Oceanography. *Bull. Amer. Meteor. Soc.,* **17,** 306–308.

Jewell, R., 1981: The Bergen school of meteorology: The cradle of modern weather-forecasting. *Bull. Amer. Meteor. Soc.,* **62,** 824–830.

______ , 1984: The meteorological judgment of Vilhelm Bjerknes. *Social Res.,* **51,** 783–807.

Jones, S.B., 1937: Stephen B. Jones to Derwent S. Whittlesey, 19 May 1937. Derwent S. Whittlesey Papers, Harvard University Archivesk, Cambridge, MA.

Kimball, H.H., 1933: Recent advances in the science of meteorology and in its practical applications. *Bull. Amer. Meteor. Soc.,* **14,** 3–7.

Koelsch, W.A., 1966: The enlargement of a world: Harvard students and geographical experience, 1840–1861. Ph.D. dissertation, University of Chicago, 347 pp.

_____ , 1980: Wallace Atwood's "Great Geographical Institute." *Ann. Assoc. Amer. Geogr.,* **70,** 567–582.

_____ , 1981a: The New England Meteorological Society, 1884–1896: A study in professionalization. *The Origins of Academic Geography in the United States,* B. W. Blouet, Ed., Shoe String Press, 43–52.

_____ , 1981b: Pioneer: The first American meteorological doctorate. *Bull. Amer. Meteor. Soc.,* **62,** 362–367.

_____ , 1983: Robert DeCourcy Ward, 1867–1931. *Geographers: Biobibliographical Studies,* **7,** 145–150.

_____ , 1986: Ben Franklin's heir: Alexander McAdie and the experimental analysis and forecasting of New England storms, 1884–1892. *New Eng. Quart.,* **59,** 523–543.

Kutzbach, G., 1979: One hundred and twenty-five years of meteorology at the University of Wisconsin. *Bull. Amer. Meteor. Soc.,* **60,** 1166–1171.

Lecuyer, C., 1992: The making of a science based technological university: Karl Compton, James Killian, and the reform of MIT, 1930–1957. *Historical Studies in the Physical and Biological Sciences,* **23,** 153–180.

Lewis, J.M., 1992: Carl-Gustaf Rossby: A study in mentorship. *Bull. Amer. Meteor. Soc.,* **73,** 1425–1438.

_____ , 1994: Cal Tech's program in meteorology: 1933–1948. *Bull. Amer. Meteor. Soc.* **75,** 69–81.

_____ , 1995: LeRoy Meisinger, Part I: Biographical tribute with an assessment of his contributions to meteorology. *Bull. Amer. Meteor. Soc.,* **76,** 33–45.

Martin, G.J., 1984: John Paul Goode, 1862–1932. *Geographers: Biobibliographical Studies,* **8,** 51–55.

McAdie, A.G., 1930: The Blue Hill Observatory, 1884–1929. *The Development of Harvard University, 1869–1929,* S.E. Morison, Ed., Harvard University Press, 549–554.

McAdie, M.R.B., 1949: *Alexander McAdie: Scientist and Writer.* [Privately published.]

Mikesell, M.W., 1984: North America. *Geography Since the Second World War,* R.J. Johnston and P. Claval, Eds., Croom Helm/Barnes and Noble, 185–213.

Miller, E.R., 1933: American pioneers in meteorology. *Mon. Wea. Rev.,* **61,** 189–193.

Millikan, R.A., 1919: Some scientific aspects of the meteorological work of the United States Army. *Proc. Amer. Philos. Soc.,* **58,** 133–149.

______ , 1920: Some scientific aspects of the meteorological work of the United States Army. *The New World of Science: Its Development during the War,* R.M. Yerkes, Ed., The Century Co., 49–62.

Mintz, Y., 1975: Jacob Bjerknes and our understanding of the atmosphere's general circulation. *Select Papers of Jacob Aall Bonnevie Bjerknes,* M.G. Wurtele, Ed., Western Periodicals, 14–15.

Morton, C., 1940: *Compendium Physicae* [c.1678]. T. Hornberger, Ed., Pubs. Col. Soc. Mass., Vol. 33, 237 pp.

Namias, J., 1980–1981: The early influence of the Bergen school on synoptic meteorology in the United States. *Pure Appl. Geophys.,* **119,** 491–500.

______ , 1982: The early history of meteorology at the Massachusetts Institute of Technology. American Geophysical Union, Committee on History of Geophysics, *Newsletter,* **2,** 112–122.

______ , 1983: The history of polar front and airmass concepts in the United States: An eyewitness account. *Bull. Amer. Meteor. Soc.,* **64,** 734–755.

______ , 1986: Autobiography. *Namias Symposium,* John O. Roads, Ed., Scripps Institution of Oceanography, 3–36.

______ , 1991: Francis W. Reichelderfer, August 6, 1895–January 26, 1983. *Natl. Acad. Sci. Biog. Mem.,* **60,** 272–291.

______ , et al. 1935: *An Introduction to the Study of Air Mass Analysis.* Amer. Meteor. Soc., 129 pp.

Nebeker, F., 1989: The 20th century transformation of meteorology. Ph.D. dissertation, Princeton University, 320 pp.

New York University, [1956]: Daniel Guggenheim School of Aeronautics. *30th Anniversary Pioneer Educator in the Air Age.* New York University College of Engineering, 93 pp.

Nielsen, C.H., 1993: Hurd Willett: Forecaster extraordinaire. *Weatherwise,* **46,** 38–44.

O'Brien, T.J., 1935: The Navy's part in modern aerological developments. *Proc. U.S. Nav. Inst.,* **61,** 385–392.

Pierce, H.F., 1930: H.F. Pierce to Committee on Admissions (Attachment), 26 August 1930. Henry F. Pierce file, Records of the Office of the Registrar, Clark University Archives, Worcester, MA.

Reichelderfer, F.W., 1928: Post graduate course in aerology and meteorology for naval officers. *Bull. Amer. Meteor. Soc.,* **9,** 149–151.

———, 1960: The Rossby memorial volume. *WMO Bull.,* **9,** 139–144.

———, 1970: The atmospheric sciences and the American Meteorological Society: The early years. *Bull. Amer. Meteor. Soc.,* **51,** 206–211.

Rossby, C.G., 1929: First annual report of the meteorological course of the Massachusetts Institute of Technology. MIT Meteorological Course, Professional Notes, No. 2, 18 pp.

———, 1939: Carl-Gustaf Rossby to Karl T. Compton, 1 February 1939 (copy). Records of the Office of the President, 1940–1946, University of Chicago Archives, Chicago, IL.

———, 1939–1940: Carl Gustaf Rossby to Karl T. Compton, 15 October 6 December 1939, 9 June 1940. Records of the Office of the President (Karl T. Compton), Institute Archives and Special Collections, Massachusetts Institute of Technology, Cambridge, MA.

Rotch, A.L., 1908: Abbott Lawrence Rotch to William Morris Davis, 7 April 1908. William Morris Davis Papers, Houghton Library, Harvard University, Cambridge, MA.

Sekera, Z., 1967: Meteorology [at UCLA]. *The Centennial Record of the University of California,* V. A. Stadtman, Ed., University of California, Berkeley, 355–356.

Stone, R.G., 1958: Charles F. Brooks. *Geog. Rev.,* **48,** 443–444.

University of California, 1977: Jack Bjerknes, 1897–1975. University of California, *In Memoriam* (May), 20–21.

Univerisity of Chicago, 1940[?]: Memorandum on Meteorology. Undated typeset manuscript. Records of the Office of the President, 1940–1946, University of Chicago Archives, Chicago, IL.

van Valkenburg, S., 1959: Charles F. Brooks, 1891–1958. *Ann. Assoc. Amer. Geogr.,* **49,** 461–465.

Varney, B.M., 1908: Early meteorology at Harvard College. *Mon. Wea. Rev.,* **36,** 140–146, 286–290.

———, 1940: Burton M. Varney to Vern O. Knudsen, 16 January 1940. Records of the Office of the Chancellor, University of California, Los Angeles, Archives, Los Angeles, CA.

von Kármán, T., 1967: *The Wind and Beyond.* Little, Brown and Co., 326 pp.

Weatherwise, 1953: Meteorological education in the United States. *Weatherwise,* **6,** 126–142.

Whitnah, D.R., 1961: *A History of the United States Weather Bureau.* University of Illinois Press, 267 pp.

———, 1966: *Safer Skyways: Federal Control of Aviation, 1926–1966.* Iowa State University Press, 417 pp.

CHAPTER 19

Meteorology Education in the United States after 1945

DAVID D. HOUGHTON

Introduction[1]

The previous chapter provides an in-depth discussion of the emergence of rigorous programs for meteorology as a physical science in American universities. Growing acceptance of meteorology as a science came with the development of theory and three-dimensional observational systems for the atmosphere, as well as with the U.S. Government's need to train people as meteorologists during World War II.

Meteorology education has continued to evolve dramatically ever since. In the post–World War II period, the applications of the meteorological profession to the needs of society were apparent, and this motivated many initiatives in the free-market U.S. educational system. In tracing these developments, the focus of this chapter will be on the general evolution of meteorology education. Only a very small sampling of all of the people and institutions that made this evolution possible will be mentioned. Fortunately, there are many historians who will be able to document and analyze the evolution more fully in years to come.

Immediate post–World War II era

As Koelsch describes in the previous chapter, World War II ended with five "major producers" of professional meteorologists in America:

---

[1]An author writing a history that includes recent events must recognize that many of the principal people he is writing about are still alive and that many perspectives abound. The summary here is from the perspective of a participant. I received my university education in meteorology in the late 1950s and early 1960s and went on to be part of the formal university educational system from 1968 to the present. My work with the American Meteorological Society (AMS) on educational matters since 1981 included six years service on the Board on Meteorological and Oceanographic Education in Universities, followed by six years service as chair of the AMS Commission on Education and Manpower (now the Education and Human Resources Commission) before I became AMS president. I am thus one of the living trees in the forest. I am grateful to all the individuals who led me to reference materials and who provided me with their personal observations.

the University of California, Los Angeles; the University of Chicago; the Massachusetts Institute of Technology; New York University; and the California Institute of Technology.[2]

Universities began to benefit from research funding from the Department of Defense, primarily through the Geophysics Research Directorate (GRD) of the Air Force Cambridge Research Center and the Office of Naval Research. GRD supported research in a broad range of areas: for instance, numerical weather prediction, general circulation, boundary-layer turbulence, and weather radar (Chan Touart and Ruth Liebowitz 1995, personal communication).

After the war, there was a major influx of people into academic meteorology programs. First, many individuals who received weather training during the war enrolled in universities, under the G.I. bill, to complete graduate-level work in meteorology (Fleagle 1994). Second, many of the German and Japanese meteorologists who emigrated to the United States went to work in universities as researchers. Together these two groups of people provided the intellectual foundation for the subsequent growth of educational programs and university-based meteorological research.

At the same time, university-based meteorological training programs developed by the U.S. Air Force and Navy continued, which further enhanced university education. As late as 1957, the air force (through the Air Force Institute of Technology) and the navy (through the Naval Weather Service) collectively enrolled over 300 of their personnel at 13 universities and the Naval Postgraduate School[3] (NAS Committee on Atmospheric Sciences 1960). At this time, the U.S. Weather Bureau also set up refresher and retraining programs at institutions such as the University of Chicago (George Platzman 1995, personal communication).

During the postwar period, the AMS played an important role in the development of education programs. Its Advisory Committee on Education gathered information on university meteorological educa-

---

[2]The California Institute of Technology dropped out of the military training program in 1943 (William Jenner 1995, personal communication) and ended its meteorology program in 1948 (Lewis 1994). A meteorology department existed at The Pennsylvania State University but in the 1940s it had yet to become involved in training programs for the military.

[3]The Naval Postgraduate School degree program originated in 1945 with the establishment of an aerology department.

tion (Beck 1953), which led to the establishment of biennial summaries of university-level academic and research programs in the United States, which continue to be published to this day. In 1953 the education and career issue of *Weatherwise,* published by the AMS, provided a summary of the 20 university programs in existence at that time (*Weatherwise* 1953) and of career opportunities in the Weather Bureau, the navy, the air force, and the private sector.

The *Compendium of Meteorology* (Malone 1951), published by the AMS and supported in part by the U.S. Air Force Geophysics Research Directorate, was a seminal resource for both research and education. It provided a comprehensive summary of the many aspects of meteorology and climatology recognized in the early 1950s and became a baseline reference for the development of both research and educational programs.

Military interest in meteorology education waned briefly after World War II but was dramatically revived in 1950 with the Korean Conflict. The U.S. Air Force found itself short of weather officers and therefore initiated a substantial university education program in 1951. At The Florida State University (FSU), for example, where a department of meteorology had been initiated in 1949, courses became fully subscribed in the third year of operation and for a number of years thereafter, because of the large number of U.S. Air Force students. The demand for meteorology education, plus increased research support from the military (as discussed herein), quickly established the department of meteorology as a significant component of FSU (Werner Baum 1995, personal communication). Similar support and development occurred on other campuses. For example, The Pennsylvania State University had about 25 students from the U.S. Air Force and Navy in 1951. Such influx was considered essential by Alfred Blackadar (Miller 1992) to reach the critical mass for the B.S. program.

Aviation forecasting strongly influenced the educational objectives and curriculum content during this post–World War II period. By 1950, the objectives had begun to broaden—a trend that has continued into the present. This broadening was facilitated by the wide range of research objectives that developed.

Nontechnical university courses in meteorology were introduced during this early post–World War II period and became popular at a number of places. The course "Weather and Man" introduced by Hans Neuberger at The Pennsylvania State University in 1948 reportedly became an "instant success" (Miller 1992). Reid Bryson (1995, personal communication) reports that before 1955 a popular meteorology course was drawing 350–400 students each semester at the University of Wisconsin—Madison.

New programs

In the late 1950s a number of coinciding factors produced new educational and research initiatives. First, the launch of the Soviet *Sputnik* in 1957, which highlighted Cold War competition in technology, enhanced national interest in many areas of science, including the atmosphere. Second, in 1957 the International Geophysical Year Program gave research activities a new sense of significance and world perspective. Third, by 1959 there were approximately as many civilians as military personnel enrolled in meteorology programs; there were 497 students in meteorology degree programs and 542 personnel in military training programs (Committee on Atmospheric Sciences 1960). Over one-half of all these people were in graduate programs, which was clear evidence of the interest in research in the atmospheric sciences and of the growing employment opportunities in meteorology.

Government funding for university research was a key facilitator for overall growth in the university sector. The 1958 report of the Committee on Meteorology of the National Academy of Sciences (NAS) and the National Research Council highlights the continuing Department of Defense support for university research after World War II, which was at the level of $3 million a year (NAS Committee on Meteorology 1958). The report urged universities to conduct basic research so that the research could "aid in determining" rather than just "fulfilling certain" military requirements. Fully 90% of university support came through the GRD before the establishment of the National Science Foundation (NSF) atmospheric sciences program in 1958. (Dave Fultz 1995, personal communication). This program solidified and further enhanced the funding base for university research (NAS Committee on Atmospheric Sciences 1960).

The 1958 report of the NAS Committee on Meteorology provided the following six far-reaching recommendations on research and education in meteorology.

1) Research support for universities should be increased immediately by 50%–100%.

2) A National Institute of Atmospheric Research should be established and operated by the nation's universities. [This was the foundation for the National Center for Atmospheric Research (NCAR) and its governing body, the University Corporation for Atmospheric Research (UCAR).]

3) The AMS should take the initiative and responsibility to increase its activities and to stimulate interest in meteorology by tenfold or more.

4) Departments of meteorology at universities should form an interuniversity committee to consider curricula, student recruitment, fellowships, and textbooks.

5) This interuniversity committee on meteorology should help the Weather Bureau in its educational and personnel development programs.

6) This interuniversity committee on meteorology should inform students who are well qualified in meteorology of available fellowships and scholarships.

Key aspects of these recommendations were attained very quickly. Two years after the report was published research funding for universities more than doubled. In 1960 UCAR was established through the focused efforts of 14 founding universities that had Ph.D. programs in meteorology. The Weather Bureau provided a small but significant increase in research funding to university scientists, which was indicative of the growing relationships between the government and academia. In addition, the NSF established its graduate fellowship program.

In 1958 the AMS defined a broad program of education initiatives and by 1960 put in place NSF-funded projects. (AMS 1960). These projects included 1) a visiting scientist program for colleges and universities, 2) a visiting foreign scientists program for colleges and universities, 3) meteorological career guidance literature for high schools, 4) a monograph series on meteorological topics of popular interest, 5) a comprehensive series of slides and motion pictures on meteorological topics for general education, and 6) precollege science centers. In 1962 the AMS also restructured its organization to establish a separate Commission on Education and Manpower, with Werner Baum as its first chairperson.

It is clear that the extensive expansion in meteorological education was made possible by the close, cooperative efforts of government and universities. The leadership was shared.

Rapid growth in the 1960s

The number of universities offering meteorological education increased rapidly during the 1960s. At the beginning of the decade there were about 24 (NAS Committee on Atmospheric Sciences 1960), 14 of which (the charter members of UCAR) offered Ph.D. degrees. By 1964, the number of universities offering meteorology education had risen to 44, and by 1970 to 62 as reported in a listing of university curricula in the atmospheric sciences (Panel on Education, NAS Committee on Atmospheric Sciences 1964; AMS 1970). The number of Ph.D.-granting institutions also increased, reaching 30 by 1969 (Fleagle 1994), with membership in UCAR reaching 24 by 1967. Overall, the number of B.S., M.S., and Ph.D. degrees awarded in meteorology per year increased rapidly during this time (Orville 1978). In the 10-year period from 1958 to 1968, federal funding in constant dollars for university research increased by a factor of 6 (NAS Committee on the Atmospheric Sciences 1971).

Many new academic programs were initiated by atmospheric scientists who moved from existing programs. In many cases "venturesome and entrepreneurial" people moved into established university departments of other disciplines and created new meteorology programs without preexisting institutional budgetary authority (Walter Saucier 1995, private communication). Some scientists even started more than one program: for example, Walter Saucier at Texas A&M University, Oklahoma University, and North Carolina State University and Reid Bryson at the University of Wisconsin—Madison and The University of Arizona.

During the 1960s, the curriculum content of university meteorology programs continued to focus on the physics and dynamics of weather systems and climatology. Other topics, described in the *Compendium of Meteorology* (Malone 1951), such as air pollution, became important at the Drexel Institute of Technology, New York University, and the University of Michigan. Other areas of study, such as agricultural meteorology, emerged at Iowa State University and Texas A&M University. The nation's university curriculum listings included courses in the upper atmosphere, plasma physics, and even solar physics (Panel on Education, NAS Committee on Atmospheric Sciences 1964). During the 1960s the boundaries of the atmospheric science field were evolving and were not well defined. In addition to meteorology departments, departments such as geology, geography, physics, astrophysics, agronomy, electrical engineering, civil engineering, astronomy, geophysics, earth sciences, and marine sciences offered some of these courses.

The emergence of NCAR in 1962 established an exciting new focal point for the university community and, at the same time, a place of perceived competition, especially for research activity in the expanding university programs. Discussion arose about the appropriate split in NSF atmospheric sciences research funding between the universities and NCAR. In practice the split seemed to be about half and half (author's recollection).

By the end of the 1960s, federal government support had a broad impact on meteorological education. University research, student fellowships, and even programs supporting school and popular education had been enhanced. At the end of this period the rapid growth in such support had ceased. Yet the cuts in research funding imposed for the fiscal year 1969 were viewed with only mild alarm by the university community because they saw the cuts as temporary (Roberts and Wolff 1969).

The last 25 years—Broadening the scope of meteorology education

The expansion in areas of application of the atmospheric sciences led to continued increase in the number of meteorology programs after 1970. Government interest in promoting and supporting meteorological education remained strong. Added to this was the incentive to support the growth of private sector needs such as TV broadcasting and environmental consulting. By 1994, according to the listing in the *1994 Curricula in the Atmospheric, Oceanic, Hydrologic, and Related Sciences,* the number of university meteorology programs had reached 109 (AMS 1994). Of these, 77 offered the Ph.D. as the highest degree, 10 offered the M.S. as the highest degree, and 22 offered only B.S. degrees. In addition, by 1994, membership in UCAR had risen to 61.

Beginning in 1970 a general broadening of university curricula in meteorology occurred. The advent of satellite systems and other remote sensing technology, such as Doppler radar and wind profilers, had a major impact on meteorology, opening up both global and mesoscale perspectives. Satellite measurement systems provided a foundation that made possible the First Global Atmospheric Research Program Global Experiment in 1978–1979, a cooperative international program. Satellite meteorology research programs developed at places such as the University of Wisconsin—Madison, Colorado State University, and The Florida State University. Satellite and radar meteorology became part of the curriculum at many universities. Atmospheric chemistry became a common curriculum component as the issues of air quality, global warming, and stratospheric ozone depletion expanded people's awareness of the serious effects human activity has on the environment. The curricula listings in the *1994 Curricula in the Atmospheric, Oceanic, Hydrologic, and Related Sciences* (AMS 1994) show that many of the originally very specialized departments in the atmospheric and related sciences had become much more generalized. In addition, curricula listings for programs such as environmental science and engineering at the Harvard School of Public Health attested to the broad connections between the atmospheric sciences and other disciplines.

Some of the academic atmospheric sciences programs, though, remained specialized in curriculum and/or level of work. As noted, the 1994 *Curricula* listing showed 22 academic institutions that offer the B.S. as the highest degree, along with other institutions that focus on training students for specific career tracks such as in the broadcast meteorology industry. Many of the expanding private sector career options, such as forensics, broadcasting, air quality, and specialized forecasting, require only the B.S. degree in meteorology for entry-level positions.

The proliferation of academic programs led to some concern about their quality in the 1980s, especially that of the small undergraduate B.S. programs. This led to dicussion within the AMS about accreditation, an issue that was eventually dropped because of lack of support. However, in 1987 the AMS did develop and publish criteria for a "minimally sound" undergraduate program (AMS 1987).

The education programs responded to the broadened disciplinary scope of the atmospheric sciences and new interdisciplinary interests in environmental and other societal issues (Fleagle 1994). Since 1990 the interdisciplinary perspective of earth system science has formalized the full interconnectedness of the atmosphere, hydrosphere, lithosphere, and biosphere. In 1991 with funding from the National Aeronautics and Space Administration (NASA), the Universities Space Research Association developed a national program to develop curricula in earth system science. By 1995, 22 universities were participating in the development of survey and senior-level courses.

Overall, federal government support for education has varied greatly during the last 25 years. Government support for education diminished in the late 1970s and early 1980s, and the NSF program for undergraduate education virtually vanished, as did support for precollege education. The graduate fellowship program continued, though. After 1985, however, there was a major resurgence of national interest in improving science education focusing on precollege and public education. The NSF budget for K–12 initiatives in science education soared.

The meteorological community was well positioned to participate in K–12 and public education enhancements because of TV weather broadcasters and the visual nature of the atmospheric sciences. The AMS, through its Education and Human Resources Commission, combined its own financial support with the energy and experience of many of its members who had been involved with K–12 activity to obtain a major grant from NSF in 1990 for a national program to improve the teaching of meteorology in the schools. The government also strongly encouraged research institutions and universities to support K–12 education in the sciences. NCAR/UCAR, which had already developed graduate-level education programs, obtained funding for K–12 education initiatives. By 1995, NCAR/UCAR was sponsoring six projects supporting students and teachers through fairs, workshops, teacher enhancement, and data provision (UCAR 1995). A 1995 listing shows that a total of 18 UCAR and 8 other universities had developed major outreach programs for K–12 meteorological education enhancement (UCAR 1995).

In recent years there has been growing concern that universities have been producing more graduates than there are jobs available, a situation exacerbated by the reduction of jobs in the government sector. The balance between the number of graduates and the number of jobs available has long been of interest. Note, for instance, the study of Ph.D.s made by Bretherton (1977). Expansion of the private sector both in size and in the scope of areas for meteorological applications has served to offset the loss of jobs in the government sector and to influence the content of university degree programs. Private sector employment had risen to an estimated 3.1% of the total profession by

1991 (Stephens and Kararosian 1992). [See chapter 15 by Spiegler, this volume.]

The university educational curriculum for the B.S. degree became overextended due to efforts to broaden programs to meet the needs of the more diverse job marketplace. This led to discussions about a five-year B.S. program or the need to consider the M.S. degree essential for entry into the profession. The 1987 AMS statement on the B.S. degree in meteorology and the atmospheric sciences was revised in 1995 after several years of deliberation (AMS 1995). The new version increased atmospheric course requirements but did not change the four-year program concept. The 1995 statement also recognized that in addition to forecasting, other career tracks were basic as well, and all B.S. program options did not need to satisfy the federal government requirements for the meteorologist classification.

With the modernization of the National Weather Service (NWS) and the technological revolution in data collection in the 1990s, it became necessary for the NWS, the air force, and the navy to establish more general training/education programs for meteorology outside universities. University programs were simply not able to cover all of the training meteorologists in operational work needed. This resulted in the establishment in 1989 of a cooperative UCAR–government program, COMET (Cooperative Operational Meteorology Education and Training). University meteorologists helped to produce some of the instructional material for COMET, and in turn, some of the COMET instructional materials were used at universities. This continued the effective cooperative government–university effort to meet meteorology education needs in the United States, which had been a tradition for over 50 years.

The future

Currently, meteorological education is well established, with over 100 universities and colleges offering degree programs in meteorology, oceanography, and related areas. Textbooks on general and specific topics appear with great frequency (see, e.g., De Souza et al. 1994; AMS 1994). Precollege education in meteorology is expanding. Many government agencies support students and educational enhancement programs in meteorology. Both the technology revolution and environmental issues require meteorology education to continue to evolve.

Academia's access to the resources of institutions such as the NAS–National Research Council, AMS, UCAR, NCAR, the NSF, and NASA should make it possible to continue joint efforts to make a strong future for meteorology education in the United States.

The major challenge for the future is to meet the educational needs for overall sustainable human development in a world where a global perspective is paramount, where our environment must be understood as the integrated entity of the earth system, and where human impact on the enrironment must be considered (Malone 1994). Educational programs must recognize and build upon the global knowledge explosion fostered by the revolution in information transfer. Meteorology education will have to expand its perspective to connect with disciplines such as the social sciences and public policy and will have to promote scientific literacy in the general population.

Acknowledgments. I wish to thank the many people who provided responses to my queries and suggestions for text material. The list includes David Atlas, Harriet Barker, Werner Baum, Reid Bryson, Horace Byers, Earl Droessler, Robert Fleagle, Richard Hallgren, George Horn, John Lewis, Ruth Liebowitz, William Jenner, William Kellogg, William Koelsch, Tom Malone, Evelyn Mazur, George Platzman, Walter Saucier, Verner Suomi, Chan Touart, and Warren Washington. Thirty-nine atmospheric science departments responded to a brief questionnaire about their history, which was very helpful.

REFERENCES

AMS, 1960: Education program of the American Meteorological Society. *Bull. Amer. Meteor. Soc.,* **41,** 496–501.

______ , 1970: *Curricula in the Atmospheric and Oceanographic Sciences—1970.* Amer. Meteor. Soc., 216 pp.

______ , 1987: The bachelor's degree in meteorology or atmospheric science. *Bull. Amer. Meteor. Soc.,* **68,** 1507.

______ , 1994: *Curricula in the Atmospheric, Oceanic, Hydrologic, and Related Sciences—1994.* Amer. Meteor. Soc. and University Corporation for Atmospheric Research, 594 pp.

______ , 1995: The bachelor's degree in atmospheric science or meteorology. *Bull. Amer. Meteor. Soc.,* **76,** 552–553.

Beck, N.C., 1953: Forward. *Weatherwise,* **6,** 116–117.

Bretherton, F.P., 1977: Training and distribution of Ph.D.'s in meteorology. *Bull. Amer. Meteor. Soc.,* **58,** 230–232.

De Souza, R.L., D.M. Matthews, R.D. Clark, and R.S. Ross, 1994: 1993 survey of textbooks used in meteorological education. *Bull. Amer. Meteor. Soc.,* **75,** 613–619.

Fleagle, R.G., 1994: *Global Environmental Change—Interactions of Science Policy, and Politics in the United States.* Praeger, 243 pp.

Lewis, J.M., 1994: Cal Tech's Program in Meteorology: 1933–1948. *Bull. Amer. Meteor. Soc.,* **75,** 69–81.

Malone, T.F., Ed., 1951: *Compendium of Meteorology.* American Meteorological Society, 1334 pp.

_____ , 1994: Sustainable human development: A paradigm for the 21st century—Challenge and opportunity for higher education. A White Paper for the National Association of State Universities and Land-Grant Colleges, Sigma Xi, Research Triangle Park, NC, 73 pp.

Miller, E.W., 1992: *The College of Earth and Mineral Sciences at Penn State.* The Pennsylvania State University Press, 291 pp.

NAS Committee on Atmospheric Sciences, 1960: The status of research and manpower in meteorology. *Bull. Amer. Meteor. Soc.,* **41,** 554–562.

_____ , 1971: *The Atmospheric Sciences and Man's Needs: Priorities for the Future.* National Academy of Sciences–National Research Council, 88 pp.

NAS Committee on Meteorology, 1958: *Research and Education in Meteorology.* National Academy of Sciences–National Research Council Publ. 479, NAS/NRC, Washington, DC, 23 pp.

Orville, R.E., 1978: Atmospheric sciences degrees in the United States, 1963–1977. *Bull. Amer. Meteor. Soc.,* **59,** 623–626.

Panel on Education, NAS Committee on Atmospheric Sciences, 1964: *Curricula in the Atmospheric Sciences—Academic Year 1963–1964.* Amer. Meteor. Soc., 124 pp.

Roberts, W.O., and E.L. Wolff, 1969. The impact of budget cuts on the atmospheric sciences in universities. *Bull. Amer. Meteor. Soc.,* **50,** 586.

Stephens, P.L., and C. Kazarosian, 1992: Results of the AMS membership survey. *Bull. Amer. Meteor. Soc.,* **73,** 486–495.

UCAR, 1995: K-12 outreach activities in weather and climate. UCAR home page for programs at UCAR, UCAR Member Institutions,

UCAR academic affiliates, and other institutions. [Available on-line from http//www.ucar.edu:8080/ucargen/education/eduhome.html.]

Weatherwise, 1953: Meteorological education in the United States. *Weatherwise,* **6,** 126–142.

Special Essays

Historical Writing on Meteorology: An Annotated Bibliography

JAMES R. FLEMING AND SIMONE L. KAPLAN

Since the pioneering work of the prolific German meteorologist, historian, and bibliographer Gustav Hellmann in the late nineteenth and early twentieth centuries, there have been a number of very good books published on the history of meteorology. The annotated bibliography that follows excludes articles and, with the exception of two of Hellmann's works published in the closing decades of the nineteenth century, discusses only books and dissertations on the history of meteorology published in the twentieth century.

Any number of good articles on the history of meteorology have been published in historical journals such as *Isis, History of Science,* and *Annals of Science* and in the scientific literature including the *World Meteorological Organization Bulletin,* the *Bulletin of the American Meteorological Society,* and the *Quarterly Journal of the Royal Meteorological Society.* Popular accounts appear everywhere but predictably so in *Weatherwise* and *Weather.* The best guides to periodical literature in the history of science are the *Isis Cumulative Bibliography* (History of Science Society 1971–1989); the *Isis Current Bibliography,* published annually in *Isis;* and the Research Libraries Information Network's (RLIN) History of Science and Technology file, available on-line (http://www-rlg.stanford.edu/rlin.html).

Reference works such as the monumental *Dictionary of Scientific Biography,* edited by C.C. Gillispie (New York 1970–1980); *The History of Modern Science: A Guide to the Second Scientific Revolution, 1800–1950,* by Stephen G. Brush (Ames 1988); and the *Companion to the History of Modern Science,* edited by R.C. Olby et al. (London 1990) are quite useful and authoritative but systematically underrepresent geophysics and meteorology. A new reference work, *The History of the Geosciences: An Encyclopedia,* edited by Gregory A. Good (Garland Press, in press), promises to correct many of these shortcomings.

And now we turn to the subject at hand—books. Rather than claiming that our list is complete (it is not) and that all of the volumes listed here are reliable histories (they are not), we offer these references and annotations as a working "historian's bookshelf."

Abbe, Truman. *Professor Abbe and the Isobars: The Story of Cleveland Abbe, America's First Weatherman.* New York: Vantage Press, Inc., 1955. 259 pp.

A work of filial piety, the book, which is based on manuscript papers and diaries, remains a valuable biography of Cleveland Abbe (1838–1916). It documents both the formative experiences of a young American scientist and the professional opportunities and struggles facing the emerging meteorological community.

Anderson, Katharine Mary. Practical science: Meteorology and the forecasting controversy in mid-Victorian Britain. Ph.D. dissertation, Northwestern University, 1994. 324 pp.

Examines the difficulties that meteorology posed for the social and cultural claims of science in Victorian Britain. Concentrates on the practical, utilitarian benefits of weather prediction and the public embarrassment caused by unsuccessful forecasts. Investigates popular notions of astrological control of the weather. Draws together several historiographic strands, including the relationship between revealed religion and science, interactions between different social interest groups, the construction of public authority, and the public's response to weather forecasting.

Arakawa, Hidetoshi. *Otenki Nihon shi* (History of Japanese Weather). Tokyo: Kawade Shobo Shinsha, 1988. 242 pp.

Japanese weather history and meteorological disasters.

Ashford, Oliver M. *Prophet—Or Professor? The Life and Work of Lewis Fry Richardson.* Boston: A. Hilger, 1985. 304 pp.

Lewis Fry Richardson, a scientist whose most significant contributions to the field of meteorology were not recognized until after his death in 1953, is appreciated as both a friend and a meteorologist in Ashford's biography. Ashford follows Richardson's early life and education, from his birth in 1881 and Quaker upbringing in Newcastle-upon-Tyne through his very diverse career, which included stints as a researcher at the National Physical Laboratory, the National Peat Industries, Ltd., and the Sunbeam Lamp Company; as Superintendent of the Eskdalemuir Observatory in Scotland; and as an ambulance driver in France in World War I. Richardson's work with W.H. Dines, one of Britain's most famous atmospheric scientists, resulted in his most famous concept, the Richardson number, which describes turbulent flows. Richardson's book, *Weather Prediction by Numerical Process* (1922), is based on his ideas (fanciful at the time) about numerical weather prediction. Richardson taught at Westminster Training College from 1920 to 1929 and at the Paisley Technical College from 1929 to 1940. In retirement he continued his interests in geophysics, especially eddy diffusion and turbulence. He remained a devout Quaker throughout his life. His major statistical studies of war and the arms race appeared posthumously.

Ashley, Bernard. *Weather Men.* London: Allman, 1970. 120 pp.

As a society, we have progressed far past the point where bad weather is seen as a sign of the anger of the gods. Now we can pick up the phone and by pressing a few buttons hear the daily forecast. Bernard Ashley tells the story of the progression of weather study in a clear, attention-grabbing, narrative fashion. The reader is yanked into tales ranging from weather lore to the first radiosonde expedition, methods of guiding planes through fog banks, and the introduction of satellites. Ashley recounts the development of atmospheric study, paying particular attention to those who made the greatest contributions, such as Edmund Halley, Buys Ballot, Napier Shaw, and James Glashier.

Atlas, David, Ed. *Radar in Meteorology: Battan Memorial and 40th Anniversary Radar Meteorology Conference.* Boston: Amer. Meteor. Soc., 1990. pp. 1–150.

Includes 18 essays on the history of radar meteorology.

Bates, Charles C., and John F. Fuller. *America's Weather Warriors, 1814–1985.* College Station: Texas A&M University Press, 1986. 360 pp.

An authoritative and well-documented history of the U.S. military weather services that includes both the human and the technical dimensions. Most of the book focuses on the period since 1940 and is based on interviews and primary sources, but earlier eras are discussed briefly.

Bolle, Bert. *Barometers.* Haarlem, 1978. Translated by J. Ramsbotham. Wappingers Falls, NY: Antique Collectors' Club Ltd., 1984. 244 pp.

More specialized than Middleton's interpretive history of the subject, Bolle documents the history of barometers from the time of Torricelli through the development of marine coastal alert devices, angle barometers, mercury wheels, aneroid barometers, and barographs. This volume is fabulously illustrated with diagrams and photographs of antique barometers, as well as formulas explaining the function of the instruments. It is a good book for those interested in the aneroid barometer, specifically, and those who collect barometers. Enthusiasts and collectors should also be aware of Edwin Banfield's *Barometers,* 3 Vols., (Baros Books, 1985) and the following booklets: Anita McConnell, *Barometers,* 2d ed., (Shire, 1994) and A.G. Thoday, *Barometers,* H.M. Stationery Office, 1978).

Brush, Stephen G., and Helmut E. Landsberg, with Martin Collins. *The History of Geophysics and Meteorology: An Annotated Bibliography.* New York: Garland Publishing Inc., 1985. 450 pp.

Compiled by a distinguished historian–scientist duo, this wide-ranging, annotated, and accessible bibliography is useful for historians of science, researchers interested in the history of their field, and students. The first part of the volume is organized into general histories, biographies, and chronologies. The second part consists of historical writings that cover the major specialties and subspecialties of geophysics and meteorology.

Burton, J.M.C. The History of the British Meteorological Office to 1905. Ph.D. dissertation. The Open University, 1988. 340 pp.

Available from the British Library Document Supply Centre, Boston Spa, United Kingdom.

Conover, John H. *The Blue Hill Meteorological Observatory: The First 100 Years, 1885–1985.* Boston: Amer. Meteor. Soc., 1990. 514 pp.

History and chronology of the observatory and a contribution to Massachusetts history.

Courain, Margaret E. Technology reconciliation in the remote sensing era of United States civilian weather forecasting, 1957–1987. Ph.D. dissertation, Rutgers University, 1991. 964 pp.

Abstract: "This dissertation seeks to advance an understanding of the management of a major technological change in meteorology. The study examines the connection between changes in production and real-time use of data products derived from remote-sensing data collection and the evolution of U.S. civilian weather forecasting 1957–1987. The role of data collection in weather forecasting throughout history is examined, giving most attention to the 1957–1987 period. Critical to the real-time use of remote-sensing data was technology reconciliation. . . . A model of the technology reconciliation is proposed which can be applied to understanding the contemporary history of other sciences. . . ."

Daniel, Howard. *One Hundred Years of International Cooperation in Meteorology (1873–1973): A Historical Review.* WMO Publication No. 345. Geneva: World Meteorological Organization, 1973. 53 pp.

Documents the formation and evolution of the International Meteorological Organization (IMO) from its origin in 1873 to its centennial in 1973. Beginning with a short history of meteorological research leading up to the first International Meteorological Congress in Vienna, Howard chronicles the growth and development of the IMO as it became well known and respected among scientists and documents its metamorphosis into the World Meteorological Organization (WMO) in 1950. Articulate and well illustrated, Daniel's booklet is aimed more toward those readers interested in the history of the WMO rather than those looking for a history of meteorology.

Davies, Arthur, and Oliver M. Ashford, eds. *Forty Years of Progress and Achievement: A Historical Review of WMO.* WMO Publication

No. 721. Geneva: World Meteorological Organization, 1990. 205 pp.

Good source of information on international coordination and projects, and on meteorology in developing countries.

DeMarrais, Gerard A. A history of air pollution meteorology through 1969. NOAA Tech. Memo., ERL ARL-74. Silver Spring, MD: Air Resources Laboratories, Environmental Research Laboratories, 1979. 78 pp.

An annotated bibliography.

Dufour, Louis. *Notes pour servir á l'histoire de la météorologie en Belgique.* 2 Vols. Uccle: Institut royal météorologique de Belgique, 1947.

Issued as part of the Instiut's *Miscellanées.*

Eisenstadt, Peter. Weather and weather forecasting in colonial America. Ph.D. dissertation, New York University, 1990. 317 pp.

Treats both weather and climate in colonial America from about 1600 to 1775. Written as an American studies dissertation, relating science to culture.

Abstract: "This study shows the ways in which changing conceptions of the weather constituted a pathway by which enlightened thought was diffused to the colonial population at large. The main conduits for the dissemination of the new conception of the weather were newspapers and magazines, sermon literature and perhaps most importantly almanacs. These various media played a crucial role in filtering and explaining an enlightened meteorology based in Newtonian science to an audience raised on a largely medieval view of the weather that emphasized the hand of God. The transition from a medieval to an enlightened understanding of the weather was a prolonged process with many intermediate stages. Yet over the eighteenth century enlightened meteorology gradually came to supplant its older rival. In the end enlightened meteorology helped redefine the colonials' understanding of their relation to nature, the passage of time, the role of God, and the meaning of America itself.

The first chapter describes the context of medieval meteorology in colonial America, especially as exemplified in physical explanations of the weather, conceptions of atmospheric effluvia, and the role of weather providence in early American culture. The second and third chapters explore aspects of weather astrology in colonial almanacs, primarily why certain almanac-makers either chose to include or scorn weather predictions. The advent of enlightened meteorology, which presaged the decline of older effluvial theories as well as new notions of providence, are discussed in the next chapter. The two subsequent chapters describe the beginnings of systematic meteorology and weather record-keeping. An extensive discussion of the scientific and ideological rationales that made weather recording attractive for so many in early America accompanies a review of the history of instrument-aided observation. The final chapter offers an overview of developments in cultural and institutional understanding of the weather in nineteenth-century America especially as it relates to communication and nationalism."

Feldman, Theodore S. The history of meteorology, 1750–1800: A study in the quantification of experimental physics. Ph.D. dissertation, University of California, Berkeley, 1983. 286 pp.

Abstract: "From a literary, qualitative pursuit physics emerged at the end of the eighteenth century as a quantitative subject based on abstract, mathematical theory and precise, systematic experiment. Scholars have generally located this development in Napoleonic France. This dissertation, which focuses on meteorology, shows quantification to have been international and to have begun as early as 1760.

In the early modern period meteorology embraced most of experimental physics. Its great popularity in the 1770s derived in part from the vogue of alpine exploration and the interest of governments both in the effects of the weather on public welfare, and in surveys, where barometric hypsometry was used to determine heights from measurements of air pressure. I have selected for this study three meteorological topics that especially preoccupied natural philosophers of the late eighteenth century: aqueous vapor, barometric hypsometry, and climatology.

The quantification of experimental physics was a revolution of styles and methods. Before 1750 natural philosophers favored qualitative, pictorial theories of vapor and experimented unsystematically. After 1770 a more abstract style and precise, systematic experimentation created a mathematical physics of vapor, in which quantity, heat and pressure were quantified for the first time.

The interaction of applied mathematics and engineering with physics also contributed to quantification. Barometric hypsometry originated in applied mathematics. But though quantitative its results were vitiated by imprecise experiment and neglect of physical factors. The adoption after 1770 of exact experimental methods and a more physical approach remedied the situation.

My examination of climatology shows that before the last third of the century the weather was perceived as a local and static event of subjective experience. After 1770 newly-founded meteorological societies assembled large collections of data and applied exact methods to their analysis. Thereby the weather was objectified and its global, dynamic character revealed."

Fierro, Alfred. *Histoire de la météorologie.* Paris: Denoel, 1991. 315 pp.

This well-written and comprehensive book (in French) traces the history of meteorology from primitive myths to the elements of popular meteorology and its birth as a science. It concentrates more on the history of meteorology in France than its development as an international science. Fierro focuses on the eighteenth and nineteenth centuries, looking at the development of instruments, synoptic meteorology, the use of the telegraph, and the conception of an international organization. Maritime and aviation meteorology are discussed, as is the development of the Bureau Central Météorologique, L'Office National Météorologique, and the current French weather service, La Météorologie Nationale.

Fleming, James R. *Guide to the History of the Atmospheric Sciences: Archives, Manuscripts, and Special Collections in the Washington, D.C. Area.* NCAR, 1989. 167 pp. [NTIS PB 89159206.]

This volume serves as a research aid for historians interested in the atmospheric sciences. It is a guide to unpublished archival and manuscript resources in the atmospheric sciences in 19 libraries and repositories in and around Washington, D.C. Included are the locations and descriptions of significant documents, personal papers, special collections, taped interviews, and historical maps and instruments found during a survey conducted in 1988. Major collections surveyed include the Library of Congress, the National Archives, and the Smithsonian Institution. Other institutions with archival holdings were surveyed by mail, with follow-up visits to those reporting relevant holdings. Although some addresses and phone numbers are out of date, especially with the recent opening of the (National) Archives II in College Park, Maryland, the description of the collections remains accurate.

Fleming, James R. *Meteorology in America, 1800–1870*. Baltimore, The Johns Hopkins University Press, 1990. 264 pp.

In the first seven decades of the nineteenth century meteorology emerged as both a legitimate science and a government service in America. Meteorologists were driven by fundamental questions about the nature of climatic change, aided by such new technologies as the telegraph, and funded by government sources expecting practical results. The observational systems they established and managed grew from small bands of isolated observers in 1800 to national and international networks seven decades later.

Between 1834 and 1859, center stage was occupied by the American storm controversy. Competing theories were developed by three prominent scientists: William Redfield, James Espy, and Robert Hare. Hotly debated issues included the cause of storms, their phenomenology, and the proper methodology for investigating them. Although it came to no clear intellectual resolution, Fleming shows how the storm controversy stimulated the development of an observational "meteorological crusade" by the American Philosophical Society, the Franklin Institute, the Army Medical Department, the Navy Department, the Smithsonian Institution, and other institutions that transformed meteorological theory and practice. Centrally located administrators organized hundreds of widely dispersed volunteer and military observers into systematic projects that covered the entire nation. Theorists then used these systems to "observe" weather patterns over large areas, making possible for the first time the compilation of accurate weather maps and charts.

When in 1870 Congress created a federal storm-warning service under the U.S. Army Signal Office, the era of amateur scientists, volunteer observers, and ad hoc organizations came to an end. But the gains had been significant, including advances in natural history, in medical geography, and in understanding the general circulation of the earth's atmosphere. This book provides a case study of complex change, cooperative institution building, and scientific controversies in early-national, Jacksonian, and Civil War America.

Fleming, James R., and Roy E. Goodman, eds. *International Bibliography of Meteorology: From the Beginning of Printing to 1889.* 4 Vols. in 1. Foreword by E. Philip Krider. Historical introduction by James Fleming. Upland, PA: Diane Publishing Co., 1994. 704 pp.

Four volumes in one: Temperature, moisture, winds, and storms. This is a new edition of the U.S. Army Signal Corps, *Bibliography of Meteorology,* 4 Vols., 1889–1891, a project one reviewer called, "the most ambitious and intensive bibliographic project ever undertaken in meteorology." Contains over 16 000 citations to meteorological literature published before 1889. Approximately two-thirds of the entries are in languages other than English, with 10% of the entries earlier than 1750.

The original volumes, produced by the U.S. Army Signal Corps between 1889 and 1891, were not easy to read or use. They were lithographed in small quantities on acidic paper. Only a few copies are still in existence and, for the most part, are now in poor condition.

This edition of the bibliography of meteorology has been completely reedited to include all the errata, additions, corrections, and supplements of the original four volumes. It is divided into subject headings and subheadings. An author index appears at the end of each volume. Articles and books that were reprinted or translated are so noted in the citations. Each citation includes the author and title of the work followed by the format, place, and date of publication. The presence of maps, charts, and plates is also indicated. This is the major bibliographic resource on the atmospheric sciences prior to the *Meteorological and Geoastrophysical Abstracts* which began in 1950.

Flohn, Hermann. *Meteorologie im Übergang: Erfahrungen und Erinnerungen, 1931–1991.* Bonn: Dummler, 1992. 81 pp.

Bonner meteorologische Abhandlungen, Heft 40.

Friedman, Robert M. *Appropriating the Weather, Vilhelm Bjerknes and the Construction of a Modern Meteorology.* Ithaca, NY: Cornell University Press, 1989. 251 pp.

Vilhelm Bjerknes (1862–1951), meteorologist and éminence gris of the Bergen School, is familiar, at least by name and reputation, to most atmospheric scientists. This is largely due to the major transformation of meteorological theory and practice, emanating from Bergen, Norway, after World War I, which brought the world, among other things, a new cyclone model, the polar front theory, airmass analysis, and a host of active, aggressive, and creative young meteorologists. During World War I the Bjerknes family residence in Bergen became the center of a new school of nonmathematical meteorological analysis. (In style it was not unlike Niels Bohr's institute for theoretical physics at his home in Copenhagen.) Believing that their system of airmass analysis could best satisfy the problems facing postwar meteorology, "apostles" and "missionaries" of the Bergen School traveled far and wide in their effort to "convert" their foreign colleagues and redefine the Norwegian periphery as a new international center of new methods for meteorological analysis and forecasting.

Friedman's book is a carefully crafted narrative analysis of the interconnected history of Bjerknes's career and the construction of a new meteorology. The book is not a biography of Bjerknes, nor is it a comprehensive history of the Bergen School's endeavors—the accomplishments of Jacob Bjerknes, of Bergeron, Rossby, Solberg, and others are presented only as they relate to Vilhelm Bjerknes's career. According to Friedman, the "emergence of the Bergen meteorology illustrates new knowledge arising through changes in practice. . . . Hard facts were not waiting in nature to be uncovered. Bergen scientists constituted their new concepts and models by drawing upon analogy, metaphor, existing theory, and ad hoc construction." The book is solidly based on archival sources, and the author avoids using today's understanding of the atmospheric sciences to comprehend the history of early twentieth century meteorology.

Frisinger, H. Howard. *The History of Meteorology to 1800.* Boston: Amer. Meteor. Soc., 1983. 148 pp.

A compact yet thorough treatment of the development of weather studies in its earliest stages, Frisinger covers two of what he calls "the three major periods in the history of meteorology." The first, "the period of speculation," starting with weather references in the Bible, is primarily colored by Aristotle's works and involves early observations by farmers and the premier scientists of the time. The next period, "the dawn of scientific meteorology," explores the invention of weather instruments and systems of observation to 1800. Last, Frisinger documents the findings of Galileo, Hadley, and Newton. Well illustrated with portraits and formulas, this book is not difficult for the amateur meteorologist to grasp.

Fujiwara, Sakuhei. *Nihon Kishogakushi.* Tokyo: Iwanami Shoten, 1951. 177 pp.

History of Japanese meteorology.

Fuller, John F. *Thor's Legions: Weather Support to the U.S. Air Force and Army, 1937–1987.* Boston: Amer. Meteor. Soc., 1990. 443 pp.

Gibbs, William J. *The Origins of Australian Meteorology.* Canberra: Australian Government Publishing Service, 1975. 32 pp.

History and biographical sketches of the founders of Australian meteorology.

Gilbert, Otto. *Die meteorologischen Theorien des griechischen Altertums.* Leipzig: B.G. Teubner, 1907. 746 pp.

Part of the *Landmarks of Science* series.

Glacken, Clarence J. *Traces on the Rhodian Shore: Nature and Culture in Western Thought from Ancient Times to the End of the Eighteenth Century.* Berkley: University of California Press, 1967. 763 pp.

This scholarly yet readable compendium traces the role of ideas and values in understanding the relationship of culture to the environment. In Western thought, at least until the end of the eighteenth century (but I suspect beyond that), concepts of the relationship of culture and nature were dominated by three types of questions: Is the earth a purposefully made creation? Have its climates and geography influenced the moral and social nature of both individuals and entire cultures? How has humanity changed the earth from its hypothetical pristine condition? The answers to these and related questions, drawn from sources covering two millennia of Western experience, should be required reading for students of the environment.

Grice, Gary K. *National Weather Services Snapshots: Portraits of a Rich Heritage.* Silver Spring, MD: National Weather Service, 1991. 93 pp.

Chiefly photographs.

Grice, Gary K., ed. *The Beginning of the National Weather Service: The Signal Service Years (1870–1891) as Viewed by Early Weather Pioneers.* Washington, DC: National Weather Service, 1991. 52 pp.

Transcribed reminiscences of former Signal Service employee selected from the records of the U.S. Weather Bureau in the National Archives.

Hellmann, Gustav (1854–1939). The following works by this prolific German author and editor contain extensive source materials on the history of meteorology. They are listed in chronological order.

Hellmann, Gustav. *Repertorium der deutchen Meteorologie. Leistungen der Deutchen in Schriften, Erfindungen und Beobachtungen auf dem Gebiete der Meteorologie und des Erdmagnetismus von den ältesten Zeiten bis zum Schlusse des Jahres 1881.* Leipzig: W. Engelmann, 1883. 996 pp.

Repertory of German meteorology. Achievements of Germans in writings, inventions, and observations in the sphere of meteorology and terrestrial magnetism from very ancient times up to the close of the year 1881.

Hellman, Gustav. *Neudrucke von Schriften und Karten über Meteorologie und Erdmagnetismus.* 1–15. Berlin: A. Asher & Co., 1893–1904. Nendeln, Liechtenstein, Kraus Reprint, 4 Vols., 1969. Contents:

1) Reynman, Leonhard. *Wetterbüchlein von wahrer Erkenntniss des Wetters. 1510.* Berlin: 1893. 56 pp.

2) Pascal, Blaise. *Recit de la grande experience de l'équilibre des liqueurs.* Paris, 1648. Berlin, 1893. 20 pp.

3) Howard, Luke. *On the Modifications of Clouds.* London, 1803. Berlin, 1893. 32 pp.

4) *Die Ältesten Karten der Isogonen, Isoklinen, Isodynamen, 1701, 1721, 1768, 1804, 1825, 1826.* Berlin: 1895. 24 pp.

5) *Die Bauern-Praktik, 1508.* Berlin: 1896. 83 pp.

6) Hadley, George. *Concerning the Cause of the General Trade-Winds.* London: 1735. Berlin: 1896. 21 pp.

7) Torricelli, Evangelista. *Esperienza dell' argento vivo.* Berlin: 1897. 15 pp.

8) *Meteorologische karten 1688, 1817, 1846, 1863, 1864. Sechs tafeln in lichtdruck mit einer einleitung.* Berlin: 1897. 12 pp., 6 charts.

9) Gellibrand, Henry. *A Discourse Mathematical on the Variation of the Magneticall Needle.* London: 1635. Berlin: 1897. 22 pp.

10) *Rara magnetica 1269–1599.* Berlin: 1898. 174 pp.

11) *Ueber Luftelektricität 1746–1753.* Berlin: 1898. 50 pp.

12) *Wetterprognosen und Wetterberichte des XV. und XVI. Jahrhunderts.* Berlin: 1899. 184 pp.

13) *Meteorologische Beobachtungen vom XIV. bis XVII. Jahrhundert.* Berlin: 1901. 212 pp.

14) *Meteorologische Optik 1000–1836 . . . Mit einer Einleitung.* Berlin: 1902. 106 pp.

15) *Denkmaler mittelalterlicher Meteorologie. Mit einer Einleitung und einem Anhang.* Berlin: 1904. 327 pp.

Hellmann, Gustav. *Beiträge zur Geschichte der Meteorologie,* 3 Vols. Veroffentlichungen des Koniglich Preussischen Meteorologischen Instituts, Nr. 273, 296, 315. Berlin: Behrend, 1914–1922. Contents:

1) Bd. Aus der Blutezeit der Astrometeorologie. (J. Stofflers Prognose für das Jahr 1524). Die ältesten instrumentellen meteorologischen Beobachtungen in Deutschland. Die älteste gedruckte Nordlichtbeschreibung. Die theologisch-meteorologische Literatur. Die Vorlaufer der Societas Meteorologica Palatina.

2) *Bd.* Entwicklungsgeschichte des meteorologischen Lehrbuches. Die Witterungsangaben in den griechischen und lateinischen Kalendern. Die Wettervorhersage im ausgehenden Mittelalter, XXI. bis XV. Jahrhundert. Wetterpropheten des XIX. und XX. Jahrhunderts. Kleine Beitrage.

3) *Bd.* Entwicklungsgeschichte des klimatologischen Lehrbuches. Geschichte des Hundert jahrigen Kalenders. Die Entwicklung unserer Kenntnisse vom Nordlicht. Die Meteorologie in ausserdeutschen Flugschriften und Flugblattern. Zur Geschichte der meteorologischen Instrumente u. Beobachtungen. Anhang: Verzeichnis meiner 1883–1922 veroffentlichen Arbeiten zur Geschichte der Meteorologie und des Erdmagnetismus.

Hellmann, Gustav. *Die Entwicklung der meteorologischen Beobachtungen in Deutschland von den ersten anfangen bis zur Einrichtung staatlicher Beobachtungsnetze.* Berlin: 1926.

On meteorological observations. From *Abhandlungen der Preussischen Akademie der Wissenschaften.* Jahrg. 1926. *Physikalisch-Mathematische Classe,* Nr. 1.

Hellmann, Gustav. *Die Entwicklung der meteorologischen Beobachtungen bis zum ende des XVIII Jahrhunderts.* Berlin: 1927. 48 pp.

On the history of meteorological observations. From *Abhandlungen der Preussischen Akademie der Wissenschaften.* Jahrg. 1927. *Physikalisch-mathematische Classe,* Nr. 1.

Heninger, S.K., Jr. *A Handbook of Renaissance Meteorology.* Durham NC: Duke University Press, 1960. 269 pp.

Heninger reconstructs Elizabethan meteorology using the scientific works of the time, such as William Fulke's *Goodly Gallerye,* Leonard Digges's *A Prognostication of Right Good Effect,* and numerous commentaries on Aristotle's *Meteorologica.* Influences from the Bible and from Protestant writings are discussed, as are ideas rooted in classical mythology. Heninger's well-documented book concludes with chapters on the meteorological imagery employed by the major creative writers of the time: Edmund Spenser, Christopher Marlowe, Ben

Johnson, George Chapman, John Donne, and William Shakespeare. An extremely comprehensive and interesting book.

Henson, Robert. *Television Weathercasting: A History.* Jefferson, NC: McFarland, 1990. 193 pp.

Robert Henson saw the need for a book documenting the history of television weathercasting and he answered it with the enthusiasm and dedication of one who loves his subject. Henson discusses the advent of weather observations in the media, from the time of "penny-papers," to radio and television broadcasts, to today's cable services. He documents the symbiotic but often stormy relationship between the National Weather Service and television stations, as well as the evolution of weather research and information technology as pertaining to public broadcasts. Severe weather warnings are highlighted. Appendixes present brief biographies of weathercasters and a listing of AMS awards and Television Seal of Approval holders. The book is well illustrated and furnished with a useful bibliography.

Hoffmann, Immanuel. *Die Anschauungen der Kirchenvater über Meteorologie: Ein Beitrag zur Geschichte der Meteorologie.* Issued as *Münchener geographische Studien* 22. München: Theodor Ackermann, 1907. 96 pp.

Meteorology in the middle ages and the fathers of the church.

Hughes, Patrick. *A Century of Weather Service: A History of the Birth and Growth of the National Weather Service, 1870–1970.* New York and London: Gordon and Breach, 212 pp.

This well-illustrated centennial volume provides an interesting narrative spanning the period from telegraphy to satellites—from the initiation of the U.S. Army Signal Office weather service in 1870 to the inauguration of the National Oceanic and Atmospheric Administration (NOAA) in 1970. Included are communications challenges, the roles of weather forecasters in the two world wars, support of aviation, and the growth of the national weather service.

Ideler, Julius Ludwig. *Meteorologia veterum Graecorum et Romanorum.* Prologomena ad novam Meteorologicorum Aristotelis editionem adornandam scripsit Iulius Ludovicus Ideler. Berolini, 1832. 254 pp.

Khrgian, A. Kh. *Meteorology: A Historical Survey,* Vol. 1, 2nd ed. rev., edited by Kh. P. Pogosyan, translated from Russian by Ron Hardin, and published for ESSA and NSF. Jerusalem: Israel Program for Scientific Translations, 1970. 385p.

This well-illustrated but rare volume covers the history of meteorology from its origins to 1950. It includes sections on instruments, observational networks, weather services, and meteorological theories. Much of the focus is on nineteenth century developments. Russian pioneers are set on an equal footing with their European and American counterparts. Fully two-thirds of the references are in Russian.

Klamkin, Charles. *Weather Vanes: The History, Design, and Manufacture of an American Folk Art.* New York: Hawthorn Books, 1973. 209 pp.

See also Crepeau, Pierre, and Pauline Portelance. *Pointing at the Wind: The Weather-vane Collection of the Canadian Museum of Civilization.* Hull, Quebec: Canadian Museum of Civilization, 1990. 84 pp.; and Westervelt, A.B., and W.T. Westervelt. *American Antique Weather Vanes: The Complete Illustrated Westervelt Catalog of 1883.* New York: Dover, 1982. 100 pp. Reprint of 1883 illustrated catalogue and price list of copper weather vanes, bannerets, and finials, manufactured by A.B. and W.T. Westervelt, New York.

Klemm, Fritz. *Die Entwicklung der meteorologischen Beobachtungen.* Offenbach am Main: Deutschen Wetterdienstes, 1973–1983.

   . . . in Franken und Bayern bis 1700. 1973. 50 pp.
   . . . in Nord- und Mitteldeutschland bis 1700. 1976. 73 pp.
   . . . in Sudwestdeutschland bis 1700. 1979. 66 pp.
   . . . in Osterreich einschliesslich Bohmen und Mahren bis zum Jahr 1700. 1983. 48 pp.

Korber, Hans Gunther. *Vom Wetteraberglauben zur Wetterforschung,* 2. Aufl. Leipzig: Edition Leipzig, 1989. 230 pp.

On the history and cultural history of meteorology.

Kraght, Peter E. *Airline Weather Services, 1931 to 1981: A History of the American Airlines Experience.* Mabank, TX: Peters Books, 1986. 182 pp.

Kutzbach, Gisela. *The Thermal Theory of Cyclones: A History of Meteorological Thought in the Nineteenth Century.* Historical Monograph Series. Boston: Amer. Meteor. Soc., 1979. 255 pp.

Kutzbach discusses the discovery of disturbances as interactions of air currents and Espy's cyclonic system theories, and takes us up to the advent of the polar front theory in 1920. Kutzbach manages to transform the somber and often stodgy characters of the time period into interesting and lively people and scientists. Easy to understand, concise, and complete, Kutzbach succeeds in providing "an insight in the particular problems and methods of problem solving in nineteenth century meteorology."

Ludlam, F.H. *The Cyclone Problem: A History of Models of the Cyclonic Storm.* Inaugural Lecture, 8 November 1966, London: Imperial College of Science and Technology, 1966. 49 pp.

Ludlam's lecture, in pamphlet form, is a concise and thorough investigation of the history of cyclonic storm models. He follows the evolution of the theory from Brandes, Bergeron, Dove, Ley, Abercrombie, to V. Bjerknes and J. Bjerknes. Furnished with diagrams representing each model's formation, as well as the formulas involved and the theories around which they revolve, this lecture is worthy of serious study.

Ludlum, David M. *Early American Hurricanes, 1492–1870.* Boston: Amer. Meteor. Soc., 1963. 198 pp.

This volume by a legendary historian of meteorology is part of the *History of American Weather* series, along with *Early American Winters, 1604–1820* (1966); *Early American Winters II, 1821–1870* (1968); and *Early American Tornadoes, 1586–1870* (1970). These volumes document the meteorological characteristics of all reported storms and winter conditions that occurred within the United States prior to 1870 and describe the intellectual efforts of American scientists of the time to understand and explain such phenomena.

McIntyre, D.P., Ed. *Meteorological Challenges: A History*. Ottawa: Information Canada, 1972, 338 pp.

The Canadian Meteorological Service, in becoming the Atmospheric Environment Service, commissioned this book of essays to commemorate its centennial year. Authors include such international scientific leaders as George Cressman, R.E. Munn, B.J. Mason, and Joseph Smagorinsky on such topics as atmospheric circulation, weather forecasting, cloud physics, air pollution, and the history of the Canadian Meteorological Service. Each essay is authoritative, well illustrated, and fairly concise. The volume was produced with the goal of documenting "the challenges faced by researchers over the past hundred years or so . . . [and the] challenges yet to be faced." In this it succeeds admirably.

Middleton, W.E.K. *A History of the Theories of Rain and Other Forms of Precipitation*. New York: Franklin Watts, 1965. 223 pp.

This book begins with early Hebrew, Greek, and Roman ideas about precipitation. The landmark invention of the barometer in the early seventeenth century changed global conceptions of water vapor and its place in the atmosphere. Middleton shows the effects of Saussure's research on theories of water vapor and precipitation and discusses seventeenth and eighteenth century theories on rain and snow. He covers expansion and compression, saline exhalations, raindrop growth, convection, and rarefaction as conceived at the time. Hutton's theory of rain leads the reader into a discussion of electrical and chemical meteorology, followed by an investigation of nineteenth century ideas on rain and hail.

Middleton, W.E.K. *The Invention of the Meteorological Instruments*. Baltimore: The Johns Hopkins University Press, 1969. 362 pp.

Middleton's books are considered to be staples in the library of any meteorologists worth their salt. Beautifully written and illustrated with diagrams of the instruments discussed, Middleton's book documents the history of meteorology as it progressed with the invention of weather instruments and how these discoveries were intertwined. The book contains formulas and explanations of how each instrument was built, how it functioned, and what it measured. See also his *History of the Barometer* (1964) and *A History of the Thermometer and Its Uses in Meteorology* (1966).

Mottelay, Paul F. *Bibliographical History of Electricity and Magnetism* London: Griffin, 1922; reprint New York: Arno Press, 1975. 673 pp.

Nebeker, Frederik. *Calculating the Weather: Meteorology in the Twentieth Century.* New York: Academic Press, 1995. 251 pp.

This volume in the international geophysics series provides a narrative account of the growth of meteorology in the twentieth century, explains how forecasting became a physics-based science, and analyzes the increasing role of calculation and digital computing in meteorology. According to the author, the main sources used in writing this history were published sources, including reports, review articles, and addresses in *Quarterly Journal of the Royal Meteorological Society* and the *Bulletin of the American Meteorological Society.*

Nihon Kisho Gakkai. *Nihon Kishogakkai Shichijugonenshi.* Tokyo: Meteorological Society of Japan, 1957. 68 pp.

A short history of the Meteorological Society of Japan in commemoration of the 75th anniversary.

O'Brien, Kaye, and Gary K. Grice. *Women in the Weather Bureau during World War II.* Washington, DC: National Weather Service, 1991. 58 pp.

Republic of India Meteorological Office, Poona. *Hundred Years of Weather Service, 1875–1975.* New Delhi: India Meteorological Department, 1976. 207 pp.

Ross, Frank X. *Weather: The Science of Meteorology from Ancient Times to the Space Age.* New York: Lothrop, Lee, & Shepard, 1965. 200 pp.

Written in a clear, articulate, narrative style, Ross discusses the major scientists involved in meteorology from Aristotle to V. and J. Bjerknes and the formation of the WMO. He then goes on to explain and explore theories and occurrences that make up the backbone of meteorology as a science, beginning with the hydrologic cycle, air masses, fronts, clouds, and storms. Meteorologists and weather tools, satellites, and methods of weather control are treated in an understandable, interesting manner. Best for the young or amateur meteorologist.

Sargent, Frederick. *Hippocratic Heritage: A History of Ideas about Weather and Human Health.* New York: Pergamon Press, 1982. 581 pp.

Follows key physicians in the evolution of human biometeorology and medical meteorology from Hippocrates through the Renaissance and into modern times. Traces 5000 years of recorded theories on the atmosphere and human health and looks at the history of human–environment relations.

Schatz, Gerald S. *The Global Weather Experiment: An Informal History.* Washington DC: National Academy of Sciences, 1978. 20 pp.

Brief treatment of the Global Atmospheric Research Program.

Schneider-Carius, Karl. *Weather Science, Weather Research: History of Their Problems and Findings from Documents during Three Thousand Years.* Munich, 1955. Translated from German, and published for NOAA and the National Science Foundation, 554 pp. New Delhi: Indian National Scientific Documentation Centre, 1975. 554 pp.

A valuable but hard to find compilation, this source book documents meteorological problems from the ancient world to the early twentieth century. Extensive quotes from original sources are provided, including Aristotle, Bacon, Boyle, Hadley, Dalton, Margules, and many others. The volume ends with a topical bibliography of over 80 pages, and a 32-page biographical index.

Shaw, Napier. *Manual of Meteorology.* Vol. 1. *Meteorology in History.* Cambridge: Cambridge University Press, 1926. 339 pp.

This volume, the first of four, is devoted to the history of the science. Seven chapters review the relations of weather and culture in antiquity. This is followed by chapters on instruments, observations, and methods of reporting the weather, including charts and graphs. A chapter is devoted to the pioneers. The volume concludes with a brief retrospective on meteorological theory and its lack of development. Rather florid but potentially a good source of quotes.

Spence, Clark C. *The Rainmakers: American Pluviculture to World War II*. Lincoln and London: University of Nebraska Press, 1980. 181 pp.

"Pluviculture," a word created by David Starr Jordan in 1925, is used to describe the formation and marketing of rainmaking ideas. Spence traces the history of rainmaking theories from the early nineteenth century, discussing efforts, both honest and fraudulent, to create rain, until 1946, when the first successful cloud-seeding experiment was completed. Set in the infancy of the Industrial Revolution, when meteorology was just beginning and farmers were buying anything they could get their hands on to end droughts. We find a thorough history of the rainmakers themselves, from "the Storm King" James Espy, to nonscientists who aided in convincing the government to experiment with cloud seeding, and the charlatans who tricked farmers out of their money. Spence concludes with a discussion of the first cloud-seeding experiments completed by the Army Air Service cooperating with scientists from Harvard and Cornell.

Sutton, Geoffrey Vincent. A science for a polite society: Cartesian natural philosophy in Paris during the reign of Louis XIII and XIV. Ph.D. dissertation, Princeton University, 1982. 489 pp.

Abstract: "The dissertation considers the reception and assimilation of Descartes' scientific system by the scientific and social elites in Paris from about 1630 to 1690. The first chapter surveys scientific practice at the Conferences of the Bureau d'adresse, by learned and literary atomists, and by mathematically inclined natural philosophers in Paris at the time of René Descartes. The second chapter sets out Descartes' scientific system, concentrating on his Meteorology to bring out the similarities and contrasts Descartes offered to the scientific styles described in the first chapter. . . ."

Takahashi, Kazuo. *Nihon bungaku to Kisho (Japanese Literature and Meteorology)*. Tokyo: Chuo Koronsha, 1978. 240 pp.

Meteorology in Japanese literature to 1868, history and criticism.

Takahashi, Koichiro, Eiji Uchida, and Takashi Nitta. *Kishogaku hyakunenshi: kishogaku no kindaishi o tankyusuru.* Tokyo: Tokyodo Shuppan, 1987. 230 pp.

One-hundred-year history of meteorology.

Tamura, Sennosuke. *Nihon kishogakushi kenkyu.* 2 Vols. Mishima-shi: Mishima Kagakushi Kenkyujo, 1979, 1981.

A historical study of Japanese meteorology; introduction in English.

Tamura, Sennosuke. *Ri Chosen kishogakushi kenkyu.* Mishima-shi: Mishima Kagakushi Kenkyujo, 1983. 399 pp.

A historical study of meteorology in the Ri Dynasty of Korea; summary in English.

Thomas, Morley K. *The Beginnings of Canadian Meteorology.* Toronto: ECW Press, 1991. 308 pp.

Follows the growth of the Canadian Meteorological Service from its origins as the Toronto Magnetic and Meteorological Observatory, founded by the British in 1839, through the career of George Templeman Kingston, the first director after confederation. Thomas documents the ongoing political battles to finance the service as it grew, as well as the innovations that shaped it. Four separate chapters document meteorological observations in each province of Canada to 1880.

USSR Academy of Sciences. *The Development of Earth Sciences in the USSR* (selected articles). 3 Vols. Translation of Razvitiye nauk o Zemle v SSSR. Wright-Patterson AFB, OH: Foreign Technology Division, Air Force Systems Command, 1969.

Volume 3 is on *Meteorology and Atmospheric Science.* Part of the series, "Fifty Years of Soviet Science and Technology," prepared by the Institut Istorii Estestvoznaniia i Tekhnikiin 1967.

Weber, Gustavus A. *The Weather Bureau: Its History, Activities and Organization.* Baltimore: The Johns Hopkins University Press, 1922. 87 pp.

Brookings Institution. Institute for Government Research. Service monographs of the United States Government, No. 9. One of a series of short sketches of government agencies.

Whitnah, Donald R. *A History of the United States Weather Bureau.* Urbana: University of Illinois Press, 1961. 267 pp.

Whitnah traces the trials and the triumphs of the Weather Bureau from its birth in 1870 to 1960. Building on the foundation laid by earlier generations of observers and theorists, a national weather service was established in the Army Signal Office in 1870. This became a civilian Weather Bureau in the U.S. Department of Agriculture in 1891. Service to aviation played a big role in decision to transfer the Weather Bureau to the Department of Commerce in 1940. Whitnah's book contains chapters on the expansion of public services and the development of research activities before 1913, the "modernization" of services before 1941, and the impact of World War II on meteorology. Whitnah also emphasizes the political battles and power struggles that the bureau weathered over the years.

Chapter 21

Historical Photo Gallery

Julie A. Burba

Jule Charney practices his photographic skills during a break between sessions at the Numerical Weather Prediction Symposium, held in 1960 in Tokyo, Japan.

Tor Bergeron (left) and Jacob Bjerknes work on the upper floor, which was designated as the work place for Bergen meteorologists, of Allegaten 33—the home of Vilhelm Bjerknes and family—in 1919.

Jacob Bjerknes working on his weather maps, circa 1920.

In 1926 staff members of the U.S. Weather Bureau forecast center drew maps of single weather elements for use by the chief weather forecaster.

Carl-Gustaf Rossby uses a rotating tank, or dishpan, built at the U.S. Weather Bureau in 1927 to study atmospheric wave motion.

U.S. Weather Bureau employees launch a pilot balloon, circa 1930, to measure wind velocities.

Navy fliers "sound" the sky over Washington, D.C., in 1934 with an aerometeorograph attached to the wing strut.

The cover photo of the Sunday supplement of the 8 November 1942 *Philadelphia Inquirer* showed Lois Coots, the only woman enrolled in the fall 1942 meteorology program at New York University (NYU), launching a pilot balloon from atop MacCracken Hall, University Heights campus of NYU, Bronx, New York. In a call to weather women nationwide to join the Civil Aeronautics Administration (CAA)/U.S. Weather Bureau program at NYU, Coots wore her khaki-colored CAA uniform for the photo.

Teletypewriter typing pool: women work in the 1940s to transmit by teletypewriter encoded analyzed weather charts to field stations.

One of the B-24M aircraft used by the 55th Reconnaissance Squadron at Guam in September 1945. On the B-24M, the psychrometer air-intake tube was located just below the weather observer's window in the nose. The radar dome was located under the rear bomb bay. This dome housed an antenna for AN/APQ-13, which was used to detect storm clouds. The aircraft was fully armed with three turrets plus normal-waist and lower-blister positions.

Irving Langmuir (left) and Bernard Vonnegut (upper right) watch Vincent Schaefer blow into a freezer to make a supercooled cloud in preparation for seeding, circa 1947, at the General Electric Research Laboratory, Schenectady, New York.

Erwin Biel, Jule Charney, and Carl-Gustaf Rossby relaxing on the Chicago lakefront in 1948.

The first numerical weather prediction team—the Princeton Gang—in 1949: John Freeman (seated) and, from left to right, Gilbert Hunt, Jule Charney, Joe Smagorinsky, Margaret Smagorinsky (computer programer), Ellen Eliassen (research assistant), and Arnt Eliassen.

Sexploitation in a Times Facsimile Corporation ad from 1952. At this time, most facsimiles were used to transmit weather data, not pinups, as one would believe from this ad.

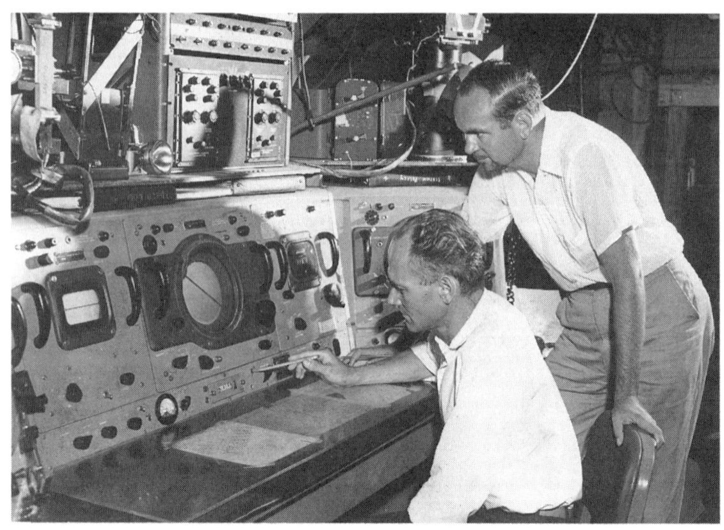

U.S. Air Force meteorologists Roger Lhermitte (seated) and Ralph Donaldson observe the radar display at the console of the new 3.2-cm CPS-9 weather surveillance radar located on the Great Blue Hill, Milton, Massachusetts, 1956.

Robert Bungaard hand plotted weather data for the Air Defense Command located in Cheyenne Mountain, Colorado, in 1959, using calculations from the Philco 2000.

Members of the Signal Corps Meteorological Team launched pillow balloons that were used to determine dispersion at the Army Electric Proving Ground, Fort Huachuca, Arizona, in 1954.

The weather center of United Air Lines at Denver, Colorado, was unique among privately operated weather sevices in the late 1950s. It received more raw data than any weather service other than the U.S. Weather Bureau's National Weather Analysis Center.

The U.S. Weather Bureau's first experimental mobile 3-cm CW Doppler radar began operation in 1957. From the beginning of its operation to 1961, the mobile Doppler radar was located in Texas, Kansas, and Indiana.

Designed in 1957 specifically for weather surveillance, the first WSR-57 was commissioned on 26 June 1959 in Miami, Florida.

This room-sized Philco Transac S-2000 computer was used in 1960 by United Aircraft Corporation's Weather System Center in East Hartford, Connecticut.

The duty forecaster, Albert E. Boyer, presented his forecast over the Travelers Broadcasting Service Station WTIC-TV, Hartford, Connecticut, circa 1960. The weekday evening two 5-minute telecasts were sponsored by the Travelers Insurance Company.

Dick Ogilvie analyzes a map in the System Forecast Unit at Eastern Air Lines in the Atlanta Terminal Building in 1961. This room was segregated from all traffic so that the forecaster was able to work uninterrupted.

Joe Smagorinsky (left), chief of the General Circulation Research Laboratory in Washington, D.C., and Francis Reichelderfer, chief of the U.S. Weather Bureau, examine an elevation contour map of the Northern Hemisphere printed out in 1962 by the STRETCH computer, which appears in the background.

Paul Frenzen (left) of the Argonne National Laboratory, Argonne, Illinois, demonstrates a wind recording instrument to Jim Culver, a participant in the 89th Annual Wheatland Plowing Match held in September 1966. Culver permitted Frenzen to set up a 25-ft tower in one of his fields. The tower held six wind recording instruments at varying levels to measure changes in wind velocity.

Does milk spoil if stored in a metal container in the hot sun? Donald F. Gatz, meteorologist with the Radiological Physics Division of Argonne National Laboratory, Argonne, Illinois, loads sample containers into a special automated rain collector, which stored rain samples in increments of 0.03 in. for later chemical analysis. Each 0.03 in. equaled about 1.5 pints of water. Gatz's 1970 research determined how thunderstorms cleanse the atmosphere of airborne particulate matter.

Pioneers in the atmospheric sciences gathered at the Conference on the Large Scale Processes in the Atmosphere held at the Massachusetts Institute of Technology on 25–28 March 1952. The conference was sponsored by the Geophysics Research Directorate of the Air Force Cambridge Research Center. Front row (left to right): Henry Samela, GRD; Hurd C. Willett, MIT; Hsiao-Lan Kuo, MIT; Sam Solot, GRD; Ralph Shapiro, GRD; Jacob Bjerknes, UCLA; Edward Lorenz, MIT; and Victor Starr, MIT. Second row (left to right): Bill Widger, GRD; Dave Fultz, U of C; Sherman Lowell, NYU; Maj. Don Martin, AWS; Seymour Hess, FSU; George Benton, JHU; Harry Wexler, USWB; and Jule Charney, IAS. Third row (left to right): Lt. Col. Edward Dolezel, ARDC; Duane Cooley, GRD; Robert White, GRD; Eberhard Wahl, GRD; Robert Long, JHU; and Yale Mintz, UCLA. Fourth row (left to right): Gunter Loeser, GRD; Thomas Keegan, GRD; Maj. Philip Thompson, MIT; Heinz Lettau, GRD; and Julius London, NYU.

Acknowledgments and photo credits

Special thanks to Mel Shapiro, Ruth Liebowitz, William Winn, and John Lewis and to Patrick Hughes for his insight and encouragement.

Photos on page 581 courtesy of Akira Kasahara; top photo on page 582 courtesy of The Norwegian Meteorological Institute; bottom photo on page 588 courtesy of Hedveg (Mrs. Jacob) Bjerknes; photos on page 583 and 584 courtesy of Patrick Hughes; photo on page 585 courtesy of the *Philadelphia Inquirer;* photo on page 587 courtesy of the GE Research and Development Center; photos on page 588 courtesy of Arnt Eliassen; top photos on both page 591 and 596 courtesy of the Geophysics Directorate (U.S. Air Force). All other photos are from the American Meteorological Society photo archives.

NOTES ON CONTRIBUTORS

Roscoe R. Braham Jr. is a scholar in residence at North Carolina State University and professor emeritus at the University of Chicago. He is a meteorologist and educator and the author of numerous articles on thunderstorms, cloud physics, weather modification, and lake-effect snow storms. He received a B.S. in geology from Ohio University, served as a pilot and weather officer in the U.S. Air Force (1942–1946), and earned an M.S. (1948) and Ph.D. (1951) in meteorology from the University of Chicago. For his work on the Thunderstorm Project (1946–1949) he received the Losey Award from the Institute of Aeronautical Sciences and the U.S. Department of Commerce Silver Medal. He was a member of the initial planning team for the University Corporation for Atmospheric Research (UCAR) (1958) and later served on its Board of Trustees and in numerous other capacities. He is a Fellow of the AMS and has served the Society as councilor, associate editor, commissioner, and president (1988). He received the AMS Rossby Medal (1981) and Charles Franklin Brooks Award (1987), and the Schaefer Award of the Weather Modification Association (1987). He is a member of Phi Beta Kappa and several scientific and professional societies. He married Mary Ann Moll in 1945; they have four children.

Julie A. Burba, born, reared, and educated in small towns amid green seas of corn in Illinois, moved to Boston, Massachusetts, in 1988. Eager to enter the publishing industry, Burba joined the staff of the American Meteorological Society in 1990 as a copy editor. Since then, she has provided her services to AMS as senior copy editor and now as news editor of the *Bulletin of the American Meteorological Society.* Prior to working at AMS, Burba worked in the public relations industry for a New England–based department store and for an agricultural membership-based organization located in Illinois. Burba received her B.A. in mass communications from Western Illinois University in 1987. She is a member of the Women's Educational and Industrial Union, the Boston Women Communicators, and the American Association for the Advancement of Science (AAAS).

Gordon D. Cartwright joined the Weather Bureau in 1929 at its office in Pittsburgh, Pennsylvania. He was hired under the terms of the

Air Commerce Act of 1926 to make three-hourly aviation-type observations for the Pittsburgh–Cleveland airmail route. From 1932 to 1939, he served as airways weather observer in Cleveland, Dallas, and Fort Worth, where he took on apprentice forecasting duties. During this period, Mr. Cartwright also ran the U.S. Army Air Corps airplane observation program at Barksdale Airbase in Shreveport, Louisiana. In 1939 he transferred to the Weather Bureau's Aviation Forecast Office at the new LaGuardia Field in Queens, New York, and worked on both domestic routes and later on the trans-Atlantic forecast team. In the mid-1940s he was granted a Weather Bureau scholarship to attend one of the special meteorology courses at New York University (NYU). The course was organized to speed up the training of weather forecasters for the expanding U.S. Army Air Corps. After a brief period as analyst under Professor Jurgen Holmboe at NYU, Cartwright was transferred to the Weather Bureau's Central Office in Washington, D.C., to assist in wartime planning, and later he served as head of the bureau's Special Services Branch, which included the Aviation, Hurricane, River and Flood, Fire Weather, and Fruit Frost Services. At the end of the war, Cartwright was selected to join the secretariat of the newly formed International Civil Aviation Organization in Montreal, Quebec, Canada; he served as secretariat officer in the MET Division until 1952. Returning to the Weather Bureau, he was assigned as supervising meteorologist for the Pacific area, based in Honolulu. In 1954 Cartwright was back in Washington, D.C., as head of the Observations and Facilities Division for two years. He was selected by the U.S. Committee for the International Geophysical Year to serve as liaison scientist at the main base in MIRNYY of the Second Soviet Antarctic Expedition. He returned to the United States in 1958, and after a brief period as chief of the International Division of the Weather Bureau in Washington, was asked by the then chief, Dr. Robert M. White, to take up a liaison position for the U.S. Government in Geneva and was assigned as science officer at the U.S. Mission, a post he held until retiring in 1983, after 54 years of service in the U.S. Government. Since that time, Cartwright has been a consultant in Geneva to the National Oceanic and Atmospheric Administration (NOAA) and the State Department.

Stanley A. Changnon has been involved in weather-related research for 45 years. He directed the atmospheric research program involving 70 scientists and engineers at the Illinois State Water Survey for 20 years and served as the survey's chief for six years. Today he is one of the survey's principal scientists and the head of his own consulting firm, which specializes in climate applications research. He conceived major national programs including METROMEX, a massive 10-year study of how large cities influence weather and change the local climate, and he planned the nation's regional climate centers program and developed the first center, the Midwestern Climate Center at the University of Illinois. His research interests include investigations of climate change and climatic variability in space and time; studies of how weather and climate impact agriculture, water resources, and policy; investigations of both inadvertent and planned weather modification; investigations of flood and droughts; and studies of severe weather. He has performed numerous projects for several state and federal agencies; received research grants totaling more than $20 million; served on numerous advisory groups including those of the National Academy of Sciences, the National Science Foundation (NSF), the Environmental Protection Agency, and the Department of Energy; been a consultant to the weather insurance industry for 35 years; provided congressional testimony on many issues; authored 600 papers and reports and five books; and received national awards for his research accomplishments from the AMS, the American Water Resources Association, the American Agricultural Economics Association, and the American Geophysical Union (AGU). He is a Fellow of the AMS and the AAAS, and past president of the American Association of State Climatologists and of the Weather Modification Association.

Kenneth C. Crawford became director of the Oklahoma Climatological Survey and state climatologist for Oklahoma in 1989. At the same time, he became a member of the faculty at the University of Oklahoma, where he serves as a professor in the School of Meteorology. He also serves as director of the Oklahoma Mesonetwork, a world-renowned environmental monitoring network composed of 111 automated stations placed in each of Oklahoma's 77 counties. Dr. Crawford received a B.S. (1966) from the University of Texas at Austin, an M.S. (1967) from The Florida State University, and a Ph.D. (1977) from the University of Oklahoma. His professional career with the National Weather Service (NWS) began in 1961. He worked as an operational forecaster at several NWS forecast offices and spent over four years in the early phases of the Doppler radar program at the National Severe

Storms Laboratory. He last served the NWS as their Oklahoma area manager (1982–1989), where he played a critical role in the successful testing of the prototype NEXRAD (Next Generation Weather Radar) system.

George P. Cressman entered The Pennsylvania State College as a physics major in 1937, where he met Professor Helmut Landsberg and took the few meteorology courses available at the time. This included the study of dynamic meteorology using Exner's book (which was in German) as a text. Upon graduation in 1941, Cressman entered the air force's graduate meteorology program and was sent to the study program at New York University, where Athelstan Spilhaus was department chairman. After completing the course, Cressman was retained as an instructor for the next class and afterward sent to the University of Chicago as instructor, where Carl-Gustav Rossby was department chairman. After forecasting assignments in the western states and then in Florida, at the end of the war, Cressman was invited by Rossby to return to Chicago for graduate studies and to be responsible for the graduate synoptic laboratory. At Chicago, professors and graduate students, under Rossby's prodding, hammered out theories of the dynamics of the near-global-scale circulations seen on the daily maps. After receiving his Ph.D., Cressman was asked by Sverre Petterssen, scientific advisor of the air force's Air Weather Service, to undertake a consultant program for the major air force weather centers. While thus employed, Cressman was also sent to the Institute of Advanced Study at Princeton to participate in the new work on numerical weather prediction. Soon, the three heads of civil and military weather services decided to establish an operational numerical prediction unit; Cressman was appointed as head of the unit.

Following major reorganizations in the Weather Bureau in 1965, Cressman was persuaded to take over responsibility for the field operations and services of the Weather Bureau and later to become director of the National Weather Service, following further organizational changes. Now fully retired, he regards his work during the early days of establishing operational numerical weather prediction as the great adventure of his career and much more enjoyable and worthwhile than all administration and reorganizations put together.

Mark DeMaria became interested in tropical cyclones while growing up in Miami after experiencing Hurricanes Cleo (1964), Betsy

(1965), and Inez (1966). He received a B.S. in meteorology at The Florida State University in 1977 and completed his M.S. in 1979 and Ph.D. in 1983 at Colorado State University, where he conducted research on numerical modeling of tropical cyclones. He was a postdoctoral fellow at the National Center for Atmospheric Research (NCAR) in 1984–85 and an assistant professor in the Department of Atmospheric Sciences at North Carolina State University from 1985 to 1987. He then returned to Miami as a research meteorologist at the Hurricane Research Division of NOAA. In 1995 he became the chief of the technical support branch at the National Hurricane Center. Over the past 15 years, the focus of Dr. DeMaria's research has been on numerical and observational studies of tropical cyclones, with emphasis on the development of techniques for operational track and intensity forecasting. He has received three awards from the AMS for his research and has published more than 30 articles on hurricanes and tropical meteorology in AMS journals.

Edwin T. Engman received his B.S. (1959) and M.S. (1961) in Agricultural Engineering from Cornell University and his Ph.D. (1974) from The Pennsylvania State University. Much of Dr. Engman's career has been spent with the USDA's Agricultural Research Service, most recently as a research hydrologist with the Hydrology Lab. Since 1990 he has been the head of the Hydrological Services Branch at the NASA/ Goddard Space Flight Center. He is particularly interested in the applications of remote sensing to hydrologic models, in monitoring the spatial variability of hydrologic parameters, and in the development and use of remotely sensed soil moisture.

James Rodger Fleming is associate professor and director of the Science and Technology Studies Program at Colby College in Maine. His research involves the history of science in America and the history of the geophysical sciences, especially meteorology. His teaching is dedicated to building bridges between the humanities and sciences. He has earned degrees in both science and history, including a B.S. in astronomy from The Pennsylvania State University (1971), an M.S. in atmospheric science from Colorado State University (1973), and an M.A. and Ph.D. in the history of science from Princeton University (1984, 1988). He is the author and editor of several books, including *Meteorology in America, 1800–1870* (1990), the *International Bibliog-*

raphy of Meteorology, 4 volumes in 1 (1994), and *Science, Technology and the Environment: Multidisciplinary Perspectives* (1994). Professor Fleming has held fellowships from the Smithsonian Institution and the National Endowment for the Humanities. He was a visiting scholar at the Massachusetts Institute of Technology (MIT) and Yale, and a visiting professor at The Pennsylvania State University's Earth System Science Center. He serves on the editorial advisory board of the Joseph Henry Papers at the Smithsonian Institution. He is a founding member of Geo–Clio (1993). From 1990 to 1995 he was the history editor of *Eos: Transactions of the American Geophysical Union.* He is a member of AMS and has served the Society in several capacities as a historical consultant and editor.

David D. Houghton is professor of Atmospheric and Oceanic Sciences at the University of Wisconsin—Madison and is currently president of the AMS. He began his professional work in 1963 as a research scientist at NCAR and was an exchange scientist in the Soviet Union and a visiting scientist at the Courant Institute of Mathematical Sciences. He joined the faculty of the University of Wisconsin—Madison in 1968 and reached full professor in 1972. He was chair of the department 1976–1979 and 1991–1994. In 1972 Dr. Houghton served as Secunded Scientist for the International Scientific and Management Group for the GARP (Global Atmospheric Research Program) Atlantic Tropical Experiment. In 1980 and 1989 he was an invited scholar in China. In 1988 he served as visiting senior UCAR scientist at the National Meteorological Center (renamed the National Centers for Environmental Prediction in 1995). He has published numerous scientific papers and was the editor of the *Handbook of Applied Meteorology.* He has served on committees for NSF, the National Research Center (NRC), UCAR, and the AMS. He has received several distinguished honors, including selection to Phi Beta Kappa and Sigma Xi, and awards of fellowships from the NSF and Phi Kappa Phi.

Simone L. Kaplan is a junior at Colby College in Waterville, Maine, where she is an English major and environmental studies minor. Her post-Colby plans include attending graduate school in pursuit of a degree in science journalism and beginning work on a book of short stories. She is interested primarily in environmental writing rather than technical journalism and hopes to find work in that field. Originally from the Boston area, she is a coxswain at Community Rowing and a member of the Colby crew team.

Edwin Kessler graduated from Columbia College, New York City, in 1950 and received his Sc.D. in meteorology from MIT in 1957. He was at the Weather Radar Branch, Air Force Cambridge Research Laboratories until 1961. From then until 1964 he was at the Travelers Research Center, Hartford, Connecticut, where he was first a senior scientist and later director of the Atmospheric Physics Division. From 1964 until 1986, he was director of the National Severe Storms Laboratory in Norman, Oklahoma. He is author of an AMS meteorological monograph, editor of three books on thunderstorms, and author or coauthor of more than 200 other reports and publications on radar meteorology, precipitation physics, climatology, alternative energy, foreign trade, agriculture, and environment. He received the Cleveland Abbe Award of the AMS in 1989. At this writing he is an adjunct professor of meteorology at the University of Oklahoma and chair of Common Cause Oklahoma. He also operates a 220-acre farm in McClain County, Oklahoma.

William A. Koelsch is professor of history and geography in the Graduate School of Geography at Clark University, where he has also served as university archivist (1972–1982) and university historian (1982–1990). He was born in 1933, graduated from Bucknell University (Sc.B., summa cum laude, 1955), Clark University (M.A., geography, 1959), and the University of Chicago (Ph.D., history, 1966). He has also studied or taught at Eckerd College, Case Western Reserve University, the University of Trier, and The University of Arizona. He is the author or editor of seven books, including *Clark University, 1887–1987,* and some 75 articles and reviews, including four papers on the history of meteorology and climatology. His twin research interests in the history of higher education and the history of geography and related fields came together in his essay for this volume. He is currently working on a series of papers on the relations of geography and the classics in Great Britain and the United States over the last 200 years.

E. Philip Krider received his B.A. degree from Carleton College in 1962 and his M.S. and Ph.D. degrees in physics from The University of Arizona in 1965 and 1969. During 1969–1971 he held a National Academy of Sciences (NAS) postdoctoral appointment in space physics at the NASA Manned Spacecraft Center in Houston, Texas. Dr. Krider has been on the faculty of The University of Arizona since 1971 and he is now director of the Institute of Atmospheric Physics and head of the Department of Atmospheric Sciences. Initially, Dr. Krider worked in elementary and cosmic ray physics; for the past 25 years, he has focused primarily on problems in atmospheric electricity. He is a Fellow of the AMS and former cochair of an NAS study on "The Earth's Electrical Environment." He has also served on an NAS panel that reviewed meteorological support for NASA space operations, and the U.S. Air Force Lightning Review Committee. He is a former associated editor of the *Journal of Geophysical Research,* former co-chief editor and editor of the *Journal of the Atmospheric Sciences,* and is currently president of the International Commission on Atmospheric Electricity. He is the author or coauthor of over 100 reviewed publications and holds eight patents. In 1985 he received the AMS Award for Outstanding Contributions to the Advance of Applied Meteorology for his development and application of lightning detection instrumentation.

John E. Kutzbach is the Plaenert-Bascom Professor of Liberal Arts at the University of Wisconsin—Madison, where he is professor and former chair of the Department of Atmospheric and Oceanic Sciences, and professor and director of the Center for Climatic Research in the Institute for Environmental Studies. He teaches courses in atmospheric science, climate and climate change, and the evolution of climate in the geologic past. His climate research is interdisciplinary and involves collaboration with geologists, ecologists, and archaeologists. It covers a broad spectrum of geologic time ranging from thousands to hundreds of millions of years in the past. He uses supercomputer models of the earth's climate to simulate the changes of climate produced by changes in the locations of continents and oceans, the height of mountains, the composition of the atmosphere, and the earth's orbit about the sun. The computer simulations are compared with observation of past environments. He is the author of over 100 publications. Dr. Kutzbach is from Reedsburg, Wisconsin. He earned his B.S., M.S., and Ph.D. at the University of Wisconsin—Madison. He has worked as a NATO (North Atlantic Treaty Organization) research fellow in England, as a research consultant to the United Nations–

sponsored World Meteorological Organization in Geneva, as a research awardee of the Alexander von Humboldt Foundation at the University of Bonn, and as a summer research visitor at NCAR. He is a Fellow of the AMS and the AGU.

Roy Leep is the executive director of weather services at WTVT Channel 13, Tampa, Florida, the nation's largest television meteorological department. The station boasts the largest privately owned Doppler radar system in the world, the 240-ft SkyTower radar with a range of 450 miles. For 38-plus years Leep has been part of Tampa Bay area television—longer than anyone else. He also directs WeatherVision, the first private weather forecasting service in the state of Florida. Leep's meteorological education includes the Weather Bureau, where he was an intern from 1946 to 1950; the U.S. Air Force weather school, where he graduated with honors and served as an instructor supervisor; and The Florida State University, where he began his radio broadcasting career. In 1957 he joined WTVT, first as a staff meteorologist, then as chief meteorologist (1959), director of the station's weather service (1963), and finally executive director (1988). Leep is active in charitable activities and is the recepient of numerous professional and civic awards and honors. He is a Fellow of the AMS and was the first meteorologist in Florida (and the 10th in the nation) to receive the AMS Seal of Approval for weathercasting.

Edward N. Lorenz first became involved with dynamic meteorology during World War II, when he took leave from his studies in mathematics at Harvard University and entered a special program at MIT, designed to meet the growing need of the armed services for weather forecasters. He subsequently served as a forecaster overseas, returning to MIT after the war and receiving the Sc.D. in meteorology in 1948. Since then, he has remained at MIT, where he was head of the meteorology department from 1977 to 1981 and is now professor emeritus. In addition to numerous articles on various aspects of atmospheric science, predominantly theoretical, he has written two books: one a comprehensive treatment of the general circulation of the atmosphere, and one an exposition of the phenomenon of chaos. He lives with his wife Jane in Cambridge, Massachusetts.

W. Paul Menzel received his B.S. in physics (with high honors) from the University of Maryland—College Park in 1967. His M.S.

(1968) and Ph.D. (1974) in theoretical solid-state physics were awarded by the University of Wisconsin—Madison. In 1975 he joined the Space Science and Engineering Center at Madison, where he investigated the possibilities for remote sensing of the atmosphere from a geosynchronous spacecraft. Specifications for the vertical temperature and moisture sounding capabilities of the Visible Infrared Spin Scan Radiometer Atmosphere Sounder (VAS) were the result of that work. In 1983 he joined the National Environmental Satellite Data and Information Service (NESDIS) to head up the Advanced Satellite Products Project, where he is still responsible for developing, testing, and evaluating procedures for derivation of new satellite products. This activity focuses on transferring advances in the research laboratory to the operational forecast arena. For the past four years, as the NESDIS Geostationary Observational Environmental Satellite (GOES) program scientist, he has been organizing the product assurance from the new GOES instruments and working toward evolving the geostationary capabilities of NOAA. Concurrently as a team member of the NASA Earth Observing System, he has been using passive infrared radiometers on ER2 aircraft to investigate cloud properties with multispectral data of high spatial resolution. Dr. Menzel is an adjunct professor in the Department of Atmospheric and Oceanic Sciences at the University of Wisconsin—Madison.

Frederik Nebeker received B.A. and M.A. degrees in mathematics from, respectively, Pomona College and the University of Wisconsin. He received an M.A. in history of science from the University of Wisconsin and a Ph.D. in history of science at Princeton University. While at Princeton, he was the editor of a major oral history project, *The Princeton Mathematics Community in the 1930s,* and he worked as an instructor in the history department. As a postdoctoral researcher at the American Philosophical Society Library he completed a bibliographical monograph, *Astronomy and the Geophysical Tradition in the United States in the 19th Century* (1991). He worked as a historian at the Center for the History of Physics at the American Institute of Physics, where he studied the role of engineering in the history of high-energy physics. Since 1990 he has been research historian at the IEEE Center for the History of Electrical Engineering at Rutgers University. He is the author of *Sparks of Genius: Portraits of Electrical Engineering Excellence* (1994) and coauthor of *The Evolution of Elec-*

trical Engineering: A Personal Perspective (1994). A revised version of his dissertation, *Calculating the Weather: Meteorology in the 20th Century,* was published in 1995.

Harold D. Orville is a distinguished professor of meteorology at the South Dakota School of Mines and Technology (SDSM&T), Rapid City. He holds a B.A. in political science from the University of Virginia, an M.S. in meteorology from The Florida State University, and a Ph.D. in meteorology from The University of Arizona. His specialties are in cloud physics, cloud dynamics, weather modification, and the numerical modeling of clouds. His career in teaching and research has been spent at SDSM&T, with sabbaticals taken to NOAA Headquarters in Rockville, Maryland, in 1972 and to the World Meteorological Organization (WMO) in Geneva, Switzerland, in 1982. He is currently the chairman of the WMO's Executive Council Panel on the Physics and Chemistry of Clouds and Weather Modification Research. He has been a Councilor of the AMS and has served on its Executive Committee as well as on many of the Society's other committees. He is currently the Scientific and Technical Activities Commissioner of the AMS.

James F.W. Purdom is chief of the NESDIS Regional and Mesoscale Meteorology Branch located at the Cooperative Institute for Research in the Atmosphere (CIRA) at Colorado State University. He received an A.B. in mathematics and physics from Transylvania College (1965) and graduate degrees in atmospheric science from St. Louis University (M.S., 1968) and Colorado State University (Ph.D., 1986). From 1966 to 1972 he was an officer in the U.S. Air Force. He joined NOAA/NESDIS as a research meteorologist in 1980. Dr. Purdom is a member of AMS and served in 1993–1994 as chair of its Committee on Satellite Meteorology and Oceanography. He was cochair of the GOES-I Technical Advisory Council and a member of the Science Advisory Committee of the U.S. Weather Research Program. He was elected Fellow of CIRA in 1981 and serves on CIRA's Advisory Council. In 1994 he received the Department of Commerce Silver Medal Award.

R.R. Rogers studied meteorology at the University of Texas, MIT, and New York University. His introduction to radar meteorology was at the MIT Weather Radar Project in the mid-1950s. After working briefly at the National Advisory Committee for Aeronautics at Langley Field, Virginia, and completing further graduate study at NYU, he joined the Applied Physics Departments of Cornell Aeronautical Laboratory in Buffalo, New York, in 1959. There he resumed working in radar meteorology and completed a Ph.D. thesis in 1964 on meteorological applications of Doppler radar. He joined the Stormy Weather Group of McGill University in 1966 and served as chair of the McGill Meteorology Department from 1978 to 1987. Rogers teaches courses in physical meteorology and does research on cloud physics and atmospheric remote sensing. He has served on the AMS committees on radar meteorology and cloud physics and was a member of the AMS Council from 1990 to 1992.

Robert J. Serafin is the director of the National Center for Atmospheric Research, a position he has held since 1989. He is an expert in meteorological instrumentation, including the areas of aviation meteorology, radar meteorology, signal processing, and remote sensing. He earned degrees in electrical engineering from the University of Notre Dame (B.S., 1958), Northwestern University (M.S., 1961), and Illinois Institute of Technology (Ph.D., 1972). In 1973 he joined NCAR as manager of its Field Observing Facility and became director of the Atmospheric Technology Division in 1981. Dr. Serafin is chair of the NRC Committee on National Weather Service Modernization and chair of the Technical Advisory Committee on NEXRAD for the National Weather Service, Federal Aviation Administration, and U.S. Air Force Weather Service. He is a chief scientist in the U.S.–People's Republic of China cooperative program in atmospheric science and technology. He is a member of the National Academy of Engineering, a Fellow of the AMS, a senior member of the Institute of Electrical and Electronics Engineers, and a member of Sigma Xi. He has served as a member of over 25 advisory committees and national panels and was the founder and first editor of the *Journal of Atmospheric and Oceanic Technology*. He is the author or coauthor of over 50 publications and holds three patents.

Paul L. Smith is a native of Missouri. He received B.S., M.S., and Ph.D. degrees from the Carnegie Institute of Technology, where he also served for several years on the faculty of the Department of Electrical Engineering. He then joined the Midwest Research Institute, where he was introduced to radar meteorology through work on the EMAC probe. In 1966, he moved to the Institute of Atmospheric Sciences at the South Dakota School of Mines and Technology and he was appointed director of the institute in 1981. His major research interests are radar meteorology, cloud physics, and weather modification. He was an NSF postdoctoral fellow in meteorology at McGill University in 1964 and returned to McGill as visiting professor of meteorology during the 1969–1970 academic year. He served as chief scientist at Air Weather Service Headquarters, Scott Air Force Base, during 1974–1975, and as visiting scientist at the Alberta Research Council in 1984–1985. In the spring semester of 1986 he was a Fulbright Lecturer in Radar Meteorology at the University of Helsinki. He was elected to the International Commission on Clouds and Precipitation in 1988 and currently serves as director of the South Dakota Space Grant Consortium. He is the author of a widely used set of lecture notes in radar meteorology and chaired the AMS Committee on Radar Meteorology in the 1970s and again in the 1990s.

David B. Spiegler, Certified Consulting Meteorologist and Fellow of the AMS, has 35 years of proven experience in various technical, management, and marketing roles in the private sector of meteorology. He is founder and president of DBS Associates, Inc. (DBS), and Impact Weather, Inc. DBS provides technical and marketing expertise across a broad spectrum of activities involving meteorology, oceanography, climatology, and the environment, including forensic meteorology. Impact Weather, Inc., is developing multimedia weather, climate, and environment products for industry. Mr. Spiegler earned a bachelor's and a master's degree in meteorology from New York University. In the first 12 years of his career, he was a research scientist at the Travelers Research Center and a senior scientist at Allied Research Associates, Inc., where he developed objective analysis and prediction techniques for the jet stream and stratospheric temperatures and winds, and for operational snow prediction along the eastern seaboard. He also developed global four-dimensional models of the atmosphere used by NASA in the planning of the space program. In the 1970s, Mr. Spiegler held various senior-level management positions at Environmental Research and Technology, Inc. (now ENSR). In the 1980s, he was a consulting meteorologist at The MITRE Corporation, where he helped develop the Automated Weather Distribution System for the U.S. Air Force. He has published over 60 technical papers and reports and has served on many AMS committees and boards. He now devotes full time to DBS and Impact Weather, Inc.

Charles H. Sprinkle is the International Affairs Officer of the NWS. He previously served for many years as the chief of the NWS's Aviation Services Branch. Mr Sprinkle received his B.S. in meteorology from The Pennsylvania State University in 1959. Prior to being assigned to NWS Headquarters in 1975, he served at forecast offices in Baltimore and Cleveland and at the National Meteorological Center. Mr. Sprinkle is a member of the AMS and the American Institute of Aeronautics and Astronautics (AIAA), and is a licensed pilot. In 1986 he was elected vice president of the WMO Technical Commission for Aeronautical Meteorology (CAeM). He has served as CAeM's president since 1990. Mr. Sprinkle also actively represents the United States at the International Civil Aviation Organization, headquartered in Montreal, Quebec, Canada. Among his many awards, Mr. Sprinkle received the 1990 Edgar S. Gorrell Award of the Air Transport Association of America for his outstanding contributions to enhancing the reliability and safety of air transportation. He also received the 1991 Losey Atmospheric Sciences Award of the AIAA for his outstanding contributions to the atmospheric sciences as applied to the advancement of aeronautics and astronautics.

Warren M. Washington is a senior scientist in the Climate and Global Dynamics Division at NCAR. Born in Portland, Oregon, Dr. Washington earned a B.S. in physics and an M.S. in meteorology from Oregon State University. After completing his Ph.D. in meteorology at The Pennsylvania State University, he joined NCAR in 1963 as a research scientist. His research specialty is computer modeling of the earth's climate. He has published more than 100 papers in professional journals. His book, *An Introduction to Three-Dimensional Climate Modeling,* coauthored with Claire Parkinson, is a standard reference on climate modeling. Dr. Washington is a consultant and advisor on climate system modeling to a number of government officials and committees. From 1978 to 1984 he served on the President's National Advisory Committee on Oceans and Atmospheres. He participated in several panels of the National Research Council and chaired the Advisory Panel for *Climate Puzzle* (1986), part of the PBS television series *Planet Earth.* He was a member of the Secretary of Energy's Advisory

Board from 1990 to 1993 and has been on the Secretary of Energy's Health and Environmental Research Advisory Committee since 1990. He serves on the Modernization Transition Committee of the NWS. In 1995 he was appointed by President Clinton to the National Science Board for a six-year term. The National Science Board helps oversee the NSF and advises the Executive Branch and Congress on science-related matters. Dr. Washington is a Fellow of the AMS, the AAAS, and the African Science Institute. From 1991 to 1995 he was on the Board of Directors of the AAAS. He is a Distinguished Alumnus of The Pennsylvania State University, Alumni Fellow of Oregon State University, and president of the Black Environmental Science Trust, a nonprofit effort to increase the numbers of African-Americans in the environmental sciences. Dr. Washington was president of the AMS in 1994 and presided over its 75th anniversary meeting.

INDEX